Advances in Organic Synthesis
(Volume 5)

Editor

Atta-ur-Rahman, *FRS*

Kings College
University of Cambridge
Cambridge
UK

Bentham Science Publishers
Executive Suite Y - 2
PO Box 7917, Saif Zone
Sharjah, U.A.E.
subscriptions@benthamscience.org

Bentham Science Publishers
P.O. Box 446
Oak Park, IL 60301-0446
USA
subscriptions@benthamscience.org

Bentham Science Publishers
P.O. Box 294
1400 AG Bussum
THE NETHERLANDS
subscriptions@benthamscience.org

CONTENTS

PREFACE

This eBook should be of interest to students and researchers interested in the art of organic synthesis. The chapters describe how challenging molecules have been synthesised by experts in the field and how new methodologies have been developed and applied where conventional chemistry had failed.

In the first article, Lönnberg describes the synthesis of glycoconjugates by the strong binding of carbohydrates with protein receptors. In another review, Alfonso highlights the development of chiral molecular receptors based on enantiopure *trans*-cyclohexane-1,2-diamine. In an interesting contribution, Pineschi and coworkers review the selective opening of asymmetric ring to allylic epoxides with nucleophiles of different kinds. Next, García-Martínez and colleagues focus on the main strategies for the synthesis of poly(phenylenevinylene)- and poly(phenyleneethynylene)-based dendrimers as well as approaches to hybrid dendritic structures that combine both scaffolds. In another review, Wu *et al.* have reviewed organic synthesis of α-activated cross conjugated cycloalkenone systems. Boechat and Bastos provide updates on the synthetic methods developed for the direct introduction of a trifluoromethyl group in carbonyl compounds. Avendaño and Cuesta discuss synthetic chemistry with n-acyliminium ions derived from piperazine-2,5-diones and related compounds. In the last chapter of this eBook, Jana and her coworkers review advancements related to the synthesis of organometallic amino acids and analogues.

I am thankful for the time and efforts made by the editorial personnel at Bentham Science Publishers, Mr. Mahmood Alam, Taimur A. Khan and Ms. Sarah A. Khan.

Atta-ur-Rahman, FRS
Kings College
University of Cambridge
Cambridge
UK

List of Contributors

Harri Lönnberg

Department of Chemistry, University of Turku, FIN -20014 Turku, Finland.

Ignacio Alfonso

Departamento de Química Biológica y Modelización Molecular, Instituto de Química Avanzada de Cataluña, Consejo Superior de Investigaciones Científicas, IQAC -CSIC, Jordi Girona, 18 -26, E -08034, Barcelona, Spain.

Mauro Pineschi

Dipartimento di Farmacia, Università di Pisa, *Via* Bonanno 33, 56126 Pisa, Italy.

Ferruccio Bertolini

Dipartimento di Farmacia, Università di Pisa, *Via* Bonanno 33, 56126 Pisa, Italy.

Valeria Di Bussolo

Dipartimento di Farmacia, Università di Pisa, *Via* Bonanno 33, 56126 Pisa, Italy.

Paolo Crotti

Dipartimento di Farmacia, Università di Pisa, *Via* Bonanno 33, 56126 Pisa, Italy.

Joaquín C. García -Martínez

Facultad de Farmacia, Universidad de Castilla -La Mancha, Avda. de los Estudiantes s/n, 02071 -Albacete, Spain.

Enrique Díez -Barra

Facultad de Ciencias y Tecnologías Químicas, Universidad de Castilla -La Mancha, Avda. Camilo José Cela, 10, 13071 -Ciudad Real, Spain.

Julián Rodríguez -López

Facultad de Ciencias y Tecnologías Químicas, Universidad de Castilla -La Mancha, Avda. Camilo José Cela, 10, 13071 -Ciudad Real, Spain.

Yen -Ku Wu

Department of Chemistry, National Tsing Hua University, Hsinchu, 30013, R.O.C. Taiwan.

Tai Wei Ly

Axikin Pharmaceuticals, Inc., 10835 Road to the Cure, Suite 250, San Diego, California 92121, USA.

Kak -Shan Shia

Institute of Biotechnology and Pharmaceutical Research, National Health Research Institutes, Miaoli County, 35053, R.O.C. Taiwan.

Nubia Boechat

Fundação Oswaldo Cruz, Farmanguinhos, Rua Sizenando Nabuco, 100 CEP 21041 -250 Manguinhos - Rio de Janeiro, RJ, Brazil.

Monica Macedo Bastos

Fundação Oswaldo Cruz, Farmanguinhos, Rua Sizenando Nabuco, 100 CEP 21041 -250 Manguinhos - Rio de Janeiro, RJ, Brazil.

Carmen Avendaño

Departamento de Química Orgánica y Farmacéutica, Facultad de Farmacia, Universidad Complutense, 28040 Madrid, Spain.

Elena de la Cuesta

Departamento de Química Orgánica y Farmacéutica, Facultad de Farmacia, Universidad Complutense, 28040 Madrid, Spain.

Poulami Jana

Department of Chemical Sciences, Indian Institute of Science Education and Research – Kolkata, Mohanpur, West Bengal - 741252, India.

Sibaprasad Maity

Department of Chemical Sciences, Indian Institute of Science Education and Research – Kolkata, Mohanpur, West Bengal - 741252, India.

Debasish Haldar

Department of Chemical Sciences, Indian Institute of Science Education and Research – Kolkata, Mohanpur, West Bengal - 741252, India.

2

CHAPTER 1

Solid-Supported Synthesis of Glycoconjugates

Harri Lönnberg[*]

Department of Chemistry, University of Turku, FIN-20014 Turku, Finland

Abstract: Carbohydrate-protein interactions play a major-role in recognition of cells by external macromolecules. Single saccharide ligands, however, bind only weakly to their protein receptors. Sufficient affinity is achieved by appropriate mutual orientation of the saccharide epitopes and their protein receptors, which allows multiple simultaneous interactions. In other words, high affinity binding is achieved through multivalency, where the high number of simultaneous binding events compensates the lack of strength of an individual interaction. Solid-supported methods allowing synthesis of multiantennary glycoconjugates have, hence, been under active development. The present paper is a review of the solid-phase protocols useful for preparation of glycoconjugates.

Keywords: Solid-support, synthesis, conjugation, 1,3-dipolar cycloaddition, oximation, peptide, oligonucleotide, cluster effect, monosaccharide, protecting group, linker, glycoconjugate, glycocluster, sugar, carbohydrate.

1. INTRODUCTION

In striking contrast to other biopolymers, carbohydrates form highly branched structures, within which the monomeric units of which may be bound to each other *via* different types of linkages. This structural diversity of carbohydrates is exploited in many biological recognition events [1, 2]. Oligosaccharides covalently linked to proteins and lipids play, for example, a role in fertilization, immune defence, cell-cell adhesion and degradation of blood clots. In addition, a wide range of human diseases are initiated through protein-carbohydrate recognition. Early phases of the infectious cycles of many bacteria, viruses and parasites involve carbohydrate-mediated recognition of host by the pathogen, as do early events in some metastatic processes. Individual monosaccharide ligands, however, bind only weakly to their protein receptors. High affinity binding is achieved by

*Address correspondence to Harri Lönnberg: Department of Chemistry, University of Turku, FIN-20014 Turku, Finland; Tel: +358 2 333 6770; Fax +358 2 333 6700; E-mail: harlon@utu.fi

proper mutual orientation of the saccharide epitopes and their protein receptors, which allows simultaneous interaction of several sugar ligands with the protein. Carbohydrate binding proteins, called lectins, usually occur as higher-order oligomeric structures that exhibit several identical binding sites [3]. In other words, the high affinity binding and, hence, cell- and tissue specificity is achieved through multivalency, where the high number of simultaneous binding events compensates the lack of strength of an individual interaction. This phenomenon is often referred to as the 'cluster glycoside effect' [4-7].

Usually, multiantennary carbohydrates anchored to proteins or lipids of cell membranes recognize extracellular proteins, but the opposite is also possible [2, 8, 9]. Hyaluronic acid, for example, is the main ligand for a transmembrane glycoprotein CD44 that is over-expressed in many cancers [10-12], galactosylated oligonucleotide phosphorothioates show enhanced uptake *via* the galactose-specific asialoglycoprotein receptor-mediated endycytosis [13, 14], antigenic peptides have been targeted to dendritic cells by glycoconjugation [15], and galactosylated dendrimers are able to inhibit at nanomolar concentrations cellular binding of pathogens bearing galactose-specific membrane proteins [16]. In addition, multivalent carbohydrate–carbohydrate interactions play a role in cell adhesion [17]. Glycoconjugates, hence, are valuable tools for glycobiology and show considerable therapeutic potential [2, 6, 8, 9, 18-25].

Workable synthesis of structurally defined glycoconjugates is a prerequisite for biological studies elucidating the recognition events. Both chemical and enzymatic methods have been applied, but chemical synthesis usually offers a greater flexibility: constructs containing natural and unnatural sugar building blocks can be assembled using natural and unnatural scaffolds and linkages. The present review deals with solid-phase synthesis of glycoconjugates, in spite of the fact that the majority of glycoconjugate synthesis has so far been carried out in solution [26]. The solid-phase synthesis, however, exhibits some advantages that expectedly make it the method of choice in future. Firstly, repeated purifications during multistep synthesis are replaced by simple washings. Secondly, parallel synthesis of combinatorial libraries is less laborious on a solid support than in solution, since at least most of the steps, if not all, may be carried out by automated machine-assisted synthesis. Synthetic glycoconjugates fall into three major categories: multiantennary glycoclusters, mimetics of linear glycopolymers, and glycopeptides. In addition, carbohydrates themselves are useful scaffolds of glycoconjugate synthesis, since they inherently exhibit various ring sizes and configurations [27]. Chemical synthesis of glycopeptides has been recently reviewed [28-30] and, hence, this subject is discussed only as far as cyclic peptide scaffolds or short peptides bearing large glycoclusters are concerned. Main attention is paid to solid-phase synthesis of multiantennary glycoclusters, mimetics of linear glycooligomers, oligonucleotide glycoconjugates and conjugates containing a sugar-derived scaffold.

2. BRANCHED MULTIANTENNARY GLYCOLUSTERS CONTAINING IDENTICAL SUGAR LIGANDS

Several alternative methods for the solid-supported synthesis of branched multiantennary glycolusters containing identical sugar units have been reported. Often lysine is used for branching. A typical example is offered by the assembly of triantennary clusters containing a photoreactive benzophenone moiety and a reporter group in addition to three α-D-manno- or α-D-fuco-pyranosyl groups [31]. Scheme **1** outlines the procedure applied. The synthesis was carried out on a sulfonamide safety-catch linker to which a 4,7,10-trioxatridecane-1,13-diamine spacer was attached *via* a succinyl linker. The chain was then elongated with Fmoc-Lys(Boc)-OH and Fmoc-Lys(Fmoc)-OH. The acid-labile Boc protection was removed and the exposed amino function was acylated with photoreactive Boc-Ser(4-BzBn)-OH. Both the Fmoc and Boc protecting groups were then removed and the exposed primary amino functions were acylated with succinic anhydride. The sugar units were finally coupled as 2-aminoethyl glycosides to the carboxy groups of the succinyl moieties. Alkylation of the sulfonamide nitrogen with iodoacetonitrile converted the linker to a good leaving group that could be displaced by an attack of amine-functionalized reporter group on the sulfonamide-bound carbonyl carbon.

A closely related strategy has been applied to obtain denderimeric octa-antennary glycopeptides bearing $(Tn)_2$-antigen (**1**) termini [32]. The tree-like structure was generated by three successive couplings of Lys or by coupling of Lys followed by two generations of *N*-lysyl-4-aminobyric acid. The $(Tn)_2$ groups were then tethered to the α- and ε-amino groups of the outmost sphere *via* a 4-aminobutyric linker. The synthesis was carried out on amino-functionalized Tentagel. The dendrimeric conjugates were not released from the support, but used as such for biological screening.

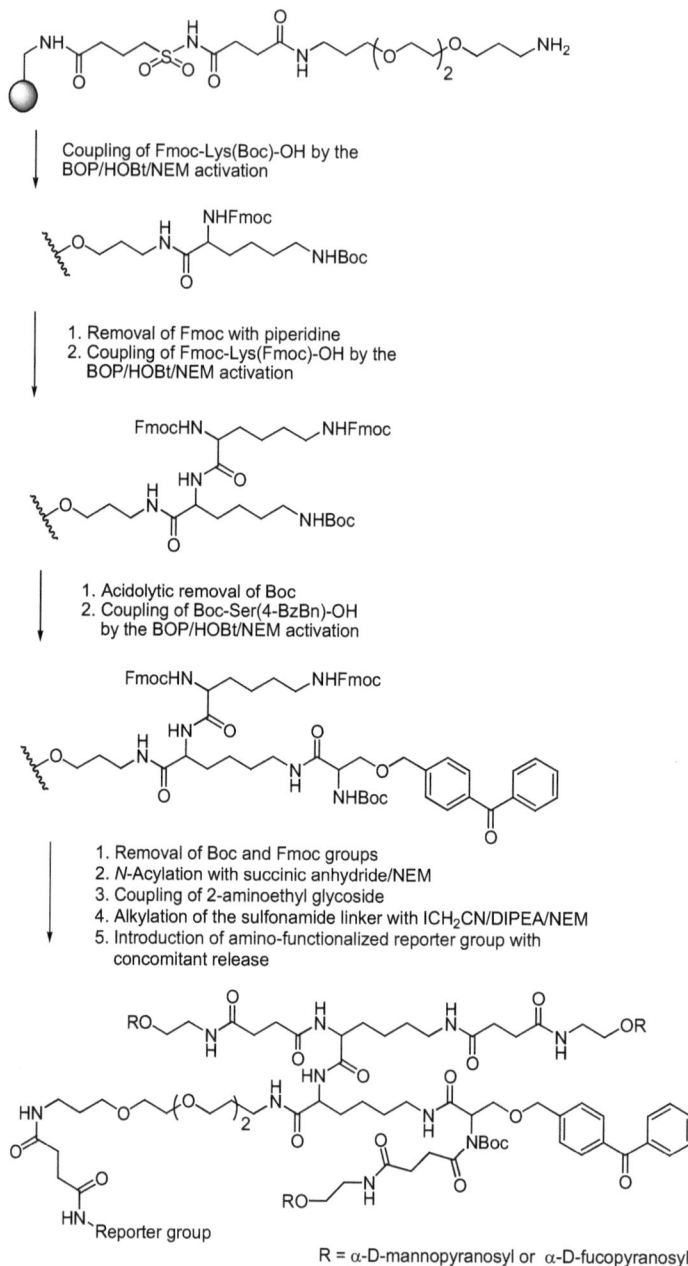

Scheme 1: Exploitation of lysine as a branching unit for construction of glycoclusters.

An even more highly branched construct bearing 16 2-thiosialic acid residues (**2**) has been prepared on a β-Ala derivatized Wang resin [33]. Four successive couplings of Lys afforded 16 primary amino groups which were then acylated with the benzotriazolyl ester of *N*-(chloroacetyl)glycylglycine. Treatment with acetyl protected 2-thiosialic acid methyl ester gave the desired conjugate. This was finally deprotected with methanolic sodium methoxide followed by sodium hydroxide (aq), and then released acidolytically from the support.

A library of fucosylated tetra-antennary peptides have been obtained by acylating the *N*-termini of tetra-antennary support-bound peptides with acetyl protected α-L-fucopyranosylacetic acid (**3**) [34]. The peptides were assembled on an amino-functionalized support by the Fmoc chemistry using the BOP/DIPEA activation. Three lysines as branching units at every second, third or fourth position afforded the tetra-antennary structure. The side-chain protections of the amino acid residues were removed by a TFA/TIS treatment and the acetyl protections of the sugar moieties by ammonolysis in MeOH. The conjugates were not released but assayed on particles.

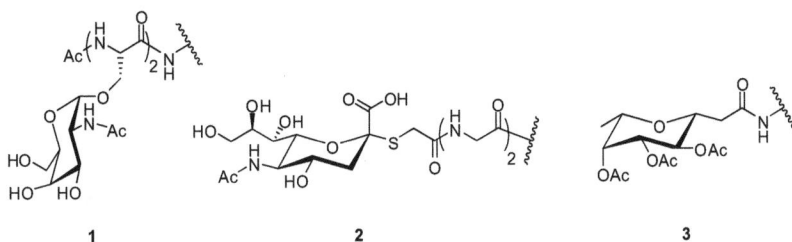

When orthogonally scissile linkers are incorporated in the tree-like scaffold, sub-libraries may be released in addition to the main library, exhibiting the maximal branching. Scheme **2** shows an interesting example of this approach. Three different linkers, *viz.* a photo-labile nitrobenzyl linker, an ester linker and a propargyl safety-catch linker, were used to construct an octa-antennary structure bearing terminal amino functions [35, 36]. These amino groups were acylated with bromoacetic acid and the sugar units were attached as 2-mercaptoethyl glycosides by acyl substitution. In addition to the octa-antennary glycoclusters, four and two antennary sub-structures were released.

Besides appropriately functionalized monosaccharides, pre-fabricated glycoclusters have been used for decoration of peptide scaffolds. For example, a tri-antennary cluster of unprotected sugars has been coupled to the *N*-terminus of a support-bound peptide by the HATU/DIPEA activation in DMF (Scheme **3**) [37].

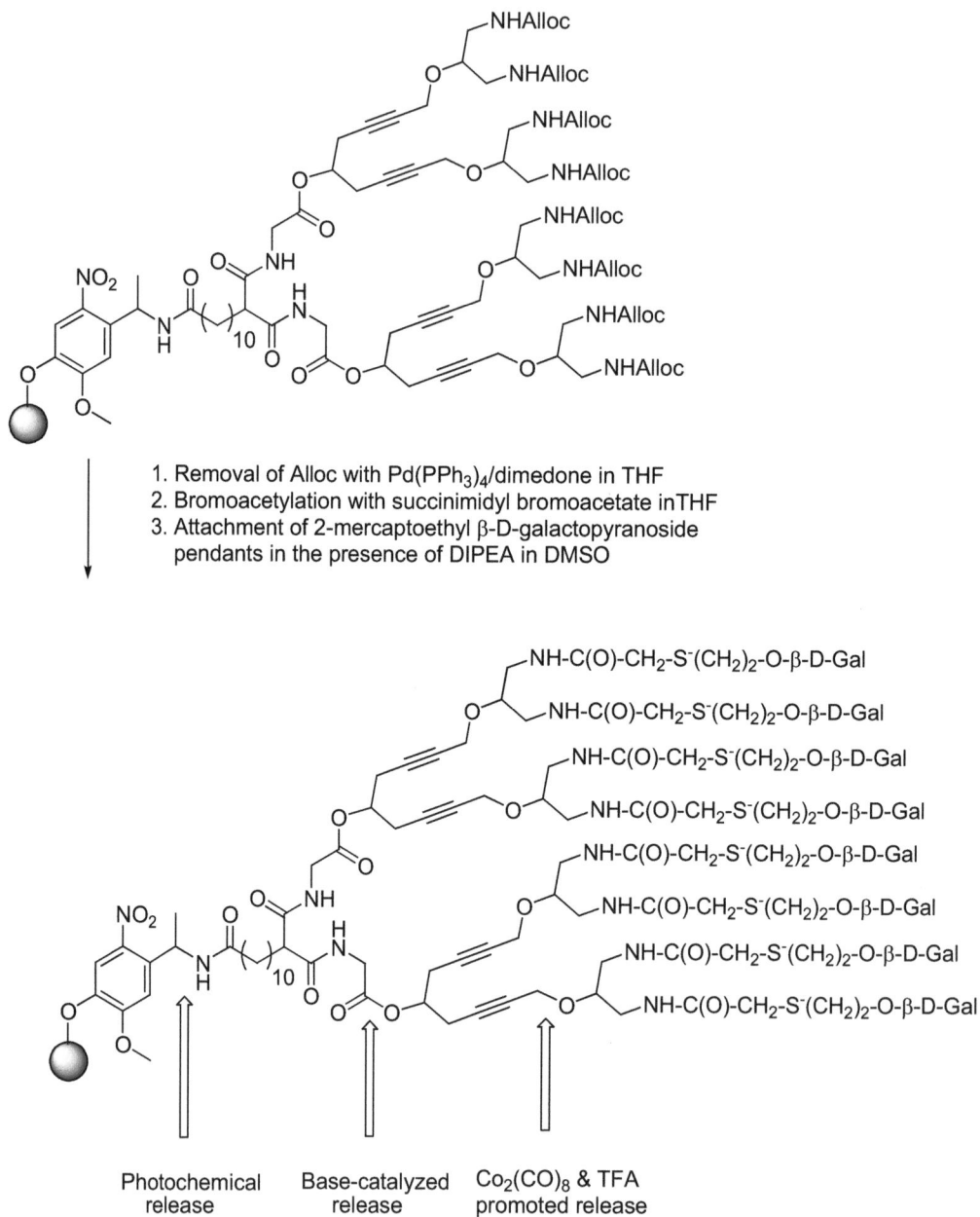

1. Removal of Alloc with Pd(PPh$_3$)$_4$/dimedone in THF
2. Bromoacetylation with succinimidyl bromoacetate inTHF
3. Attachment of 2-mercaptoethyl β-D-galactopyranoside pendants in the presence of DIPEA in DMSO

Photochemical release Base-catalyzed release Co$_2$(CO)$_8$ & TFA promoted release

Scheme 2: Utilization of orthogonally scissile linkers that allow release of several different sub-libraries.

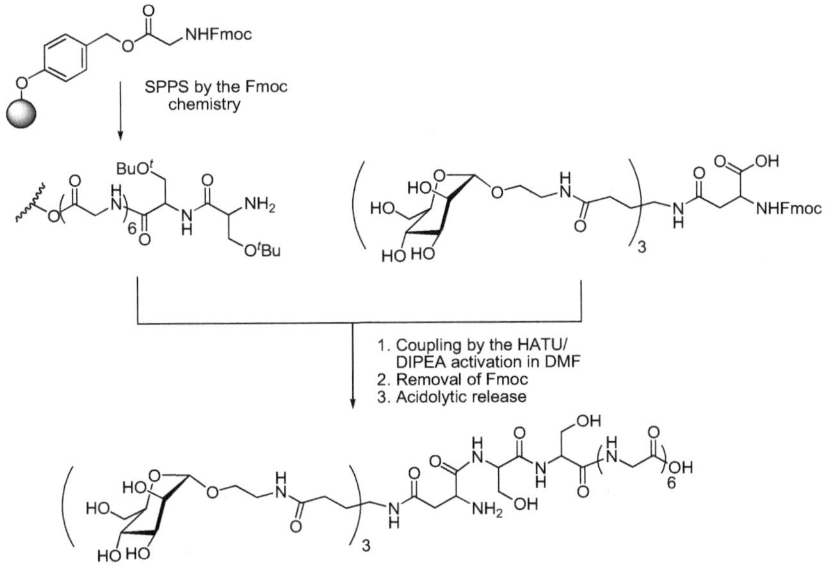

Scheme 3: Decoration of a peptide scaffold with pre-fabricated glyco clusters.

Recently, phosphodiester scaffolds have been utilized in synthesis of multi-antennary glycoclusters. Tetragalactosyl clusters have been constructed by assembling the phosphodiester backbone on a solid support from propargyl derivatized phosphoramidite building-blocks and attaching the azide-functionalized galactosyl pendants to this backbone with the aid of microwave-assisted click reaction (Scheme 4) [38], *i.e.* a Cu(I)-promoted 1,3-dipolar cycloaddition of alkylazide to alkyne [39, 40]. Another related example is offered by synthesis of fucosylated pentaerythrityl phosphodiester oligomers [41]. The backbone has been assembled from DMTrO-protected phosphoramidites derived from 2,2-bis(propargyloxymethyl)propane-1,3-diol (**4**) and the support-anchored oligomer has then been subjected to click reaction with fully acetylated 8-azido-3,6-dioxaoctyl β-D-fucopyranoside (**5**). This has allowed attachement of 10 sugar units. In both cases, ammonolytic cleavage of the succinyl linker and concomitant removal of the acetyl groups give the cluster.

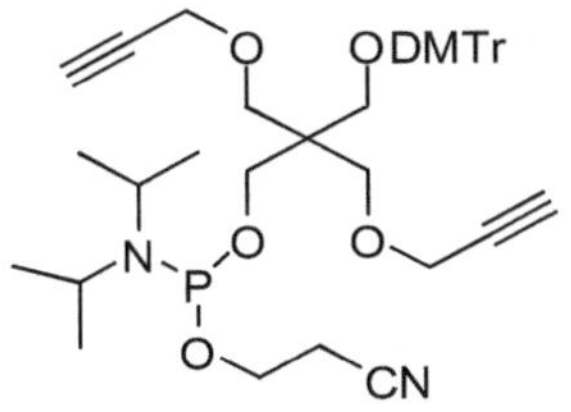

4 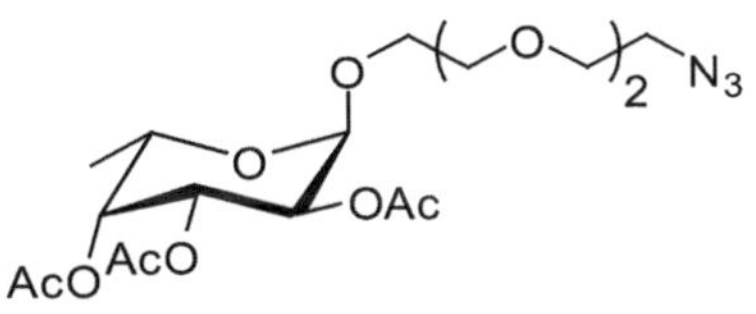   **5**

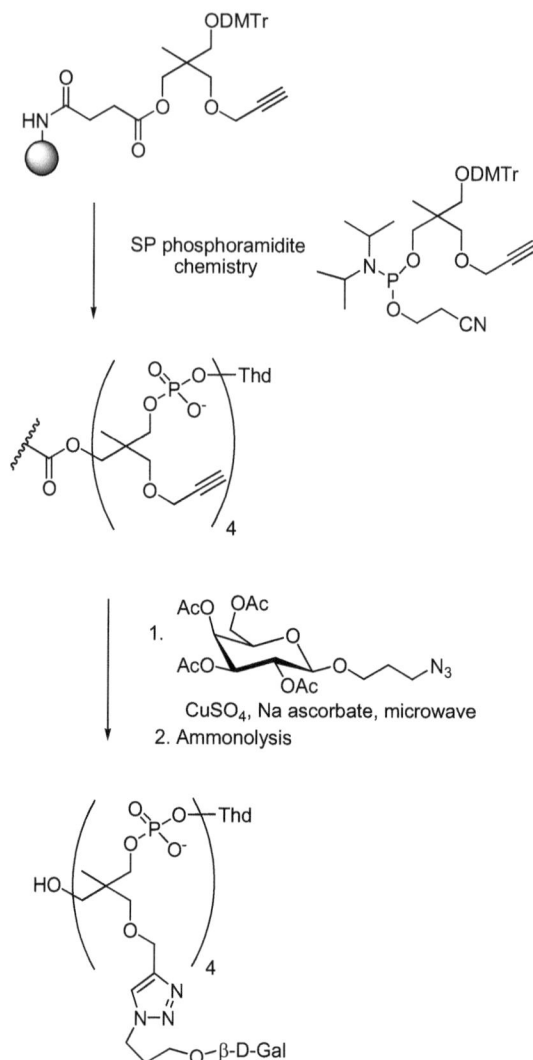

Scheme 4: Construction of glycol clusters by utilization of phosphoramidite chemistry for the assembly of the scaffold and click chemistry for its post-synthetic decoration.

3. BRANCHED MULTIANTENNARY GLYCOLUSTERS CONTAINING TWO OR MORE DIFFERENT SUGAR LIGANDS

Since lectins usually are specific for a single type of sugar, homoclusters have been the main object of synthetic efforts. There are, however, some indications

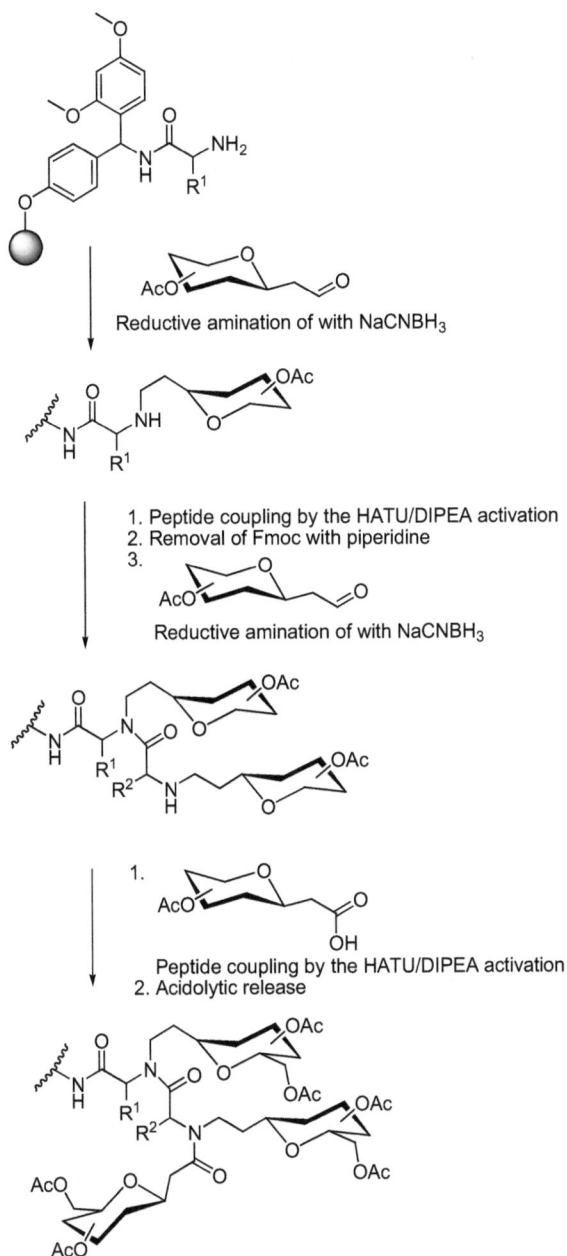

Scheme 5: Exploitation of peptide chemistry for construction of a linear glycocluster containing three different sugars.

that in special cases, heteroclusters containing more than one type of sugars bind more tightly than the corresponding homoclusters [42]. For example, mixed-type of α-Man, β-Glc and α-Man, β-Lac heteroglycoclusters exhibit 8 times higher affinity to mannose specific Con A lectin than α-Man homoclusters of identical valency [43]. Recent studies on of β-cyclodextrin-scaffolded glycoclusters decorated with α-mannosyl and β-lactosyl antennas verify the existence of such a synergic heterocluster effect) [44]. It has also been shown that antibodies are able to interact simultaneously with several glycans, and the affinity is higher with heterogeneous glycans [45]. Accordingly, synthesis of glycoconjugates containing more than one type of sugar ligands also is of interest. As an example of such a synthesis, preparation of acyclic peptide conjugates containing up to three different sugars is outlined in Scheme **5** [46-48]. The *C*-terminal amino acid was immobilized to a Rink amide linker and a fully acetylated glycosylacetaldehyde was attached to the α-amino function by reductive alkylation. The next amino acid was coupled to the secondary amino group and, after removal of the amino protection, another glycosylacetaldehyde was attached. The third sugar unit was then introduced by acylation of the resulting secondary amine by a glycosylacetic acid and the conjugate was released acidolytically.

Another approach for preparation of branched peptide-based glycoconjugates containing three different sugars utilizes as a branching unit 2,2-bis(aminomethyl)-β-alanine, the amino groups of which are protected with orthogonally removable Fmoc, Boc and Alloc groups (Scheme **6**) [49]. This building block was coupled to a glycine derivatized SCAL-linker. This linker withstands a variety of conditions, but by reduction of the sulfoxide groups to sulfide groups it is converted acid-labile [50]. The protecting groups were removed in the order Fmoc, Boc, Alloc and an *O*-glycosylated serine block was introduced as a pentafluorophenyl ester after each removal. The Fmoc groups were replaced with acetyl groups before the release of the conjugate into solution. The sugar hydroxyl groups were deacetylated in solution phase.

An even larger conjugate has been obtained by solid-supported conjugation of two different pre-fabricated sugar clusters to a peptide scaffold (Scheme **7**) [51]. *N*-(2-Azidoethyl)-*N*-(*N*-Boc-2-aminoethyl)glycine (**6**) was used as a branching unit that was coupled to a glycine derivatized sulfamylbutyryl safety-catch linker. The Boc protection was removed and an acetyl protected triantennary isocyanate sugar

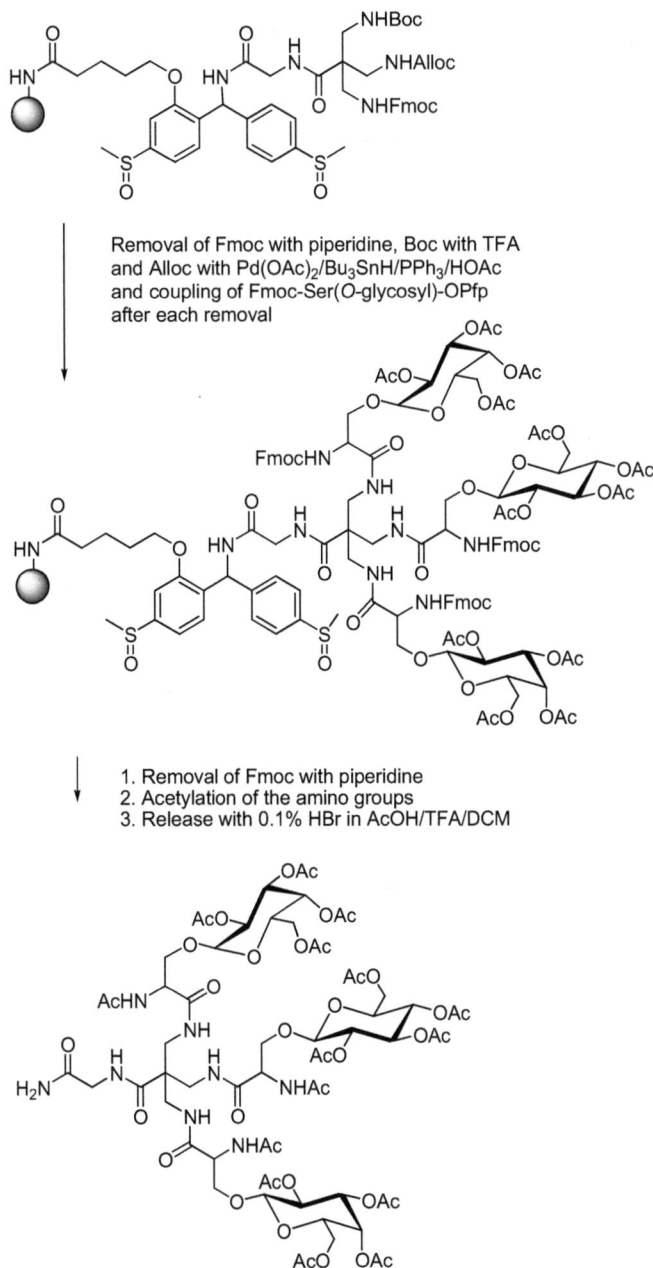

Scheme 6: Exploitation of peptide chemistry for construction of a branched glycocluster containing three different sugars.

conjugate (**7**) was attached. The azido function was then reduced to an amino group and the second sugar block (**8**) was introduced. The sulfonamide linkage was alkylated with iodoacetonitrile and the conjugate was released by acyl substitution with benzyl amine.

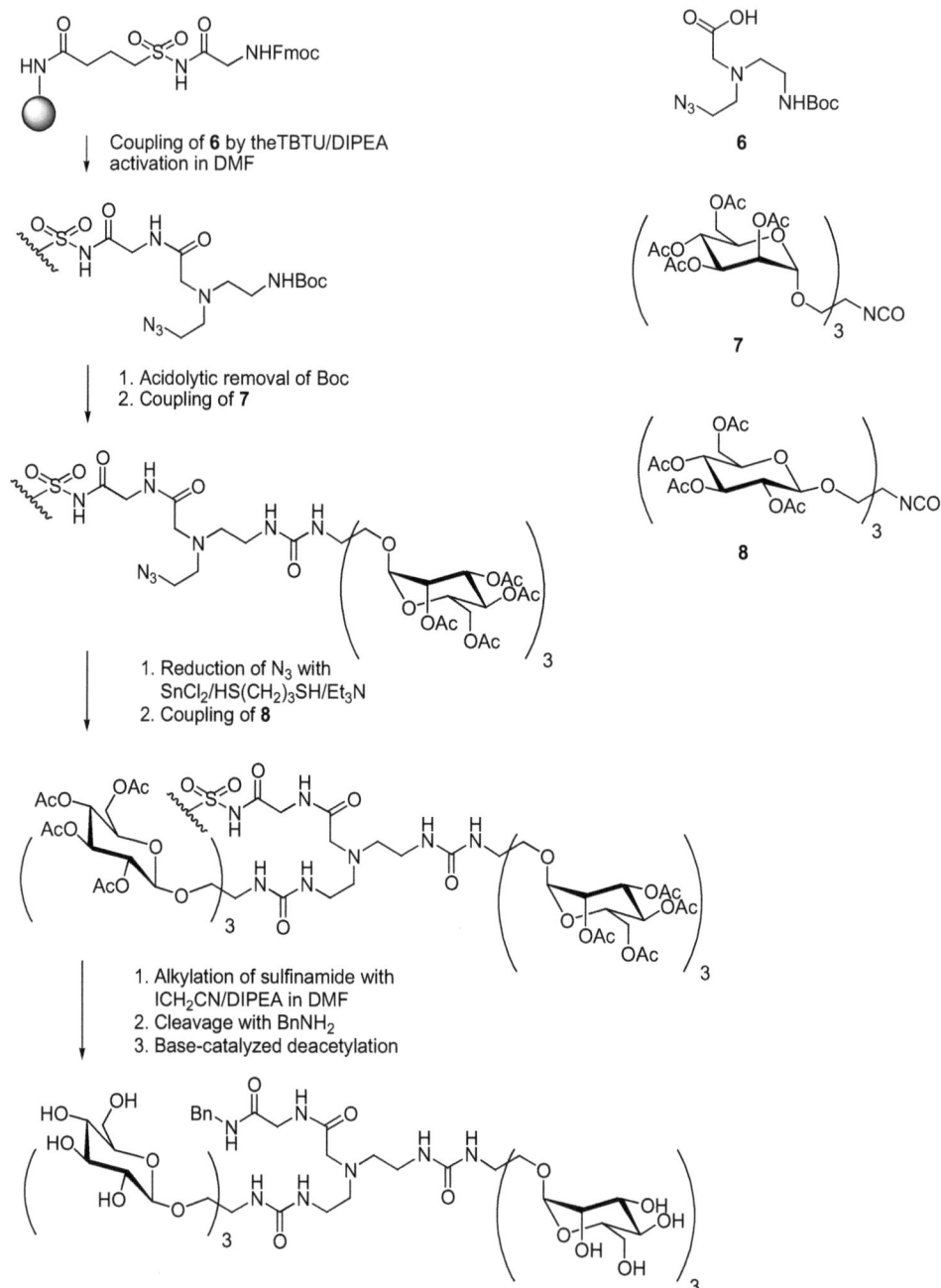

Scheme 7: Conjugation of two different pre-fabricated glycoclusters to a peptide scaffold.

Conjugates containing one nucleoside in addition to two dissimilar sugars have been prepared by using N^α-Fmoc-N^ε-Boc-lysine as a scaffold (Scheme **8**) [52].

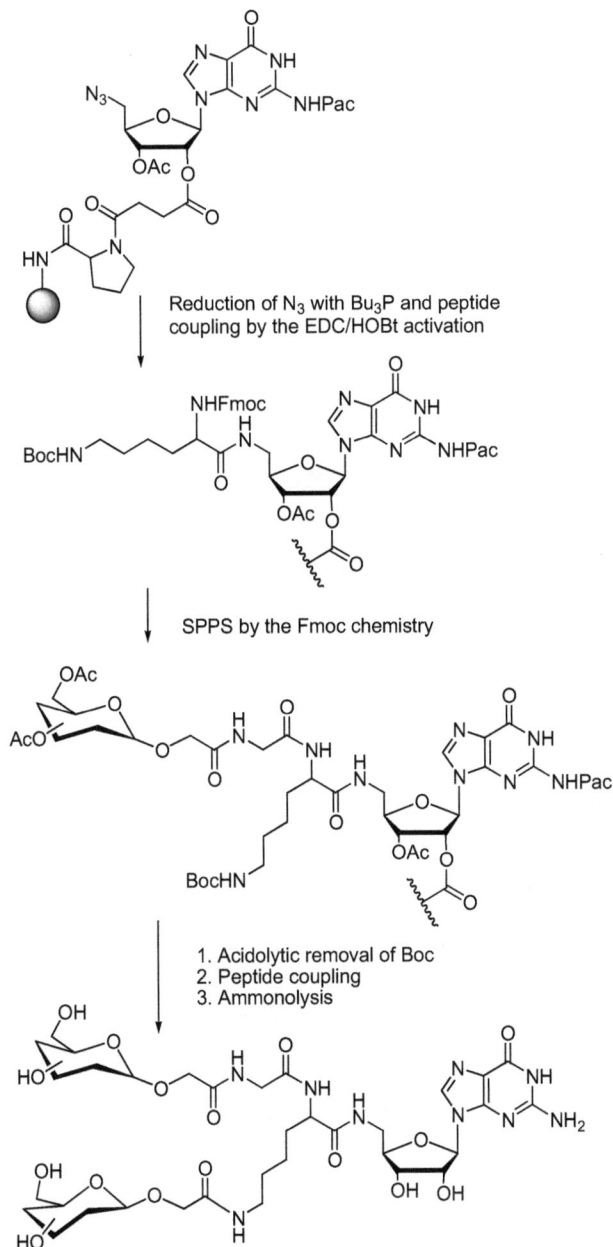

Scheme 8: Synthesis of a heterotopic cluster containing a nucleoside in addition to two dissimilar sugars.

Acyl protected 5′-azido-5′-deoxyguanosine was immobilized through a 2′-*O*-succinyl linker to a proline functionalized support. The azido group was reduced to an amino group and this was then acylated with the orthogonally protected lysine block. The Fmoc group was removed and the exposed amino group was subjected to chain elongation by glycine and the first acetylated 2-hydroxy-2-oxoethyl glycoside. The Boc group was removed and the second glycosidic unit was coupled. The conjugate was released and deprotected by ammonolysis.

4. CYCLIC MULTIANTENNARY GLYCOLUSTERS

Appropriately designed cyclic peptides have received interest as scaffolds that orient the saccharide ligands to emanate from a single face of the ring structure. A cyclic decapeptide scaffold has, for example, been used for this purpose [53]. A linear decapeptide containing four N^ε-Dde protected Lys residues was first assembled by the Fmoc/^{t}Bu strategy on the amino function of an allyl glutamate handle attached to a Rink amide linker through its γ-carboxylic function (Scheme **9**). The allyl and *N*-terminal Fmoc protections were removed and the chain was cyclized by PyAOP/DIPEA activation. The lysine side chains were deprotected and elongated with serine residues, which were subsequently oxidized to glyoxalyl residues. Oximation of these groups with aminooxy-functionalized sugars gave the desired tetra-antennary conjugate. The conjugates were not released in solution but used as support-anchored probes for screening of lectin binding.

Cyclic octapeptides bearing three identical sugar ligands have been obtained by a closely related approach. The linear precursor incorporating three N^ω-[1-(4,4-dimethyl-2,6-dioxocyclohexylidene)-3-methylbutyl] (Ddv) protected Lys or Orn residues and an *N*-terminal N^ε-Boc-Lys was assembled by the Fmoc chemistry on a β-alanine derivatized Sieber linker and cyclized [54]. The Ddv protections were then removed and the exposed amino functions were reacted with 4-nitrophenoxycarbonyl activated 4-hydroxybuten-2-yl glycoside per acetates. The resulting urethane linkages withstood acidolytic release of the conjugate with 1% TFA. The acetyl protections of the sugar moieties were removed in solution.

Solid-phase chemistry has also been successfully applied to preparation of cyclic peptide conjugates bearing three different sugars. In this case, a cyclic β-peptide

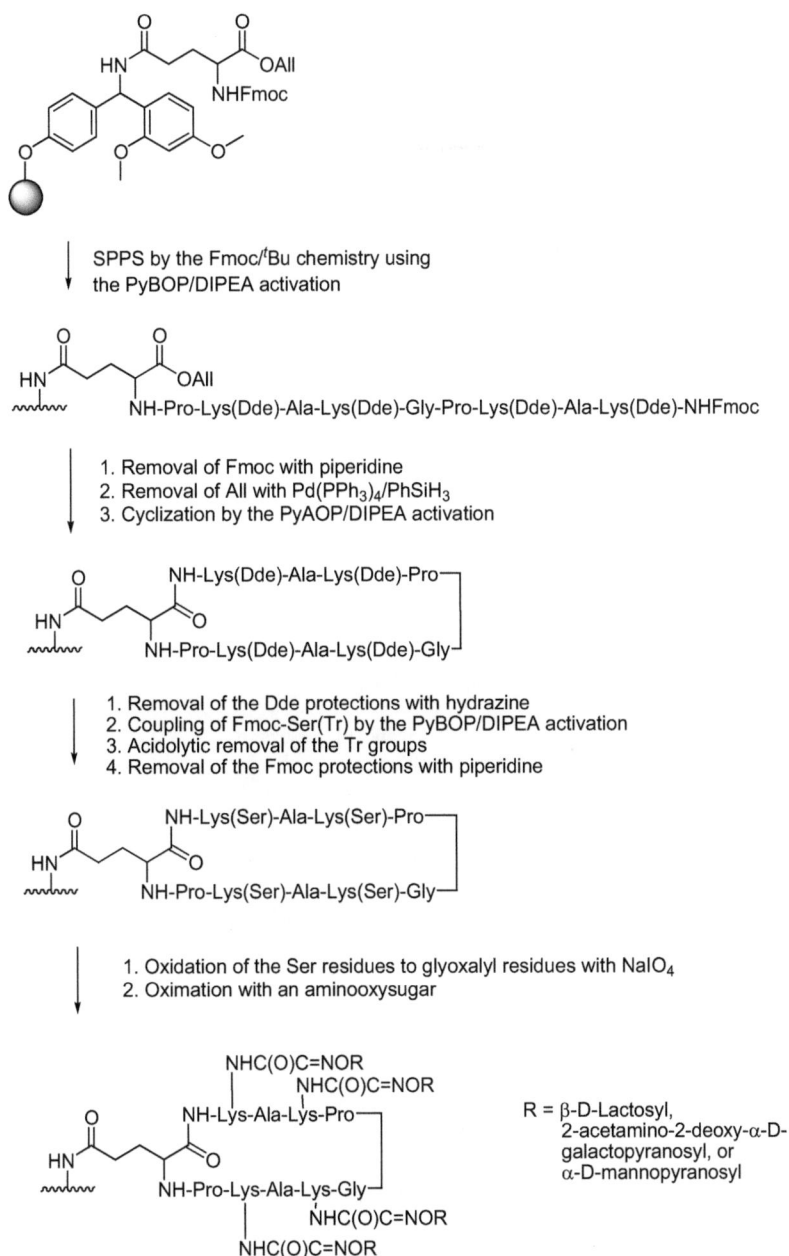

Scheme 9: Synthesis of a homotopic glycocluster on a cyclic peptide scaffold.

structure has been used to orient the sugar ligands (Scheme **10**) [55]. For the peptide assembly, β-alanine and three differently protected β-amino-L-alanines

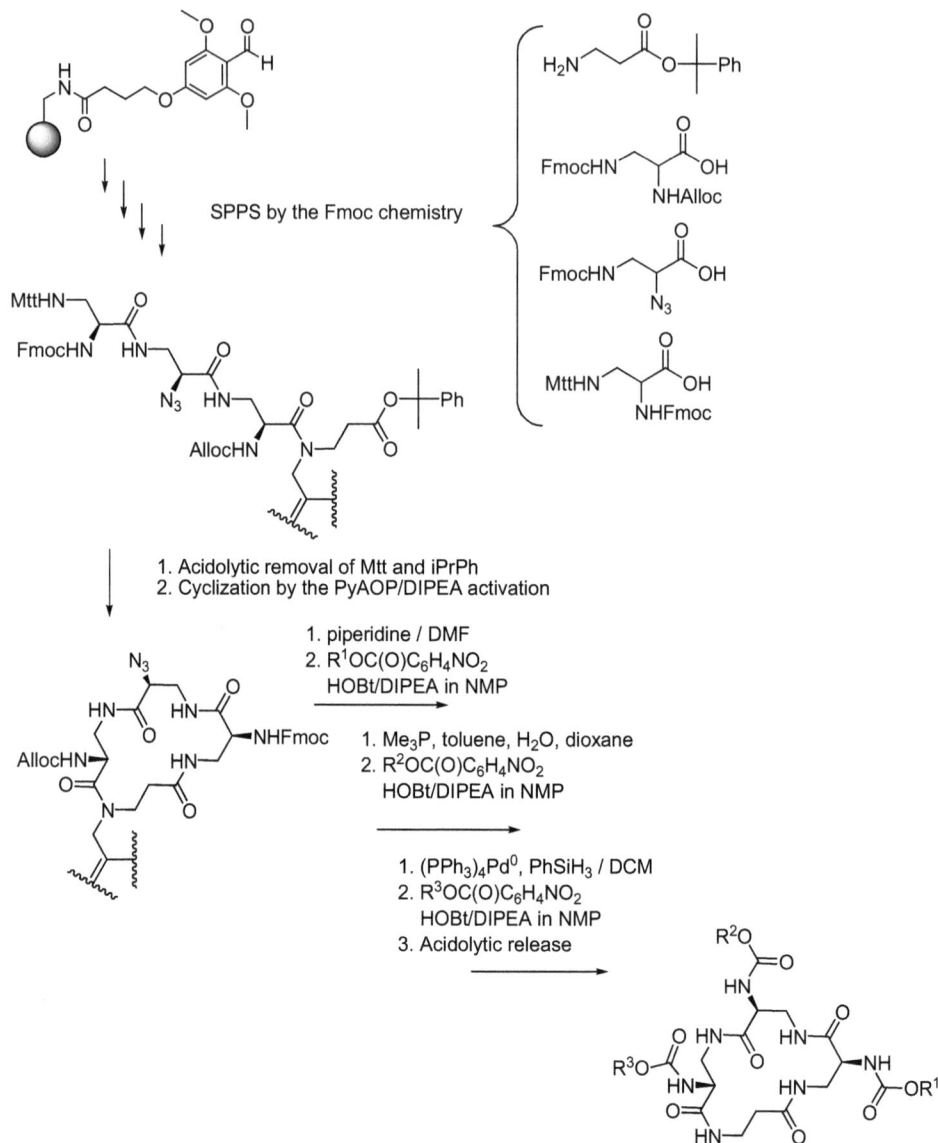

Scheme 10: Synthesis of a heterotopic glycocluster on a cyclic peptide scaffold.

have been used. β-Alanine protected as an acid-labile 1-methyl-1-phenylethyl ester was first immobized by reductive amination to a BAL-linker, and the chain of the orthogonally protected β-amino-L-alanines was assembled on the secondary nitrogen. The acid labile protections from the terminal carboxy and amino functions were removed, and cyclization was achieved by a PyAOP/DIPEA treatment. The α-amino functions were exposed in a stepwise manner by removing the Fmoc group, reducing the azido group to amino group and removing the Alloc group, in this order. Immediately after each de-masking, a 4-nitrophenylcarbonate activated glycoside was coupled. Acidolytic cleavage released the conjugate. The sugar ester protections were removed in solution by a hydrazine treatment.

5. CONSTRUCTS INCORPORATING SUGAR AMINO ACIDS AND THEIR CONGENERS

In spite of the remarkable progress made during the past decade, the solid-phase synthesis of oligosaccharides is still not a routine method [56-61]. The efficiency of glycosidation is not always sufficient to allow rapid solid-supported chain assembly and *cis*-1,2-gycosyl donors tend to yield anomeric mixtures. Hence, to overcome these problems, increasing interest has been paid to construction of oligosaccharide mimetics by achiral linkages, above all by peptide bonds. A tetrasaccharide analog consisting of benzyl 2-amino-2-deoxy-α-D-glucopyranosiduronic acid units [62] and a trisaccharide analog derived from (5-amino-5-deoxy-β-D-arabinofuranosyl)carboxylic acid [63] have been assembled by the Fmoc chemistry on a Rink-amide linker (Scheme **11**). The monomeric units were coupled as azido derivatives by DIC/HOBt activation and the azido group was after each coupling step reduced to an amino function with DTT/DIPEA treatment in DMF. Before acidolytic release with 50% TFA in DCM, the hydroxyl functions were acetylated. Numerous related pseudo-oligosaccharides have been prepared and the advances achieved in this type of chemistry has been recently reviewed [64].

Cyclic structures incorporating both amino acid and sugar units constitute an interesting class of glycoconjugates. This kind of structural analogs of Gramicidin S, a naturally occurring antimicrobial cyclic peptide that lyse microbial cells, offer an illustrative example. This peptide contains two Phe-Pro fragments serving as reverse turns within the cyclic β-sheet structure. Analogs having these dipeptidic moieties replaced with sugars have been prepared by solid-supported assembly of the linear precursor, followed by cyclization in solution (Scheme **12**) [65]. In

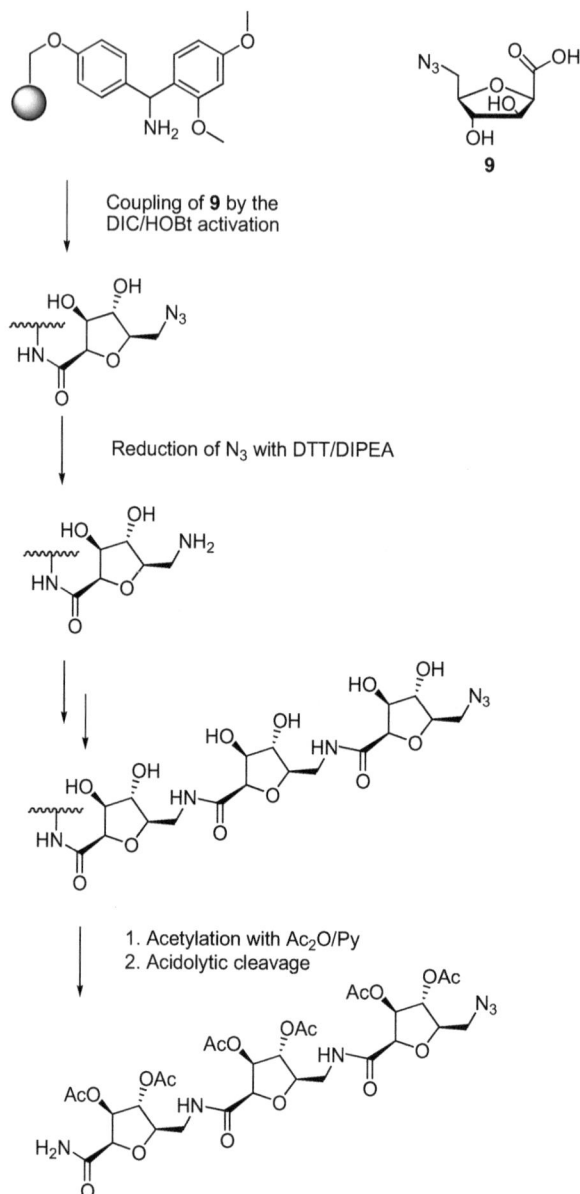

Scheme 11: Assembly of a linear cluster from sugar amino acids.

more detail, the peptide chain was assembled by the Fmoc chemistry on a HMPB-linker attached to MBHA-functionalized polystyrene using the BOP/HOBt/DIPEA activation. The sugar amino acid building blocks, however, contained azido functions instead of a Fmoc protected amino group and, hence, the protecting group removal was at the respective stages of synthesis replaced with reduction by trimethylphosphine. The acyclic precursor still bearing Boc protections on the side chain amino functions was released from the support by mild acid treatment and cyclized in solution through the terminal carboxy and amino groups by a PyBOP/HOBt/DIPEA treatment in DMF. Finally the Boc protections were removed.

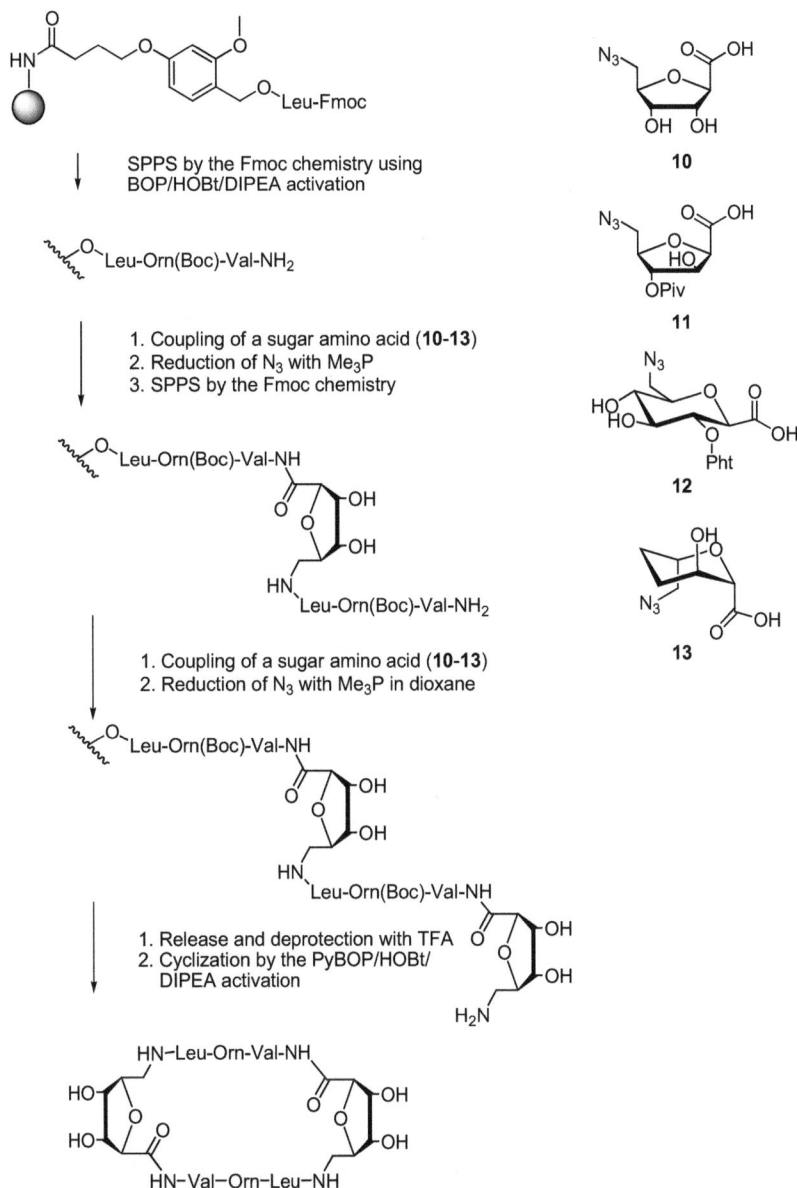

Scheme 12: Synthesis of cyclic structures containing sugar amino acids.

Essentially the same chemistry has been applied to obtain cyclic nonamers having only one of the Phe-Pro fragments replaced with a sugar amino acid, either **14** or **15** [66]. A cyclic tetramer containing two pyranoid sugar amino acids linked together by Gly residues (**16**) [67] has been synthesized by introducing the sugar units as *O*-benzylated Fmoc protected amino acids, not as azides. A bicyclic sugar amino acid derived from arabinofuranose has been incorporated into a cyclic tetrapeptide structure (**17**) by assembling the acyclic precursor on a Sasrin linker and cyclizing it in solution [68]. A noteworthy step in this synthesis is simultaneous reduction of the azido function of the sugar block with tributylphosphine and DIC/HOBt promoted coupling of the next amino acid. Compound **17** is a selective antagonist of $\alpha_v\beta_3$ integrins.

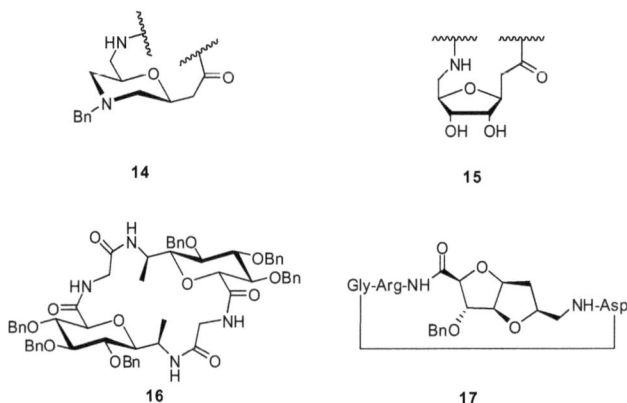

14

15

16

17

In addition to sugar amino acids, glycosylmethylamines have been used to incorporate sugars into cyclic peptide structures. Solid-supported synthesis of cyclic tetramers containing two regular amino acids and two sugar units represent this kind of an approach (Scheme **13**) [69]. *N*-Fmoc protected glutamic acid allyl ester was coupled through the γ-carboxy function to SCAL-linker. The Fmoc protection was removed and an *N*-Boc-protected glycosylmethylamine was coupled as a 4-nitrophenyl activated sugar carbonate. Boc protection was required, since removal of the Fmoc protection was observed to be accompanied by migration of toluoyl groups from sugar hydroxyls to the amino group. The Boc group was removed and the chain was elongated with an *N*-Boc protected amino acid and the second sugar block. The terminal Boc and Alloc protections were removed and cyclization was carried out on-support by using PyBOP as an activator. Reduction of the sulfoxide substituents of the SCAL linker under acidic conditions released the sugar protected conjugate, which was then deprotected in solution by methoxide ion catalyzed transesterification.

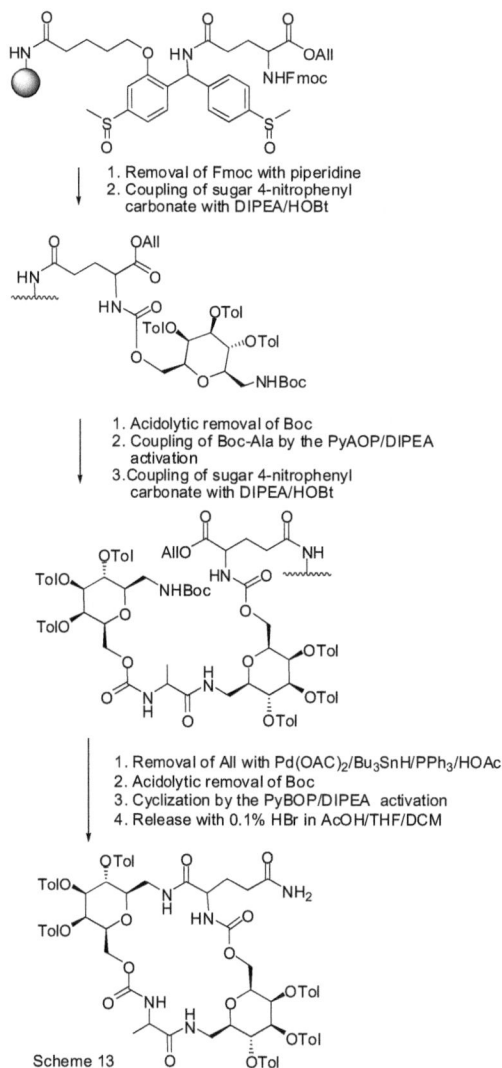

Scheme 13: Synthesis of cyclic structures containing amino acids derived from *C*-glycosides.

6. OLIGONUCLEOTIDE GLYCOCONJUGATES

Conjugation with glycoclusters has been regarded as a viable approach to affect the cellular uptake, cell-type specificity and pharmacokinetics of therapeutic oligonucleotides, including both antisense oligomers and siRNA [18, 70, 71].

Solid-supported synthesis of oligonucleotide glycoconjugates has, however, been rather recently reviewed [72] and, hence, only a short overview of this subject is presented here.

The most straightforward method for incorporation of sugars into oligonucleotides is the use of pre-fabricated building blocks of nucleoside glycoconjugates, such as **18** and **19**, in conventional oligonucleotide synthesis by the phosphoramidite strategy [73-77]. When an exceptionally long coupling time is used, this approach may be applied to introduction of even rather complex carbohydrate groups, such as neomycin, an aminoglycoside antibiotic (**20**) [78]. Alternatively, thymidine analogs bearing an *N*-hydroxysuccinimido functionalized side-arm instead of the 5-methyl group may be introduced in desired positions within the oligonucleotide chain and then post-synthetically derivatized with aminofunctionalized oligosaccharides [79].

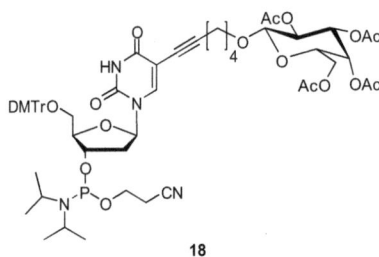

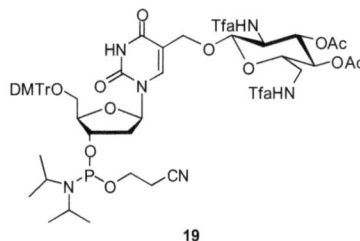

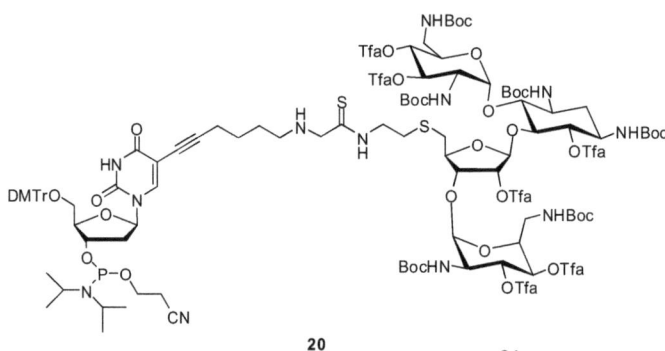

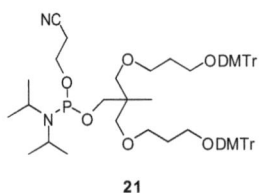

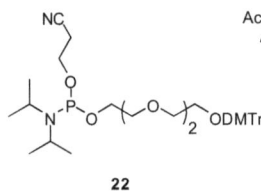

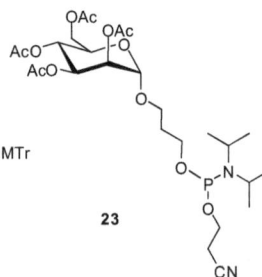

Another simple way to provide oligonucleotide with an additional carbohydrate moiety is introduction of a pre-fabricated acyl-protected sugar phosphoramidite to the 5′-terminus of the support-bound oligonuclotide as the last step of the chain assembly [81-83]. As an extension of this methodology, a tree-like tetramannosyl conjugate has been prepared by using phosphoramidate **21** for branching, **22** for elongation of the branches and **23** for the terminal mannosylation [84]. A similar approach has more recently been applied to obtain related oligonucleotide glycoconjugates [85]. Even more impressively, mannose has been used as the branching unit [86]. This allows tetraantennary branching and, hence, constructs bearing 16 terminal galactopyranosides have been obtained. In more detail, propargyl α-D-mannopyranoside was first immobilized by Cu(I) promoted 1,3-dipolar cycloaddition to an ω-azidoalkyl arm of a solid-supported branched linker. The hydroxyl groups were phosphitylated with pent-4-yn-1-yl phosphoramidite and subjected to another click reaction with propargyl α-D-mannopyranoside. The phosphitylation cycle was repeated and the terminal sugar units were introduces as peracetylated 4-(azidomethyl)cyclohexylmethyl β-D-galactopyranosides. Finally, the oligonucleotide chain was assembled on another arm of the branched linker. A similar strategy has been used to obtain fucosyl clusters containing up to 8 sugar units [87].

3′-terminal sugar conjugates are, in turn, obtained by utilizing an *O*-acetyled glycoside as a handle immobilized to the support *via* a succinyl linker [80, 81, 88]. The oligonucleotide is the assembled on one of the sugar hydroxyl groups. 3′-Conjugates of even structurally complex aminoglycoside antibiotics have been prepared in a similar manner [89]. Trigalactosylated tetrahydroxycholane has been introduced into the 3′-terminus with the aid of a branched linker (**24**) and to the 5′-terminus as a phosphoramidite reagent (**25**) [71]. 3′-(β-Cyclodextrin) conjugates have been obtained by assembling the oligonucleotide chain in the reversed 5′→ 3′ and conjugating an amino-functionalized cyclodextrin to the terminal 3′-hydroxy function activated as a 4-nitrophenyl carbonate [90]. 3′-Conjugates derived from an amide-linked tetrasaccharide mimetic have been prepared by sequential peptide and phosphoramidite coupling on a single support [91]. The 3′-sugar tail may also be elongated by coupling one or more sugar phosphoramidites before the first nucleosidic building blocks [92].

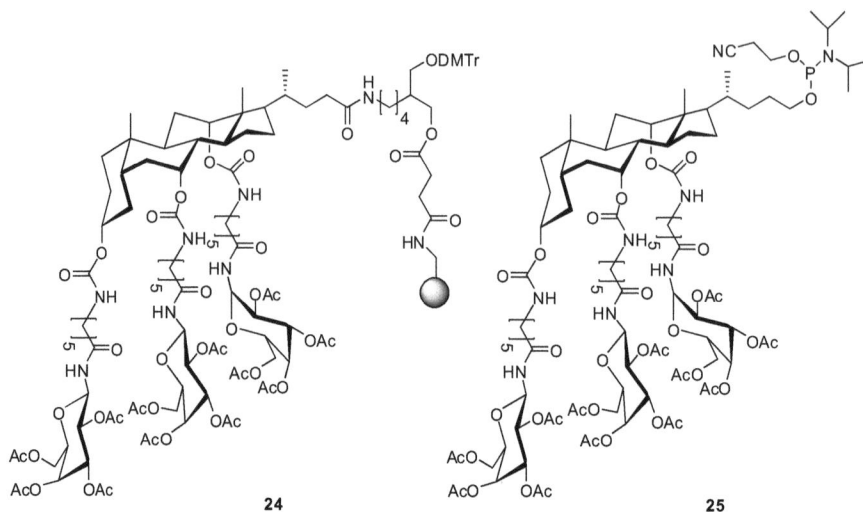

24   **25**

Sugar phosphoramidites also allow introduction of sugar moiety into intrachain positions in oligonucleotides [93]. Ribostamycin, an aminoglycoside antibiotics, has been introduced into an intrachain position by inserting an abasic ribofuranose phosphoramidite bearing two levulinoyloxyalkyl side-arms into the oligonucleotide chain, removing the levulinoyl protections on-support and coupling the ribostamycin phosphoramidite reagent to the exposed hydroxyl functions (Scheme **14**) [94].

Phosphoramidite chemistry has also been exploited in construction of oligonucleotide glycoconjugates bearing three different sugars. For this purpose, a non-nucleosidic building block bearing three orthogonally removable hydroxy protections was coupled to the 5′-terminus of a solid-supported oligonucleotide (Scheme **15**) [95]. The hydroxyl protections were removed in a stepwise manner and the sugar units were introduced as phosphoramidite reagents. Methyl protected phosphoramidites and a polymer support had to be used for the chain assembly, since the 2-cyanoethyl groups and CPG did not withstand removal of the fluoride ion labile TBDPS group. The methyl protections were removed from the phosphodiester likages before standard ammonolysis.

Click chemistry is another conjugation procedure extensively applied to preparation of oligonucleotide glycoconjugates. Sugar azides are usually conjugated to a support-anchored alkynylated oligonucleotide. 3′-Terminal conjugates bearing up to three β-D-galactopyranosyl or α-D-mannopyranosyl groups have been obtained by

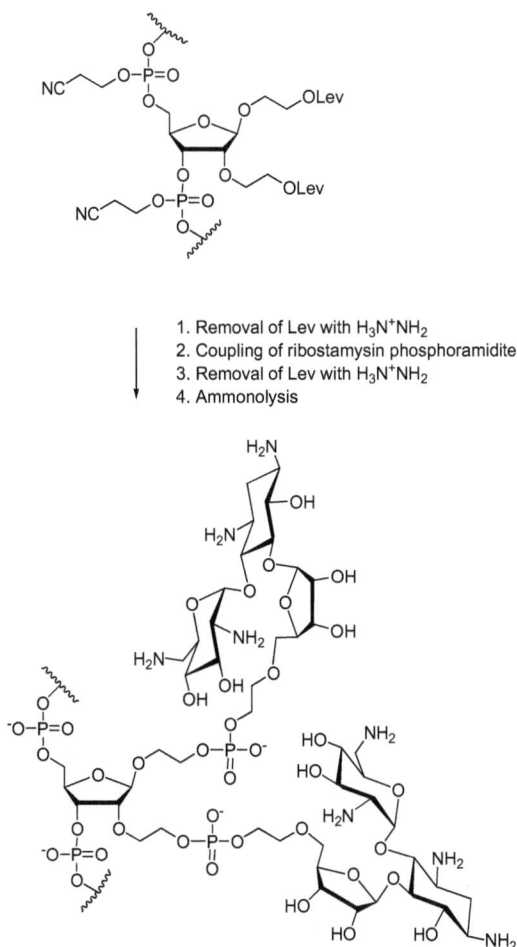

Scheme 14: Incorporation of aminoglycoside antibiotic structures into oligonucleotides.

consecutive application of oxidative amidation and microwave assisted click chemistry, as outlined in Scheme **16** [96, 97]. A non-nucleosidic tail was first assembled by the H-phosphonate chemistry, the H-phosphonate diester linkages were subjected to oxidation with propargylamine, and the azido-functionalized sugars were then attached by the click reaction under microwave irradiation. The oligonucleotide sequence was finally assembled by the phosphoramidate chemistry on the hydroxyl function of the last non-nucleosidic unit and conventional ammonolysis completed the synthesis. Azido-functionalized sugars have also been conjugated to 4´-(4-pentyn-1-yl)aminomethylthymidines introduced into several intrachain positions [98]. Introduction of a cyclooctyne derived non-nucleosidic phosphoramidite (**26**) in desired positions within the oligonucleotide chain has enabled conjugation of azido-functionalized sugars by copper-free 1,3-cycloaddition [99].

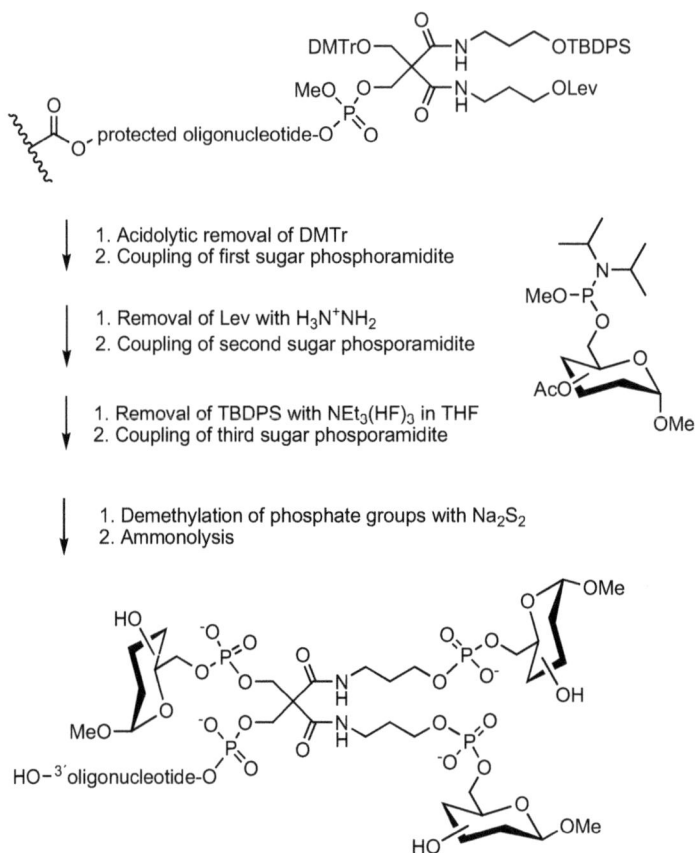

Scheme 15: Synthesis of oligonucleotide glycoconjugates containing three different sugars.

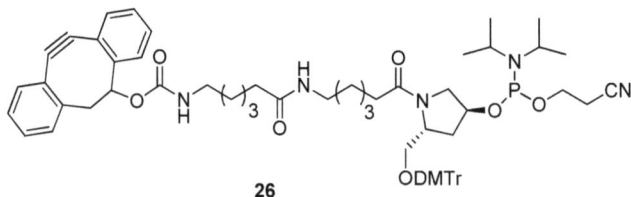

26

The reaction of alkynylated sugars with support-bound azido-functionalized oligonucleotides is, however, also possible [100]. A 5′-DMTr protected 3′-*H*-phosphonate monomer has been inserted into the oligonucleotide by the *H*-phosphonate chemistry and propyne derivatized sugars have then been attached

(Scheme **17**). It is worth nothing that the corresponding phosphoramidite building block cannot be prepared, owing to intramolecular Staudinger reaction between the azido and phosphoramidite moieties. Introduction of nucleosidic [101] or non-nucleosidic [102] building blocks bearing terminal alkynyl and ω-haloalkyl groups into the oligonucleotide chain has allowed introduction of two different sugars in selected positions by the click reaction. First, an azido-functionalized sugar is conjugated and then the halogen substituent is exchanged to azido on-support and an alkynyl derivatized sugar is coupled.

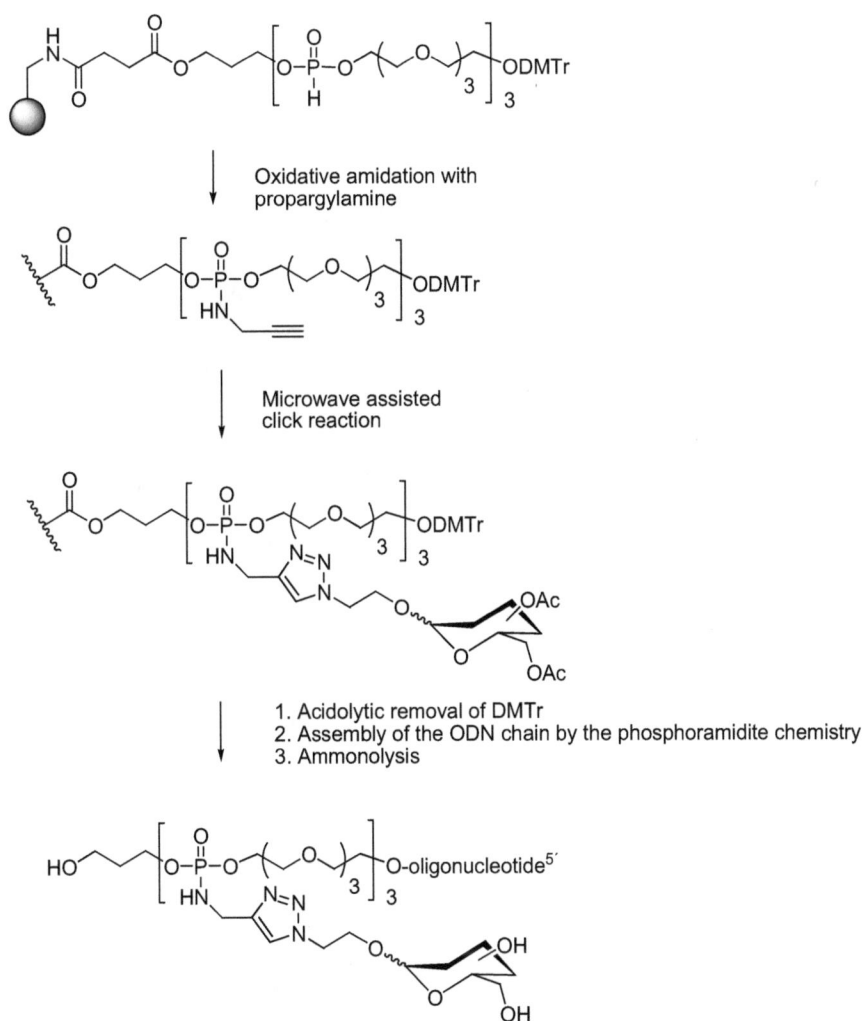

Scheme 16: Conjugation of azido-functionalized sugars to an ω-alkynyl-functionalized phosphodiester scaffold.

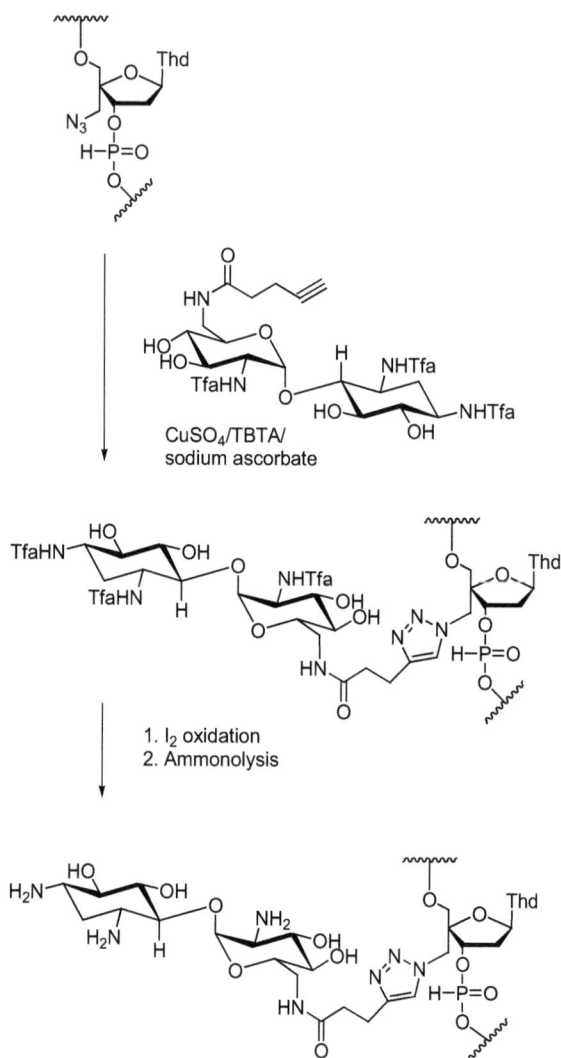

Scheme 17: Conjugation of ω-alkynyl-functionalized sugars to an azido-functionalized oligonucleotide.

Besides phosphoramidite coupling and click reaction, solid-supported oximation has been applied to preparation of oligonucleotide glycoconjugates. Conjugates bearing up to 6 identical mono- [103] or di-saccharides [104] were obtained as depicted in Scheme **18**. Three non-nucleosidic phosphoramidite building blocks, each bearing two phthaloyl protected aminooxy functions, were inserted into the oligonucleotide chain by the phosphoramidite chemistry. The phthaloyl groups were removed on-support and aldehyde-functionalized sugars introduced by oximation (Scheme **18**). Solid-supported oximation has also utilized for the conjugation of pre-fabricated Glu-clusters to aminooxy functionalized oligonucleotides [105]. The oxime linkages withstood standard ammonolysis, which also quantitatively removed the acetyl protections from the sugar hydroxyl functions.

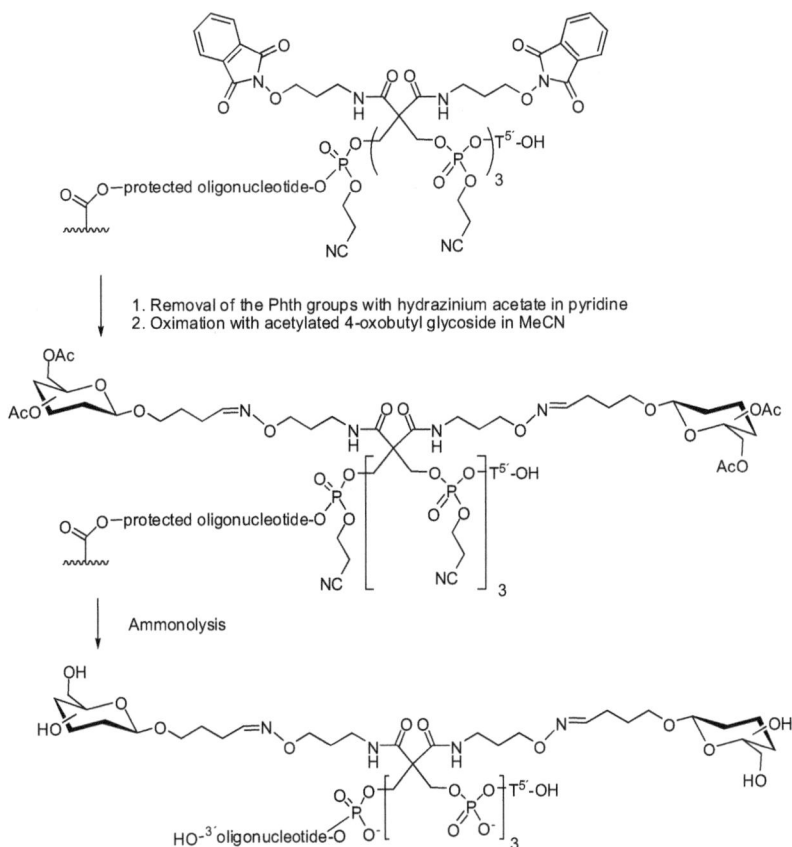

Scheme 18

Scheme 18: Exploitation of solid-supported oximation for construction of oligonucleotide glycoconjugates.

7. SUGARS AS SCAFFOLDS

Since the density of functional groups in sugars is high and their mutual orientation depends on the sugar ring configuration, monosaccharides are attractive candidates for scaffolds, with which the spatial orientation of the conjugate groups attached to the sugar core may be systematically tuned [106-108]. The feasibility of this approach has been demonstrated by using orthogonally protected D-glucopyranose [109, 110], its 2-amino analog [111] and D-galactopyranose [112] as support-anchored scaffolds. Scheme **19** outlines, as an illustrative example, the use of 3-*O*-allyl-4-*O*-(1-ethoxyethyl-6-*O*-TBDMS-1-deoxy-1-thio-β-D-galactopyranose as a multipodal handle. This handle was immobilized as a carboxy-functionalized

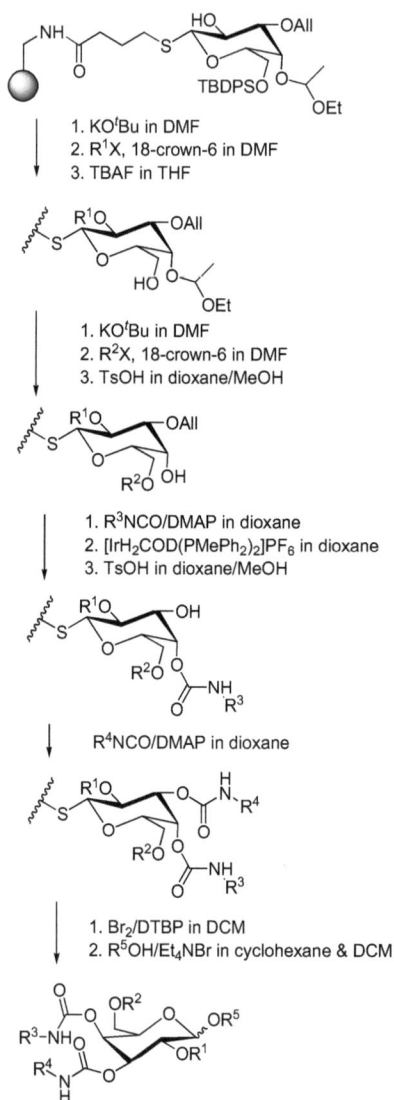

Scheme 19: An orthogonally protected monosaccharide as a scaffold for construction of glycoclusters.

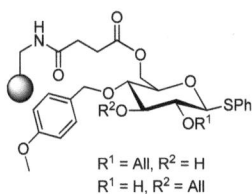

27

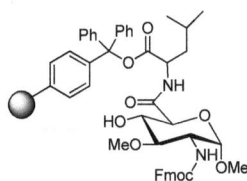

28

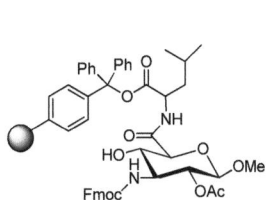

29

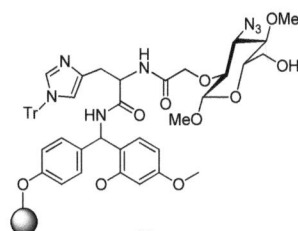

30

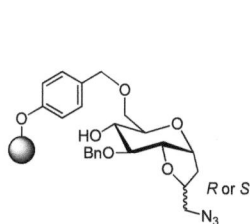

31

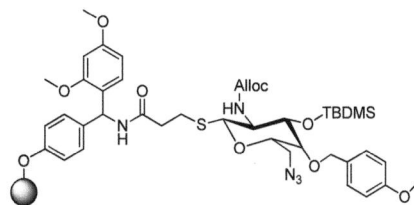

32

thioglycoside through an amide linkage to an amino-functionalized support. The unprotected 2-hydroxy function was alkylated and the other three hydroxyl groups were deblocked and derivatized in a stepwide manner. Accordingly, the TBDPS group was removed with fluoride ion and the exposed primary hydroxyl group was alkylated. The acid-labile 1-ethoxyethyl group was then removed and the exposed 4-OH was subjected to carbamoylation with isocyanates. Finally, the allyl group was removed by sequential treatments with an iridium complex and *p*-toluenesulfonic acid and then carbamoylated. The conjugate was released from the support by bromine promoted alcoholysis, which gave the product as an anomeric mixture. A very similar approach has been applied to obtain glucopyranose conjugates containing a permanent propyl ether protection instead of removable allyl group [109, 113]. In addition to derivatization of 4-OH and 6-OH with isocyanates, 1,1′-carbonyldiimidazole and trichloromethyl carbonochloridate activation followed by treatment with an amine has been applied to carbamoylation of these functions [114]. Esterification of the same groups with Boc-Gly-OH has allowed further derivatization by peptide coupling [114]. A thiophenyl D-glucopyranoside scaffold bearing orthogonally removable 4-methoxybenzyl and allyl group at 4-OH and at 2-OH or 3-OH, respectively, has been immobilized to an amino-functionalized support through a 6-*O*-linker (27), but not used for library any synthesis [115].

2,3-Anhydrolevoglucosan has been tethered to the Rink amide linker through 4-OH and functionalized by an attack of a variety of oxygen, nitrogen and sulfur nucleophiles on C2. Finally, the exposed 3-OH has been alkylated (Scheme 20) [116]. Products containing a iodophenyl group on the C2- or C3-substituent, or incorporated in the linker structure, have further been deriatized by Pd-promoted alkynylation or arylation.

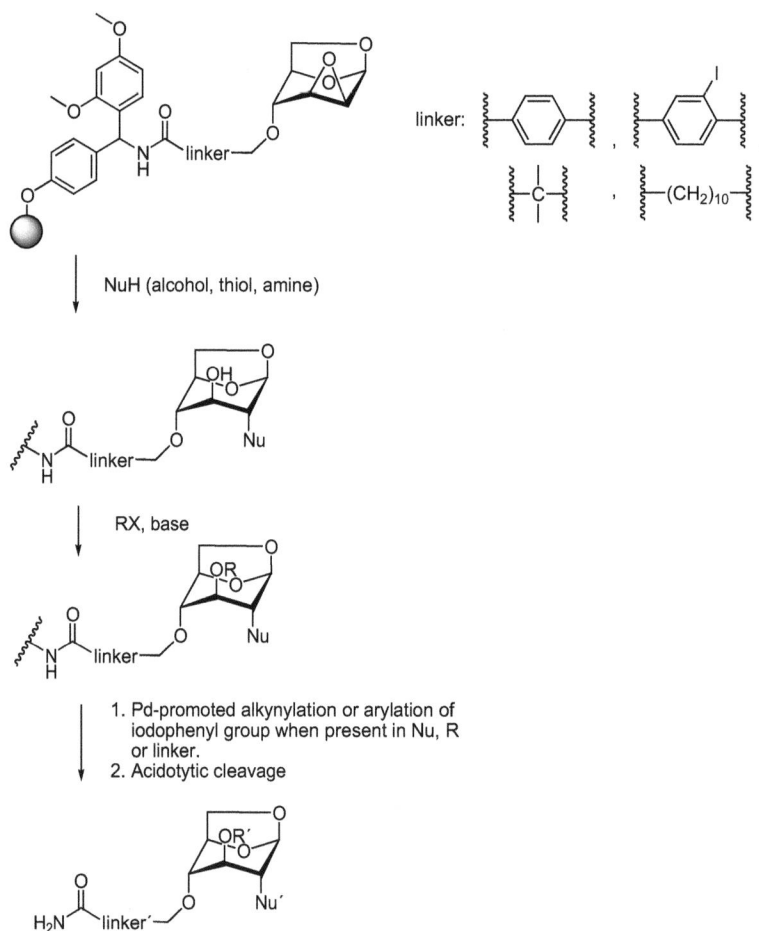

Scheme 20: 2,3-Anhydrolevoglucosan as a scaffold for construction of glycoclusters.

Besides hexopyranoses, their amino congeners have been used as scaffolds. Library syntheses have been carried out on trityl-linked supports (**28**) and (**29**) by first reacting the unprotected hydroxyl group with isocyanates and then acylating the deprotected amino group [117]. A related procedure involves coupling of methyl 3-azido-3-deoxy-2-*O*-carboxymethyl-4-*O*-methyl-β-D-glucopyranoside scaffold to a His(Tr)-Rink-amide linker (**30**), carbamoylation of the 6-hydroxy group with isocyanates, reduction of the azido function and acylation of the resulting amino group [118]. A similar methodology has more recently been applied to decoration of *cis*-fused perhydrofuropyrans (**31**) on a Wang linker [119]. After acylation or carbamoylation of 4-OH, the azido function is reduced to amino group, which is then converted to an amido, sulfonamido, urea, thiourea, or *N*-alkylamido function. An alternative approach for preparation of bicyclic sugar-derived compounds involves solid-phase functionalization of a monosaccharide followed by ring-closing methathesis (Scheme **21**) [120].

A scaffold bearing four orthogonally, not only selectively, removable protecting groups has been prepared from 2,6-diamino-2,6-dideoxy-D-galactose (**32**) [121]. Any of the groups may be removed leaving the others intact: Alloc with Pd(PPh$_3$)$_4$ in dioxane, TBDMS with tetrabutylammonium hydrogen difluoride in MeCN, 4-methoxybenzyl with DDQ in wet DCM and the azido function undergoes Staudinger reduction with Bu$_3$P in THF. Among the exposed functionalities, the hydroxyl groups have been subjected to carbamoylation with isocyanates and the amino and 3-OH groups to acylation [122]. The conjugates have been released by acidolytic cleavage of the Rink amide linker giving 4-amino-4-oxobutyl thioglycosides.

A library of lipidated disaccharides mimicking the structure of a disaccharidic degradation product of Moenomycin A has been generated by solid-phase chemistry. Meonomycin A is a pentasaccharide bearing a phosphoglycerate linked long chain lipid moiety [123]. It inhibits transglycosylases playing role in the synthesis of bacterial cell wall peptidodlycans, and the disaccharide core structure partly retains this inhibitory activity [124]. As indicated in Scheme **22**, an appropriately protected disaccharide thioglycoside was attached *via* the C6-carboxy function to a photo-labile linker, the levulinoyl protected 3-OH of the

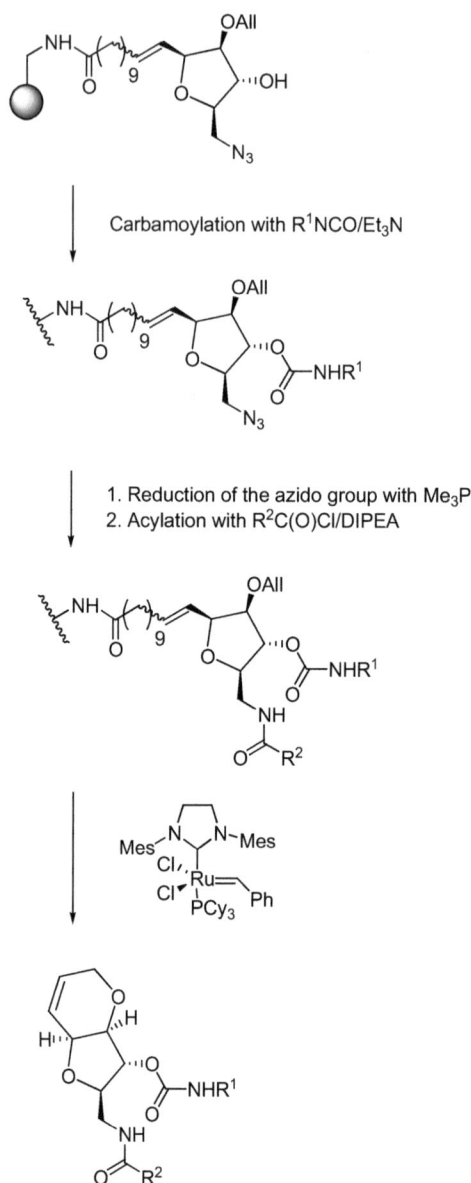

Scheme 21: An alkene linker in glycocluster synthesis: cleavage by ring-closing methathesis.

same sugar unit was deblocked and subjected to carbamoylation, and the thioglycosidic bond was hydrolyzed [125]. The anomeric hydroxyl group was then phosphitylated with the desired lipid phosphoramidite. Alternatively, a C3 azido function was exploited. The group was reduced to an amino group and acylated before the phosphitylation step.

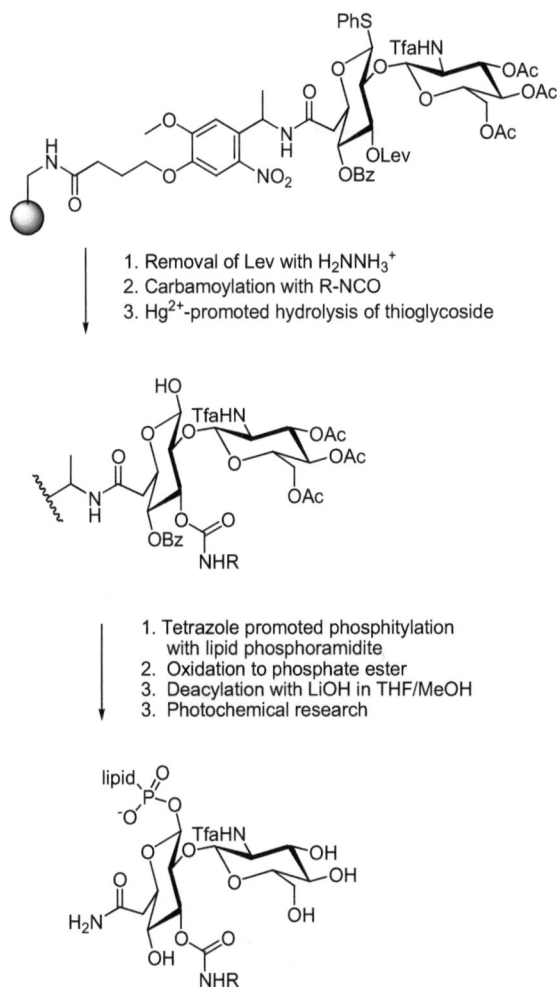

Scheme 22: Synthesis of lipidated glycoclusters.

A four-helix peptide bundle has been assembled using solid-supported galactopyranoside as a core structure (Scheme **23**) [126].

Bleomycin A$_5$ (**38**), an anticancer antibiotic isolated from *Streptomyces verticillus*, and its analogs containing a monosaccharide unit instead of the naturally occurring disaccharide, have also been prepared on a solid support [127]. As outlined in Scheme **24**, the synthesis was carried out on a Dde-based linker derivatized with spermidine, the secondary amino group of which was protected

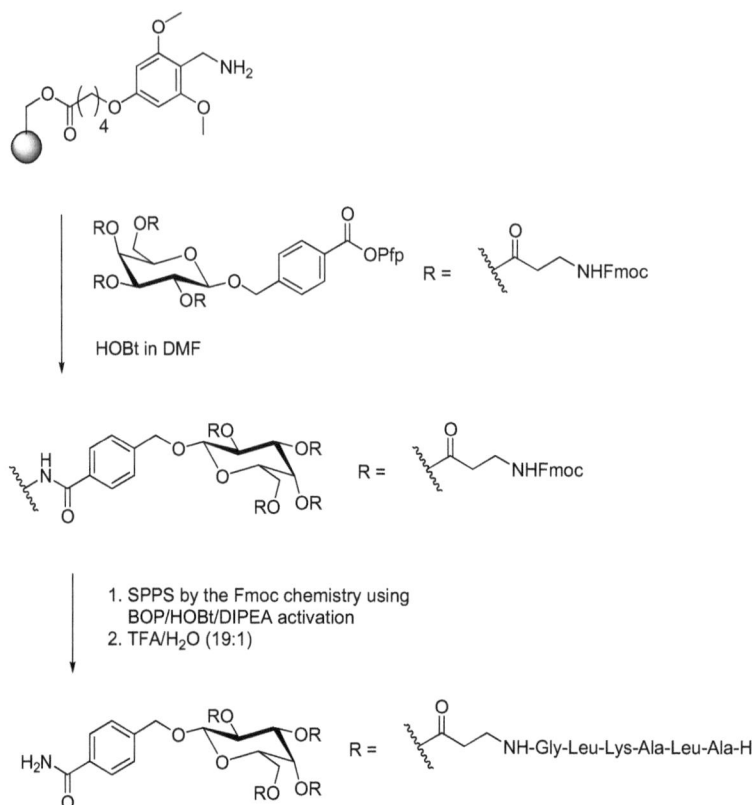

Scheme 23: Preparation of a four-helix bundle on a galactopyranoside scaffold.

with a Boc or NBS group. Fmoc protected 2'-(2-aminoethyl)-2,4'-bithiazole-4-carboxylic acid (**33**), N^α-Fmoc-(*S*)-threonine (**34**) and N^α-Fmoc-(2*S*,3*S*,4*R*)-4-amino-3-hydroxy-2-methylpentanoic acid (**35**) were attached in this order by peptide coupling and the sugar moieties were then added as pre-fabricated histidine conjugates (**36**), again by peptide coupling. Finally, the *N*-Boc protected pyrimidoblamic acid (**37**) moiety was introduced, the remaining protecting groups were acidolytically removed and the conjugate was released by a hydrazine treatment.

ACKNOWLEDGEMENT

Declared none.

CONFLICT OF INTEREST

The author(s) confirm that this chapter content has no conflict of interest.

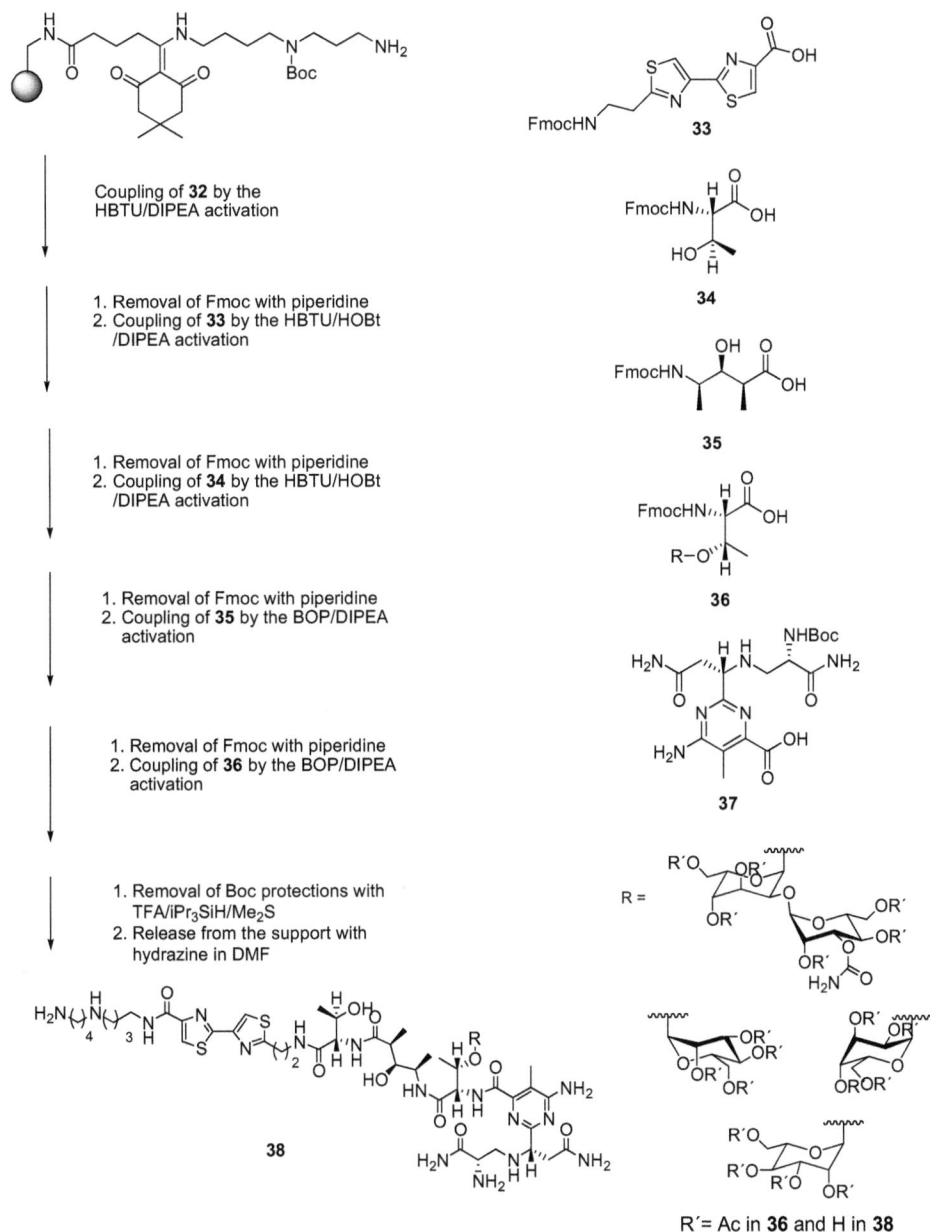

Scheme 24: Synthesis of Bleomycin A₅ and its congeners.

DISCLOSURE

The chapter submitted for series eBook titled **"Advances in Organic Synthesis, Volume 5"** is an update of our article published in **CURRENT ORGANIC**

SYNTHESIS, Volume 6, Number 4, November Issue 2009, with additional text and references.

ABBREVIATIONS

Ac   = Acetyl

All   = Allyl

Alloc   = Allyloxycarbonyl

CPG   = Controlled pore glass

Cy   = Cyclohexyl

Dde   = 1-(4,4-Dimethyl-2,6-dioxocyclohexylidene)ethyl

Ddv   = 1-(4,4-Dimethyl-2,6-dioxocyclohexylidene)-3-methylbutyl

DIC   = Diisopropylcarbodiimide

DMTr   = 4,4′-Dimethoxytrityl

DTBP   = 2,6-Di-*tert*-butylpyridine

Boc   = *tert*-Butoxycarbonyl

BOP   = (Benzotriazol-1-yloxy)tris(dimethylamino)phosphonium hexafluorophosphate

Bz   = Benzoyl

Bn   = Benzyl

DIPEA   = *N*,*N*-Diisopropylethylamine

DCM   = Dichloromethane

DDQ   = 2,3-Dichloro-5,6-dicyanobenzoquinone

DMAP   = 4-(Dimethylamino)pyridine

DMF   = *N,N*-dimethylformamide

DTT   = Dithiotreitol

EDC   = 1-Ethyl-3-(3-dimetylaminopropyl)carbodiimide

Fmoc   = Fluoren-9-ylmethoxycarbonyl

HATU   = *O*-(7-azabenzotriazol-1-yl)-1,1,3,3-tetramethyluronium
            hexafluorophosphate

HBTU   = *O*-(Benzotriazol-1-yl)-1,1,3,3-tetramethyluronium
            hexafluorophosphate

HMPB   = 4-(4-Hydroxymethyl-3-methoxyphenoxy)butanoic acid

HOBt   = 1-Hydroxybenzotriazole

iPrPh   = 1-Methyl-1-phenylethyl

Lev   = Levulinoyl

MBHA   = 4-Methylbenzhydrylamine

Mes   = Mesitylene

Mtt   = 4-Methyltrityl

NBS   = 2-Nitrobenzenesulfonyl

NEM   = *N*-Ethylmorpholine

Pac   = Phenoxyacetyl

Pfp   = Pentafluorophenyl

Phth   = Phthaloyl

PyAOP = 7-Azabenzotriazol-1-yloxytris(pyrrolidino)phosphonium Hexafluorophosphate

PyBOB = Benzotriazol-1-yloxytris(pyrrolidino)phosphonium hexafluorophosphate

SCAL = Safety catch linker

SP = Solid phase

SPPS = Solid phase peptide synthesis

TATU = *O*-(7-azabenzotriazol-1-yl)-1,1,3,3-tetramethyluronium tetrafluoroborate

TBAF = Tetrabutylammonium fluoride

TBDPS = *tert*-Butyldiphenylsilyl

TBDMS = *tert*-Butyldimethylsilyl

TBTA = Tris[(1-benzyl-1,2,3-triazol-4-yl)methyl]amine

TBTU = *O*-(Benzotriazol-1-yl)-1,1,3,3-tetramethyluronium tetrafluoroborate

TFA = Trifluoroacetic acid

Tfa = Trifluoroacetyl

Thd = Thymidine

THF = Tetrahydrofuran

TIS = Triisopropylsilane

Tr = Trityl

Ts = Tosyl

REFERENCES

[1]   Dwek, R. A. Glycobiology: Toward understanding the function of sugars. *Chem. Rev.* **1996**, *96*, 683-720.

[2]   Gabius, H. –J. Siebert, H. –C.; André, S.; Jiménez-Barbero, J.; Rűdiger, H. Chemical biology of the sugar code.*ChemBioChem* **2004**, *5*, 740-764.

[3]   Drickamer, K. Increasing diversity of animal lectin structures. *Curr. Opin. Struct. Biol.* **1995**, *5*, 612-616.

[4]   Lee, Y. C.; Lee, R. T. Carbohydrate-protein interactions: basis of glycobiology. *Acc. Chem. Res.* **1995***, 28*, 321-327.

[5]   Mammen, M.; Choi, S. –K.; Whitesides, G. M. Polyvalent interactions in biological systems: implications for design and use of multivalent ligands and inhibitors. *Angew. Chem. Int. Ed.* **1998**, *37*, 2754-2794.

[6]   Lundquist, J. J.; Toone, E. J. The cluster glycoside effect. *Chem. Rev.* **2002**, *102*, 555-578.

[7]   Collins, B. E.; Paulson, J. C. Cell surface biology mediated by low affinity multivalent protein–glycan interactions. *Curr. Opin. Chem. Biol.* **2004**, *8*, 617-625.

[8]   Doores, K. J.; Gamblin, D. P.; Davis, B. G. Exploring and exploiting the therapeutic potential of glycoconjugates. *Chem. Eur. J.* **2006**, *12*, 656-665.

[9]   Kiessling, L. L.; Gestwicki, J. E.; Strong, L. E. Synthetic multivalent ligands as probes of signal transduction. *Angew. Chem. Int. Ed.* **2006**, *45*, 2348-2368.

[10]   Toole, B. P. Hyaluronan: from extracellular glue to pericellular cue. *Nat. Rev. Cancer* **2004**, *4*, 528-539.

[11]   De Stefano, I.; Battaglia, A.; Zannoni, G. F.; Prisco, M. G.; Fattorossi, A.; Travaglia, D.; Baroni, S.; Renier, D.; Scambia, G.; Ferlini, C.; Gallo, D. Hyaluronic acid–paclitaxel: effects of intraperitoneal administration against CD44(+) human ovarian cancer xenografts. *Cancer Chemother. Pharmacol.* **2011**, *68*, 107-116.

[12]   Veiseh, M.; Turley, E. A. Hyaluronan metabolism in remodeling extracellular matrix: probes for imaging and therapy of breast cancer. *Integr. Biol.* **2011**, *3*, 304–315.

[13]   Zhu, L.; Ye, Z.; Cheng, K.; Miller, D. D.; Mahato, R. I. Site-specific delivery of oligonucleotides to hepatocytes after systemic administration. *Bioconjugate Chem.* **2008**, *19*, 290-298.

[14]   Zhu, L.; Mahato, R. I. Targeted delivery of siRNA to hepatocytes and hepatic stellate cells by bioconjugation. *Bioconjugate Chem.* **2010**, *21,* 2119–2127.

[15]   Srinivas, O.; Larrieu, P.; Duverger, E.; Boccaccio, C.; Bousser, M.-T.; Monsigny, M.; Fonteneau, J.-F.; Jotereau, F.; Roche, A.-C. Synthesis of glycocluster-tumor antigenic peptide conjugates for dendritic cell targeting. *Bioconjugate Chem.* **2007**, *18*, 1547-1554.

[16]   Rendle, P. M.; Seger, A.; Rodrigues, J.; Oldham, N. J.; Bott, R. R.; Jones, J. B.; Cowan, M. M.; Davis, B. G. Glycodendriproteins: a synthetic glycoprotein mimic enzyme with branched sugar-display potently inhibits bacterial aggregation. *J. Am. Chem. Soc.* **2004**, *126*, 4750-4751.

[17]   Lorenz, B.; de Cienfuegos, L. A.; Oelkers,M.; Kriemen, E.; Brand, C.; Stephan, M.; Sunnick, E.; Yu¨ksel, D.; Kalsani, V.; Kumar, K.; Werz, D. B.; Janshoff, A. Model system for cell adhesion mediated by weak carbohydrate–carbohydrate interactions. *J. Am. Chem. Soc.* **2012**, *134*, 3326–3329.

[18]   Yan, H.; Tram, K. Glycotargeting to improve cellular delivery efficiency of nucleic acids. *Glycoconj. J.* **2007**, *24*, 107–123.

[19] Kiessling, L.; Gestwicki, J. E.; Strong, L. E. Synthetic multivalent ligands in the exploration of cell-surface interactions. *Curr. Opin. Chem. Biol.* **2000**, *4*, 696-703.

[20] Pratt, M. R.; Bertozzi, C. R. Synthetic glycopeptides and glycoproteins as tools for biology. *Chem. Soc. Rev.* **2005**, *34*, 58-68.

[21] Bertozzi, C. R.; Kiessling, L. L. Chemical glycobiology. *Science* **2001**, *291*, 2357-2364.

[22] Lindhorst, T. K. Artificial multivalent sugar ligands to understand and manipulate carbohydrate-protein interactions. *Top. Curr. Chem.* **2002**, *218*, 201-235.

[23] Ernst, B.; Magnani, J. L. From carbohydrate leads to glycomimetic drugs. *Nat. Rev. Drug Discovery* **2009**, *8*, 661-677.

[24] Bernardi, A.; Cheshev, P. Interfering with the sugar code: design and synthesis of oligosaccharide mimics. *Chem. Eur. J.* **2008**, *14*, 7434-7441.

[25] Imberty, A.; Chabre, Y. M.; Roy, R. Glycomimetics and glycodendrimers as high affinity microbial anti-adhesins. *Chem. Eur. J.* **2008**, *14*, 7490 – 7499.

[26] Deniaud, D.; Julienne, K.; Gouin, S. G. Insights in the rational design of synthetic multivalent glycoconjugates as lectin ligands. *Org. Biomol. Chem.*, **2011**, *9*, 966-979.

[27] Schweizer, F. ; Hindsgaul, O. Combinatorial synthesis of carbohydrates. *Curr. Opin. Chem. Biol.* **1999**, *3*, 291-298.

[28] Haase, C.; Seitz, O. Chemical synthesis of glycopeptides. *Top. Curr. Chem.* **2007**, *267*, 1-36.

[29] Specker, D.; Wittmann, V. Synthesis and application of glycopeptide and glycoprotein mimetics. *Top. Curr. Chem.* **2007**, *267*, 65-107.

[30] Boltje, T. J.; Buskas, T.; Boons, G.-J. Opportunities and challenges in synthetic oligosaccharide and glycoconjugate research. *Nat. Chem.* **2009**, *1*, 611-622.

[31] Lee, M.-R.; Jung, D.-W.; Williams, D.; Shin, I. Efficient solid-phase synthesis of trifunctional probes and their application to the detection of carbohydrate-binding proteins. *Org. Lett.* **2005**, *7*, 5477-5480.

[32] Jezek, J.; Velek, J.; Veprek, P.; Velkova, V.; Trnka, T.; Pecka, J.; Ledvina, M.; Vondrasek, J.; Pisacka, M. Solid phase synthesis of glycopeptide dendrimers with Tn antigenic structure and their biological activities. Part I. *J. Pept. Sci.* **1999**, *5*, 46-55.

[33] Roy, R.; Zanini, D.; Meunier, S. J.; Romanowska, A. Solid-phase synthesis of dendritic sialoside inhibitors of influenza A virus haemagglutinin. *J. Chem. Soc. Chem. Commun.* **1993**, 1869-1872.

[34] Kolomiets, E.; Johansson, E. M. V.; Renaudet, O.; Darbre, T.; Reymond, J.-L. Neoglycopeptide dendrimer libraries as a source of lectin binding ligands. *Org Lett.* **2007**, *9*, 1465-1468.

[35] Amaya, T.; Tanaka, H.; Takahashi, T. Combinatorial synthesis of carbohydrate cluster on tree-type linker with orthogonally cleavable parts. *Synlett* **2004**, 497-502.

[36] Amaya, T.; Tanaka, H.; Takahashi, T. Solid-phase synthesis of carbohydrate cluster on tree-type linker with three types of orthogonally cleavable part. *Synlett* **2004**, 503-507.

[37] Shaikh, H. A.; Sönnichsen, F. D.; Lindhorst, T. K. Synthesis of glycocluster peptides. *Carbohydr. Res.* **2008**, *343*, 1665-1674.

[38] Pourceau, G.; Meyer, A.; Vasseur, J.-J. ; Morvan, F. Combinatorial and automated synthesis of phosphodiester galactosyl cluster on solid support by click chemistry assisted by microwaves. *J. Org. Chem.* **2008**, *73*, 6014-6016.

[39] Tornoe, C. W.; Christensen, C.; Meldal, M. Peptidotriazoles on solid phase: [1,2,3]-triazoles by regiospecific copper(I)-catalyzed 1,3-dipolar cycloadditions of terminal alkynes to azides. *J. Org. Chem.* **2002**, *67*, 3057-3064.

[40]   Rostovtsev, V. V.; Green, L. G.; Fokin, V. V.; Sharpless, K. B. A stepwise Huisgen cycloaddition pr ocess: Copper(I)-Catalyzed Regioselective "Ligation" of Azides and Terminal Alkynes. *Angew. Chem. Int. Ed.* **2002**, *41*, 2596-2599.

[41]   Morvan, F.; Meyer, A.; Jochum, A.; Sabin, C.; Chevolot, Y.; Imberty, A.; Praly, J.-P.; Vasseur, J.-J.; Souteyrand, E.; Vidal, S. Fucosylated pentaerythrityl phosphodiester oligomers (PePOs): automated synthesis of DNA-based glycoclusters and binding to *Pseudomonas aeruginosa* lectin (PA-IIL). *Bioconjugate Chem.* **2007**, *18*, 1637-1643.

[42]   Ambrosi, M.; Cameron, N. R.; Davis, B. G. Lectins: tools for the molecular understanding of the glycocode. *Org. Biomol. Chem.* **2005**, *3*, 1593–1608.

[43]   Gomez-Garcia, M.; Benito, J. M.; Rodriguez-Lucena, D.; Yu, J.-X.; Chmurski, K.; Ortiz Mellet, C.; Gutierrez Gallego, R.; Maestre, A.; Defaye, J.; Garcia Fernandez, J. M. Probing secondary carbohydrate–protein interactions with highly dense cyclodextrin-centered heteroglycoclusters: the heterocluster effect. *J. Am. Chem. Soc.* **2005**, *127*, 7970-7971.

[44]   Gomez-Garcia, M.; Benito, J. M.; Butera, A. P.; Ortiz Mellet, C.; Garcia Fernandez, J. M.; Jimenez Blanco, J. L. Probing carbohydrate-lectin recognition in heterogeneous environments with monodisperse cyclodextrin-based glycoclusters. *J. Org. Chem.* **2012**, *77*, 1273–1288.

[45]   Liang, C.-H.; Wang, S.-K.; Lin, C.-W.; Wang, C.-C.; Wong, C.-H.; Wu, C.-Y. Effects of neighboring glycans on antibody–carbohydrate interaction. *Angew. Chem. Int. Ed.* **2011**, *50*, 1608-1612.

[46]   Kutterer, K. M. K.; Barnes, M. L.; Arya, P. Automated, solid-phase synthesis of *C*-neoglycopeptides: coupling of glycosyl derivatives to resin-bound peptides. *J. Comb. Chem.* **1999**, *1*, 28-31.

[47]   Arya, P.; Kutterer, K. M. K.; Barkley, A. Glycomimetics: a programmed approach toward neoglycopeptide libraries. *J. Comb. Chem.* **2000**, *2*, 120-126.

[48]   Arya, P.; Barkley, A.; Randell, K. D. Automated high-throughput synthesis of artificial glycopeptides. Small-molecule probes for chemical glycobiology. *J. Comb. Chem.* **2002**, *4*, 193-198.

[49]   Katajisto, J.; Karskela, T.; Heinonen, P.; Lönnberg, H. An orthogonally protected α,α-bis(aminomethyl)-β-alanine building block for the construction of glycoconjugates on a solid support. *J. Org. Chem.*, **2002**, *67*, 7995-8001.

[50]   Patek, M.; Lebl, M. Safety-catch anchoring linkage for synthesis of peptide amides by Boc/Fmoc strategy. *Tetrahedron Lett.* **1991**, *32*, 3891-3894.

[51]   Diaz-Moscoso, A.; Benito, J. M.; Ortiz Mellet, C.; Garcia Fernandez, J. M. Efficient use of Ellman safety-catch linker for solid-phase assisted synthesis of multivalent glycoconjugates. *J. Comb. Chem.* **2007**, *9*, 339-342.

[52]   Filippov, D. V.; van den Elst, H.; Tromp, C. M.; van der Marel, G. A.; van Boeckel, C. A. A.; Overkleeft, H. S.; van Boom, J. H. Parallel solid phase synthesis of tricomponent bisubstrate analogues as potential fucosyltransferase inhibitors. *Synlett* **2004**, 773-778.

[53]   Renaudet, O.; Dumy, P. On-bead synthesis and binding assay of chemoselectively template-assembled multivalent neoglycopeptides. Org. Biomol. Chem. **2006**, *4*, 2628-2636.

[54]   Wittmann, V.; Seeberger, S. Combinatorial solid-phase synthesis of multivalent cyclic neoglycopeptides. *Angew. Chem. Int. Ed.* **2000**, *39*, 4348-4352.

[55]   Virta, P.; Karskela, M.; Lönnberg, H. Orthogonally protected cyclo-β-tetrapeptides as solid-supported scaffolds for the synthesis of glycoclusters. *J. Org. Chem.* **2006**, *71*, 1989-1999.

[56]   Plante, O. J.; Palmacci, E. R., Seeberger, P. H. Automated solid-phase synthesis of oligosaccharides. *Science* **2001**, *291*, 1523-1527.

[57]   Palmacci, E. R.; Plante, O. J.; Hewitt, M. C.; Seeberger, P. H. Automated solid-phase synthesis of oligosaccharides. *Helv. Chim. Acta* **2003**, *86*, 3975-3990.

[58]   Ratner, D. M.; Swanson, E. R.; Seeberger, P. H. Automated synthesis of a protected *N*-linked glycoprotein core pentasaccharide. *Org. Lett.* **2003**, *5*, 4717-4720.

[59]   Routenberg Love, K.; Seeberger, P. H. Automated solid-phase synthesis of protected tumor-associated antigen and blood group determinant oligosaccharides. *Angew. Chem. Int. Ed.* **2004**, *43*, 602-605.

[60]   Ojeda, R.; Terenti, O.; de Paz, J.-L.; Martin-Lomas, M. Synthesis of heparin-like oligosaccharides on polymer supports. *Glycoconjugate J.* **2004**, *21*, 179-195.

[61]   Jonke, S.; Liu, K.-g., Schmidt, R. R. Solid-phase oligosaccharide synthesis of a small library of N-glycans. *Chem. Eur. J.* **2006**, *12*, 1274-1290.

[62]   Müller, C.; Kitas, E.; Wessel, H. P. Novel oligosaccharide mimetics by solid-phase synthesis. *J. Chem. Soc. Chem. Commun.* **1995**, 2425-2426.

[63]   Long, D. D.; Smith, M. D.; Marquess, D. G.; Claridge, T. D. W.; Fleet, G. W. J. A solid phase approach to oligomers of carbohydrate amino-acids: secondary structure in a trimeric furanose carbopeptoid. *Tetrahedron Lett.* **1998**, *39*, 9293-9296.

[64]   Jimenez Blanco, J. L.; Ortega-Caballero, F.; Ortiz Mellet, C.; Garcia Fernandez, J. M. (Pseudo)amide-linked oligosaccharide mimetics: molecular recognition and supramolecular properties. *Beilstein J. Org. Chem.* **2010**, *6*, No. 20.

[65]   Grotenbreg, G. M.; Kronemeijer, M.; Timmer, M. S. M.; El oualid, F.; van Well, R. M.; Verdoes, M.; Spalburg, E.; van Hooft, P. A. V.; de Neeling, A. J.; Noort, D.; van Boom, J. H.; van der Marel, G. A.; Overkleeft, H. S.; Overhand, M. A practical synthesis of Gramicidin S and sugar amino acid containing analogues. *J. Org. Chem.* **2004**, *69*, 7851-7859.

[66]   Grotenbreg, G. M.; Christina, A. E.; Buizert, A. E. M.; van der Marel, G. A.; Overkleeft, H. S.; Overhand, M. Synthesis and application of carbohydrate-derived morpholine amino acids. *J. Org. Chem.* **2004**, *69*, 8331-8339.

[67]   Raunkjaer, M.; El Oualid, F.; van der Marel, G. A.; Overkleeft, H. S.; Overhand, M. Alkylated sugar amino acids: A new entry toward highly functionalized dipeptide isosters. *Org. Lett.* **2004**, *6*, 3167-3170.

[68]   Peri, F.; Bassetti, R.; Caneva, E.; de Gioia, L.; La Ferla, B.; Presta, M.; Tanghetti, E.; Nicotra, F. Arabinose-derived bicyclic amino acids: synthesis, conformational analysis and construction of an $\alpha_v\beta_3$-selective RGD peptide. *J. Chem. Soc. Perkin Trans 1* **2002**, 638-644.

[69]   Katajisto, J.; Lönnberg, H. Solid-phase synthesis of cyclic *C*-glycoside/amino acid hybrids by carbamate coupling chemistry and on-support cyclization. *Eur. J. Org. Chem.* **2005**, 3518-3525.

[70]   Zatsepin, T. S.; Oretskaya, T. S. Synthesis and applications of oligonucleotide-carbohydrate conjugates. *Chem. Biodiv.* **2004**, *1*, 1401-1416.

[71]   Maier, M. A.; Yannopoulos, C. G.; Mohamed, N.; Roland, A.; Fritz, H.; Mohan, V.; Just, G.; Manoharan, M. Synthesis of antisense oligonucleotides conjugated to a multivalent carbohydrate cluster for cellular targeting. *Bioconjugate Chem.* **2003**, *14*, 18-29.

[72]   Lönnberg, H. Solid-phase synthesis of oligonucleotide conjugates useful for delivery and targeting of potential nucleic acid therapeutics. *Bioconjugate Chem.* **2009**, *20*, 1065-1094.

[73]   Matsuura, K.; Hibino, M.; Ikeda, T.; Yamada, Y.; Kobayashi, K. Self-organized glycoclusters along DNA: Effect of the spatial arrangement of galactoside residues on cooperative lectin recognition. *Chem. Eur. J.* **2004**, *10*, 352-359.

[74]   Matsuura, K.; Hibino, M.; Yamada, Y.; Kobayashi, K. Construction of glycol-clusters by self-organization of site-specifically glycosylated oligonucleotides and their cooperative amplification of lectin recognition. *J. Am. Chem. Soc.* **2001**, *123*, 357-358.

[75]   Turner, J. J.; Meeuwenoord, N. J.; Rood, A.; Borst, P.; van der Marel, G. A.; van Boom, J. H. Reinvestigation into the synthesis of oligonucleotides containing 5-(β-D-glucopyranosyloxymethyl)-2′-deoxyuridine. *Eur. J. Org. Chem.* **2003**, 3832-3839.

[76]   Tona, R.; Bertolini, R.; Hunziker, J. Synthesis of aminoglycoside-modified oligonucleotides. *Org. Lett.* **2000**, *2*, 1693-1696.

[77]   Hunzinker, J. Synthesis of 5-(2-amino-2-deoxy-β-Dglucopyranosyloxymethyl)-2′-deoxyuridine and its incorporation into oligothymidylates. *Bioorg. Med. Chem. Lett.* **1999**, *9*, 201-204.

[78]   Charles, I.; Xi, H.; Arya, D. P. Sequence-specific targeting of RNA with an oligonucleotide-neomycin conjugate. *Bioconjugate Chem.* **2007**, *18*, 160-169.

[79]   Schlegel, M. K.; Hütter, J.; Eriksson, M.; Lepenies, B.; Seeberger, P. H. Defined presentation of carbohydrates on a duplex DNA scaffold. *ChemBioChem* **2011**, *12*, 2791 – 2800.

[80]   D′Onofrio, J.; Petraccone, L.; Martino, L.; Di Fabio, G.; Iadonisi, A.; Balzarini, J.; Giancola, C.; Montesarchio, D. Synthesis, biophysical characterization and anti-HIV activity of glycol-conjugated G-quadruplex-forming oligonucleotides. *Bioconjugate Chem.* **2008**, *19*, 607-616.

[81]   D′Onofrio, J.; Erra, E.; Di Fabio, G.; Iadonisi, A.; Petraccone, L.; De Napoli, L.; Barone, G.; Balzarini, J.; Giancola, C.; Montesarchio, D. Synthesis and biophysical characterization of G-rich oligonucleotides conjugated with sugar-phosphate tails. *Nucleosides Nucleotides Nucleic Acids* **2007**, *26*, 1225-1229.

[82]   Akhtar, S.; Routledge, A.; Patel, R.; Gardiner, J. M. Synthesis of mono- and dimannoside phosphoramidite derivatives for solid-phase conjugation to oligonucleotides. *Tetrahedron Lett.* **1995**, *36*, 7333-7336.

[83]   Wang, Y.; Sheppard, T. L. Chemoenzymatic synthesis and antibody detection of DNA glycoconjugates. *Bioconjugate Chem.* **2003**, *14*, 1314-1322.

[84]   Dubber, M.; Fréchet, J. M. J. Solid-phase synthesis of multivalent glycoconjugates on a DNA synthesizer. *Bioconjugate Chem.* **2003**, *14*, 239-246.

[85]   Ugarte-Uribe, B.; Perez-Rentero, S.; Lucas, R.; Avino, A.; Reina, J. J.; Alkorta, I.; Eritja, R.; Morales, J. C. Synthesis, cell-surface binding, and cellular uptake of fluorescently labeled glucose-DNA conjugates with different carbohydrate presentation. *Bioconjugate Chem.* **2010**, *21,* 1280–1287.

[86]   Pourceau, G.; Meyer, A.; Chevolot, Y.; Souteyrand, E.; Vasseur, J.-J.; Morvan, F. Oligonucleotide carbohydrate-centered galactosyl cluster conjugates synthesized by click and phosphoramidite chemistries. *Bioconjugate Chem.* **2010**, *21,* 1520–1529.

[87]   Gerland, B.; Goudot, A.; Pourceau, G.; Meyer, A.; Dugas, V.; Cecioni, S.; Vidal, S.; Souteyrand, E.; Vasseur, J.-J.; Chevolot, Y.; Morvan, F. Synthesis of a library of fucosylated glycoclusters and determination of their binding toward pseudomonas aeruginosa Lectin B (PA-IIL) using a DNA-based carbohydrate microarray. *Bioconjugate Chem.* **2012**, *23*, 1534−1547.

[88]   Adinolfi, M.; De Napoli, L.; Di Fabio, G.; Iadonisi, A.; Montesarchio, D. Modulating the activity of oligonucleotides by carbohydrate conjugation: solid phase synthesis of sucrose-oligonucleotide hybrids. *Org. Biomol. Chem.* **2004**, *2*, 1879-1886.

[89]   Kiviniemi, A.; Virta, P. Synthesis of aminoglycoside-3′-conjugates of 2′-O-methyl oligoribonucleotides and their invasion to a ^{19}F labeled HIV-1 TAR model. *Bioconjugate Chem.* **2011**, *22*, 1559–1566.

[90]   Habus, I.; Zhao, Q.; Agrawal, S. Synthesis, hybridization properties, nuclease stability, and cellular uptake of the oligonucleotide-amino-β-cyclodextrins and adamantane conjugates. *Bioconjugate Chem.* **1995**, *6*, 327-331.

[91]   D'Onofrio, J.; de Champdoré, M.; De Napoli, L.; Montesarchio, D.; Di Fabio, G. Glycomimetics as decorating motifs for oligonucleotides: Solid-phase synthesis, stability, and hybridization properties of carbopeptoid–oligonucleotide conjugates. *Bioconjugate Chem.* **2005**, *16*, 1299-1309.

[92]   Adinolfi, M.; De Napoli, L.; Di Fabio, G.; Iadonisi, A.; Montesarchio, D.; Piccialli, G. Solid phase synthesis of oligonucleotides tethered to oligo-glucose phosphate tail. *Tetrahedron* **2002**, *58*, 6697-6704.

[93]   Sheppard, T. L.; Wong, C.-H.; Joyce, G. F. Nucleoglycoconjugates: design and synthesis of a new class of DNA-carbohydrate conjugates. *Angew. Chem. Int. Ed.* **2000**, *39*, 3660-3663.

[94]   Ketomäki, K.; Virta, P. Synthesis of aminoglycoside conjugates of 2′-O-methyl oligoribonucleotides. *Bioconjugate Chem.* **2008**, *19*, 766-777.

[95]   Katajisto, J.; Heinonen, P.; Lönnberg, H. Solid-phase synthesis of oligonucleotide glycoconjugates bearing three different glycosyl groups: orthogonally protected bis(hydroxymethyl)-*N,N′*-bis(3-hydroxypropyl)malondiamide phosphoramidite as key building block. *J. Org. Chem.* **2004**, *69*, 7609-7615.

[96]   Chevolot, Y.; Bouillon, C.; Vidal, S.; Morvan, F.; Meyer, A.; Cloarec, J.-P.; Jochum, A.; Praly, J.-P.; Vasseur, J.-J.; Souteyrand, E. DNA-based carbohydrate biochips: a platform for surface glyco-engineering. *Angew. Chem. Int. Ed.* **2007**, *46*, 2398-2402.

[97]   Bouillon, C.; Meyer, A.; Vidal, S.; Jochum, A.; Chevolot, Y.; Cloarec, J.-P.; Praly, J.-P.; Vasseur, J.-J.; Morvan, F. Microwave assisted click chemistry for the synthesis of multiple labeled-carbohydrate oligonucleotides on solid support. *J. Org. Chem.* **2006**, *71*, 4700-4702.

[98]   Kiviniemi, A.; Virta, P.; Lönnberg, H. Solid-supported synthesis and click conjugation of 4′-C-alkyne functionalized oligodeoxyribonucleotides. *Bioconjugate Chem.* **2010**, *21*, 1890-1901.

[99]   Jayaprakash, K. N.; Peng, C. G.; Butler, D.; Varghese, J. P.; Maier, M. A.; Rajeev, K. G.; Manoharan, M. Non-nucleoside building blocks for copper-assisted and copper-free click chemistry for the efficient synthesis of RNA conjugates. *Org. Lett.* **2010**, *12*, 5410-5413.

[100] Kiviniemi, A.; Virta, P.; Lönnberg, H. Utilization of intrachain 4′-C-azidomethylthymidine for preparation of oligodeoxyribonucleotide conjugates by click chemistry in solution and on a solid support. *Bioconjugate Chem.* **2008**, *19*, 1726-1734.

[101] Kiviniemi, A.; Virta, P.; Drenichev, M. S.; Mikhailov, S. N.; Lönnberg, H. Solid-supported 2′-O-glycoconjugation of oligonucleotides by azidation and click reactions. *Bioconjugate Chem.* **2011**, *22*, 1249–1255.

[102] Pourceau, G.; Meyer, A.; Vasseur, J.-J.; Morvan, F. Synthesis of mannose and galactose oligonucleotide conjugates by bi-click chemistry. *J. Org. Chem.* **2009**, *74*, 1218-1222.

[103] Katajisto, J.; Virta, P.; Lönnberg, H. Solid-phase synthesis of multiantennary oligonucleotide glycoconjugates utilizing on-support oximation. *Bioconjugate Chem.* **2004**, *15*, 890-896.

[104] Karskela, M.; Virta, P.; Malinen, M.; Urtti, A.; Lönnberg, H. Synthesis and cellular uptake of fluorescently labeled multivalent hyaluronan disaccharide conjugates of oligonucleotide phosphorothioates. *Bioconjugate Chem.* **2008**, *19*, 2549-2558.

[105] Karskela, M.; Helkearo, M.; Virta, P.; Lönnberg, H. Synthesis of oligonucleotide glycoconjugates using sequential click and oximation ligations. *Bioconjugate Chem.* **2010**, *21*, 748-756.

[106] Meutermans, W.; Le, G. T.; Becker, B. Development of delivery methods for carbohydrate-based drugs: controlled release of biologically-active short chain fatty acid-hexosamine analogs. *ChemMedChem* **2006**, *1*, 1164-1194.

[107] Velter, I. ; La Ferla, B. ; Nicotra, F. Carbohydrate-based molecular scaffolding. *J. Carbohydr. Chem.* **2006**, *25*, 97-138.

[108] Cipolla, L. ; Peri, F. ; La Ferla, B. ; Redaelli, C. ; Nicotra, F. Carbohydrate scaffolds for the production of bioactive compounds. *Curr. Org. Synth.* **2005**, *2*, 153-173.

[109] Opatz, T., Kallus, C., Wunberg, T., Schmidt, W., Henke, S., and Kunz, H. D-Glucose as a multivalent chiral scaffold for combinatorial chemistry. *Carbohydr. Res.* **2002**, *337*, 2089-2110.

[110] Opatz, T.; Kallus, C.; Wunberg, T.; Schmidt, W.; Henke, S.; Kunz, H. D-Glucose as a pentavalent chiral scaffold. *Eur. J. Org. Chem.* **2003**, 1527-1536.

[111] Thanh, G. L.; Abbenante, G.; Adamson, G.; Becker, B.; Clark, C.; Condie, G.; Falzun, T.; Grathwohl, M.; Gupta, P.; Hanson, M.; Huynh, N.; Katavic, P.; Kuipers, K.; Lam, A.; Liu, L.; Mann, M.; Mason, J.; McKeveney, D.; Muldoon, C.; Pearson, A.; Rajaratnam, P.; Ryan, S.; Tometzki, G.; Verquin, G.; Waanders, J.; West, M.; Wilcox, N.; Wimmer, N.; Yau, A.; Zuegg, J.; Meutermans, W. A versatile synthetic approach toward diversity libraries using monosaccharide scaffolds. *J. Org. Chem.* **2010**, *75*, 197–203.

[112] Kallus, C.; Opatz, T.; Wunberg, T.; Schmidt, W.; Henke, S.; Hunz, H. Combinatorial solid-phase synthesis using D-galactose as a chiral five-dimension-diversity scaffold. *Tetrahedron Lett.* **1999**, *40*, 7783-7786.

[113] Wunberg, T.; Kallus, C.; Opatz, T.; Henke, S.; Schmidt, W.; Kunz, H. Carbohydrates as multifunctional chiral scaffolds in combinatorial synthesis. *Angew. Chem. Int. Ed.* **1998**, *37*, 2503-2505.

[114] Opatz, T.; Kallus, C.; Wunberg, T.; Kunz, H. Combinatorial synthesis of amino acid- and peptide-carbohydrate conjugates on solid phase. *Tetrahedron* **2004**, *60*, 8613-8626.

[115] Peri, F.; Nicotra, F.; Leslie, C. P.; Micheli, F.; Seneci, P.; Marchioro, C. D-glucose as a regioselectively addressable scaffold for combinatorial chemistry on solid phase. *J. Carbohydr. Chem.* **2003**, *22*, 57-71.

[116] Brill, W. K.-D.; De Mesmaeker, A.; Wendeborn, S. Solid-phase synthesis of levoglucosan derivatives. *Synlett* **1998**, 1085-1090.

[117] Sofia, M. J.; Hunter, R.; Chan, T. Y.; Vaughan, A.; Dulina, R.; Wang, H.; Gange, D. Carbohydrate-based small-molecule scaffolds for the construction of universal pharmacophore mapping libraries. *J. Org. Chem.* **1998**, *63*, 2802-2803.

[118] Jain, R.; Kamau, M.; Wang, C.; Ippolito, R.; Wang, H.; Dulina, R.; Anderson, J.; Gange, D.; Sofia, M. J. 3-Azido-3-deoxy-glycopyranoside derivatives as scaffolds for the synthesis of carbohydrate-Based universal pharmacophore mapping libraries. *Bioorg. Med. Chem. Lett.* **2003**, *13*, 2185-2189.

[119] Cervi, G.; Peri, F.; Battistini, C.; Gennari, C.; Nicotra, F. Bicyclic carbohydrate-derived scaffolds for combinatorial libraries. *Bioorg. Med. Chem.* **2006**, *14*, 3349-3367.

[120] Timmer, M. S. M.; Verdoes, M.; Sliedregt, L. A. J. M.; van der Marel, G. A.; van Boom, J. H.; Overkleeft, H. S. The use of a mannitol-derived fused oxacycle as a combinatorial scaffold. *J. Org. Chem.* **2003**, *68*, 9406-9411.

[121] Hünger, U.; Ohnsmann, J.; Kunz, H. Carbohydrate scaffolds for combinatorial syntheses that allow selective deprotection of all four positions independent of the sequence. *Angew. Chem. Int. Ed.* **2004**, *43*, 1104-1107.

[122] Madalinski, M.; Stoll, M.; Dietrich, U.; Kunz, H. A selectively deprotectable 2,6-diaminogalactose scaffold for the solid-phase synthesis of potential RNA ligands. *Synthesis* **2008**, 1106-1120.

[123] Welzel, P. Transglycosylase inhibition, In *Antibiotics and Antiviral Compounds –Chemical Synthesis and Modification*, Krohn, K.; Kirst, H. A.; Maag, H.; Eds., VCH: Weinheim, 1993, pp 373-378.

[124] Riedel, S.; Donnerstag, A.; Hennig, L.; Welzel, P. Synthesis and transglycosylase-inhibiting properties of a disaccharide analogue of moenomycin A lacking substitution at C-4 of unit F. *Tetrahedron* **1999**, *55*, 1921-1936.

[125] Sofia, M. J.; Allanson, N.; Hatzenbuhler, N. T.; Jain, R.; Kakarla, R.; Kogan, N.; Liang, R.; Liu, D.; Silva, D. J.; Wang, H.; Gange, D.; Anderson, J.; Chen, A.; Chi, F.; Dulina, R.; Huang, B.; Kamau, M.; Wang, C.; Baizman, E.; Branstrom, A.; Bristol, N.; Goldman, R.; Han, K.; Longley, C., Midha, S.; Axelrod, H. R. Discovery of novel disaccharide antibacterial agents using a combinatorial library approach. *J. Med. Chem.* **1999**, *42*, 3193-3198.

[126] Jensen, K. J.; Barany, G. Carbopeptides: carbohydrates as potential templates for *de novo* design of protein models. *J. Peptide Res.* **2000**, *56*, 3-11.

[127] Thomas, C. J.; Chizhov, A. O.; Leitheiser, C. J.; Rishel, M. J.; Konishi, K.; Tao, Z.-F.; Hecht, S. M. Solid-phase synthesis of bleomycin A_5 and three monosaccharide analogues: exploring the role of the carbohydrate moiety in RNA cleavage. *J. Am. Chem. Soc.* **2002**, *124*, 12926-12927.

Send Orders of Reprints at bspsaif@emirates.net.ae
Advances in Organic Synthesis, Vol. 5, 2013, 51-100

CHAPTER 2

Chiral Molecular Receptors Based on *Trans*-Cyclohexane-1,2-Diamine

Ignacio Alfonso[*]

Departamento de Química Biológica y Modelización Molecular, Instituto de Química Avanzada de Cataluña, Consejo Superior de Investigaciones Científicas, IQAC-CSIC, Jordi Girona, 18-26, E-08034, Barcelona, Spain

Abstract: This review highlights the development of new receptors based on enantiopure *trans*-cyclohexane-1,2-diamine, a chiral building block presenting unique structural and conformational properties. Different architectures sharing this structural motif, which were prepared following several synthetic approaches, have been summarized from the recent literature.

Keywords: *Trans*-cyclohexane-1,2-diamine, chiral receptor, nitrogen, molecular recognition, macrocycles.

1. INTRODUCTION

Trans-cyclohexane-1,2-diamine (**1**) (Fig. **1**) is one of the most widely used chiral diamines in modern chemistry. The fields of application of this molecule range from the preparation of chiral catalysts for asymmetric synthesis [1], the synthesis of supramolecular receptors [2] or the preparation of chiral stationary phases for separation science [3]. This cyclic diamine has unique structural properties which make it very useful for the induction of a chiral environment (chiral ligands) as well as for the development of new synthetic strategies, taking advantage of its geometrical preorganization. In this paper, we will review the recent literature regarding the synthesis of new receptors bearing this common structural motif. A seminal review paper from Hanessian [4] about the use of this diamine in asymmetric synthesis and, to a lesser extent, in molecular recognition should be an excellent source for the important references prior to 1997. Some

*Address correspondence to Ignacio Alfonso: Departamento de Química Biológica y Modelización Molecular, Instituto de Química Avanzada de Cataluña, Consejo Superior de Investigaciones Científicas, IQAC-CSIC, Jordi Girona, 18-26, E-08034, Barcelona, Spain; Tel: +34 934 006 100; Fax +34 932 045 904; E-mail: ignacio.alfonso@iqac.csic.es

Atta-ur-Rahman (Ed)

more recent reviews, not directly dealing with this diamine but covering closely related fields [5], could serve as complementary literature regarding this topic. The main intention of this review is to try to highlight the importance of this scaffold in modern chemistry. Moreover, the examples devoted to the preparation of new receptors to be used in the supramolecular chemistry field will be preferentially selected and accordingly cited.

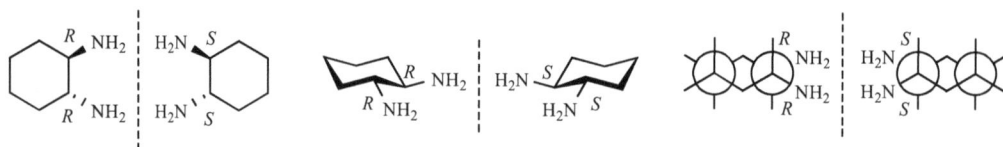

Figure 1: Different representations of the enantiomers of chiral *trans*-cyclohexane-1,2-diamine (**1**).

For a clearer presentation of the data, a logical classification has been carried out attending to the structure of the final compounds, with increasing complexity. Thus, we have made the corresponding divisions dealing with: 1) open-chain, 2) macrocyclic and 3) macropolycyclic, interlocked and cage compounds. Besides, two additional sections highlighting the structural peculiarities of diamine (**1**) and the corresponding methods for its resolution have been added. Obviously, these two short sections will be discussed before the main body of the review.

2. STRUCTURAL PECULIARITIES OF *TRANS*-CYCLOHEXANE-1,2-DIAMINE

Most of the practical applications of this diamine in synthesis, catalysis and molecular recognition are closely related to its three dimensional structure. Therefore, a short paragraph in this regard should be pertinent. The six member ring of the molecule with a *trans*-1,2-substitution renders a C_2 symmetry and a rigid scaffold with very precisely defined disposition of the two amino groups. Regarding the cyclohexane geometry, at least three different conformers can be proposed: two chair-like (diaxial and diequatorial) conformers and a twisted boat, which are all of them minima of energy (Fig. **2**). Theoretical calculations (DFT at B3LYP//6-31G*) [6] shows that the diequatorial conformer is highly favored in the gas phase due to the minimization of 1,3-diaxial repulsions and to the possibility of establishing an intramolecular N-H···N hydrogen bond. The corresponding diaxial conformer is 4.67 kcal/mol higher in energy than the diequatorial one, although this geometry has been experimentally observed in some specific situations [7]. The twisted boat is even less stable, being 9.90 kcal/mol higher in energy than the diequatorial chair conformation. Thus, for most of the compounds bearing this diamine, the diequatorial chair conformation should be the most favorable. Within this geometry, the nitrogen atoms are in *gauche* disposition, setting the two C-N bonds at an angle close to 60°, which is a key feature regarding the reactivity of the system. For instance, this fact has been used to explain the differences in basicity between (**1**) [8] and the non-cyclic 1,2-ethanediame [9]. The cyclohexanediamine shows a more basic amino group for the first protonation process, but a lesser basic nitrogen for the addition of the second proton. The geometrical disposition of the nitrogens allows the stabilization of the monoprotonated form by a strong bifurcated [HN···H$^+$···NH] hydrogen bond, while their close proximity disfavors the setting of two positive charges after the second protonation process. Other consequences of the geometrical characteristics are the very good ability as ligand for transition metals [10] and, in some cases, a much better crystallinity of the corresponding compounds when compared with other closely related 1,2-diamines [11]. Regarding the structural characterization of the compounds derived from (**1**), some peculiar features must be highlighted. The symmetry and rigidity of the cyclohexane ring make the systems to display very characteristic ^{1}H NMR spectra with clear axial and equatorial positions for all the protons within the ring. This fact induces a large anisochrony of all the methylenes of the cycle and a characteristic coupling constants pattern, with larger J values for geminal and axial-axial vicinal coupling constants when compared to the axial-equatorial or the equatorial-equatorial vicinal ones. Another very useful technique for the characterization of derivatives of (**1**) is circular dichroism (CD) [12]. When suitable chromophores are attached to the nitrogen atoms, very representative signatures are obtained in the CD spectra, showing the characteristic bisigned signal corresponding, very usually, to a large split Cotton effect [13].

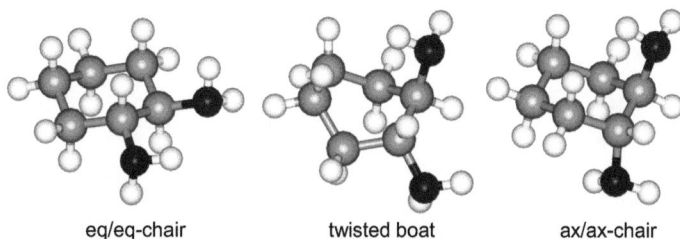

eq/eq-chair   twisted boat   ax/ax-chair

Figure 2: Minimum conformations for the *trans*-cyclohexane-1,2-diamine (**1**).

3. ENANTIOPURE *TRANS*-CYCLOHEXANE-1,2-DIAMINE: CHEMICAL AND BIOCATALYTIC RESOLUTIONS

In order to synthesize chiral receptors with high enantiopurity, the starting diamine should be available in enantioenriched (ideally enantiopure) form. Currently, both enantiomeric forms of this diamine are commercially available, although for historical reasons, some words regarding the different methodologies for its resolution are pertinent. Two main approaches have been used until now: the chemical and the biocatalytic one. The most important chemical approach takes advantage of the diamine base character and its high trend to induce stable crystal structures and solid conglomerates. Initially, Galsbøl *et al.* reported how diastereomeric crystals can be obtained from this diamine and tartaric acid [14] yielding, after recrystallization, the enantiopure form of the compound as a tartrate salt (Scheme 1). Later, Jacobsen and co-workers reported an improved version of this methodology [15], which did not require crystallization steps (by simple precipitation) and could be performed with the *cis/trans* mixture of the ligand. A comprehensive study of the crystal organization of this salt was further reported [16], showing that the chiral recognition is mainly due to the formation of helical supramolecular structures in the solid state. Helical structures of this type have been also observed for solid state supramolecular complexes of (1) with different chiral diols [17] and diacids [18] highlighting how the three dimensional geometry of this diamine has an important impact on the supramolecular complexes obtained through non-covalent contacts. Even more recently, a deep study from Díaz-Díaz, Marrero-Tellado and co-workers showed how the [(1) · tartrate] salt can self-assemble into supramolecular fibers in different organic media, which trap the solvent molecules to finally render the formation of stable organogels [19]. A preparative resolution of this diamine by preferential crystallization of the corresponding citrate salt has been also recently reported, as well as the structural study of the formed conglomerate [20].

Scheme 1: Chemical resolution of (1) by precipitation/crystallization of the corresponding tartrate salt.

Another simple general methodology to obtain chiral amines and diamines is the enzymatic resolution of the corresponding racemate, which has been highly developed in the last decades [21]. Within this field, we used *Candida antarctica* lipase (CAL-B) for the sequential kinetic resolution (Scheme **2**) of *trans*-cylohexane-1,2-diamine (**1**) [22]. In this reaction, both the substrate (**1**) and products (**2a-b**) were obtained in enantiopure forms and used for the synthesis of chiral polyazamacrocycles with applications in chiral anion binding (see below).

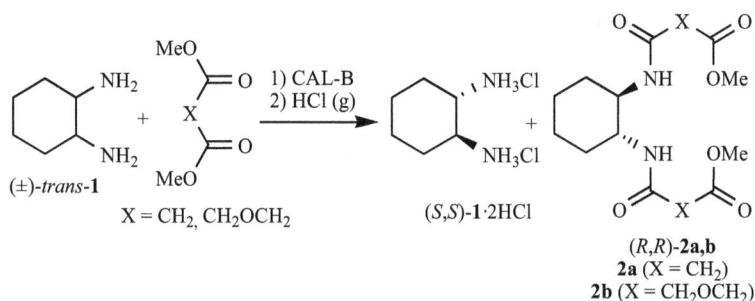

Scheme 2: Enzymatic resolution of *trans*-cyclohexane-1,2-diamine.

Recently, a more sophisticated biocatalytic approach has been reported [23], in which *N,N*-dialkylated enantiopure derivatives of (**1**) can be prepared with high selectivity and enantiomeric excesses from non-expensive cyclohexene oxide (**3**). This methodology was used for the efficient synthesis of the analgesic U-(−)-50,488.

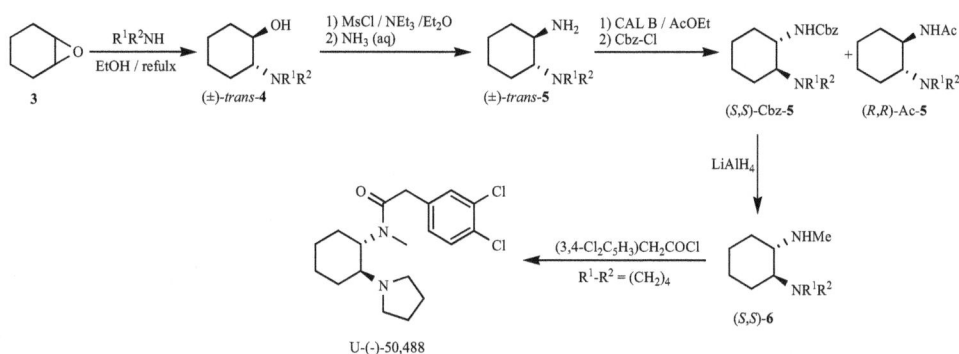

Scheme 3: Synthesis of *N,N*-dialkylated enantiopure derivatives of (**1**).

4. OPEN-CHAIN MOLECULES

The corresponding open chain receptors derived from diamine (**1**) have very specific conformational properties due to the presence of this building block. Thus, they are less flexible than conventional open chain receptors and are highly preorganized in a well defined conformation, thanks to the *trans*-diequatorial disposition of both nitrogen atoms of (**1**). One of the seminal papers using this moiety in the preparation of a molecular receptor is that reported by Hamilton *et al.* [24] They coupled the diamine (*R,R*)-(**1**) with two molecules of *N*-protected amino acids, leading to the preparation of a family (**7a-c**) of simple C_2 symmetrical enantiopure open chain hosts for carboxylate anions, which resemble the ristocetin binding site. The ^{1}H NMR titration studies rendered that the serine derivative is the most efficient due to the synergic establishing of two additional OH hydrogen bonds, apart form those from the amide and carbamate NH groups (Fig. **3**).

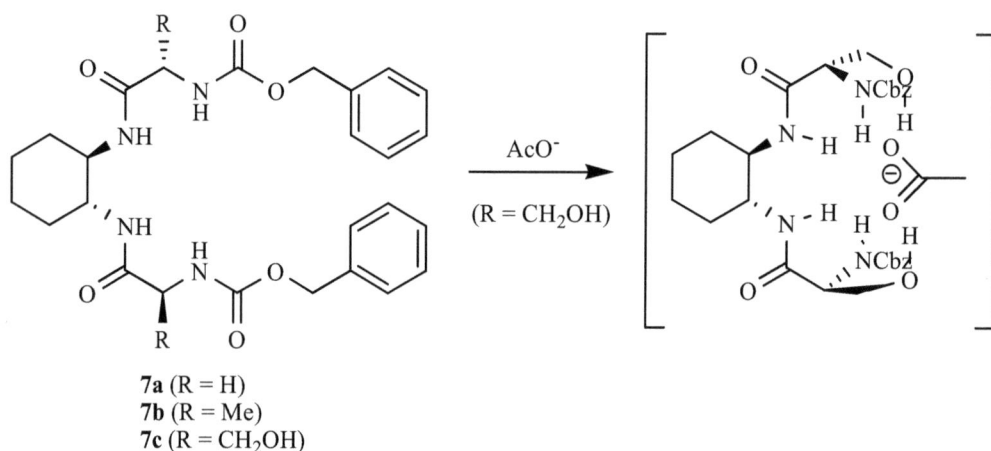

7a (R = H)
7b (R = Me)
7c (R = CH$_2$OH)

Figure 3: A chiral receptor for carboxylate anions that resembles ristocetin binding site.

A simpler design was reported by Fabbrizzi and co-workers (Fig. (**4**)), who prepared a bis(thiourea) derivative (**8**) by conventional coupling of (**1**) with the corresponding thiocianate [25]. The obtained compound (**8**) is able to form four very stable hydrogen bonds with anions, especially with carboxylates and phosphates, as shown by UV titration experiments. The chiral scaffold in (**8**) prompted the authors to check its potential enantioselectivity in the binding of a chiral polyanion with biological interest, namely D-2,3-diphosphoglycerate. Although the receptor showed strong binding by UV titration in DMSO, only a modest enantioselectivity ($\Delta\log K = 0.35$) was observed when comparing the interaction with each enantiomer of the host (**8**). The observed binding was additionally supported by ^{1}H and ^{31}P NMR experiments.

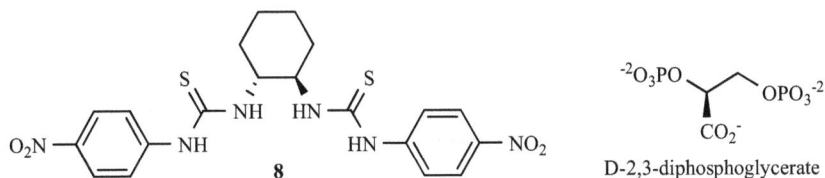

Figure 4: Structure of an enantioselective host (**8**) for a chiral phosphate.

We have also prepared an open-chain bis(amidoamine) compound, bearing two units of (**1**) in its structure (Scheme **4**) [26]. The key enantiopure diamino derivative (*R,R*)-(**9**) was obtained as depicted in Scheme **3**. The fundamental structural features of the final compound (**10**) are the presence, within a C_2 symmetry, of two amides and two tertiary amines connected through a 2,6-bis(carbonyl)pyridine unit. This last moiety stabilizes a pincer-like U-shaped conformation of the compound, as shown by NMR and molecular modeling. The basic nature of the system allowed the interaction with different chiral carboxylic acids in solution, leading to the formation of (**10**): acid supramolecular complexes with 1:2 stoichiometry. In these complexes, a splitting of the ^{1}H NMR signals of the acids was observed and, therefore, this compound can be used as an efficient chiral solvating agent (CSA) for the determination of the enantiomeric excesses of carboxylic acids by NMR. A more detailed study using high resolution NMR experiments (including intermolecular 1D ROESY) and molecular modeling allowed us to propose a reasonable model for the supramolecular complex, stabilized by electrostatic, H-bonding and π-π interactions. Other interesting CSAs based on monosubstituted (**1**) [(1*R*,2*R*)-1-(1',8'-naphthalimide)-2-aminocyclohexane and its corresponding 4'-derivatives, (**11a-c**)] have been recently reported (Scheme **4**) [27]. Despite their very simple structures, they also displayed good NMR enantiodiscrimination of chiral carboxylic acids, with promising practical applications.

One of the most interesting topics in current supramolecular chemistry is the study of the self-assembling properties of peptide-like molecules [28]. A recent paper from Hanessian *et al.* [29] reported on the preparation and the study of non-cyclic peptide-like systems, where the U-shaped folding conformation was induced by the cyclohexane moiety (Fig. **5**). The authors found quite intriguing effects of the stereochemical relationships between the chiral centers of the cyclohexane framework and those of the corresponding amino acids of the peptidic arms. They coupled two D/L-alternated sequences of tripeptides to (**1**) moiety ((**12a,b**) in Fig. **5**) and they realized that both diastereomers displayed quite different macroscopic behavior. While the compound (**12a**) bearing (*R,R*)-(**1**) formed crystals suitable for X-ray diffraction analysis, its diastereomer (**12b**) formed a supramolecular gel. Analysis of the solid-state structure of (**12a**) revealed a highly H-bonded helical open-ended tubular superstructure, with the tripeptide strands intertwining like a pair of self-embracing arms. Structurally related bis-tripeptides with different amino acids showed extensive intramolecular H-bonding by NMR studies. Overall, this paper demonstrates the ability of this diamine to induce a helical and highly chiral environment, with very promising applications in the preparation of biologically interesting self-assembling systems.

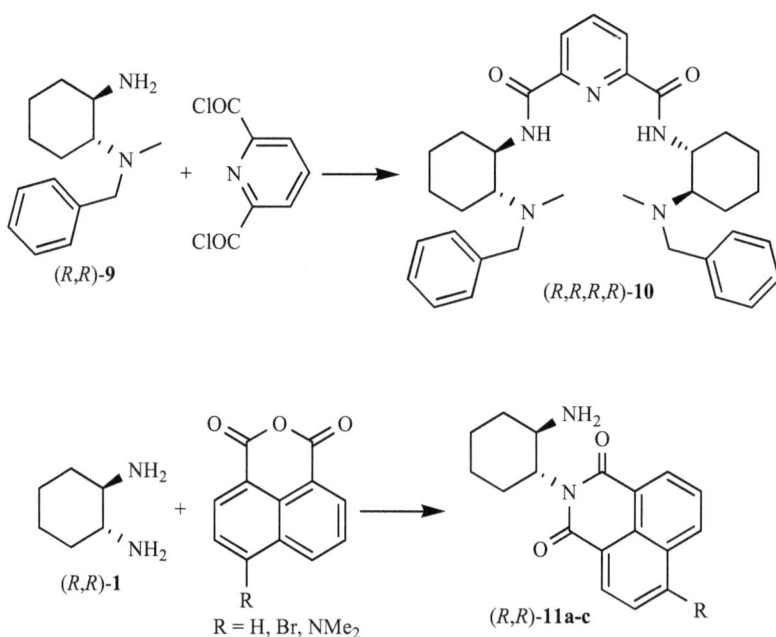

Scheme 4: Synthesis of a chiral solvating agents (**10**) and (**11a-c**) based on diamine (**1**).

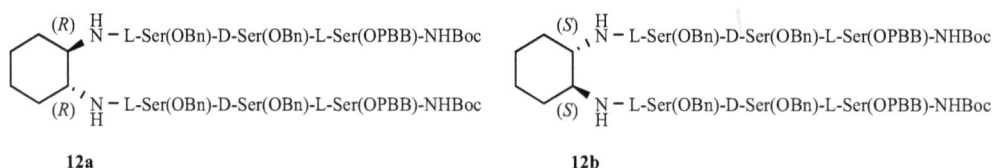

Figure 5: Peptide-coupled derivatives of (**1**).

Another field of application of diamine (**1**) is for the preparation of optically active open chain polyamines. Attracted by the potential applications of this family of compounds [30], we decided to prepare several open-chain chiral polyamines bearing *trans*-cyclohexane-1,2-diamine in their structures. The key step for this synthetic strategy was the solvent dependent selective alkylation of the corresponding bis(sulfonamide) (**13**) (Scheme **5**) [31]. In this case, the conformational constriction of the cyclohexane ring, as well as the selection of the right combination of protecting groups and reaction solvent allowed the selective preparation of either the mono- (**14**) or dialkylated (**15**) derivative of the cyclohexanebis(sulfonamide) frame (**13**) [32]. Both intermediates were successfully used for the modular synthesis of a family of polyamines (**16a-c**),

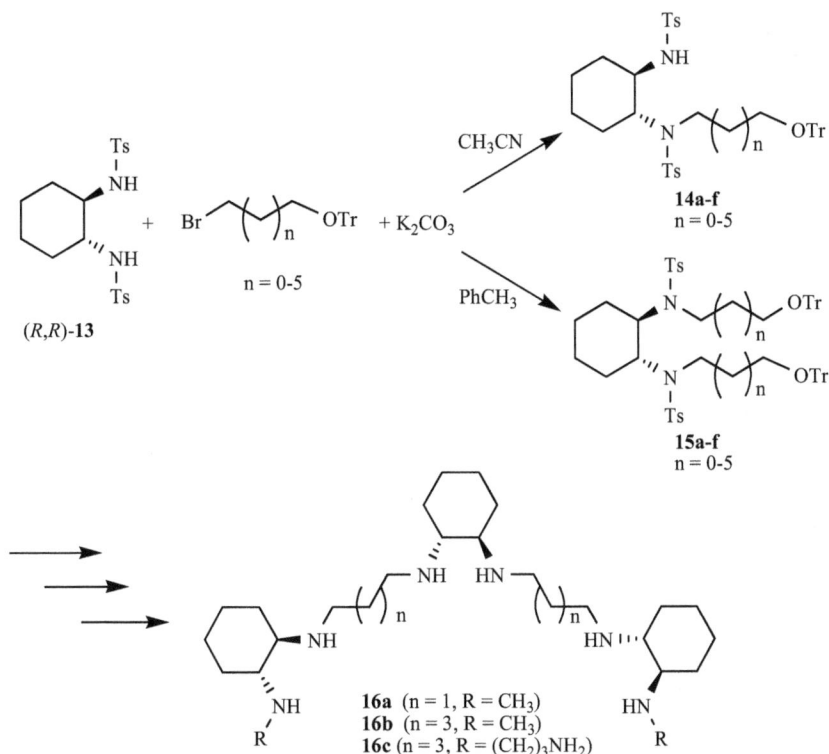

Scheme 5: Synthesis of enantiopure open-chain polyamines bearing (**1**) moieties in their structures.

changing their structural variables, such as the spacer (n), configuration of the chiral centers or substitution of the terminal amino groups (R). Some of these polyamines have shown to bind DNA molecules in a stereoselective manner at physiological conditions [33]. Melting studies with model oligonucleotides showed that polyamines (**16a**) having *all-R* configuration of their chiral centers bind stronger to the AT-rich oligonuclotide ($\Delta\Delta T_m \approx 1.5$ °C), while the *all-S* enantiomer clearly preferred the CG-rich sequence ($\Delta\Delta T_m \approx 5$ °C). Interestingly, this selectivity is practically eliminated with the more flexible derivatives (**16b**). A reasonable explanation for this trend could be the different helix handness of the oligonucleotides in the presence of a charged electrolyte, giving rise to left- or right-handed helix depending on the CG-base content.

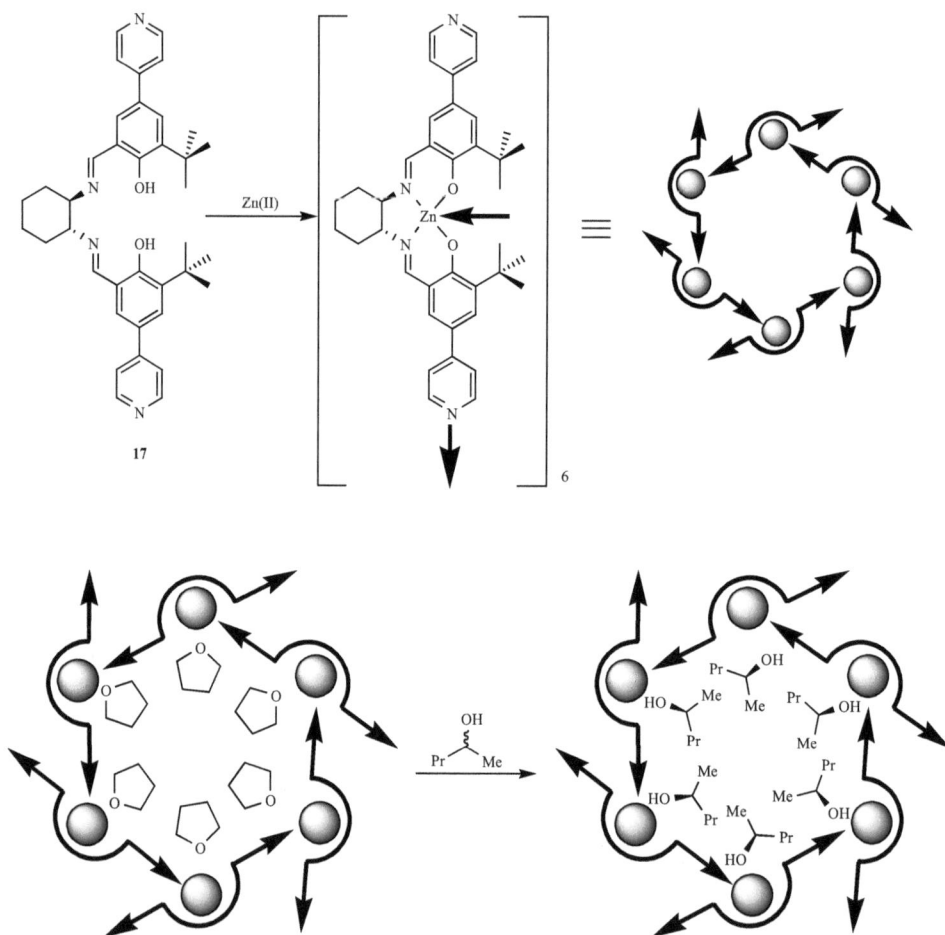

Figure 6: A metal ligand able to self-assemble into a hexameric porous material.

Inclusion compounds are highly interesting in separation science, and molecules derived from (**1**) have been also studied in this context. Recently, an efficient system based on a salen-type metal complex has been developed for the enantioselective separation of polar molecules (Fig. **6**) [34]. In the presence of Zn(II), the ligand (**17**) is able to self-assemble in a hexameric nanotubular solid state supramolecular structure with a large porous cavity, filled by six solvent (THF) molecules. Soaking these crystals in racemic 2-pentanol in a sealed vial at 40 °C produced the exchange of the included molecules. Further analysis of the alcohol molecules within the cavity (by chiral GC) rendered a 99.5% ee, demonstrating the ability of the tubular hollow nanostructures of (**17**) as chiral selector. Although assays with other chiral alcohols proved to be unsuccessful, these preliminary results are very promising within the field of separation science of racemic mixtures.

As a summary of this section, it can be concluded that this diamine is a very useful building block for preparing open chain receptors, still showing some degree of conformational preorganization, which is highly valuable for molecular recognition or self-assembling processes.

5. MACROCYCLIC MOLECULES

When reviewing the chiral receptors derived from (**1**), macrocyclic compounds are, by far, the most abundant structures. This fact is due to several reasons. First of all, supramolecular chemists normally prefer to use macrocyclic hosts, since they display large cavities for the molecular recognition to take place, and a high degree of structural preorganization. Moreover, in the case of diamine (**1**), additional characteristics have made an enormous explosion of its usage for the preparation of macrocycles. Thus, as previously commented, the cylohexane ring induces a folding of the molecule in a U-shaped conformation which facilitates the macrocyclization process. The cyclization step is, in most cases, hampered by low yields or by the formation of untreatable mixtures of oligomers of different sizes. In the case of cyclohexane-1,2-diamine ring, it is possible to highly control the geometrical parameters leading to more satisfactory yields and purer crude materials. This rationale has been exploited for the synthesis of different macrocyclic oligoimines directly through the condensation between (**1**) and different dialdehydes. The disposition of the C-N bonds in (**1**) forming an angle

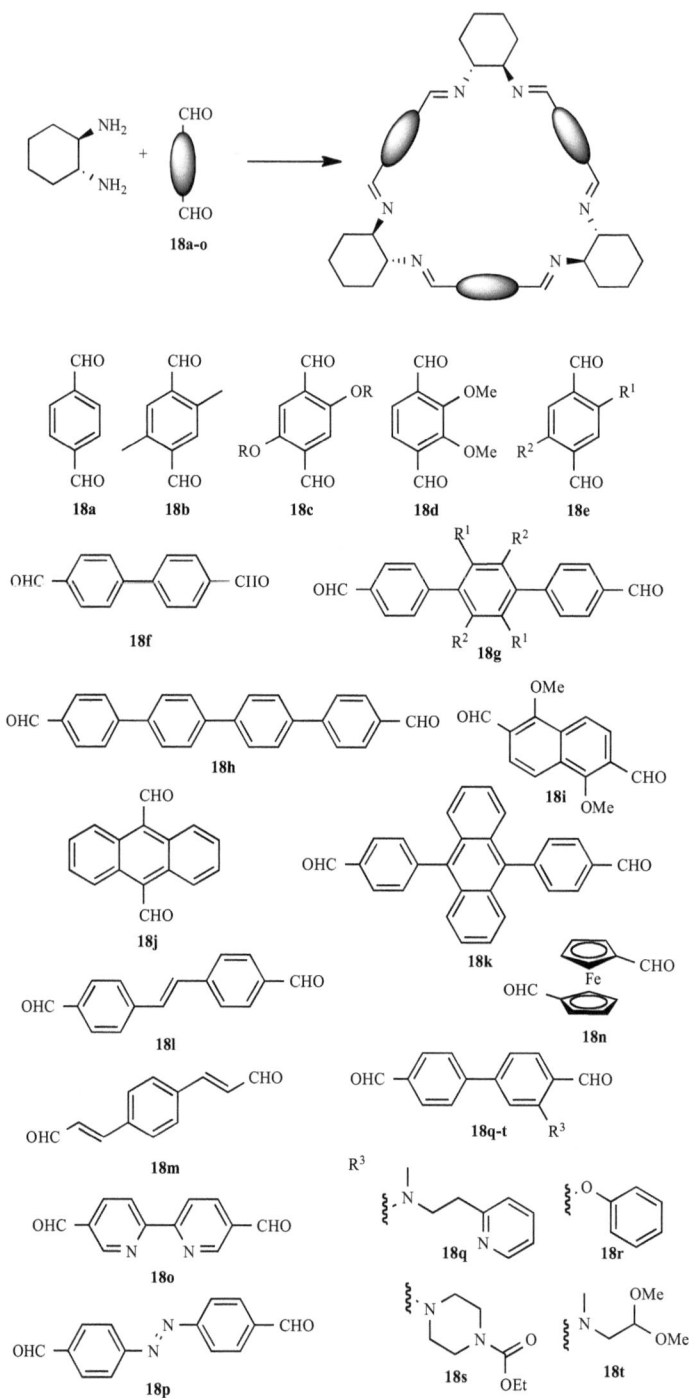

Scheme 6: One-pot synthesis of *trianglimines* derived from (**1**) and aromatic dialdehydes setting the CHO groups at 180°.

close to 60° is essential for the success of the reaction. Thus, when the diamine is mixed with aromatic dialdehydes (**18a-o**) setting their CHO groups in a planar disposition and at 180°, the polyimine tends to form a triangle for purely geometrical reasons, leading to the [3+3] condensation product (Scheme **6** and Table **1**) [35-45]. For obvious reasons these macrocyclic compounds have been generically named *trianglimines* by several authors.

Table 1: Isolated yields and corresponding references for the one-pot synthesis of trianglimines derived from (**1**) and aromatic dialdehydes setting the CHO groups at 180°

Dialdehyde	R	Yield (%)	References
18a	-	90	[35]
18b	-	97	[36]
18c	R = Me	90	[35, 37]
	R = Et	18	[38]
	R = Pr	17	[38]
	R = Bu	32	[38]
	R = Bn	23	[38]
	R = p-FBn	29	[38]
18d	-	25	[36]
18e	R^1 = Et; R^2 = OMe	8	[37]
	R^1 = Bu; R^2 = OMe	13	[37]
	R^1 = $(CH_2)_2$OMe; R^2 = OMe	5	[37]
	R^1 = Me; R^2 = H	46	[37]
18f	-	32-67	[39, 40]
18g	R^1 = Me; R^2 = H	18	[41]
	R^1 = R^2 = F	55	[42]
	R^1 = R^2 = H	99	[40]
18h	-	57	[41]
18i	-	14	[41]
18j	-	80	[39]
18k	-	99	[40]
18l	-	68	[40]
18m	-	58	[40]
18n	-	13.8	[41]
18o	-	79-84	[43]
18p	-	91	[44]
18q-t	R^3 (see scheme **6**)	4-37 (different regioisomers)	[45]

The reaction usually proceeds in methylene chloride at room temperature, although some authors found better results in more polar solvents and/or at slightly higher temperatures. In most cases, the crude reaction product is very clean, showing a good yield of the non-purified [3+3] compound. However, a recrystallization process is necessary in some examples, usually leading to a dramatic decrease of the isolated yield. This fact is more likely due to the dynamic nature of the imine bond. Thus, the formation of a given macrocycle is in equilibrium with its hydrolysis or recombination into smaller or larger cyclic or open chain oligomers. For the simplest aromatic dialdehyde (**18a**), the effect of performing the reaction either with enantiopure (**1**) or with its racemic mixture was thoroughly investigated [35b]. The configurations of the corresponding diastereomeric products resulting in the individual reactions were examined by ^{1}H and ^{13}C NMR spectroscopy. Unambiguous proof was also obtained by X-ray crystal structure analysis of both alternative diastereomers, revealing an interesting stereoselective stacking of the triangles into microporous chiral columns [34b]. The most interesting feature of this procedure is the high degree of the control over the cavity size of the corresponding macrocyclic *trianglimine*, by controlling the structure of the dialdehyde (see, for instance, the large cavity obtained with dialdehyde (**18h**)). Moreover, other characteristics, such as the polarity, the photochemical (**18f-k, 18p**) and the electrochemical (**18n**) activities, or the presence of additional functional groups with potential binding sites (**18o**) have been successfully implemented within this general procedure.

On the other hand, when the angle formed by the two CHO groups of the dialdehyde is smaller than 180°, the condensation is not so selective, mainly leading to a mixture of [2+2] and [3+3] macrocycles (Scheme **7**). The ratio between them usually depends on the geometrical disposition of the CHO groups in the dicarbonyl compound. Since the [2+2] compound shows binary symmetry and the [3+3] cycle displays ternary symmetry, in most of the cases, both compounds display very similar NMR spectral data. Therefore, they can be detected and distinguished by mass spectrometry, the most common technique being ESI-MS, as it allows semiquantitative determination of the proportion between cyclic oligomers (Table **2**). After careful purification, the obtained major compound was, in most of the cases, the [2+2] macrocycle. In some examples, the results were rationalized attending to geometrical reasons, since an angle lower than 180° would disfavor the formation of the [3+3] triangle. Moreover, some authors supported their conclusions with a detailed ^{1}H NMR analysis [41], combined with theoretical calculations [36, 42, 47, 48], Besides, the corresponding crystal structure of the [2+2] macrocycle [46] further supported the conclusions. Once again, attending to the geometrical shape of the [2+2] compounds, these macrocycles have been called *rhombimines*.

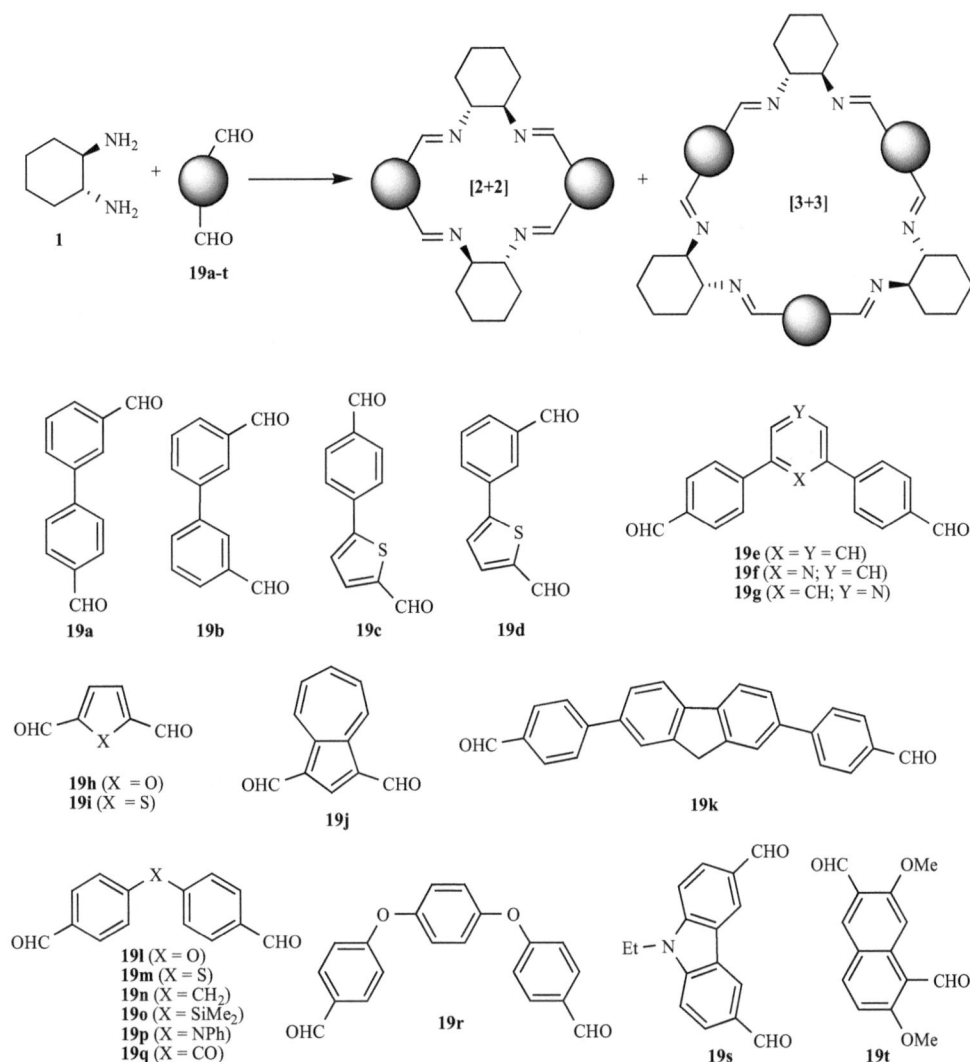

Scheme 7: One-pot synthesis of macrocyclic imines derived from (**1**) and aromatic dialdehydes (**19a-t**) setting the CHO groups at less than 180°.

Table 2: Proportions of oligomers, isolated yields and references for the condensation between (**1**) and aromatic dialdehydes (**19a-t**) with the CHO groups at an angle <180°

Dialdehyde	[2+2]: [3+3] Ratio	Isolated Yield (%)	References
19a	97: 3	67 ([2+2])	[42]
19b	98: 2	72 ([2+2])	[42]
19c	95: 5	72 ([2+2])	[42]
19d	98: 2	66 ([2+2])	[42]
19e	90: 10	79 ([2+2])	[42]
19f	90: 10	78 ([2+2])	[42]
19g	88: 12	81 ([2+2])	[42]
19h	~0: 100	14 ([3+3])	[42]
19i (in MeOH)	50: 50	-	[46]
19i (in CH$_2$Cl$_2$)	-	32 ([3+3])	[36]
19j	20: 80	90 ([2+2] + [3+3])	[36]
19k	85: 15	34 ([2+2])	[42]
19l	[2+2] + ([3+3] and [4+4] traces)	68 ([2+2])	[47, 48]
19m	not determined	>90 ([2+2])	[46]
19n	not determined	82 ([2+2])	[46]
19o	not determined	90 ([2+2])	[46]
19p	not determined	64 ([2+2])	[46]
19q	mixture of oligomers	-	[46]
19r	[2+2] + ([3+3] traces)	42 ([2+2])	[45]
19s	-	30 ([2+2])	[36]
19t	-	18 ([3+3])	[41]

The effect of microwave irradiation (M.W.) in these types of multicomponent condensations has been also investigated by Srimirugan and co-workers [49-51]. They studied the reaction between (**1**) and aromatic dialdehydes (**20a-g**) in polar protic solvents (usually ethanol: water mixtures) and under M.W. conditions (Scheme **8**) Table **3**. They obtained the corresponding [2+2] cycle as the mayor product, sometimes unpurified by other larger cyclic oligomers [49]. For some derivates, they were able to grow single crystals suitable for X-ray diffraction analysis [50] showing calixarene-like structures with oval-shaped cavities in the solid state. Regarding the minor compounds sometimes obtained in the cyclization reaction, further studies with (**20g**) [51] demonstrated that the [2+2] compound was accompanied with the [6+6] giant-size 72-member ring macrocycle. Remarkably, the intermediate oligomers going from [2+2] to [6+6] (namely [3+3], [4+4] or [5+5]) were not detected in the ESI-MS spectra of the crude reaction. Overall, it seems that when the dialdehyde has some flexibility and the angle between the CHO groups is lower than 180° but higher than 120°, the [3+3] macrocycle can be obtained under kinetic control (short reaction times and using solvents with low polarity). However, in most cases, when using polar solvents and higher temperatures (or M.W.), the thermodynamically controlled compound was obtained, usually being the [2+2] macrocycle. Very remarkably, the observation of a 72-membered ring as a stable thermodynamic product suggests that a fine tuning of the geometry of the dialdehyde could be used to prepare very sophisticated receptors based on diamine (**1**).

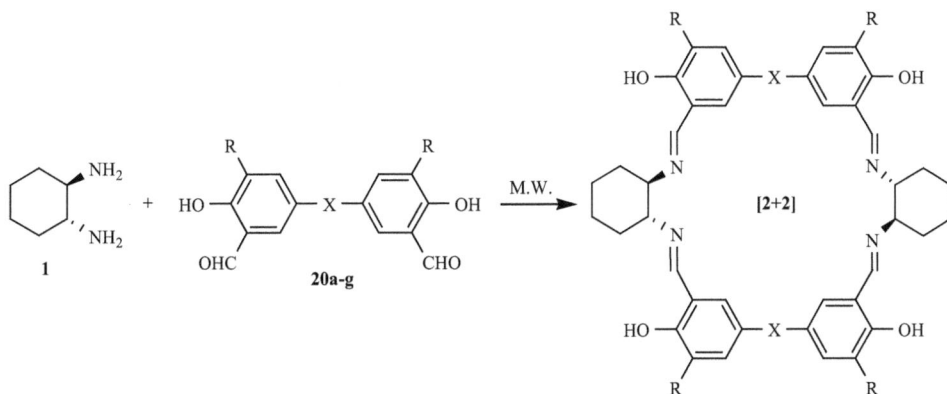

Scheme 8: Synthesis of macrocycles by microwave irradiation promoted condensation of aromatic dialdehydes and (**1**).

Table 3: Microwave irradiation promoted condensation of dialdehydes and (**1**)

Dialdehyde	X	R	Yield (%)	References
20a	CH_2	H	55 (mixture)	[49]
20b	CH_2	Br	45 ([2+2])	[49]
20c	$C(Me)_2$	Br	41 ([2+2])	[49]
20d	CH_2	NO_2	not determined	[49]
20e	SO_2	H	47 ([2+2])	[49, 50]
20f	$C(Me)_2$	H	78 (mixture)	[49]
20g	-	H	36 ([2+2]) + 14 ([6+6])	[49-51]

Also, the effect of supercritical CO_2 on the simple condensation of (**1**) with (**18a**) (Scheme **6**) or (**21a**) (Scheme **9**) has been also recently studied as a greener approach to the reaction, in the absence of additional solvent or catalysts [52].

A particular case of aromatic dialdehydes is the family of isophthalic-like derivatives, for which the angle is exactly 120°. Their corresponding imine condensations with (**1**) have been extensively studied (Scheme **9**) with different substitutions on the aromatic ring. The results (Table **4**) suggest that the [3+3] macrocycle was obtained under kinetic control, but the [2+2] macrocycle is the thermodynamically favored compound. Actually, upon extensive heating, the [3+3] macrocycle undergo cycle contraction to the [2+2] derivative [36].

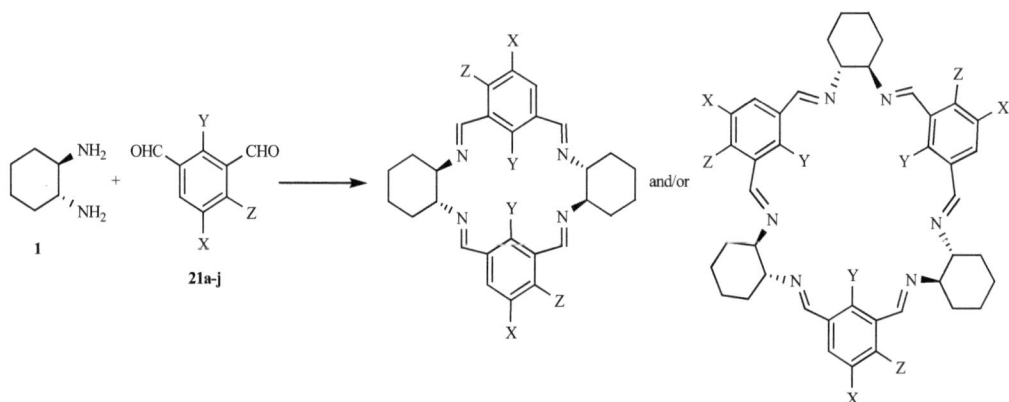

Scheme 9: One-pot synthesis of macrocyclic oligoimines derived from (**1**) and aromatic isophthaldehydes setting the CHO groups at 120°.

Table 4: Synthesis of oligoimines derived from (**1**) and aromatic isophthaldehydes

Dialdehyde	X	Y	Z	Yield (%)	References
21a	H	H	H	90 ([3+3])	[35a]
21b	Me	H	H	75 ([3+3])	[36, 37]
21c	H	Me	H	67 ([3+3])	[36]
21d	OMe	OMe	OMe	70 ([3+3])	[36]
21e	Me	OH	H	78-80 ([3+3])	[53, 54, 55]
21f	Me	H	OH	80 ([3+3])	[53]
21g	Me	OTs	H	59 ([3+3])	[38]
21h	Me	OCH_2CO_2Me	H	6 ([3+3])+N. R.[a] ([2+2])	[38]
21i	H	OH	H	N. R.[a] ([2+2])	[56]
21j	tBu	OH	H	N. R.[a] ([2+2])	[55]
21b [b]	Me	H	H	~100 ([2+2])	[36]
21c [b]	H	Me	H	~100 ([2+2])	[36]

[a]N. R.: yield not reported; [b]Upon refluxing (12-72 h) in CH_2Cl_2, the [3+3] cycle undergoes ring contraction to the [2+2] cycle.

More recently, Lisowski and co-workers have reported an interesting template effect in the condensation of **(1)** and **(21j)**: by controlling the amount of divalent zinc ion, it is possible to obtain either the [2+2] (with 1 equivalent of Zn), or the [3+3] (with 0.5 equivalent of Zn) macrocycle [57]. A deep structural study allowed the authors to explain this effect by the formation of different multinuclear metal complexes.

Despite the large number of examples showing multicomponent ([n+n] with n ≥ 2) condensation of **(1)** with aromatic dialdehydes, when the geometry of the dicarbonylic compound allows the [1+1] cyclization, the corresponding product is mainly formed. This fact has been reported for flexible **(22a-d)** [58] or concave rigid **(23)** [59] aromatic dialdehydes (Scheme **10**).

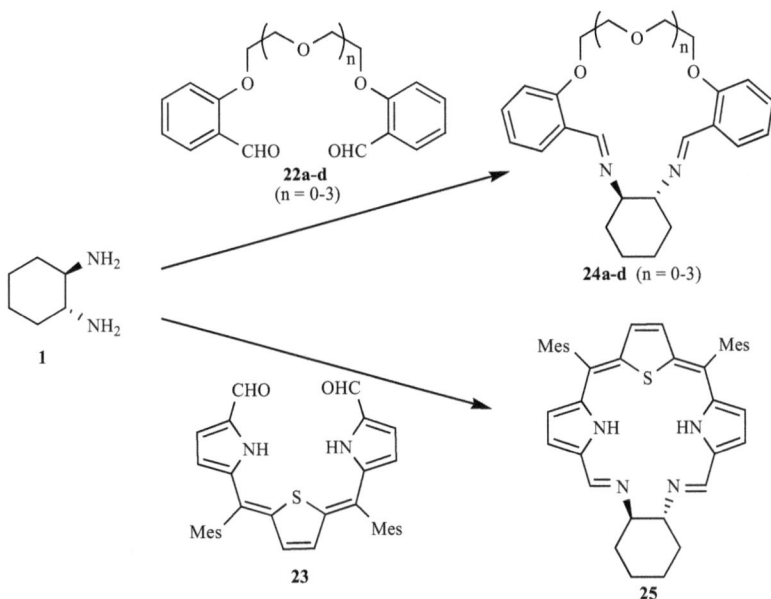

Scheme 10: [1+1] Macrocyclization reactions of **(1)**.

Regarding the corresponding reactions with aliphatic dialdehydes, a lesser number of papers have been reported so far. Different reasons could explain this fact: *a priori*, aliphatic dialdehydes are difficult to synthesize, they are usually less stable and, accordingly, the corresponding imines hydrolyze easier than the aromatic ones. Moreover, they often produce side reactions by aldol-type auto-condensation processes. However, Gawroński, de Meijere and co-workers [60] used a very clever approach to the preparation of large size, highly stable macrocyclic [2+2] and [3+3] oligoimines from aliphatic dialdehydes (Scheme **11**). To this aim, they used highly rigid aliphatic dialdehydes containing a cyclohexane (**25**), a bicycle [2.2.2]octane (**26**) or a [7]triangulane (**27**) skeleton. Following the trends previously commented, the dialdehydes with the CHO groups at 180°, (**25**) and (**26**), yielded the corresponding [3+3] macrocycles, (**28**) and (**29**), in 97% and 94% yields, respectively. However, the dialdehyde (**27**), which sets the CHO groups at ~160° led to the unique formation (93%) of the [2+2] oligoimine (**30**). Interestingly, the enantiomerically pure dialdehyde (**27**) showed a remarkable diastereoselectivity in the condensation with the two enantiomers of (**1**): only the (*R,R*) enantiomer gave a [2+2] macrocyclization product. All the final macrocycles were deeply characterized by NMR, CD and X-ray diffraction of the corresponding single crystals.

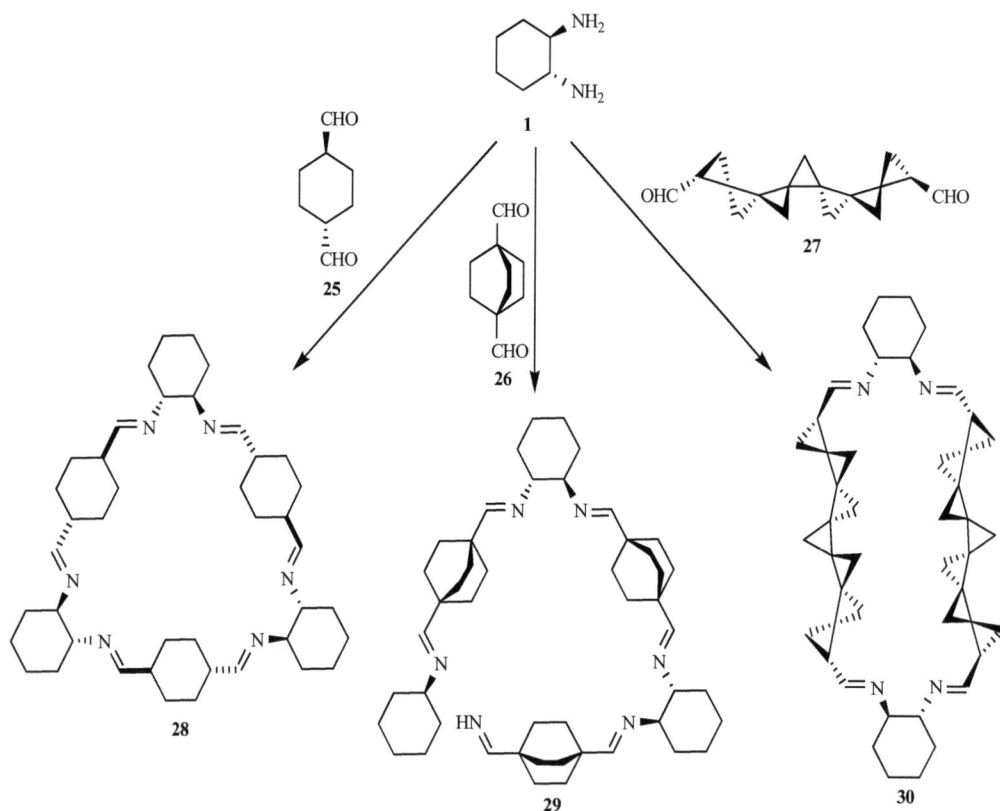

Scheme 11: One-pot synthesis of macrocyclic oligoimines derived from (**1**) and constrained aliphatic dialdehydes.

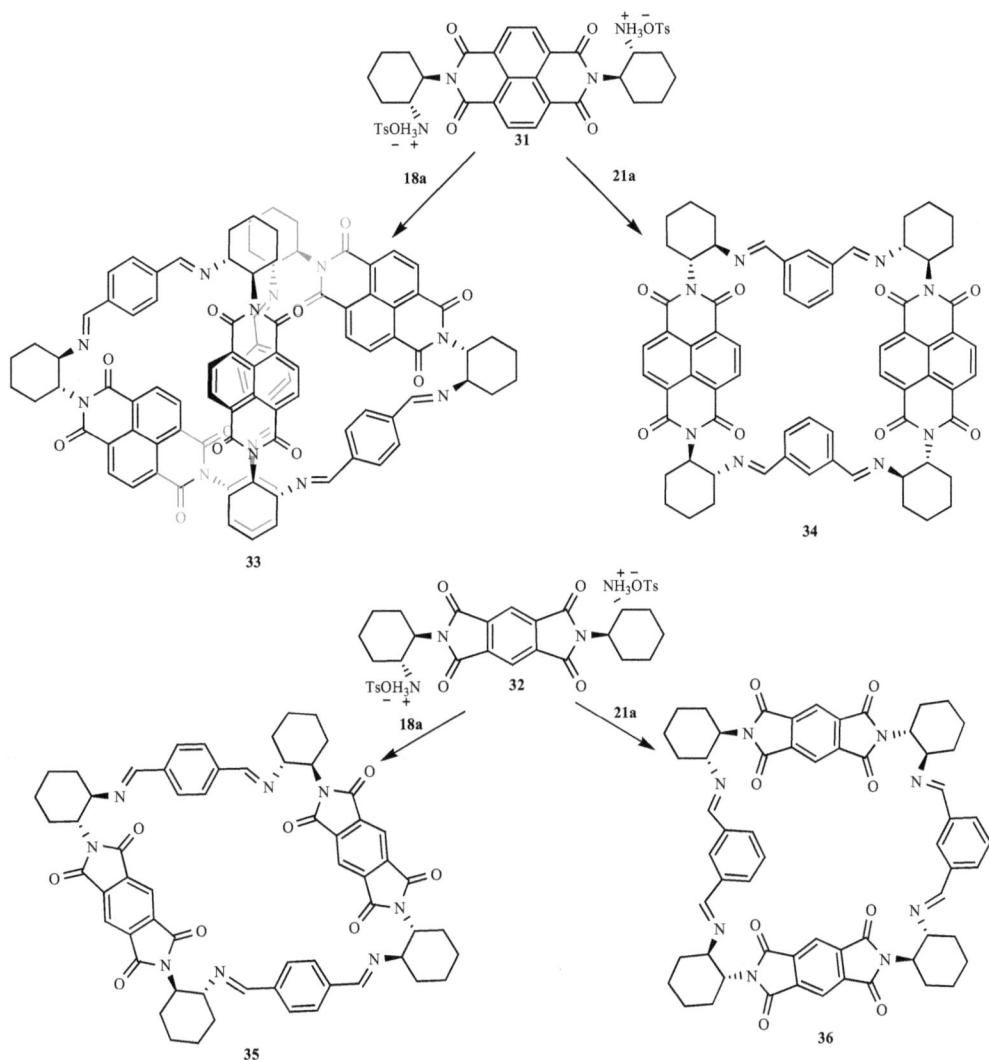

Scheme 12: Synthesis of one *loopimine* (**33**) and three *rectanglimines* (**34-36**) by [3+3] or [2+2] cyclocondensation reactions, respectively.

Another approach to obtain molecularly diverse large macrocycles has been recently reported by Gawroński (Scheme **12**) [61]. In this paper, the authors prepared two building blocks consisting of two units of (**1**) co-linearly joined by a rigid aromatic dimide, leading to compounds (**31**) and (**32**). Then, they carried out the imine condensation with the simplest aromatic dialdehydes: terephthaldehyde (**18a**) and isophthaldehyde (**21a**). Interestingly, (**31**) produced the [3+3] cycloaduct (**33**) with (**18a**), but the [2+2] product (**34**) when using (**21a**), in 82%

and 60% yields, respectively. However, (32) always led to the [2+2] macrocycles, (35) and (36), in 78% and 65% yields, respectively. After the corresponding structural analysis by CD and with the help of molecular modeling, the authors determined the shapes and conformations of the macrocycles, calling *loopimine* to (33) and *rectanglimines* to (34), (35) and (36), for analogy with the common nomenclature used in the field.

Since the imine bond formation is reversible, despite the high stability of the described macrocycles, when carrying out the usual purification operations (crystallization, chromatography) some of them recombined to other oligomers in different conditions (solvent and temperature). Thus, in order to get kinetically and thermodynamically stable cyclic compounds, the described oligoimines have been easily reduced (Scheme 13) to the corresponding macrocyclic polyamines in very good yields [35-39, 44, 46-48, 51, 57, 62]. Moreover, the reduction step can be performed *in situ*, leading to a one pot two-steps reductive amination macrocyclization, as an excellent general synthetic methodology. Despite that, in some cases, the isolation of the oligoimine intermediate and further reduction rendered a cleaner crude and better isolated overall yields [42]. The macrocyclic polyamines thus synthesized display unique and interesting structural properties with high potential in supramolecular chemistry. They display large sizes of the macrocyclic cavities, with several nitrogen atoms available for the interactions with different substrates. These amino groups, as free bases, are able to bind cationic transition metals. On the other hand, when these nitrogen atoms are partially or totally protonated, the obtained species strongly bind anions, even in highly competing protic solvents (see below). Moreover, some of these macrocycles have been alkylated (Scheme 13), leading to the preparation of a broad family of macrocyclic polyazamacrocycles (*tranglamines*) [62]. The structural and chiroptical properties of some of these azacrown molecules have been studied using a combination of CD, molecular modeling and X-ray diffraction of single crystals, when available. Recently, Savoia *et al.* have reported a very interesting approach to a new family of trianglamines, from the corresponding [3+3] trianglimines by the addition of organolithium reagents (Scheme 13) [63]. Very remarkably, the six newly formed C-C bonds were produced in a completely stereoselective fashion.

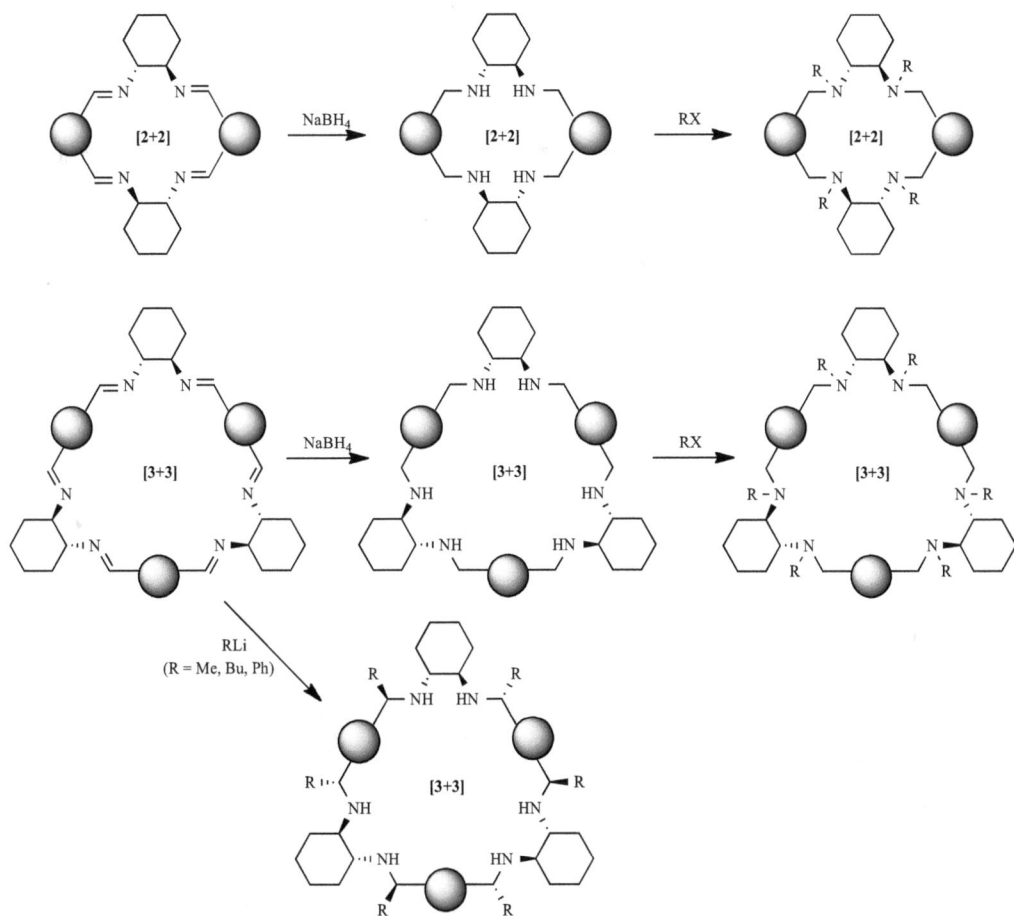

Scheme 13: General synthesis of chiral macrocyclic polyamines from the corresponding polyimines.

As previously mentioned, these trianglamines are partially protonated in aqueous solution close to neutral pH, displaying a highly positive charge density. This property has allowed these compounds to exert anion binding abilities in aqueous solution [64]. Regarding that, the simplest trianglamine derived from terephthaldehyde (**18a**) has been studied for the binding of isomeric benzenetricarboxylic acids (BTC) and 1,3,5-benzenetriacetic acid (1,3,5-BTA) [65]. Potentiometric titrations revealed a high degree of self-complementarity in the host-guest binding event, displaying the best fit structure with 1,3,5-BTC, as expected from the molecular symmetry and size of the receptor (Fig. **7**). Detailed studies by NMR (including intermolecular NOEs) and molecular modeling strongly supported a supramolecular structure as depicted in Fig. **7**, where the 1,3,5-BTC is included inside the macrocyclic cavity, setting electrostatic H-bonding and aryl-aryl [C-H$\cdots\pi$] host-guest interactions.

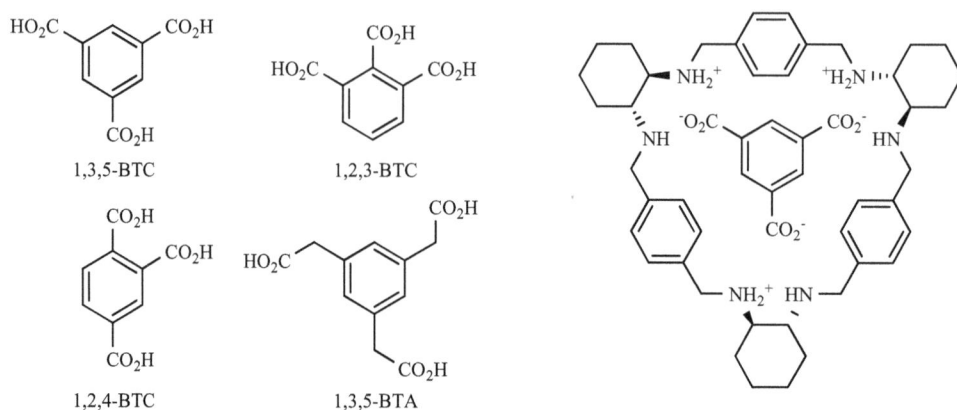

Figure 7: Binding of benzenetricarboxylic (BTC) and benzenetriacetic (BTA) acids by a macrocyclic *trianglamine*.

These compounds have been used as effective CSAs for secondary alcohols and carboxylic acids [66]. Interestingly, in some cases, the presence of several H-donor and H-acceptor functions allowed to get very good NMR enantiodiscrimination using substoichiometric amounts of the CSA.

In the last couple of years, an effort has been made in order to translate the binding selectivity into useful applications, such as the development of fluorescent sensors able to yield differential signal depending on the enantiomeric composition of the substrate [67]. For instance, Pu *et al.* (Fig. **(8)**)prepared the fluorescent macrocycles **(37)** and **(38a,b)** using the reductive amination synthetic approach since both precursors, diamine and dialdehyde, are conformationally constrained [68]. These macrocycles share a 1,1'-binaphthol moiety which acts as a chiral reporter and also as a fluorescent antenna. The obtained receptors are able to display a highly enantioselective fluorescent response to mandelic acid derivatives. Recently, Zhu, Cheng and co-workers used a chiral macrocycle derived from **(1)** for a cascade recognition of Cu(II) and α-amino acids in protic solutions [69].

Very recently, a conceptually different methodology for the discovery and synthesis of new receptors has been proposed: the dynamic combinatorial chemistry (DCC) approach [70]. We wondered if we could use this procedure to prepare new chiral azamacrocycles [71]. With this aim, we focused our attention on the imine formation with a pyridine derived dialdehyde (**39**). Thus, the condensation between diamine (**1**) and dialdehyde (**39**) in methanol led to an equilibrium mixture containing different cyclic and linear oligomers (Scheme **14**) [71, 72]. Inspection of many different cationic templates rendered interesting results. Thus, the system is shifted to the dimeric species (**40**) in the presence of Ba(II) but to the trimer (**41**) in the presence of Cd(II), although cadmium ion is smaller than barium. Deep NMR, mass spectrometry, UV and modeling studies demonstrated the formation of dinuclear metal complexes between the trimeric oligoimine (**41**) and Cd(II). *In situ* reduction of these complexes led to the selective and efficient preparation of either the dimeric (**42**) or trimeric (**43**) polyazamacrocyclic compounds, both in very high isolated yields.

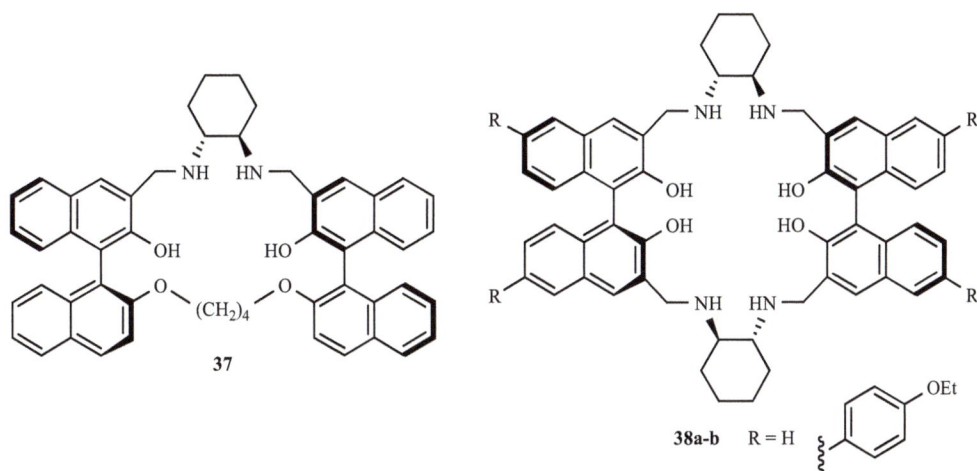

Figure 8: Macrocyclic polyamines used as enantioselective fluorescent sensors.

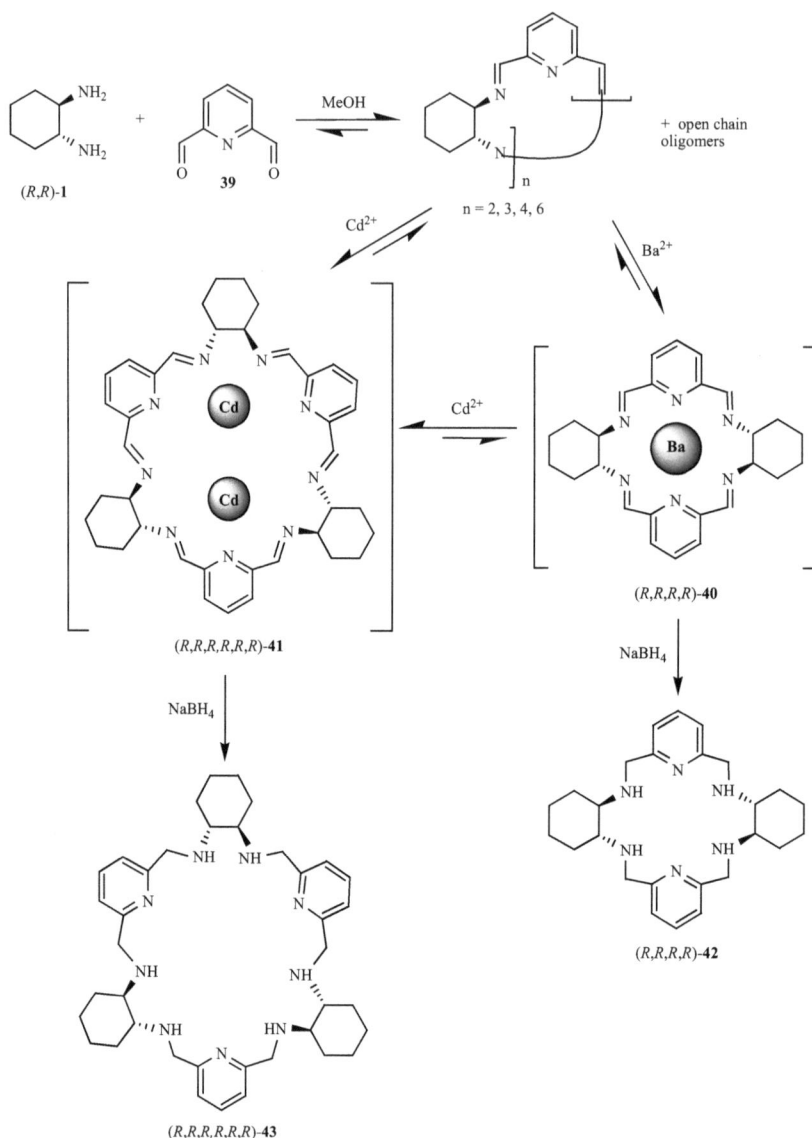

Scheme 14: Syntheses of enantiopure macrocyclic polyamines from a DCL of oligoimines.

When the untemplated reaction was performed with the racemic mixture of (**1**) and dialdehyde (**39**), a large number of different oligomers and stereoisomers are *a priori* possible (Scheme **15**). Detailed structural characterization of the oligoimines mixture showed the major formation of the heterochiral [2+2] C_{2h}-*meso* (hetero-**40**) and the [4+4] D_{2d} (heteroB-**44**) compounds, with a small amount of the C_2 symmetrical trimer (hetero-**41**) (for assumed terminology, See Scheme

15) [73]. This is most likely due to a higher stability of alternating configurations in the diamine moieties and the lower solubility of the obtained species.

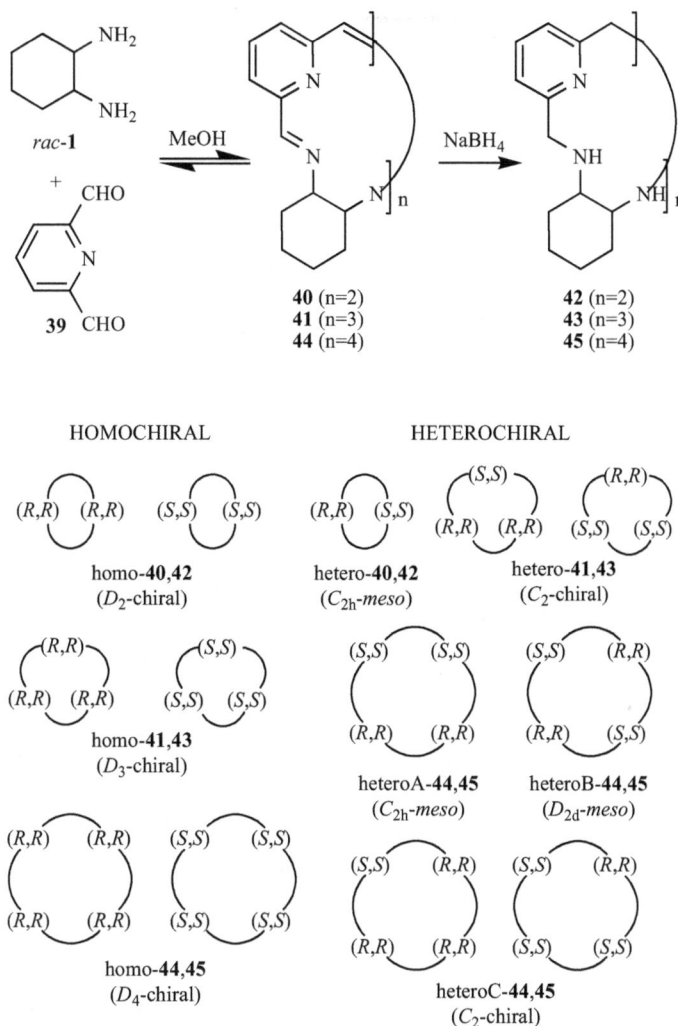

40 (n=2)
41 (n=3)
44 (n=4)

42 (n=2)
43 (n=3)
45 (n=4)

HOMOCHIRAL   HETEROCHIRAL

homo-**40,42**
(D_2-chiral)

hetero-**40,42**
(C_{2h}-*meso*)

hetero-**41,43**
(C_2-chiral)

homo-**41,43**
(D_3-chiral)

heteroA-**44,45**
(C_{2h}-*meso*)

heteroB-**44,45**
(D_{2d}-*meso*)

homo-**44,45**
(D_4-chiral)

heteroC-**44,45**
(C_2-chiral)

Scheme 15: Schematic representation of the possible products of the reaction between racemic (**1**) and (**39**).

The effect of cationic templates on the racemic DCL of oligoimines has been also investigated (Scheme **16**) [73]. The presence of Ba(II) led to the unique formation, after reduction, of the [2+2] macrocycle. A detailed analysis showed a statistical mixture of isomers, (hetero-**42**) and (homo-**42**), in a 1: 1.6 mole ratio, respectively. More remarkably, the Cd(II) template promoted the unique formation of the [3+3] cyclic oligomer, as a single isomer (hetero-**43**), showing a highly diastereoselective host amplification effect.

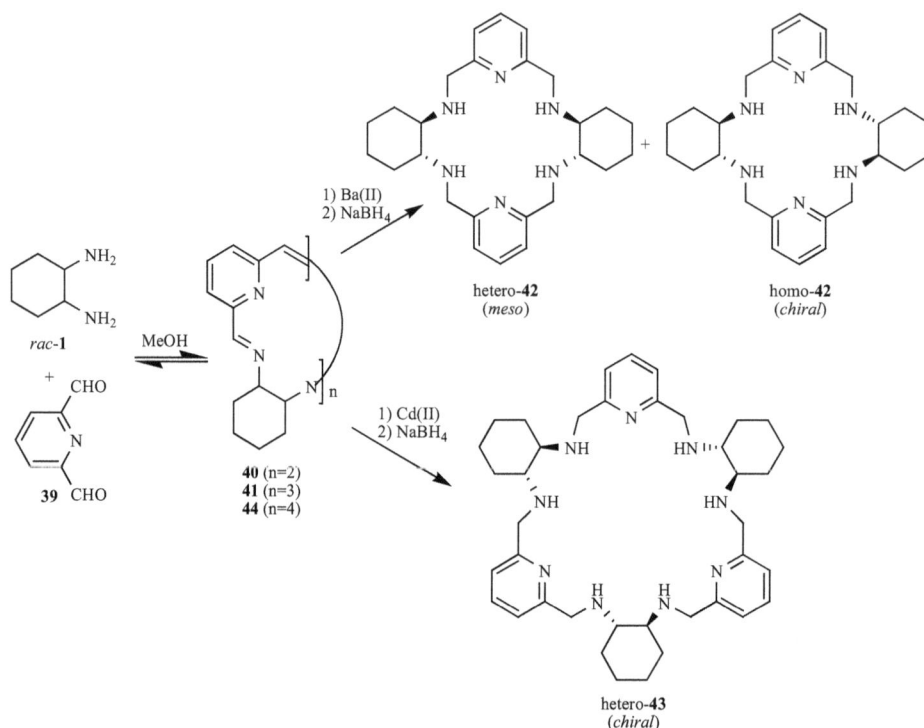

Scheme 16: Metal assisted amplifications from a DCL of racemic (**1**).

The homochiral compounds containing several units of pyridine and diamine (**1**), in enantiopure forms, have been used as receptors for biologically important chiral dicarboxylates at physiological conditions. For example, (all-*R*-**42**) showed to be a highly enantioselective receptor (up to $\Delta\Delta G = 6.12$ kJ/mol) for malate dianion in water, over a wide pH range and in the presence of the competing chloride anion, as in biomolecular media [74]. Further studies with other dicarboxylates and using a multidisciplinary approach allowed proposing a model for the selectivity displayed, in which the twisted helical conformation of the receptor plays a fundamental role [75]. Besides, on the other hand, compound (all-*S*-**42**) is an excellent CSA for chiral carboxylic acids in CDCl₃ [76].

Regarding the larger macrocycle (all-*R*-**43**), it displayed a complicated binding equilibrium pattern with the same dicarboxylates, but with weak interaction, probably due to a poor size complementarity. However, this compound has proved to be very useful for metal binding studies. The large size of the marocyclic cavity of (all-*R*-**43**) is able to host an uncommon multi-metallic $[Cu_3(OH)_2]^{4+}$ cluster, which showed ferromagnetic properties due to the electronic communication between the three Cu(II) ions [77]. These clusters are very interesting as synthetic models of metalloenzymes. On the other hand, the lanthanide complexes of enantiopure homochiral (homo-**43**) form a helical wrapping around the rare earth metal cations, with very interesting time-dependent chiroptical properties sourcing from the transition between *P* and *M* helix signs [78]. On the other hand, the corresponding complexes with the heterochiral isomer (hetero-**43**) also showed a dynamic behavior in solution, although the presence of diastereomers in this case is due to a change in the direction of the helix axis [79].

Although the reductive amination multicomponent macrocyclization is a very efficient synthetic strategy to obtain large macrocycles derived from (**1**), it is mainly restricted to aromatic or constrained aliphatic dialdehydes. Accordingly, in order to get simpler and more flexible aliphatic macrocycles, another strategy has to be used. A synthetic procedure based on the direct macrocyclization through dialkylation of *N,N'*-diisopropyl derivative of (**1**) has been reported [80], although the structural variability of the accessible macrocycles seems to be limited, once again, to aromatic moieties. However, probably one of the most popular and efficient procedure for the synthesis of polyazamacrocycles is that described by Ritchman and Atkins [81], which consists of a reaction between a bis(sulfonamide) and an appropriate α,ω-dielectrophilic system (Scheme **17**) in the presence of an excess of a base (typically potassium or cesium carbonate). In this procedure, the preorganization leading to the macrocyclization is obtained with the sulfonamide protecting group, which makes the linear precursor to fold in polar solvents (typically acetonitrile or DMF). Besides, the sulfonamide group increases the acidity of the terminal nitrogen atoms and protects the secondary amine of the linear polyamine precursor. Following this general scheme, one, two or even more fragments of (**1**) can be implemented in the final structure of the macrocyclic polyamine (Scheme **17**).

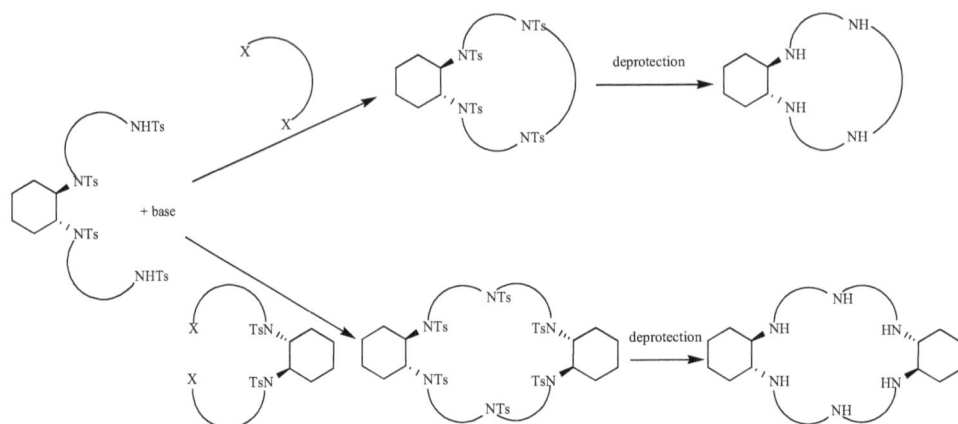

Scheme 17: General Richman-Atkins procedure for the preparation of polyazamacrocycles containing diamine (**1**).

Following that general strategy, we have used a chemo-enzymatic approach to synthesize different polyaza [32, 82] and oxaazamacrocycles [83] in enantiopure forms (Fig. **9**), combining enzymatic aminolysis resolutions with the Richman-Atkins procedure. Some of these macrocycles have shown to be enantioselective receptors for dicarboxylates in aqueous solution [83, 84]. For instance, protonated hexaazamacrocycle (**46**) showed a moderate D-selectivity towards tartrate dianion, whereas (**47**·H_6^{6+}) exhibited a good preference for *N*-Ac-D-aspartate (K_{ass}(D)/K_{ass}(L) = 5.89). The most surprising results were obtained with the *N*-Ac derivative of glutamate anion, which forms very stable complexes with both (**46**) and (**47**). The stoichiometry of these complexes can be 1: 1 or 1: 2 (receptor/anion), depending on the number of the protons and also on the enantiomer of the anion. For this last anion, both azamacrocycles exhibited a very good D-preference. Regarding the oxaazamacrocycles, the triprotonated species of (**49**) showed a moderate L-enantioselectivity with malate and tartrate (bearing OH groups). On the other hand, the receptor (**48**) displayed a good D-selectivity towards *N*-Ac-aspartate ($\Delta\Delta G$ = 3.83 kJ/mol). Despite the lack of enantioselectivity of this compound with hydroxyacids, a very good diastereopreference towards the *meso* form of tartrate was observed ($\Delta\Delta G$ = 7.5 kJ/mol), which helped us to propose a reasonable mode of binding, based on electrostatic and H-bonding interactions. Finally, the smaller cyclam derivative (**50**) behaved as an excellent ligand for transition metals, and, when partially protonated, also as a good host for small inorganic anions [85].

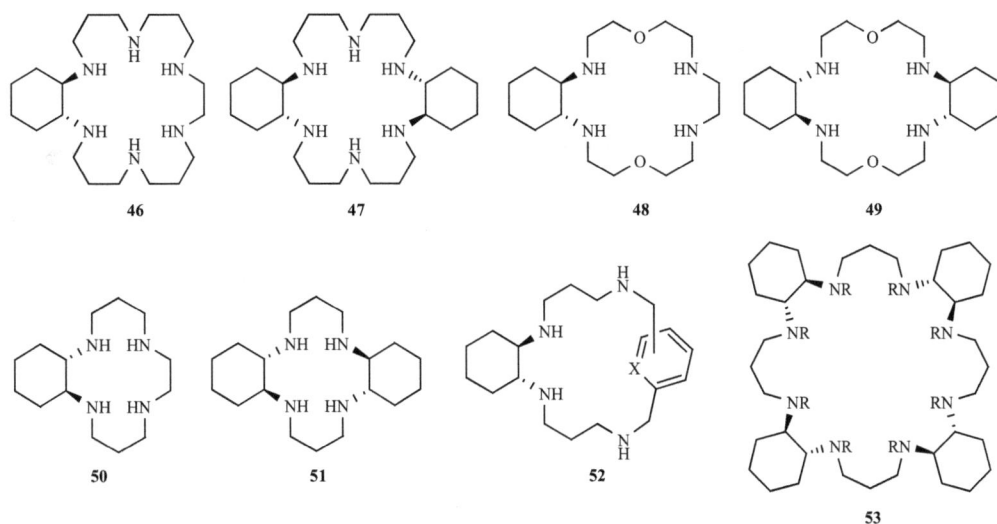

Figure 9: Molecular structures of enantiopure macrocyclic polyamines prepared by the Richman-Atkins procedure.

Other interesting macrocyclic receptors derived from (**1**) have been described, bearing amide [86, 87], sulfonamide [88] and thiourea [89] functionalities in their structure (Fig. **10**). Tetraamide (**54a**) has been studied in its ability to display NMR enantiodiscrimination as CSA for chiral substrates with H-bonding and π-acidic groups [86]. The structurally related silica-supported compound (**54b**) has been used as chiral stationary phase for the HPLC separation of racemates [87]. Recently, Kilburn and co-workers reported on the multi-step synthesis of amide/sulfonamide macrocycles (**55a-b**) which bind acetate anion in organic solvents [88]. Besides, the solid state structure of (**55a**) showed interesting H-bonding pattern. The same author described the preparation of amide/thiourea based macrocycles (**56a-b**) which showed a good enantiodiscrimination in the interaction with *N*-Boc-glutamate dianion, although the binding studies were complicated by the insolubility of the compounds and the presence of multiple equilibria of different host: guest stoichiometries [89].

As previously stated, probably the main problem to prepare macrocyclic receptors is the low selectivity often found for the cyclization step. This usually leads to low yields and tedious purification steps. Considering the potential applications of macrocycles derived from amino acids in the molecular recognition field [90], we

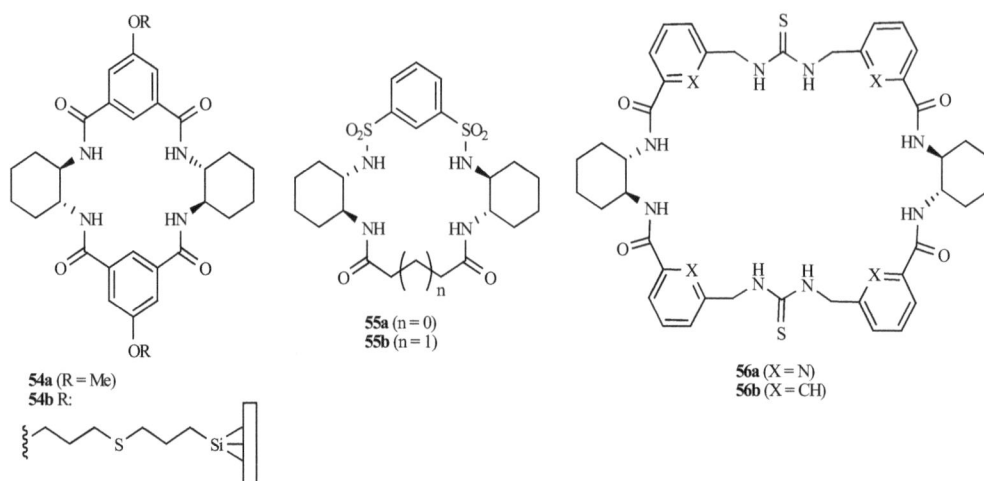

54a (R = Me)
54b R:

55a (n = 0)
55b (n = 1)

56a (X = N)
56b (X = CH)

Figure 10: Hydrogen bonding chiral receptors based on (**1**).

envisioned to carry out a reductive amination reaction between the corresponding open-chain pseudopeptidic bis(amidoamines) (**57a-i**) and the aromatic dialdehydes (**18a**) and (**21a**) (Scheme **18**). When the reaction was performed with the flexible ethylenic spacer (**57a,b**), it always led to a complicated mixture of open chain oligomers, with no formation of the intended macrocycle (**58a,b**) [91]. However, when the (*R,R*)-cyclohexane-1,2-diamine derivatives (**58c-i**) were used, the process nicely led to the corresponding [2+2] macrocyclic products (**58c-g**, *m-***58c**), easily isolable in moderate to high yields after the *in situ* reduction (Table **5**) [91, 92]. The tetraimine intermediate (Scheme **18**) was thoroughly studied by NMR, circular dichroism and theoretical calculations. Interestingly, the observed preorganization for the cyclization is highly configurationally driven, as it is shown by the *match*/*mismatch* relationships between the absolute configurations of both the chiral diamine and the amino acid. Thus, when the reaction was performed with a bis(amidoamine) containing (*R,R*) configuration of the cyclohexane moiety and *R* configuration of the amino acid moieties [compounds (**57h**) and (**57i**)], it also led to a complicated mixture of compounds. A more detailed structural analysis with open chain model systems showed that the *mismatch* combination of stereocenters produced a less reactive diamine and led to a more constrained corresponding diimine [92]. Moreover, the tetra-HCl salt of one the final macrocycles (**58d**) showed a very interesting porous structure in the solid state, mimicking a synthetic chloride channel [92].

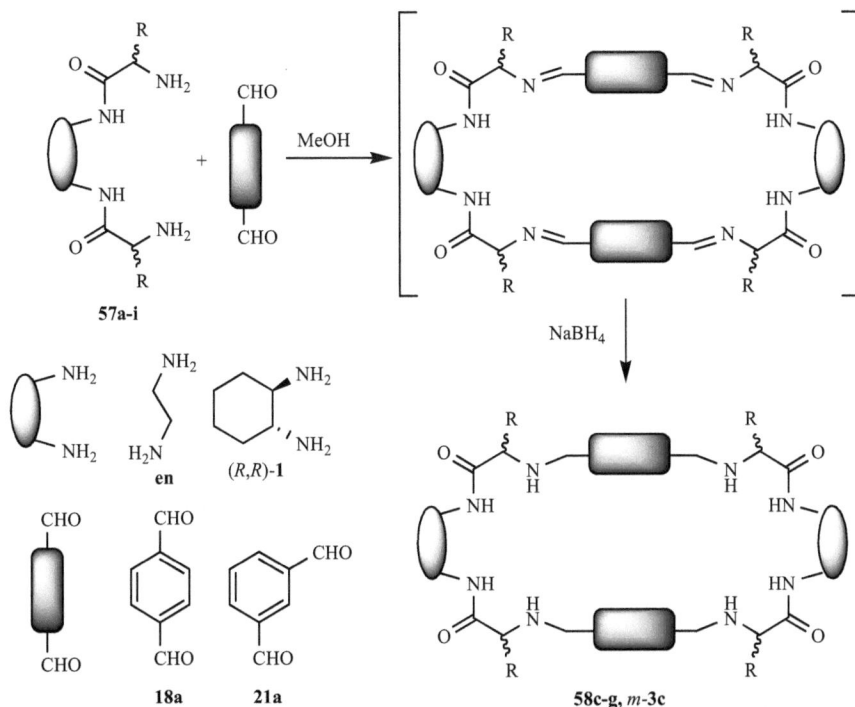

Scheme 18: Synthesis of chiral pseudopeptidic macrocycles through a configurationally driven preorganization.

Table 5: Synthesis of chiral pseudopeptidic macrocycles

Sust.	Diamine	R (Cα-Conf.)	Dialdehyde.	Prod.	Yield [a]
57a	en	iPr (*S*)	**18a**	**58a**	-[b]
57b	en	CH₂Ph (*S*)	**18a**	**58b**	-[b]
57c	(*R,R*)-1	iPr (*S*)	**18a**	**58c**	67
57d	(*R,R*)-1	CH₂Ph (*S*)	**18a**	**58d**	55
57c	(*R,R*)-1	iPr (*S*)	**21a**	*m*-**58c**	35
57e	(*R,R*)-1	iBu (*S*)	**18a**	**58e**	58
57f	(*R,R*)-1	sec-Bu (*S*)	**18a**	**58f**	41
57g	(*R,R*)-1	(CH₂)₂CONHTr (*S*)	**18a**	**58g**	17
57h	(*R,R*)-1	iPr (*R*)	**18a**	**58h**	-[b]
57i	(*R,R*)-1	CH₂Ph (*R*)	**18a**	**58i**	-[b]

[a]Isolated pure yield; [b]A complicated mixture of oligomers was obtained.

Recently, we have found that the mismatch combination of sterocenters (**57h**) and (**57i**) can be also biased toward the formation of the macrocycles by the suitable anion templation [93]. The addition of the appropriate dicarboxylate produced the [2+2] macrocycle under thermodynamic control, but mixtures of larger macrocycles under kinetic control.

6.   MACROPOLYCYCLIC,   INTERLOCKED   AND   CAGE-LIKE MOLECULES

The diamine (**1**) has been also used to prepare macropolycyclic, interlocked and cage-like compounds. Within this topic, seminal papers were published by Still and co-workers [94] about the preparation of an amide macrotricycle (**59**) which was studied for the binding of peptidic molecules (Fig. **11**). The synthesis of (**59**) can be carried out either stepwise in 39% overall yield, or through a direct one-step condensation of (**1**) with 1,3,5-benzenetris(carbonylchloride) in basic medium (19% yield). This receptor showed an excellent side chain selectivity (up to 3 kcal/mol) and enantioselectivity (>99% ee) in the interaction with different peptidic sequences [94a]. Comparative NMR and molecular modeling studies suggested that the supramolecular complexes were stabilized by several H-bonding interactions and by the inclusion of the guest side chain (R) within the cavity of the host. This macrotricyclic scaffold was additionally used as the starting point for the development of chiral stationary phases for HPLC [95] or for the preparation of fluorescent chemosensors [96].

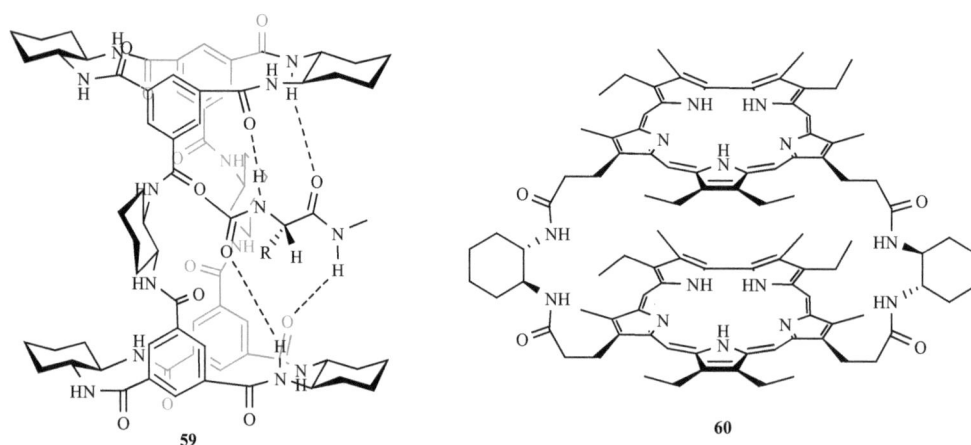

Figure 11: Large receptors from (**1**) able to bind peptides and amino acids.

Another classical example is the work reported by Sessler *et al.* [97]. They synthesized a family of sapphyrin-based receptors for dicarboxylate anions.

Binding studies with *N*-protected aspartic and glutamic dicarboxylates showed that the receptor (**60**) bearing the chiral cyclohexanediamine moiety was the most enantioselective (up to $\Delta\Delta G^{\circ} = 0.84$ kcal/mol for the binding of *N*-Cbz-D/L-Glu).

Interlocked molecules are very interesting compounds from the structural point of view, as well as for their potential applications as molecular machines and in materials science [98]. The preparation of an interlocked catenane based on diamine (**1**) and β-cyclodextrin (β-CD) has been also reported [99]. The authors followed the methodology shown in Scheme **19**. They initially prepared the inclusion complex between β-CD and terephthaldehyde (**18b**) taking advantage of the hydrophobic cavity of the macrocyclic oligosaccharide. Then, imine condensation of 1 equivalent of this complex with three equivalents of (**1**) and 2 equivalents of (**18a**) rendered the expected *trianglimine*-β-CD-[2]-catenane, which was reduced *in situ* to yield the corresponding *trianglamine*-β-CD-[2]-catenane (**61**). The reaction can be carried out with both enantiomers of the diamine leading to the corresponding diastereomeric compounds. The interlocked structure of (**61**) was unambiguously demonstrated by NMR and mass spectrometry. Although the obtained overall yield was somewhat low (18%), the easiness of the procedure opens a way to the preparation of topologically interesting chiral catenane molecules bearing diamine (**1**) in their structure.

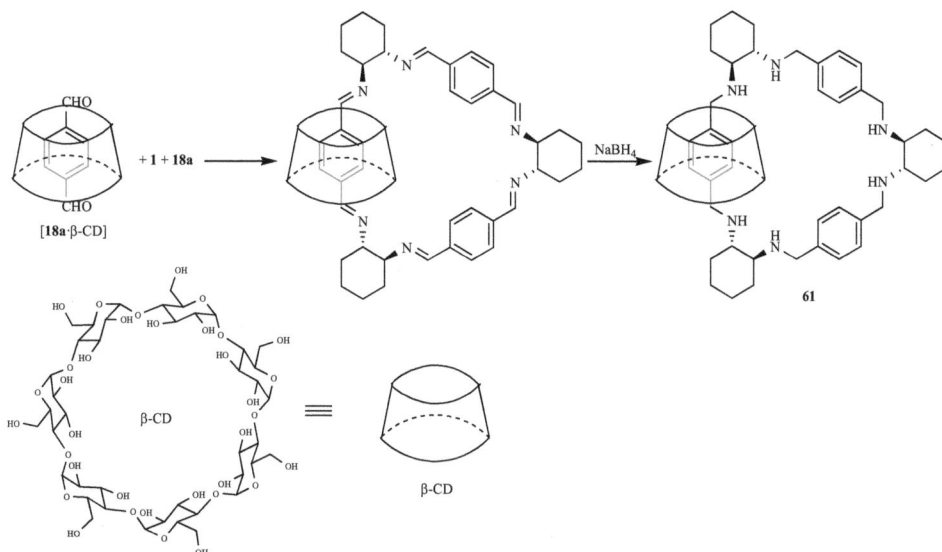

Scheme 19: Synthesis of a *trianglamine*-β-cyclodextrin-[2]-catenane (**61**).

Another group of interesting macropolycyclic molecules are those derived from spherical or polyhedral geometries [100]. Also in this field, the specific characteristics of (1) have found a space for applications. Regarding that, Gawroński recently reported the very efficient synthesis of a chiral iminospherand with unprecedented tetrahedral symmetry [101]. The compound was obtained by the direct condensation between diamine (1) (D in Scheme 20) and the simplest aromatic trialdehyde, 1,3,5-triformylbenzene (A), to yield the corresponding [6+4] multicomponent cyclization. The identity of the D_6A_4 compound (62) was established by mass spectrometry and the connectivity rendering the tetrahedral symmetry was ascertained by NMR and supported by molecular modeling. Although the inner space in this iminospherand is rather small, the authors claimed that more useful chiral containers should be accessible using a more elaborated trialdehyde (A). More recently, the group of Cooper has deeply studied this family of tetrahedral cages as porous materials with the ability to entrap (and release) different guests [102].

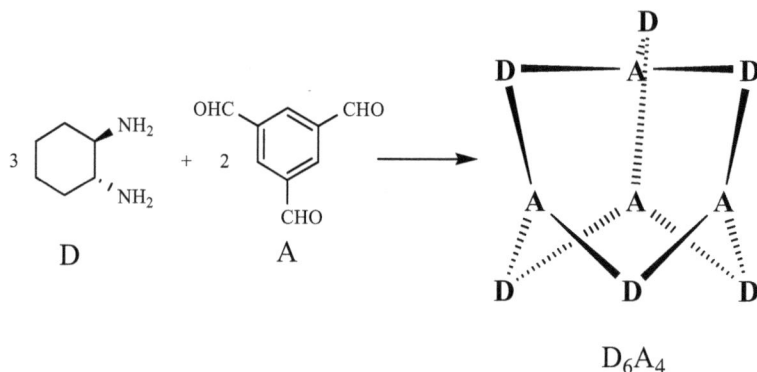

Scheme 20: One-pot synthesis of a chiral iminospherand (D_6A_4) (62) with tetrahedral symmetry.

Also within the field of nanometer-sized molecular polyhedron synthesis, a very clever application of diamine (1) has been recently reported [103]. In this paper, the authors intended to synthesize a molecular chiral nanocube by a dynamic covalent approach, based on imine bond formation (Scheme 21). In order to do that, they needed to prepare the cyclotriveratrylene trialdehyde (63) in enantiopure form. They solved that problem by using a dynamic kinetic resolution of (63) with diamine (R,R)-(1) in CHCl₃ containing a catalytic amount of TFA (12 h; 80 °C). Under those conditions, the trialdehyde is able to racemize and the (P)-(63)

isomer reacts with (*R,R*)-(**1**) to yield the pure cryptophane (*P,P,R,R,R,R,R,R*)-(**64**) in 92% yield. Precipitation of (*P,P,R,R,R,R,R,R*)-(**64**) and hydrolysis with TFA/water gave (*P*)-(**63**) (92% yield, >99% ee) and allowed recovery of (*R,R*)-(**1**). This example represents a very remarkable application of (**1**) in the macropolycyclic chemistry field.

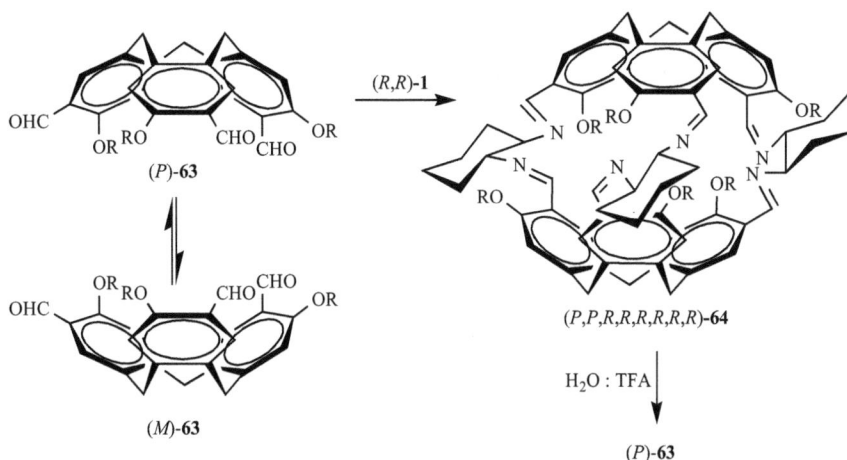

Scheme 21: Synthesis of a cryptophane (**64**) for the dynamic kinetic resolution of the trialdehyde (**63**).

Even more recently, a straightforward preparation of a chiral cage has been communicated [104]. The authors also used diamine (**1**) as the chiral scaffold and for its geometrical propensity to yield [3+3] imine cyclocondensations (Scheme **22**). For the construction of the bowl-shaped "ceiling" of the intended cage, they synthesized the corresponding tripodal hexaaldehyde (**67**) using a triple copper catalyzed Huisgen cycloaddition of (**65**) and (**66**). Trianglimine formation with (**1**) and (**67**) yielded the intended cage almost quantitatively. Very remarkably, the overall coupling (imine formation + Huisgen cycloaddition) between the three components can be also performed in one pot yielding, after purification, the same cage (**68**) in 70% yield. The authors suggested that the dynamic nature of the imine functions allows the auto-repair of any eventual "mistake" and drives the reaction to the thermodynamically stable compound. Preliminary studies suggested that (**68**) shows a good selectivity in the binding of Ni(II) over other metallic cations, which could be very promising for applications of this or related molecules in catalysis or as biomimetic synthetic systems.

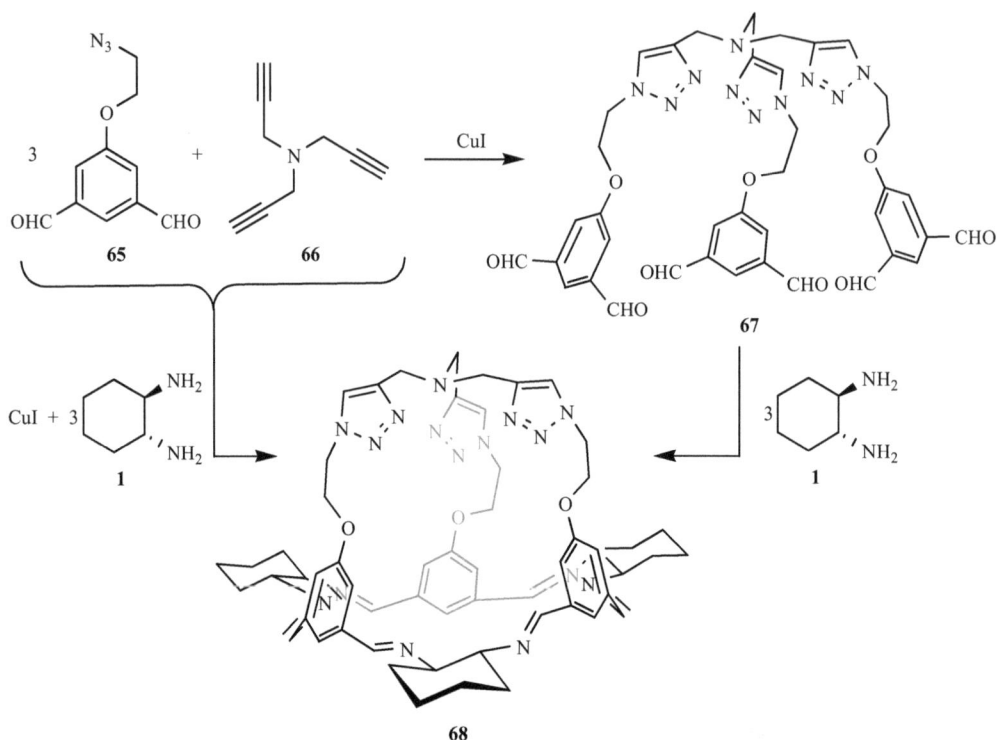

Scheme 22: Imine-Huisgen multicomponent reaction for the synthesis of a chiral cage compound (**68**).

We have recently reported the synthesis of pseudopeptidic molecular cages [105], as a further application of our dynamic covalent procedure for the preparation of pseudopeptidic macrocycles (Scheme **23**). Thus, the condensation of the pseudopeptidic bis(amidoamines) with 1,3,5-triformylbenzene, followed by *in situ* reduction, led to the corresponding cages (**69a-c**) in very good overall isolated yields (30-59%).

Although rigorously not being a macrocyclic cage, the tripodal receptor (**70**) designed by Roelens, Jiménez-Barbero and co-workers (Fig. **12**) [106] is better classified in this section, since the detailed structural analysis [107] rendered a cup-like conformation in solution, able to cage sugar derivatives by a combination of H-bonding and CH···π interactions. The compound (**70**) resulted to be a very stereoselective receptor for mannosides and also more efficient than the corresponding open-chain tripodal derivative (**71**).

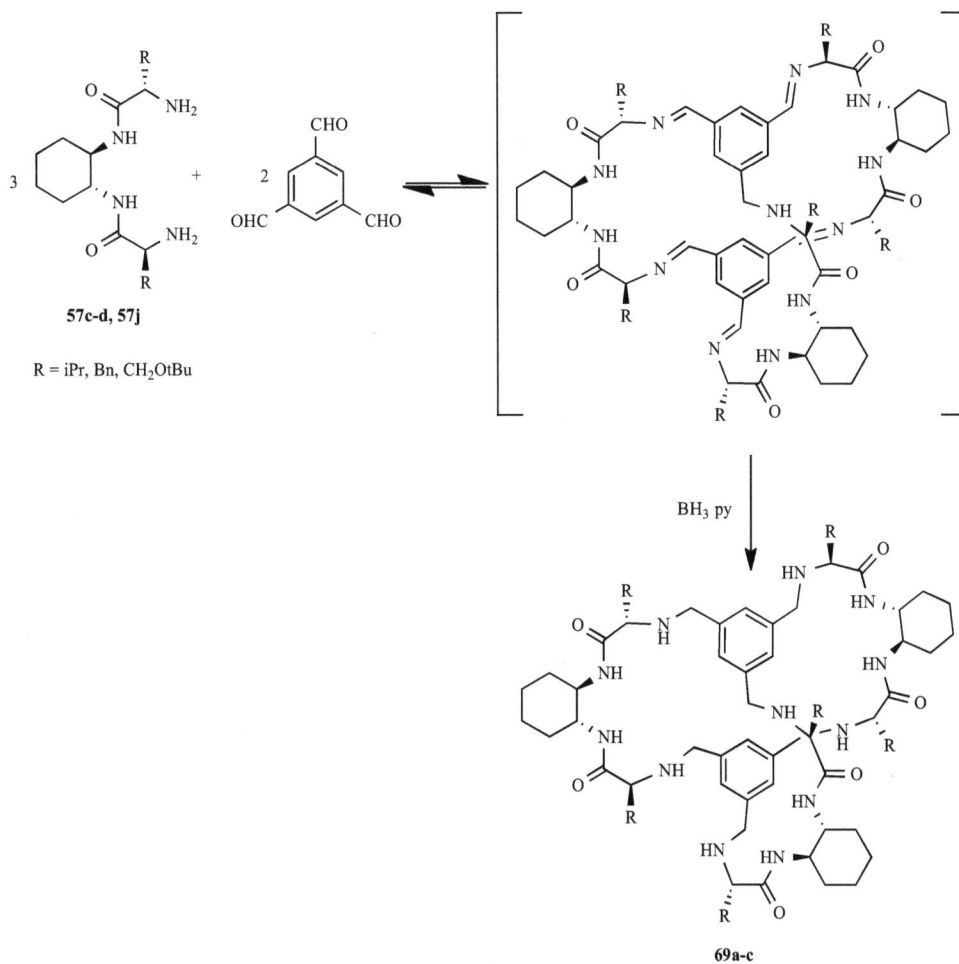

Scheme 23: Synthesis of pseudopeptidic molecular cages.

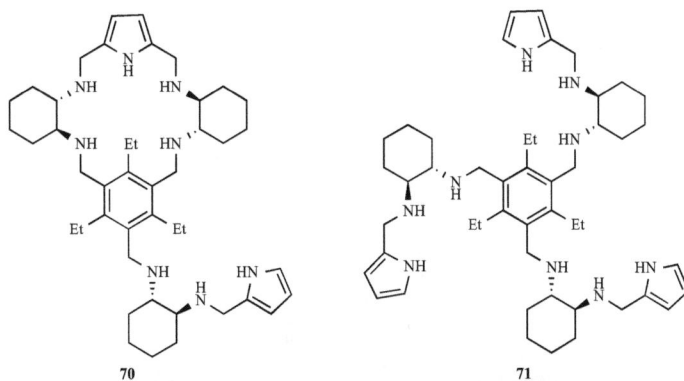

Figure 12: Chiral tripodal receptors for the enantioselective recognition of β-manosides.

Finally, Cooper and co-workers have recently reported the most outstanding example of macrobicyclization reaction from (**1**). The direct condensation of the chiral diamine with the simple tripodal trialdehyde (**72**) rendered the formation of a [8+12] cage (**73**) by a multicomponent reaction where 20 building blocks are specifically connected in one pot (Scheme **24**) [108]. The cage can be represented by a highly distorted cube (A_8B_{12}) where the 8 triphenyl amine moieties (A) occupy the vertices, while the 12 diamines occupy the edges. The crystal structure of the obtained cage showed the giant inner cavity, which is filled with solvent molecules. Actually, a detailed molecular modeling study illustrated how the cage should collapse in a tightly folded structure when these inner solvent molecules are removed. To date, this example represents the largest cage built from the chiral diamine (**1**) and foresees a still bright future of this interesting building block in the field of molecular cages and organic porous materials.

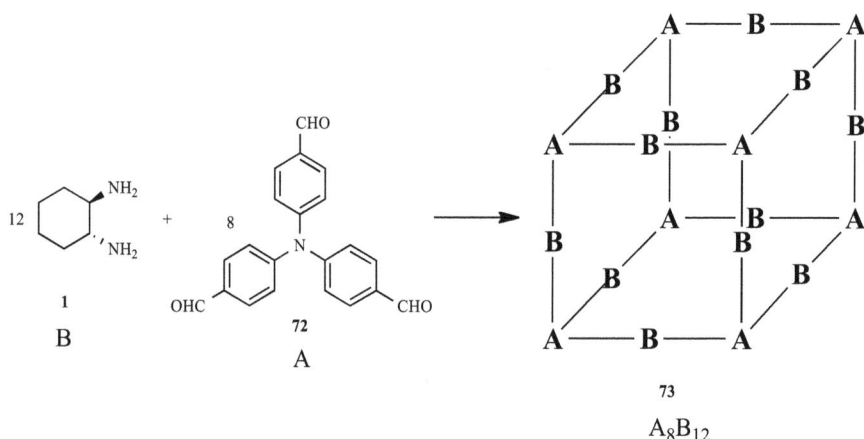

Scheme 24: Synthesis of a large A_8B_{12} organic molecular cage.

CONCLUSIONS AND OUTLOOK

In this paper, the applications of diamine (**1**) for the preparation of chiral receptors have been reviewed. The described compounds ranged from open-chain to macrocyclic and macropolycyclic molecules, all of them with a well defined (and predictable) conformation due to the rigidity of the cyclohexane scaffold. Many different synthetic procedures have been reported, either under kinetic or under thermodynamic control conditions. The availability of diamine (**1**) in enantiopure forms, its well defined three-dimensional structure and the characteristic spectroscopic features of its derivatives make the compounds derived from (**1**) very appealing to be used in supramolecular chemistry. The results reported to date showed that the hosts bearing this moiety in their structures are able to exert high degrees of selectivity for the binding of biologically interesting guests. Moreover, they often display a highly enantioselective interaction with chiral substrates, with very promising applications for separation devices, sensors and catalysts. Despite the large number of potential receptors already described, much more research regarding their molecular recognition and self-assembling properties is still necessary. Accordingly, in spite of the research already done in the field, the increasing use of (**1**) for the synthesis and study of new molecular architectures is highly envisioned for the next future.

ACKNOWLEDGEMENTS

Financial support from CSIC-I3 (2007801001) and the Spanish Ministry of Science and Innovation (CTQ2009-14366-C02-02) is gratefully acknowledged.

CONFLICT OF INTEREST

The author(s) confirm that this chapter content has no conflict of interest.

DISCLOSURE

DISCLOSURE: The chapter submitted for series eBook titled **"Advances in Organic Synthesis, Volume 5"** is an update of our article published in **CURRENT ORGANIC SYNTHESIS, Volume 7, Number 1, February Issue 2010,** with additional text and references.

REFERENCES

[1]   (a) Jacobsen, E. N.; Zhang, W.; Muci, A. R.; Ecker, J. R.; Deng, L. Highly enantioselective epoxidation catalyists derived from 1,2-diaminocyclohexane. *J. Am. Chem. Soc.* **1991**, *113*, 7063-7064. (b) Deng, L.; Jacobsen, E. N. A practical, highly enantiolselective synthesis of the taxol side-chain *via* asymmetric catalysis. *J. Org. Chem.* **1992**, *57*, 4320-4323. (c) Chang, S.; Lee, N. H.; Jacobsen, E. N. Regioselective and enantioselective catalytic epoxidation of conjugated polyenes – Formal synthesis of LTA(4) methyl-ester. *J. Org. Chem.* **1993**, *58,* 6939-6941. (d) Jacobsen, E. N. Asymmetric catalysis of epoxide ring opening reactions. *Acc. Chem. Res.* **2000**, *33*, 421-431. (e) Cozzi, P. G. Metal–Salen Schiff

base complexes in catalysis: practical aspects. *Chem. Soc. Rev.* **2004**, *33*, 410-421. (f) Katsuki, T. Unique asymmetric catalysis of cis-β metal complexes of salen and its related Schiff-base ligands. *Chem. Soc. Rev.* **2004**, *33*, 437-444.

[2]   (a) Still, W. C. Discovery of sequence-selective peptide binding by synthetic receptors using encoded combinatorial libraries. *Acc. Chem. Res.* **1996**, *29*, 155-163. (b) Stibor, I.; Zlatuskova, P. Chiral recognition of anions. *Top. Curr. Chem.* **2005**, *255*, 31-63.

[3]   (a) Pirkle, W. H.; Pochapsky, T. C. Considerations of chiral recognition relevant to the liquid-chromatographic separation of enentiomers. *Chem. Rev.* **1989**, *89*, 347-362. (b) Gasparrini, F.; Lunazzi, L.; Misiti, D.; Villani, C. Organic stereochemistry and conformational-analysis from enantioselective chromatography and dynamic Nuclear-Magnetic Resonance measurements. *Acc. Chem. Res.* **1995**, *28*, 163-170. (c) Burguete, M. I.; Fréchet, J. M. J.; García-Verdugo, E.; Janco, M.; Luis, S. V.; Svec, F.; Vicent, M. J.; Xu, M. New CSPs based on peptidomimetics: efficient chiral selectors in enantioselective separations. *Polym. Bull.* **2002**, *48*, 9-15.

[4]   Bennani, Y. L.; Hanessian, S. *trans*-1,2-Diaminocyclohexane Derivatives as Chiral Reagents, Scaffolds, and Ligands for Catalysis: Applications in Asymmetric Synthesis and Molecular Recognition. *Chem. Rev.* **1997**, *97*, 3161-3195.

[5]   (a) Pu, L. 1,1'-Binaphthyl dimers, oligomers, and polymers: Molecular recognition, asymmetric catalysis, and new materials. *Chem. Rev.* **1998**, *98*, 2405-2494. (b) Borisova, N. E.; Reshetova, M. D.; Ustynyuk, Y. A. Metal-Free Methods in the Synthesis of Macrocyclic Schiff Bases. *Chem. Rev.* **2007**, *107*, 46-79. (c) Savoia, D.; Gualandi, A. Chiral Perazamacrocycles: Synthesis and Applications. Part 1. *Curr. Org. Synth.* **2009**, *6*, 102-118. (d) Savoia, D.; Gualandi, A. Chiral Perazamacrocycles: Synthesis and Applications. Part 2. *Curr. Org. Synth.* **2009**, *6*, 119-142.

[6]   Performed with Spartan `06, Wavefunction, Inc. Irvine, CA.

[7]   Alfonso, I.; Astorga, C.; Rebolledo, F.; Gotor, V.; García-Granda, S.; Tesouro, A. Structural studies of two protonated forms of a C_2 symmetrical optically active cyclam derivative. *J. Chem. Soc., Perkin Trans. 2* **2000**, 899-904.

[8]   Bertsc, C.; Fernelius, W. C.; Block, P. B. Pennsylvania State Univ. Contract No AT(30-1)-907, April 30, **1956**.

[9]   Barbucci, R.; Paoletti, P.; Vacca, A. Predictions of enthalpies of protonation of amines – log K, ΔH, and Δ values for protonation of ethylenediamine and tri-methylenediamne, tetra-methylenediamine, penta-methylenediamine and hexa-methylenediamine. *J. Chem. Soc. (A)* **1970**, 2202.

[10] (a) Derwahl, A.; Wasgestian, F.; House, D. A.; Robinson, W. T. The crystal structures and doublet emission properties of hexaam(m)ine Cr(III) complexes. *Coord. Chem. Rev.* **2001**, *211*, 45-67. (b) P. Cole, A. P.; Mahadevan, V.; Mirica, L. M.; Ottenwaelder, X.; Snack, T. D. P. Bis(μ-oxo)dicopper(III) Complexes of a Homologous Series of SimplePeralkylated 1,2-Diamines: Steric Modulation of Structure, Stability, and Reactivity. *Inorg. Chem.* **2005**, *44*, 7345-7364. (c) Wang, Y.; Fu, H.; Shen, F.; Sheng, X.; Peng, A.; Gu, Z.; Ma, H.; Ma, J. S.; Yao, J. Distinct *M* and *P* Helical Complexes of H_2O and Metal Ions Ni^{II}, Cu^{II}, and Zn^{II} with Enantiomerically Pure Chiral Bis(pyrrol-2-ylmethyleneamine)cyclohexane Ligands: Crystal Structures and Circular Dichroism Properties. *Inorg. Chem.* **2007**, *46*, 3548-3556.

[11] Sawada, T.; Yoshizawa, M.; Sato, S.; Fujita, M. Minimal nucleotide duplex formation in water through enclathration in self-assembled hosts *Nature Chem.* **2009**, *1*, 53-56.

[12]   (a) Berova, N.; Di Bari, L.; Pescitelli, G. Application of electronic circular dichroism in configurational and conformational analysis of organic compounds *Chem. Soc. Rev.* **2007**, *36*, 914-931. (b) Berova, N.; Nakanishi, K.; Woody, R. W. *Circular Dichroism. Principles and Applications*, Wiley-VCH, 2000, New York.

[13]   (a) Gawroński, J.; Brzostowska, M.; Kacprzak, K.; Kołbon, H.; Skowronek, P. Chirality of Aromatic Bis-Imides from Their Circular Dichroism Spectra. *Chirality* **2000**, *12*, 263-268. (b) Gawroński, J.; Brzostowska, M.; Gawrońska, K.; Koput, J.; Rychlewska, U.; Skowronek, P.; Nordén B. Novel Chiral Pyromellitdiimide (1,2,4,5-Benzenetetracarboxydiimide) Dimers and Trimers: Exploring Their Structure, Electronic Transitions, and Exciton Coupling. *Chem. Eur. J.* **2002**, *8*, 2484-2494.

[14]   Gasbøl, F.; Steenbøl, P.; Seremen, B. S. The Preparation, Separation and Characterization of the lel₃ and ob₃ isomers of Tris(*trans*-1,2-cyclohexanediamine)rhodium(III) Complexes. *Acta Chem. Scad.* **1972**, *26*, 3605-3610.

[15]   Larrow, J. F.; Jacobsen, E. A Practical Method for the Large-Scale Preparation of [*N,N'*-Bis(3,5-di-*tert*-butylsalicylidene)-1,2-Cyclohexanediaminato(2-)]manganese (III) Chloride, a Highly Enantioselective Epoxidation Catalyst. *J. Org. Chem.* **1994**, *59*, 1939-1942.

[16]   Rychlewska, U. Chiral recognition in salts of *trans*-1,2-diaminocyclohexane and optically active tartaric acids: crystal structure of 1:2 salt of (*S,S*)-diaminocyclohexane with (*R,R*)-tartaric acid. *J. Mol. Struct.* **1999**, *474*, 235-243.

[17]   (a) Hanessian, S.; Gomtsyan, A.; Simard, M.; Roelens, S. Molecular Recognition and Self-Assembly by "Weak" Hydrogen Bonding: Unprecedented Supramolecular Helicate Structures from Diamine/Diol Motifs. *J. Am. Chem. Soc.* **1994**, *116*, 4495-4496. (b) Hanessian, S.; Simard, M.; Roelens, S. Molecular Recognition and Self- Assembly by Non-amidic Hydrogen Bonding. An Exceptional "Assembler" of Neutral and Charged Supramolecular Structures. *J. Am. Chem. Soc.* **1995**, *117*, 7630-7645. (c) Ratajczak-Sitarz, M.; Katrusiak, A.; Gawrońska, K.; Gawroński, J. Racemate resolution *via* diastereomeric helicates in hydrogen-bonded co-crystals: the case of BINOL–diamine complexes. *Tetrahedron: Asymmetry* **2007**, *18*, 765-773.

[18]   Imai, Y.; Kawaguchi, K.; Murata, K.; Sato, T.; Kuroda, R.; Matsubara, Y. Tunable Two-Component Host System with Multiple Chiral Points Composed of Cyclohexanediamine and 1,1'-Binaphthyl-2,2'-dicarboxylic Acid. *Crystal Growth & Design* **2007**, *7*, 1676-1678.

[19]   García-Velázquez, D.; Díaz-Díaz, D.; Gutiérrez-Ravelo, A.; Marrero-Tellado, J. J. Instantaneous Low Temperature Gelation by a Multicomponent Organogelator Liquid System Based on Ammonium Salts. *J. Am. Chem. Soc.* **2008**, *130*, 7967-7973.

[20]   Galland, A.; Dupray, V.; Lafontaine, A.; Berton, B.; Sanselme, M.; Atmani, H.; Coquerel, G. Preparative resolution of (±)-trans-1,2-diaminocyclohexane by means of preferential crystallization of its citrate monohydrate. *Tetrahedron: Asymmetry*, **2010**, *21*, 2212-2217.

[21]   (a) Alfonso, I.; Gotor, V. Biocatalytic and biomimetic aminolysis reactions: useful tools for selective transformations on polyfunctional substrates. *Chem. Soc. Rev.* **2004**, *33*, 201-209. (b) Gotor-Fernández, V.; Gotor, V. Enzymatic aminolysis and ammonolysis processes in the preparation of chiral nitrogenated compounds. *Curr. Org. Chem.* **2006**, *10*, 1125-1143. (c) Gotor-Fernández, V.; Busto, E.; Gotor, V. *Candida antarctica* lipase B: An ideal biocatalyst for the preparation of nitrogenated organic compounds. *Adv. Synth. Catal.* **2006**, *348*, 797-812.

[22]   Alfonso, I.; Astorga, C.; Rebolledo, F.; Gotor, V. Sequential biocatalytic resolution of (±)-*trans*-cyclohexane-1,2-diamine. Chemoenzymatic synthesis of an optically active polyamine. *Chem. Commun.* **1996**, 2471-2472.

[23]   González-Sabin, J.; Gotor, V.; Rebolledo, F. Chemoenzymatic preparation of optically active trans-cyclohexane-1,2-diamine derivatives: An efficient synthesis of the analgesic U-(-)-50,488. *Chem. Eur. J.* **2004**, *10*, 5788-5794.

[24]   Albert, J. S.; Hamilton, A. D. Synthetic analogs of the ristocetin binding site: Neutral, multidentate receptors for carboxylate recognition. *Tetrahedron Lett.* **1993**, *34*, 7363-7366.

[25]   Amendola, V.; Boiocchi, M.; Esteban-Gómez, D.; Fabbrizzi, L.; Monzani, E. Chiral receptors for phosphate ions. *Org. Biomol. Chem.* **2005**, *3*, 2632-2639.

[26]   Peña, C.; González-Sabín, J.; Alfonso, I.; Rebolledo, F.; Gotor, V. Cycloalkane-1,2-diamine derivatives as chiral solvating agents. Study of the structural variables controlling the NMR enantiodiscrimination of chiral carboxylic acids. *Tetrahedron* **2008**, *64*, 7709-7717.

[27]   Yang, X.; Wang, G.; Zhong, C.; Wua, X.; Fu, E. Novel NMR chiral solvating agents derived from (1*R*,2*R*)-diaminocyclohexane: synthesis and enantiodiscrimination for chiral carboxylic acids. *Tetrahedron: Asymmetry* **2006**, *17*, 916-921.

[28]   (a) Cherny, I.; Gazit, E. Amyloids: Not only pathological agents but also ordered nanomaterials. *Angew. Chem. Int. Ed.* **2008**, *47*, 4062-4069. (b) Ulijn, R. V.; Smith, A. M. Designing peptide based nanomaterials. *Chem. Soc. Rev.* **2008**, *37*, 664-675. (c) Gazit, E. Self-assembled peptide nanostructures: the design of molecular building blocks and their technological utilization. *Chem. Soc. Rev.* **2007**, *36*, 1263-1269.

[29]   Hanessian, S.; Vinci, V.; Fettis, K.; Maris, T.; Viet M. T. P. Self-Assembly of Noncyclic Bis-D- and L-tripeptides into Higher Order Tubular Constructs: Design, Synthesis, and X-ray Crystal Superstructure. *J. Org. Chem.* **2008**, *73*, 1181-1191.

[30]   (a) Campbell, D. R.; Morris, D. R.; Bartos, D.; Daves, G. D., Jr.; Bartos, F. *Advances in Polyamine Research*; Raven Press: New York, 1978. (b) Goldenburg, S. H.; Algranati, J. D. *The Biology and Chemistry of Polyamines*; JCSU Press: Oxford, 1990.

[31]   Peña, C.; Alfonso, I.; Voelcker, N. H.; Gotor, V. Solvent dependent selective alkylation of a bis(sulfonamide) for the synthesis of a DNA-binding chiral polyamine *Tetrahedron Lett.* **2005**, *46*, 2783-2787.

[32]   Peña, C.; Alfonso, I.; Gotor, V. Conformationally Biased Selective Alkylation of trans-Cyclohexane-1,2-bis(sulfonamide) Assisted by Solvent-Tuned Protecting Groups: Applications to the Synthesis of a Large Optically Active Polyazamacrocycle. *Eur. J. Org. Chem.* **2006**, 3887-3897.

[33]   Peña, C.; Alfonso, I.; Tooth, B.; Voelcker, N. H.; Gotor, V. Synthesis and Stereoselective DNA Binding Abilities of New Optically Active Open-Chain Polyamines. *J. Org. Chem.* **2007**, *72*, 1924-1930.

[34]   Li, G.; Yu, W.; Cui, Y. A Homochiral Nanotubular Crystalline Framework of Metallomacrocycles for Enantioselective Recognition and Separation. *J. Am. Chem. Soc.* **2008**, *130*, 4582-4583.

[35]   (a) Gawroński, J.; Kolbon, H.; Kwit, M.; Katrusiak, A. Designing Large Triangular Chiral Macrocycles: Efficient [3 + 3] Diamine-Dialdehyde Condensations Based on Conformational Bias. *J. Org. Chem.* **2000**, *65*, 5768-5773. (b) Chadim, M.; Buděšínský, M.; Hodaňová, J.; Závada, J.; Junk, P. C. (3+3)-Cyclocondensation of the enantiopure and racemic forms of *trans*-1,2-diaminocyclohexane with terephthaldehyde. Formation of diastereomeric molecular triangles and their stereoselective solid-state stacking into microporous chiral columns. *Tetrahedron: Asymmetry* **2001**, *12*, 127-133.

[36] Kuhnert, N.; Rossignolo, G. M.; Lopez-Periago, A. The synthesis of trianglimines: on the scope and limitations of the [3+3] cyclocondensation reaction between (1*R*,2*R*)-diaminocyclohexane and aromatic dicarboxaldehydes. *Org. Biomol. Chem.* **2003**, *1*, 1157-1170.

[37] Kuhnert, N.; Lopez-Periago, A. M. Synthesis of novel chiral non-racemic substituted trianglimine and trianglamine macrocycles *Tetrahedron Lett.* **2002**, *43*, 3329-3332.

[38] Kuhnert, N.; Lopez-Periago, A.; Rossignolo, G. M. The synthesis and conformation of oxygenated trianglimine macrocycles *Org. Biomol. Chem.* **2005**, *3*, 524-537.

[39] Kuhnert, N.; Strabnig, C.; Lopez-Periago, A. M. Synthesis of novel enantiomerically pure trianglimine and trianglamine macrocycles. *Tetraherdon: Asymmetry* **2002**, *13*, 123-128.

[40] Kwit, M.; Skowronek, P.; Kolbon, H.; Gawroński, J. Synthesis, Structure, and Contrasting Chiroptical Properties of Large Trianglimine Macrocycles. *Chirality* **2005**, *17*, S93-S100.

[41] Kuhnert, N.; Burzlaff, N.; Patel, C.; Lopez-Periago, A. Tuning the size of macrocyclic cavities in trianglimine macrocycles. *Org. Biomol. Chem.* **2005**, *3*, 1911-1921.

[42] Kuhnert, N.; Patel, Ch.; Jami, F. Synthesis of chiral nonracemic polyimine macrocycles from cyclocondensation reactions of biaryl and terphenyl aromatic dicarboxaldehydes and 1*R*,2*R*-diaminocyclohexane. *Tetrahedron Lett.* **2005**, *46*, 7575-7579.

[43] Hodačová, J.; Buděšínský, M. New Synthetic Path to 2,2'-Bipyridine-5,5'-dicarbaldehyde and Its Use in the [3+3] Cyclocondensation with trans-1,2-Diaminocyclohexane. *Org. Lett.* **2007**, *9*, 5641-5643.

[44] Tanaka, K.; Fukuaoka, S.; Miyanishi, H.; Takahashi, H. Novel chiral Schiff base macrocycles containing azobenzene chromophore: gelation and guest inclusion. *Tetrahedron Lett.* **2010**, *51*, 2693-2696.

[45] Nour, H. F.; Matei, M. F.; Bassil, B. S.; Kuhnert, N. Synthesis of tri-substituted biaryl based trianglimines: formation of C3-symmetrical and non-symmetrical regioisomers. *Org. Biomol. Chem.* **2011**, *9*, 3258-3271.

[46] Gao, J.; Martell, A. E. Novel chiral N_4S_2- and N_6S_3-donor macrocyclic ligands: synthesis, protonation constants, metal-ion binding and asymmetric catalysis in the Henry reaction. *Org. Biomol. Chem.* **2003**, *1*, 2801-2806.

[47] Gawroński, J.; Brzostowska, M; Kwit, M.; Plutecka, A.; Rychlewska, U. Rhombimines-Cyclic Tetraimines of *trans*-1,2-Diaminocyclohexane Shaped by the Diaryl Ether Structural Motif. *J. Org. Chem.* **2005**, *70*, 10147-10150.

[48] Gawroński, J.; Kwit, M.; Grajewski, J.; Gajewy, J.; Długokińska, A. Structural constraints for the formation of macrocyclic rhombimines. *Tetrahedron: Asymmetry* **2007**, *18*, 2632-2637.

[49] Srimurugan, S.; Viswanathan, B.; Varadarajan, T. K.; Varghese, B. Microwave assisted cyclocondensation of dialdehydes with chiral diamines forming calixsalen type macrocycles. *Tetrahedron Lett.* **2005**, *46*, 3151-3155.

[50] Srimurugan, S.; Viswanathan, B.; Varadarajan, T. K.; Varghese, B. Synthesis and crystal structure of [2+2] calixsalens. *Org. Biomol. Chem.* **2006**, *4*, 3044-3047.

[51] Srimurugan, S; Suresh, P. Microwave assisted synthesis of 72-membered chiral hexanuclear [6+6] macrocyclic Schiff base. *J. Incl. Phenom. Macrocycl. Chem.* **2007**, *59*, 383-388.

[52] López-Pariago, A. M.; García-González, C. A.; Domingo, C. Towards the synthesis of Schiff base macrocycles under supercritical CO_2 conditions. *Chem. Commun.* **2010**, *46*, 4315-4317.

[53] Korupoju, S. R.; Zacharias, P. S. New optically active hexaaza triphenolic macrocycles: synthesis, molecular structure and crystal packing features. *Chem. Commun.* **1998**, 1267-1268.

[54] Kwit, M.; Gawroński, J. Chiral calixsalen-type macrocycles from *trans*-1,2-diaminocyclohexane. *Tetrahedron: Asymmetry* **2003**, *14*, 1303-1308.

[55] Gao, J.; Reibenspies, J. H.; Zingaro, R. A.; Woolley, F. R.; Martell, A. E.; Clearfield, A. Novel Chiral "Calixsalen" Macrocycle and Chiral Robson-type Macrocyclic Complexes. *Inorg. Chem.* **2005**, *44*, 232-241.

[56] Kim, G.-J.; Park, D.-W.; Tak, Y.-S. Synthesis and catalytic activity of new macrocyclic chiral salen complexes. *Catal. Lett.* **2000**, *65*, 127-133.

[57] Sarnicka, A.; Starynowicz, P.; Lisowski, J. Controlling the macrocycle size by the stoichiometry of the applied template ion. *Chem. Commun.* **2012**, *48*, 2237-2239.

[58] Correa, W. H.; Scout, J. L. Synthesis and Characterisation of Macrocyclic Diamino Chiral Crown Ethers. *Molecules* **2004**, *9*, 513-519.

[59] Won, D.-H.; Lee, C.-H. Thiophene-containing Schiff-base macrocycles: intermediate compounds between macroaromatics and azamacrocycles. *Tetrahedron Lett.* **2001**, *42*, 1969-1972.

[60] Kwit, M.; Plutecka, A.; Rychlewska, U.; Gawroński, J.; Khlebnikov, A. F.; Kozhushkov, S. I.; Rauch, K.; de Meijere, A. Chiral Macrocyclic Aliphatic Oligoimines Derived from *trans*-1,2-Diaminocyclohexane. *Chem. Eur. J.* **2007**, *13*, 8688-8695.

[61] Kaik, M.; Gawroński, J. Unprecedented Selectivity in the Formation of Large-Ring Oligoimines from Conformationally Bistable Chiral Diamines. *Org. Lett.* **2006**, *8*, 2921-2924.

[62] Gawroński, J.; Gawrońska, K.; Grajewski, J.; Kwit, M.; Plutecka, A.; Rychlewska, U. Trianglamines—Readily Prepared, Conformationally Flexible Inclusion-Forming Chiral Hexamines. *Chem. Eur. J.* **2006**, *12*, 1807-1817.

[63] Savoia, D.; Gualandi, A.; Stoeckli-Evans, H. Stereoselective synthesis of ring C-hexasubstituted trianglamines. *Org. Biomol. Chem.* **2010**, *8*, 3992-3996.

[64] (a) García-España, E.; Díaz, P.; Llinares, J. M.; Bianchi, A. Anion coordination chemistry in aqueous solution of polyammonium receptors. *Coord. Chem. Rev.* **2006**, *250*, 2952-2986. (b) Wichmann, K.; Antonioli, B.; Sohnel, T.; Wenzel, M.; Gloe, K.; Gloe, K.; Price, J. R.; Lindoy, L. F.; Blake, A. J.; Schroder, M. Polyamine-based anion receptors: Extraction and structural studies. *Coord. Chem. Rev.* **2006**, *250*, 2987-3003. (c) Gale, P. A.; Quesada, R. Anion coordination and anion-templated assembly: Highlights from 2002 to 2004. *Coord. Chem. Rev.* **2006**, *250*, 3219-3244.

[65] Hodačová, J.; Chadim, M.; Závada, J.; Aguilar, J.; García-España, E.; Luis, S. V.; Miravet, J. F. Shape-Complementarity in the Recognition of Tricarboxylic Acids by a [3+3] Polyazacyclophane Receptor. *J. Org. Chem.* **2005**, *70*, 2042-2047.

[66] (a) Tanaka, K.; Fukuda, N.; Fujiwara, T. Trianglamine as a new chiral shift reagent for secondary alcohols. *Tetrahedron: Asymmetry* **2007**, *28*, 2657-2661. (b) Tanaka, K.; Fukuda, N. 'Calixarene-like' chiral amine macrocycles as novel chiral shift reagents for carboxylic acids. *Tetrahedron: Asymmetry* **2009**, *20*, 111-114. (c) Quinn, T. P.; Atwood, P. D.; Tanski, J. M.; Moore, T. F.; Folmer-Andersen, J. F. Aza-Crown Macrocycles as Chiral Solvating Agents for Mandelic Acid Derivatives. *J. Org. Chem.* **2011**, *76*, 10020-10030. (d) Tanaka, K.; Nakai, Y.; Takashi, H. Efficient NMR chiral discrimination of carboxylic acids using rhombamine macrocycles as chiral shift reagent. *Tetrahedron: Asymmetry* **2011**, *22*,

178-184. (e) Gualandi, A.; Grilli, S.; Savoia, D.; Kwit, M.; Gawroński, J. C-hexaphenyl-substituted trianglamine as a chiral solvating agent for carboxylic acids. *Org. Biomol. Chem.* **2011**, *9*, 4234-4241.

[67]   Pu, L. Fluorescence of organic molecules in chiral recognition. *Chem. Rev.* **2004**, *104*, 1687-1716.

[68]   (a) Li, Z.-B.; Lin, J.; Pu, L. A Cyclohexyl-1,2-diamine-Derived Bis(binaphthyl) Macrocycle: Enhanced Sensitivity and Enantioselectivity in the Fluorescent Recognition of Mandelic Acid. *Angew. Chem. Int. Ed.* **2005**, *44*, 1690-1693. (b) Li, Z.-B.; Lin, J.; Sabat, M.; Hyacinth, M.; Pu, L. Enantioselective fluorescent recognition of chiral acids by cyclohexane-1,2-diamine-based bisbinaphthyl molecules. *J. Org. Chem.* **2007**, *72*, 4905-4916.

[69]   Yang, X.; Liu, X.; Shen, K.; Zhu, C.; Cheng, Y. A Chiral Perazamacrocyclic Fluorescent Sensor for Cascade Recognition of Cu(II) and the Unmodified α-Amino Acids in Protic Solutions. *Org. Lett.* **2011**, *13*, 3510-3513.

[70]   (a) Lehn, J. M. From supramolecular chemistry towards constitutional dynamic chemistry and adaptive chemistry. *Chem. Soc. Rev.* **2007**, *36*, 151-160. (b) Corbett, P. T.; Leclaire, J.; Vial, L.; West, K. R.; Wietor, J.-L.; Sanders, J. K. M.; Otto, S. Dynamic Combinatorial Chemistry. *Chem. Rev.* **2006**, *106*, 3652-3711.

[71]   González-Alvarez, A.; Alfonso, I.; López-Ortiz, F.; Aguirre, A.; García-Granda, S.; Gotor, V. Selective Host Amplification from a Dynamic Combinatorial Library of Oligoimines for the Syntheses of Different Optically Active Polyazamacrocycles. *Eur. J. Org. Chem.* **2004**, 1117-1127.

[72]   Gregolinski, J.; Lisowski, J.; Lis, T. New 2+2, 3+3 and 4+4 macrocycles derived from 1,2-diaminocyclohexane and 2,6-diformylpyridine. *Org. Biomol. Chem.* **2005**, *3*, 3161-3166.

[73]   González-Álvarez, A.; Alfonso, I.; Gotor, V. Highly diastereoselective amplification from a dynamic combinatorial library of macrocyclic oligoimines. *Chem. Commun.* **2006**, 2224-2226.

[74]   González-Álvarez, A.; Alfonso, I.; Díaz, P.; García-España, E.; Gotor, V. A highly enantioselective abiotic receptor for malate dianion in aqueous solution. *Chem. Commun.* **2006**, 1227-1229.

[75]   González-Álvarez, A.; Alfonso, I.; Díaz, P.; García-España, E.; Gotor-Fernández, V.; Gotor, V. A Simple Helical Macrocyclic Polyazapyridinophane as a Stereoselective Receptor of Biologically Important Dicarboxylates under Physiological Conditions. *J. Org. Chem.* **2008**, *73*, 374-382.

[76]   González-Álvarez, A.; Alfonso, I.; Gotor, V. An azamacrocyclic receptor as efficient polytopic chiral solvating agent for carboxylic acids. *Tetrahedron Lett.* **2006**, *47*, 6397-6400.

[77]   González-Álvarez, A.; Alfonso, I.; Cano, J.; Díaz, P.; Gotor, V.; Gotor-Fernández, V.; García-España, E.; García-Granda, S.; Jiménez, H. J.; Lloret, F. A Ferromagnetic $[Cu_3(OH)_2]^{4+}$ Cluster Formed inside a Tritopic Nonaazapyridinophane: Crystal Structure and Solution Studies. *Angew. Chem., Int. Ed.* **2009**, *48*, 6055-6058.

[78]   (a) Gregoliński, J.; Lisowski, J. Helicity Inversion in Lanthanide (III) Complexes with Chiral Nonaaza Macrocyclic Ligands. *Angew. Chem. Int. Ed.* **2006**, *45*, 6122-6126. (b) Gregoliński, J.; Starynowicz, P.; Hua, K. T.; Lunkley, J. L.; Muller, G.; Lisowski, J. Helical Lanthanide (III) Complexes with Chiral Nonaaza Macrocycle. *J. Am. Chem. Soc.* **2008**, *130*, 17761-17773.

[79]   Gregoliński, J.; Lis, T.; Cyganik, M.; Lisowski, J. Lanthanide Complexes of the Heterochiral Nonaaza Macrocycle: Switching the Orientation of the Helix Axis. *Inorg. Chem.* **2008**, *47*, 11527-11534.

[80]   Padmaja, M.; Periasamy, M. The synthesis of novel chiral macrocyclic and polymeric amines containing a *trans*-1,2-diaminocyclohexane system. *Tetrahedron: Asymmetry* **2004**, *15*, 2437-2441.

[81]   Richman, J. E.; Atkins, T. J. Nitrogen Analogs of Crown Ethers. *J. Am. Chem. Soc.* **1974**, *96*, 2268-2270.

[82]   (a) Alfonso, I.; Rebolledo, F.; Gotor, V. Chemoenzymatic syntheses of two optically active hexaazamacrocycles. *Tetrahedron: Asymmetry* **1999**, *10*, 367-374. (b) Alfonso, I.; Astorga, C.; Rebolledo, F.; Gotor, V. Optically active tetraazamacrocycles analogous to cyclam. *Tetrahedron: Asymmetry* **1999**, *10*, 2515-2522. (c) González-Álvarez, A.; Rubio, M.; Alfonso, I.; Gotor, V. Chemoenzymatic syntheses of new optically active C_2-symmetrical macrocyclic polyazacyclophanes. *Tetrahedron: Asymmetry* **2005**, *16*, 1361-1365.

[83]   Alfonso, I.; Rebolledo, F.; Gotor, V. Optically Active Dioxatetraazamacrocycles: Chemoenzymatic Syntheses and Applications in Chiral Anion Recognition. *Chem. Eur. J.* **2000**, *6*, 3331-3338.

[84]   Alfonso, I.; Dietrich, B.; Rebolledo, F.; Gotor, V.; Lehn, J.-M. Optically Active Hexaazamacrocycles: Protonation Behavior and Chiral-Anion Recognition. *Helv. Chim. Acta* **2001**, *84*, 280-295.

[85]   Alfonso, I.; Astorga, C.; Gotor, V. Electrospray Ionization Mass Spectrometry (ESI-MS) as a Useful Tool for Fast Evaluation of Anion and Cation Complexation Abilities of a Cyclam Derivative. *J. Incl. Phenom. Macrocycl. Chem.* **2005**, *53*, 131-137.

[86]   Uccello-Barretta, G.; Balzano, F.; Martinelli, J.; Berni, M.-G.; Villani, C.; Gasparrini, F. NMR enantiodiscrimination by cyclic tetraamidic chiral solvating agents. *Tetrahedron: Asymmetry* **2005**, *16*, 3746-3751.

[87]   Gasparrini, F.; Misiti, D.; Pierini, M.; Villani, C. A Chiral A_2B_2 Macrocyclic Minireceptor with Extreme Enantioselectivity. *Org. Lett.* **2002**, *4*, 3993-3996.

[88]   Mammoliti, O.; Allasia, S.; Dixon, S.; Kilburn, J. D. Synthesis and anion-binding properties of new disulfonamide-based receptors. *Tetrahedron* **2009**, *65*, 2184-2195.

[89]   Ragusa, A.; Rossi, S.; Haynes, J. M.; Stein, M.; Kilburn, J. D. Novel Enantioselective Receptors for *N*-Protected Glutamate and Aspartate. *Chem. Eur. J.* **2005**, *11*, 5674-5688.

[90]   Kubik, S. Amino acid containing anion receptors. *Chem. Soc. Rev.* **2009**, *38*, 585-605.

[91]   Bru, M.; Alfonso, I.; Burguete, M. I.; Luis, S. V. Efficient syntheses of new chiral peptidomimetic macrocycles through a configurationally driven preorganization. *Tetrahedron Lett.* **2005**, *46*, 7781-7785.

[92]   Alfonso, I.; Bolte, M.; Bru, M.; Burguete, M. I.; Luis, S. V. Designed Folding of Pseudopeptides: The Transformation of a Configurationally Driven Preorganization into a Stereoselective Multicomponent Macrocyclization Reaction. *Chem. Eur. J.* **2008**, *14*, 8879-8891.

[93]   Bru, M.; Alfonso, I.; Bolte, M.; Burguete, M. I.; Luis, S. V. Structurally disfavoured pseudopeptidic macrocycles through anion templation. *Chem. Commun.* **2011**, *47*, 283-285.

[94]   (a) Yoon, S. S.; Still, W. C. An Exceptional Synthetic Receptor for Peptides. *J. Am. Chem. Soc.* **1993**, *115*, 823-824. (b) Tomeiro, M.; Still, W. C. Sequence-Selective Binding of Peptides in Water by a Synthetic Receptor Molecule *J. Am. Chem. Soc.* **1995**, *117*, 5887-5888. (c) Pan, Z.; Still, W. C. Macrocyclic Oligomers of Isophthalic Acid and *trans*-1,2-Diaminocyclohexane-Building Blocks for Synthetic Peptide Receptors. *Tetrahedron Lett.* **1996**, *37*, 8699-8702.

[95]   Gasparrini, F.; Misiti, D.; Still, W. C.; Villani, C.; Wennemers, H. Enantioselective and Diastereoselective Binding Study of Silica Bound Macrobicyclic Receptors by HPLC. *J. Org. Chem.* **1997**, *62*, 8221-8224.

[96]   (a) Chen, C.-T.; Wagner, H.; Still, W. C. Fluorescent, Sequence-Selective Peptide Detection by Synthetic Small Molecules. *Science* **1998**, *279*, 851-853. (b) Chang, K.-H.; Liao, J.-H.; Chen, C.-T.; Mehta, B. K.; Chou, P.-T.; Fang, J.-M. Stereoselective Recognition of Tripeptides Guided by Encoded Library Screening: Construction of Chiral Macrocyclic Tetraamide Ruthenium Receptor for Peptide Sensing. *J. Org.Chem.* **2005**, *70*, 2026-2032.

[97]   Sessler, J. L.; Andrievsky, A.; Král, V.; Link, V. Chiral Recognition of Dicarboxylate Anions by Sapphyrin-Based Receptors. *J. Am. Chem. Soc.* **1997**, *119*, 9385-9392.

[98]   (a) Amabilino, D. B.; Stoddart, J. F. Interlocked and intertwined structures and superstructures. *Chem. Rev.* **1995**, *95*, 2725-2828. (b) Raymo, F. M.; Stoddart, J. F. Interlocked macromolecules. *Chem. Rev.* **1999**, *99*, 1643-1663.

[99]   Kuhnert, N.; Tang, B. Synthesis of diastereomeric trianglamine-β-cyclodextrin-[2]-catenanes. *Tetrahedron Lett.* **2006**, *47*, 2985-2988.

[100] (a) MacGillivray, L. R.; Atwood, J. L. Structural classification and general principles for the design of spherical molecular hosts. *Angew. Chem. Int. Ed.* **1999**, *38*, 1019-1034. (b) Leininger, S.; Olenyuk, B.; Stang, P. J. Self-assembly of Discrete Cyclic Nanostructures Mediated by Transition Metals. *Chem. Rev.* **2000**, *100*, 853-907. (c) Cram, D. J.; Cram, J. M. *Container Molecules and their Guests*; Royal Society of Chemistry: Cambridge, U.K., 1994.

[101] Skowronek, P.; Gawroński, J. Chiral Iminospherand of a Tetrahedral Symmetry Spontaneously Assembled in a [6+4] Cyclocondensation. *Org. Lett.* **2008**, *10*, 4755-4758.

[102] (a) Tozawa, T.; Jones, J. T. A.; Swamy, S. I.; Jiang, S.; Adams, D. J.; Shakespeare, S.; Clowes, R.; Bradshaw, D.; Hasell, T.; Chong, S. Y.; Tang, C.; Thompson, S.; Parker, J.; Trewin, A.; Bacsa, J.; Slawin, A. M. Z.; Steiner, A.; Cooper, A. I. Porous Organic Cages. *Nat. Mater.* **2009**, *8*, 973-978. (b) Holst, J. R.; Trewin, A.; Cooper, A. I. Porous organic molecules. *Nat. Chem.* **2010**, *2*, 915-920. (c) Jones, J. T. A.; Hasell, T.; Wu, T.; Bacsa, J.; Jelfs, K. E.; Scmidtmann, M.; Chong, S. Y.; Adams, D. J.; Trewin, A.; Schiffman, F.; Cora, F.; Slater, B.; Steiner, A.; Day, G. M.; Cooper, A. I. Modular and predictable assembly of porous organic molecular crystals. *Nature* **2011**, *474*, 367-371. (d) Hasell, T.; Schmidtmann, M.; Cooper, A. I. Molecular Doping of Porous Organic Cages. *J. Am. Chem. Soc.* **2011**, *133*, 14920-14923. (e) Hasell, T.; Chong, S. Y.; Jelfs, K. E.; Adams, D. J.; Cooper, A. I. Porous Organic Cage Nanocrystals by Solution Mixing. *J. Am. Chem. Soc.* **2012**, *134*, 588-598. (f) Hasell, T.; Schmidtmann, M.; Stone, C. A.; Smith, M. W.; Cooper, A. I. Reversible water uptake by a stable imine-based porous organic cage. *Chem. Commun.* **2012**, *48*, 4689-4691. (g) Hasell, T.; Chong, S. Y.; Schmidtmann, M.; Adams, D. J.; Cooper, A. I. Porous Organic Alloys. *Angew. Chem. Int. Ed.* **2012**, *51*, 7154-7157.

[103] Xu, D.; Warmuth, R. Edge-Directed Dynamic Covalent Synthesis of a Chiral Nanocube. *J. Am. Chem. Soc.* **2008**, *130*, 7520-7521.

[104] Steinmetz, V.; Couty, F.; David, O. R. P. One-step synthesis of chiral cages. *Chem. Commun.* **2009**, 343-345.

[105] Moure, A.; Luis, S. V.; Alfonso, I. Efficient Synthesis of Pseudopeptidic Molecular Cages. *Chem. Eur. J.* **2012**, *18*, 5496-5500.

[106] (a) Ardá, A.; Venturi, C.; Nativi, C.; Francesconi, O.; Gabrielli, G.; Cañada, F. J.; Jiménez-Barbero, J.; Roelens, S. A Chiral Pyrrolic Tripodal Receptor Enantioselectively Recognizes β-Mannose and β-Mannosides. *Chem. Eur. J.* **2010**, *16*, 414-418. (b) Nativi, C.; Francesconi, O.; Gabrielli, G.; Vacca, A.; Roelens, S. Chiral Diaminopyrrolic Receptors for Selective Recognition of Mannosides, Part 1: Design, Synthesis, and Affinities of Second-Generation Tripodal Receptors. *Chem. Eur. J.* **2011**, *17*, 4814-4820.

[107] Ardá, A.; Cañada, F. J.; Nativi, C.; Francesconi, O.; Gabrielli, G.; Jiménez-Barbero, J.; Roelens, S. Chiral Diaminopyrrolic Receptors for Selective Recognition of Mannosides, Part 2: A 3D View of the Recognition Modes by X-ray, NMR Spectroscopy, and Molecular Modeling, *Chem. Eur. J.* **2011**, *17*, 4821-4829.

[108] Jelfs, K. E.; Wu, X.; Schmidtmann, M.; Jones, J. T. A.; Warren, J. E.; Adams, D. J.; Cooper, A. I. Large Self-Assembled Chiral Organic Cages: Synthesis, Structure, and Shape Persistance. *Angew. Chem. Int. Ed.* **2011**, *50*, 10653-10656.

Send Orders of Reprints at bspsaif@emirates.net.ae

CHAPTER 3

Regio- and Stereoselective Ring Opening of Allylic Epoxides

Mauro Pineschi*, Ferruccio Bertolini, Valeria Di Bussolo, and Paolo Crotti*

Dipartimento di Farmacia, Università di Pisa, Via Bonanno 33, 56126 Pisa, Italy

Abstract: Three-membered heterocyclic rings offer a powerful combination of reactivity, stability, availability, and atom economy. In fact, the asymmetric ring opening of allylic epoxides with nucleophiles of different kinds, as will be discussed in this Review, offers the possibility of generating valuable chiral non-racemic building blocks in a very simple manner and in a stereodefined fashion.

Keywords: Allylic epoxides, nucleophilic ring opening, heteronucleophiles, carbon nucleophiles, regioselectivity, stereoselectivity, enantioselectivity.

1. INTRODUCTION

Allylic epoxides, otherwise called vinyloxiranes or vinyl epoxides, are important building blocks in synthetic organic chemistry. They can be readily accessed either by oxidation of dienes or by nucleophilic addition of activated allyl groups to aldehydes or ketones [1]. When the epoxide ring, which is a particular kind of leaving group, is flanked by a double bond, it can be considered as a peculiar class of allylic substrates in which the allylic substitution is accompanied by a ring opening process in which the leaving group is maintained in the final product. Hence, allylic epoxides combine the reactivity of epoxides with that of allylic substrates allowing a wide range of useful synthetic transformations, including reactions with nucleophiles and Lewis acid rearrangement processes [1-3].

In principle, nucleophilic attack can take place on three of the four consecutive functionalized carbon atoms of the allylic epoxide (pathways *a,b,c*, Scheme **1**). Nucleophilic attack on the oxirane carbon adjacent to the carbon-carbon double bond [carbon C(2), allylic position] affords the S_N2 addition product (*1,2-addition*

---

*Address correspondence to Mauro Pineschi and Paolo Crotti: Dipartimento di Farmacia, Università di Pisa, *via* Bonanno 33, 56126, Pisa, Italy; Tel: +39 2219 668, +39 2219 690; Fax: +39 050 2219 660; E-mails: pineschi@farm.unipi.it and crotti@farm.unipi.it.

Atta-ur-Rahman (Ed)

product), usually obtained in ring-opening reactions (path *b*). Nucleophilic addition to the least hindered oxirane carbon [carbon C(1)] affords the regioisomeric *1,2-addition product* (path *c*), though this pathway is not commonly observed. Finally, the conjugated addition of the nucleophile to vinyl carbon C(4) gives the corresponding S_N2' addition product (*1,4-addition product*, path *a*).

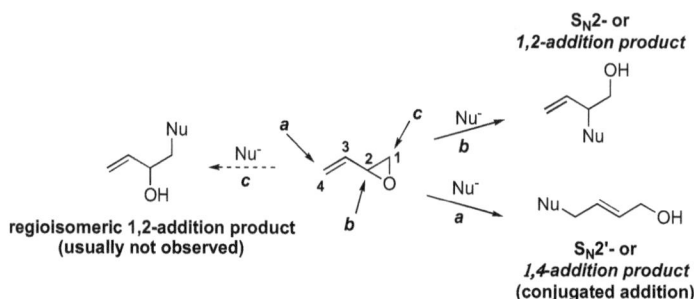

Scheme 1: Possibile reaction pathways for the nucleophilic ring-opening of an allylic epoxide.

Much work has been centred on developing conditions to obtain each of these three products selectively. The factors which favour nucleophilic attack by route *b* (direct addition or S_N2 process) and/or route *a* (conjugated addition or S_N2'process) are very sensitive to the type of reagent or catalyst used. The following classification of reactions is based on the type of reagent used as the nucleophile.

This Review resumes the nucleophilic ring opening of these systems up to July 2012, with particular emphasis on regio- and stereoselective transformations.

2. CARBON NUCLEOPHILES

2.1. Alkylative Ring Opening

In general, the regioselectivity of the ring opening of allylic epoxides with organometallic reagents depends largely on the hard or soft properties of the nucleophiles. It is generally admitted that hard alkyl metals attack under charge control seeking the most positive center of the allylic epoxide situated in the allylic position (S_N2 addition) [4]. On the other hand, softer nucleophiles should react through HOMO-LUMO interactions, resulting in an orbital-controlled reaction (S_N2' addition) [4]. For example, Grignard reagents [5-9], alkyllithium

[7, 10-12], organoaluminum [13-15], and organozinc species favour the S_N2 addition products [16], although a mixture of regioisomers is generally obtained. The 1,2-regioselectivity can be increased by the use of the RLi in the presence of stoichiometric amounts of $BF_3 \cdot Et_2O$ [17, 18]. More recently, Alexakis *et al.* showed that trialkylorganozincates and tetraalkyaluminates provide a convenient way to introduce alkyl substituents regioselectively at the allylic position of 1,3-diene monoepoxides (Scheme **2**) [19]. It was observed that a polar solvent such as THF and the presence of a substituent at the double bond, necessarily present in endocyclic allylic epoxides, favoured the S_N2 addition.

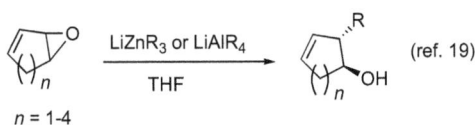

Scheme 2: Selective S_N2 addition to allylic epoxides.

Based on their previous experience with vinylsulfone chemistry, Fuchs *et al.* [20] developed stereoselective protocols for the *syn* and *anti* S_N2-type methylation of a series of enantiomerically enriched epoxyvinyl sulfones giving access to all eight possible diastereoisomer stereotetrads, seven of which are commonly found in polypropionate natural products. Also allylstannanes exclusively gave the 1,2-adduct in the presence of $BF_3 \cdot Et_2O$ with various allylic epoxides [21]. More recently, an allylboration of cyclic allylic epoxides with allyldiethylborane and (2-cyclohexenyl)dicyclohexylborane [22], and the use of diethylzinc in the presence of CF_3COOH to give the *cis-1,2-addition product* have been described [23].

On the other hand, organocopper reagents are especially well suited for the S_N2'-addition to allylic epoxides [24]. In particular, Marino *et al.* have shown that the use of soft alkylcyanocopper derivatives (RCuCNLi) with allylic epoxides affords the S_N2'-adducts exclusively with *anti*-stereoselectivity (Scheme **3**, eq. a) [25-29].

As dialkylzincs are hard alkyl nucleophiles, but usually too weak to react with these conjugate systems, Lipshutz *et al.* generated a cuprate *in situ* by a transmetallation reaction with a catalytic amount of a cyanocuprate, leading to more reactive and selective species. With this approach it is possible to transfer functionalized organozinc reagents with a competitive transfer of the methyl ligand drastically reduced (Scheme **3**, eq. b) [30].

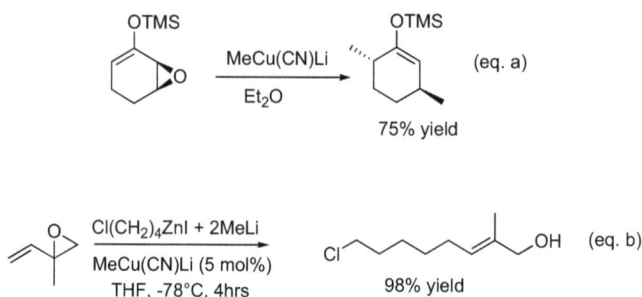

Scheme 3: Selective S_N2' addition to allylic epoxides.

In an S_N2'-type nucleophilic displacement both the *anti* and the *syn*-stereoselective reaction pathway are in principle possible [31], but normally in a copper-catalyzed reaction the *anti*-S_N2' facial selectivity is largely predominant if not exclusive. This preference, which has been qualitatively rationalized in terms of orbital symmetry [32, 33], can be overridden when a particular chelating hindered group is present close to the reaction site. For example, Marino and Fernández de la Pradilla reported sulfoxide-controlled S_N2' displacements between cyanocuprates and chiral non-racemic epoxy vinyl sulfoxides **1** and **2** [34]. As shown in Scheme **4**, the absolute configuration of the newly formed carbon-carbon bond is primarily controlled by the chiral sulphur atom, which in the non-reinforcing situation depicted in eq. a (Scheme **4**) can override the intrinsic *anti* tendency of the vinyl epoxide moiety and undergo *syn*-addition.

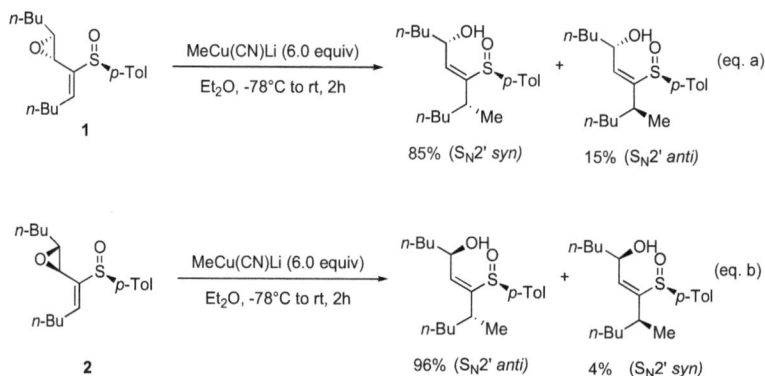

Scheme 4: Sulfoxide-controlled S_N2' displacements between cyanocuprates and epoxy vinyl sulfoxides.

By constrast, under identical reaction conditions, diastereoisomeric sulfinyl epoxide **2** gave almost exclusively the corresponding S_N2' adduct (eq. b).

By the use of different combinations of CuCN-derived cuprate reagents, solvents, and reaction conditions Dieter *et al.* developed very interesting sequential bis-allylic substitution reactions of electron-deficient (compound **3**) and electron-rich vinyloxiranes (compound **4**) to give vicinal alkane stereogenic centers with high *syn*-diastereoselectivities (Scheme **5**) [35]. For example, ethyl (*E*)-4,5-epoxy-2-hexenoate **3** afforded excellent S_N2'/S_N2 regioselectivity and *anti*-diastereoselectivity with dilakylzinc reagents in the presence of CuCN and in the subsequent allylic substitution of the resulting allylic acetate to give the corresponding *syn*-adduct **5**. When dealing with vinyloxirane **4**, best results in terms of regioselectivity and diastereoselectivity were obtained by the use of an allylic phosphate in combination with lithium alkylcyanocuprates in the second step (compound **6**, Scheme **5**).

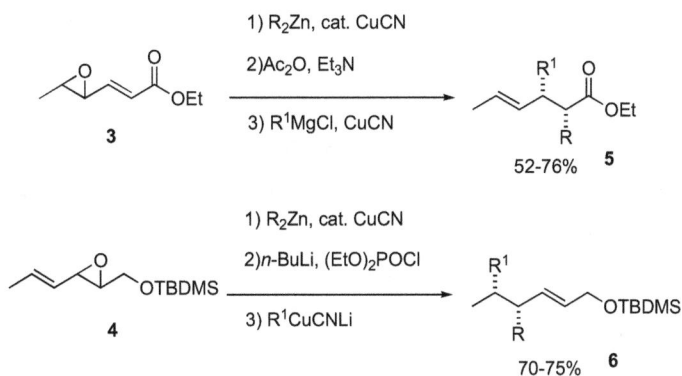

Scheme 5: Sequential bis-allylic substitution with CuCN-derived cuprates.

Further exceptions to the general trend outlined above can be obtained by varying the electronic properties of the allylic epoxides and/or the steric properties of the organometallic reagent. For example, high levels of S_N2' regioselectivity have been achieved by the use of *gem*-difluorinated allylic epoxides (such as **7**) in combination with RLi in anhydrous THF (Scheme **6**) [36]. The addition reaction affording compound **8** occurred at the highly positively charged terminal fluorine-possessing sp^2 carbon atom. This is in sharp contrast with the case of the corresponding non-fluorinated vinyloxiranes, which gave a mixture of regioisomers [5]. *Ab initio* calculation demonstrated that the charge on the fluorine-possessing carbon atom of allylic epoxide **7a** was unusually altered from –0.425 of the non-fluorinated prototype **7b** to +0.805 (Scheme **6**).

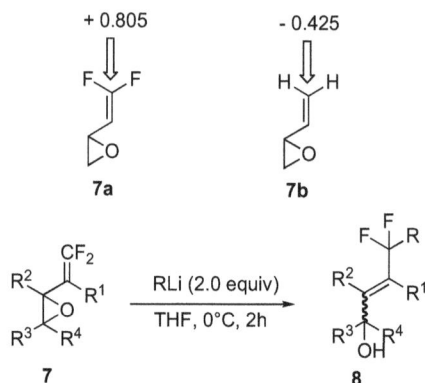

Scheme 6: *Ab initio* calculation of *gem*-difluorinated allylic epoxide **7a** and general reactivity of compounds of type **7** with alkyllithium.

As shown by Amos B. Smith III, a high chemoselectivity can be achieved in the addition of 9-derived lithium dthiane anion to vinyl epoxides by exploiting the steric nature of the dithiane. Unencumbered dithiane anions (R=H, Ph, TMS) afforded S_N2 adducts of type **10**, whereas sterically encumbered anions (R= Et, *i*-Pr, TIPS) lead to allylic alcohols of type **11** by S_N2' additions to the more accessible alkene terminus (Scheme **7**) [37]. The use of diastereoisomeric (*E*)- and (*Z*)-allylic epoxides revealed that the S_N2' dithiane addition takes place in a *syn* fashion with respect to the epoxide. Recently, this procedure based on the S_N2/ S_N2' reaction manifold of appropriate vinyl epoxides and dithiane anions was applied to the synthesis of the C(1-27) macrocyclic skeleton of the aglycone of (+)-rimocidin [38].

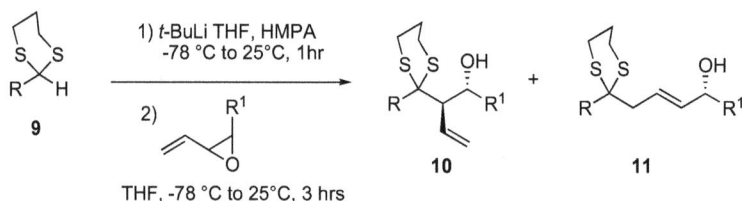

Scheme 7: Regioselectivity of dithiane addition to vinyl epoxides: steric control over the S_N2 and S_N2' manifolds.

On the other hand, particular structural reasons present in the allylic epoxide can also give a particular regioselectivity when dealing with organometallic reagents. A noteworthy result in this sense came from the examination by Crotti and co-workers of the behaviour of the D-galactal-derived vinyl oxirane **12β**, obtained *in situ* by basic cyclization of the corresponding *trans* hydroxy mesylate (Scheme **8**).

While Grignard reagents such as MeMgBr and PhMgBr gave a Grob fragmentation process, cuprates such as Me$_2$CuLi and EtMgBr in the presence of stoichiometric amount of CuCN afforded the corresponding *anti-1,2-addition product*, alcohol **13**, as the only reaction product. On the other hand, and contrary to the normal trend described above, lithium alkyls such as MeLi, BuLi, *i*-PrLi, *t*-BuLi, and PhLi gave the corresponding β-*C*-glycoside **14** with a complete 1,4-regioselectivity and complete β-stereoselectivity. Coordination of the reagent (RLi) with the oxirane oxygen through the metal, as shown in **12'**, was considered to be responsible for the regio- and stereoselectivity observed (Scheme **8**) [39, 40]. A complementary 1,4-regio- and α-stereoselectivity was obtained with the diastereoisomeric D-allal-derived allyl epoxide **12α** [41]. Application of the same protocol to the 6-deoxy derivatives **12α-Me** and **12β-Me** afforded a corresponding completely stereoselective substrate-dependent *C*-glycosylation process as with **12α** and **12β** [42].

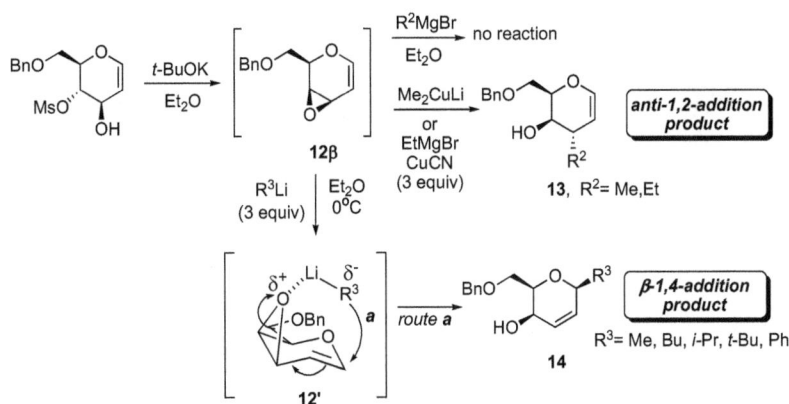

Scheme 8: Different reactivity of organometallic reagents with D-galactal-derived vinyl oxirane **12β**.

An S$_N$2'-type addition reaction between lithiated carbon nucleophiles and silylated vinyl epoxides has recently been reported by Malacria *et al.* [43]. In particular, the thorough study of these reactions demonstrated that it is possible to control the diastereoselectivity of this nucleophilic substitution by variation of the temperature, addition of salts and polarity of the solvent [44].

Cyclic vinyl diepoxides **15a,b** undergo clean S$_N$2' reactions delivering oxabridged systems of various sizes after a domino intramolecular *O*-alkylation reaction. For example, the CuCN-catalyzed reaction of Grignard reagents with vinyl diepoxide **15b** proceeded with complete S$_N$2' regioselectivity to give the allylic epoxy alcohols **16a,b** as the primary reaction products. Subsequent intramolecular *O*-alkylation occurred with complete regioselectivity at the allylic position of the remaining allylic epoxide, affording the bicyclo[4.2.1]nonane skeleton, present in **17a,b** (Scheme **9**) [45].

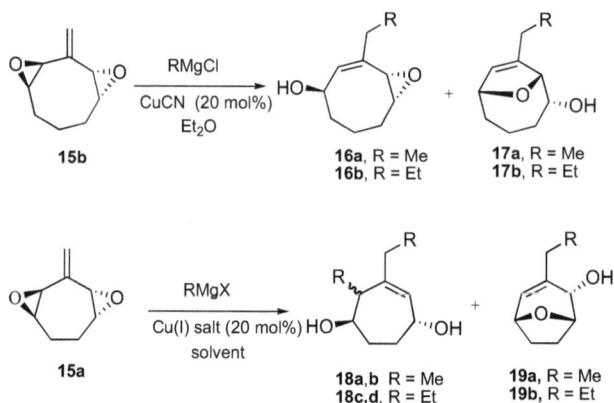

Scheme 9: Copper-catalyzed reactions of vinyl diepoxides **15a,b** with Grignard reagents.

The application of the same copper-catalyzed addition protocol of Grignard reagents to the seven-membered vinyl diepoxide **15a** proved to be strongly influenced by the type of RMgX and reaction conditions utilized. However, when the reaction was performed with a threefold excess of the Grignard reagent in THF, it invariably afforded diols **18a-d** as the major products. Interestingly, these compounds are derived from a double alkylation process in which the alkyl moiety is transferred twice, first in a S$_N$2'- then in a S$_N$2-fashion. With substrate **15a**, an intramolecular *O*-alkylation also occurred, affording regioisomeric bicyclo[3.2.1] compounds of type **19**, in which attack occurred at the less activated allylic position of the oxirane moiety (Scheme **9**) [45].

2.1.1. Enantioselective Alkylations

As regards the asymmetric ring opening (ARO) reaction of epoxides, a large variety of nucleophiles have been successfully employed, with the majority of these being hetero-atom-based, whereas the use of carbon-based nucleophiles remains quite limited [46]. In general, there are two main groups of allylic epoxides to perform ARO with: racemic allylic epoxides and *meso*-allylic epoxides. The former case is valuable for the enantioselective formation of the ring opening product and/or for the isolation of the non-reacting epoxide enantiomer. Unlike saturated epoxides, the use of a *meso* substrate is not common with allylic epoxides because these are not easy to prepare in a symmetrical form.

The remarkable results previously obtained in the addition of organozinc reagents to enones by Feringa *et al.* [47], prompted Pineschi group to verify the effectiveness of the chiral copper complexes of Binol-based phosphoramidites in the *anti*-stereoselective conjugated addition of organozinc reagents to allylic epoxides. Indeed, a remarkable ligand-accelerated catalysis effect thanks to the presence of a catalytic amount of copper-complexes with Binol-based phosphoramidites such as **20** was found (Scheme **10**). When reactions were performed in accordance with a classic kinetic resolution (*KR*) protocol (eq. a, Scheme **10**), it was possible to obtain the corresponding allylic alcohol (S_N2' pathway) with a high regio- and enantioselectivity and to recover the unreacted allylic epoxide with a high optical purity [48]. Very interestingly, a complete conversion of a racemic allylic epoxide into constitutionally different enantiomerically enriched ring-opened products, regioisomeric allylic and homoallylic alcohols of type **21** and **22**, was simply obtained by means of an excess of the dialkylzinc reagent and a longer reaction time (eq. b, Scheme **10**) [49].

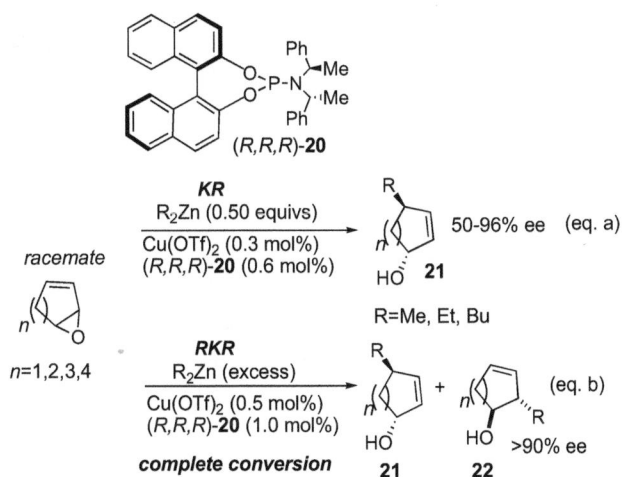

Scheme 10: Copper-phosphoramidite-catalyzed kinetic and regiodivergent kinetic resolution with dialkylzinc reagents.

This is one of the very rare examples of a regiodivergent kinetic resolution (*RKR*) by the use of a hard organometallic reagent and a chiral catalyst in a carbon-carbon bond-forming reaction [50]. More recently, the *RKR* was successfully applied to a variety of cyclic semirigid allylic epoxides and, in some specific cases, to conformationally mobile allylic epoxides [51]. Recently, in vinyl epoxides **carba-12α** and **carba-12β**, the carba analogs of glycal-derived epoxides **12α** and **12β** (see Scheme **8**), high levels of anti-1,4-addition were observed in their reactions with MeMgBr/CuCN and Me$_2$Zn in the presence of catalytic amounts of Cu(OTf)$_2$/phosphoramidite ligands (Scheme **11**) [52]. A complementary reaction of the enantiomers of the starting epoxide was found for epoxide **carba-12β**, even if it was inferior to that previously observed, under the same reaction conditions, for the enantiomers of 1,3-cyclohexadiene monoepoxide [49, 50]. Moreover, a significant detrimental effect exherted by the adjacent benzyloxymethyl moiety, as tentatively shown in **A** (Scheme **11**), on the S$_N$2' reactivity and on the chiral recognition process of epoxide **carba-12α** was found [52].

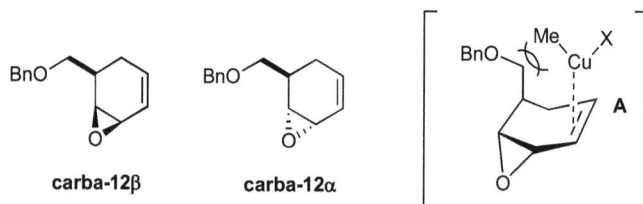

carba-12β   carba-12α

Scheme 11: Influence of a substituent at the C-6 position on copper-catalyzed *RKR* of carba analogs of 6-O-(benzyl)-D-allal- and -D-galactal-derived allyl epoxides.

Quite recently, Alexakis and Equey have shown that cyclic allylic epoxides can be alkylated also by organoaluminum reagents in combination with copper salts and chiral phosphoramidites [53]. They found that the solvent played a central role in the conjugated addition with organoaluminum reagents. In fact, the reaction must be carried out in THF, whereas the use of less coordinating solvents such as toluene and CH$_2$Cl$_2$ gave only the formation of oligomeric products. More recently, the same author developed a copper-catalyzed kinetic resolution of 1,3-cyclohexadiene monoepoxide with Grignard reagents. Among the several chiral ligands screened, chiral ferrocenyl diphosphines gave the best results in terms of regio- and enantioselectivity [54].

Alexakis *et al.* also performed the kinetic resolution of cyclic allylic epoxides with Grignard reagents by means of copper catalysts with chiral phosphorus ligand of the SimplePhos family [55]. In this reaction conditions it was possible to obtain good levels of enantioselectivities of the S_N2' adducts also with hindered and functionalized Grignard reagents such as $TMSCH_2MgCl$. The allylic alcohols were obtained in up to 96% *ee* albeit with a moderate conversion of 37% using *c*-hexMgCl.

To avoid the inherent limitations of resolution processes, Pineschi *et al.* studied the reactions of several symmetrical bis-allylic epoxides with dialkylzinc reagents in combinations of chiral copper complexes with phosphoramidite **20** (Scheme **12**).

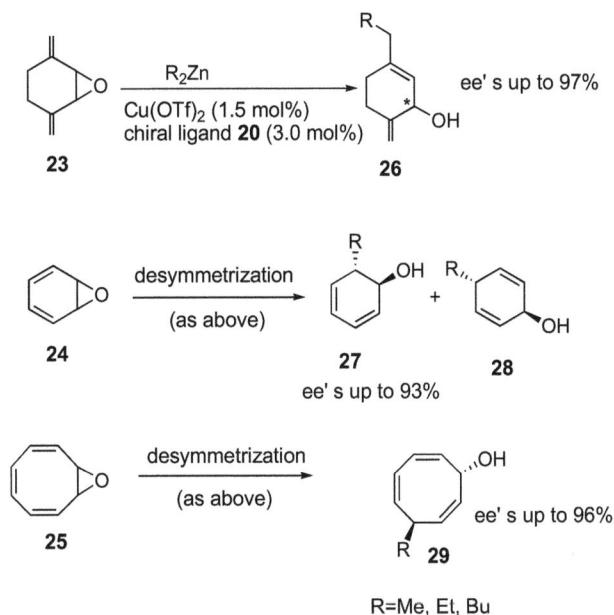

Scheme 12: Catalytic and enantioselective desymmetrization of symmetrical allylic epoxides **23**-**25**.

For example, *meso*-methylidene cycloalkane epoxide **23**, afforded the corresponding bis-allylic alcohols of type **26** in good yields and with high regio- and enantioselectivities [56]. The effective chiral recognition of the two enantiotopic faces was effected by the same chiral catalyst, even with very reactive arene oxides. In this framework, the unprecedented catalytic enantioselective trapping of benzene oxide (**24**) with dialkylzincs gave a crude reaction mixture of novel regioisomeric dienols **27** (*anti*-α-adduct) and **28** (achiral γ-adduct) (Scheme **12**) [57]. The monoepoxide **25** derived from 1,3,5,7-cyclooctatetraene (COT) is also a symmetrical molecule, and it has several distinctive features not observed in symmetrical vinyl oxirane substrates examined. In fact, COT-monoepoxide has a special structure imposed by three consecutive double bonds where the double bonds and the epoxide ring are not in the same plane (ca. 60° deviation). Moreover, ring contraction isomerization to the seven-membered trienyl carboxaldehyde, which in turn added the organometallic reagents delivering substituted cyclohepta-trienyl alcohols, was the most common reaction observed when organometallic reagents were employed with this substrate [58]. This catalytic system allowed a highly enantioselective desymmetrization of COT-monoepoxide with dialkylzinc reagents to give the corresponding trienyl alcohol **29** with a high yield, complete conjugated regioselectivity and high enantioselectivity [59, 60].

2.2. Arylation and Alkenylation

In contrast with the high regioselectivity displayed by alkyl cyanocuprates, sp^2-hybridized cyanocuprates are more prone to give mixtures of *trans*-1,2- and *trans*-1,4-adducts when allowed to react with allylic epoxides [29, 34]. For example, the reaction of phenyl cyanocuprate with cyclopentadiene monoepoxide was devoid of the high regioselectivity observed with cyclohexadiene monoepoxide [29]. However, descriptions of total synthesis based on the nucleophilic ring opening of allylic epoxides with aryl copper reagents have been reported [61, 62].

A highly regio- and stereoselective S_N2' reaction of an aryl cyanocuprate **31** with silyl enol ether epoxide **30** has recently been described as the key step for the total synthesis of (+)-machaeriol D (Scheme **13**) [63].

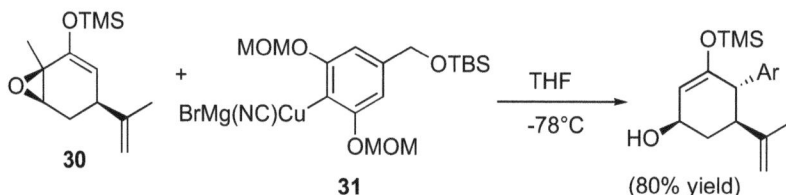

Scheme 13: Regio- and stereoselective S_N2' reaction of an aryl cyanocuprate.

In the case of alkenyl-based cuprates, the degree of regioselectivity depends largely on the substitution pattern of the vinylic moiety attached to the copper. Thus, whereas the reaction of vinylcyanocuprate with cyclopentadiene monoepoxide showed a complete lack of regioselectivity [29], the cuprate derived from (*E*)-1-iodo-3-[*t*-butyldimethylsilyl)oxy]-1-octene provided a 1:4 mixture of 1,4- **32** and 1,2-adduct **33**, in a 80% overall yield. Subsequently, allylic alcohol **32** was used for a stereocontrolled synthesis of prostaglandins (Scheme **14**) [64].

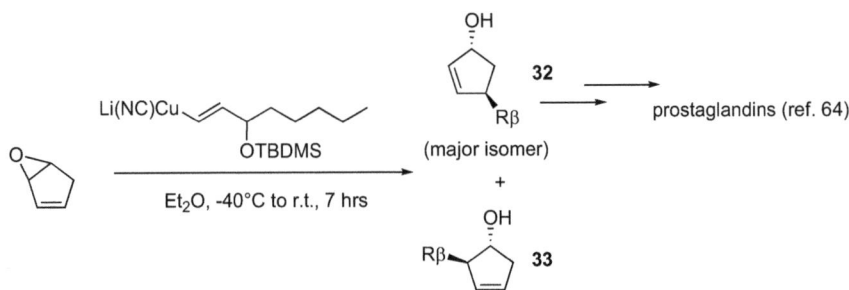

Scheme 14: Regio- and stereoselective S_N2' reaction of an alkenyl cyanocuprate in the total synthesis of prostaglandins.

The reaction of functionalized (*Z*)-vinylcuprates obtained from the corresponding (*Z*)-vinyl tellurides has also been reported to give a mixture of regioisomers when treated with allylic epoxides [65].

Allylic epoxides undergo mild, rapid oxidative addition to a variety of palladium(0) complexes, generating a cationic η^3-palladium species. Several carbon nucleophiles have been used in this reaction. For example, the reaction of aryl- and vinylmercurials (RHgCl) with allylic epoxides promoted or catalyzed by Li_2PdCl_4 provided the corresponding allylic alcohol (S_N2' addition) in high yields only for aliphatic allylic epoxides [66].

Organostannanes are especially well suited for delivering aryl or alkenyl moieties; they readily couple with allylic epoxides at ambient temperature, and add regioselectively, giving predominantly the *1,4-addition product*. For example, the reaction of Me_3SnPh catalyzed by $(CH_3CN)_2PdCl_2$ (4.0 mol%) with isoprene monoepoxide in the presence of 10 equiv of H_2O gave the coupled product **34** with an excellent regioselectivity, albeit as an *E/Z* mixture (eq. a, Scheme **15**). On the other hand, in coupling reactions with cyclic allylic epoxides, the reaction is poorly regioselective but the reaction is completely *anti*-stereoselective, with the organic group from the tin partner coupling *trans* to the alcohol function (eq. b, Scheme **15**) [67].

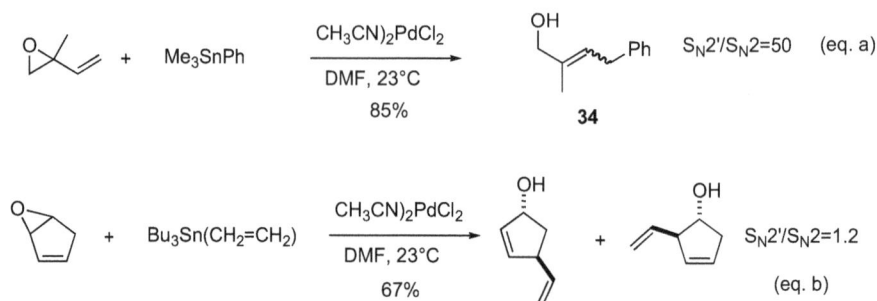

Scheme 15: Palladium-catalyzed coupling of vinyl epoxides with organostannanes.

More recently, the selective activation of a substrate such as **35**, bearing two potential nucleophilic sites, has been achieved by using different palladium catalysts. Hence, the reaction of **35** with isoprene monoepoxide, catalyzed by palladium complexes without strongly coordinated ligands, favoured Pd/Sn transmetallation, giving product **36** with a moderate to good yield (Scheme **16**) [68].

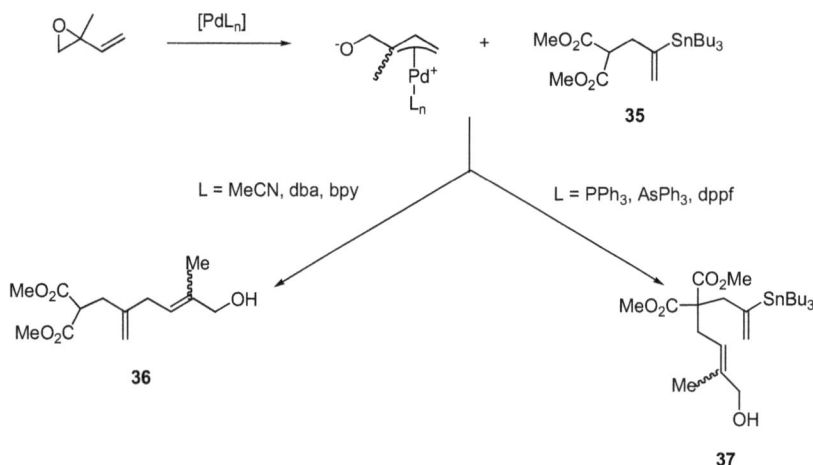

Scheme 16: Reaction of isoprene monoepoxide with palladium-switchable bisnucleophiles.

On the other hand, when the electrophilicity of the (η^3-allyl)palladium complex was reduced using a phosphine or arsine ligand, nucleophilic attack of the malonate type anion on the complex occurred, giving products of type **37**. Allylic alcohols **36** and **37** can be further manipulated by intramolecular allylic alkylation to give five-membered carbocycles.

Organoboron compounds occupy a privileged position among organometallic reagents, owing to their ease of synthesis, stability, commercial availability and synthetic versatility. Surprisingly, the metal-catalyzed reaction of aryl- and alkenyl boronic acids with allylic epoxides is scarce. In a seminal paper, Suzuki and Miyaura reported a cross-coupling of 1-alkenylboranes with butadiene monoepoxide catalyzed by nickel or palladium complexes, occurring with a low regioselectivity [69].

A major advance in this field has quite recently been reported by Szabó and co-workers with the use of palladium pincer complexes **38** [70]. The reaction occurred in mild conditions using 0.5-2.5 mol% pincer complex-catalyst affording allylic alcohols with a high S_N2' regioselectivity and good to excellent yields (Scheme **17**). It should be noted that there is a substantial difference between the mechanism of the pincer complex-catalyzed reaction and the corresponding palladium(0)-catalyzed processes.

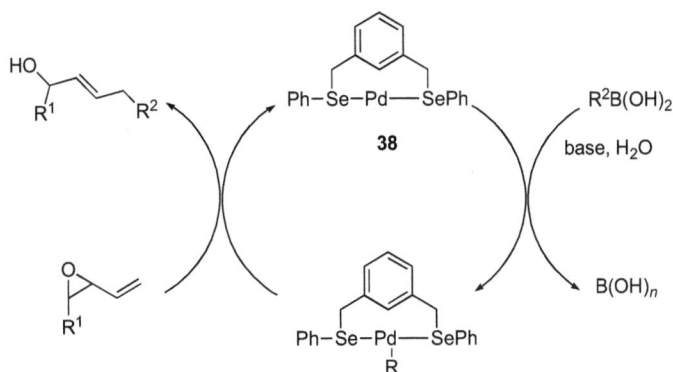

Scheme 17: Palladium pincer complex-catalyzed cross-coupling of vinyl epoxides with organoboronic acids

In this case, the oxidation state of the palladium atom is +2 under the catalytic process, and according to the authors, a (η^3-allyl)palladium complex was not formed. The initial step of the reaction is presumed to be the transmetallation of the organoboronic acid to the pincer complex, followed by an S_N2' type transfer process to give the corresponding allylic alcohol (Scheme **17**). However, a high S_N2' regioselectivity was obtained only with allylic epoxides unsubstituted on the double bond, while the use of 1,3-cyclohexadiene monoepoxide afforded a 2:1 mixture of regioisomeric alcohols.

Very recently, a key step for the synthesis of montanine-like alkaloids was realized by arylation of the pyrrolidine ring using a regioselective and *anti*-stereoselective palladium-catalyzed ring opening of vinyl epoxide **39** to give compound **40** with 85% yield (Scheme **18**) [71]. Unlike Szabò's reaction condition, the reaction was successful in anhydrous THF and it is presumed to occurr *via* a (η^3-allyl)palladium(II) complex by opening of the epoxide ring.

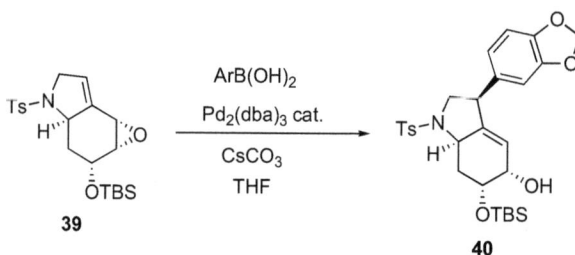

Scheme 18: Palladium-catalyzed stereoselective arylation of a cyclic vinyl epoxide *via* organoboronic acids.

A very interesting activation of organosiloxane reagents with catalytic amounts of fluoride containing N-heterocyclic carbene-copper complexes (such as **41**, Scheme **19**) was recently reported by Ball *et al.* [72].

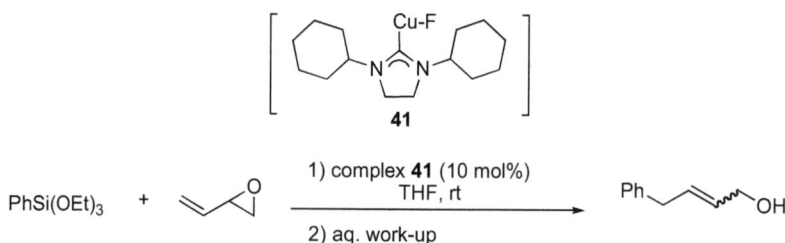

Scheme 19: N-Heterocyclic carbene-copper complexes catalyzed coupling of organosiloxanes with allylic epoxides.

The arylation reaction appears general for aryl and heteroaryl silanes with good functional group tolerance. The corresponding allylic alcohols were isolated as a mixture of *E*/*Z* isomers in acyclic cases, whereas with 1,3-cyclohexadiene monoepoxide a mixture of regioisomers was obtained.

A different way of generating arylcopper species was recently reported by Cheng *et al.* using a cooperative copper- and palladium-catalyzed three-component coupling of *in situ* generated benzene with allylic epoxides and terminal alkynes [73]. The three-component reaction also proceeded in the presence of the palladium complex alone (dppp= diphenylphosphino propane) without copper(I) salt, but the dual-metal catalyst system greatly enhances the regioselectivity of the reaction (Scheme **20**).

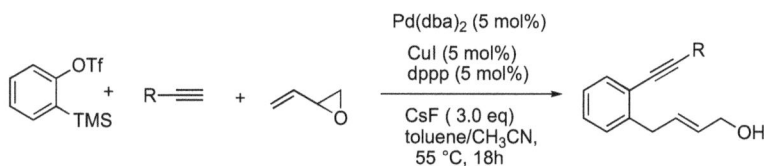

Scheme 20: Three-component coupling of benzynes, allylic epoxides, and terminal alkynes.

2.3. Reactions with Enolates

The nucleophilic addition in neutral conditions of stabilized enolates, such as those derived from malonates, β-diketones, β-ketoacid or β-ketosulfones, to (π-allyl)palladium complexes derived from allylic epoxides was independently developed by Tsuji [74] and Trost [75] in 1981. The regiocontrol for the formation of the new C-C bond distal to the hydroxy group (*i.e.* 1,4-addition) was very high in many cases and can be explained by the electronic effect of the epoxide oxygen. As regards stereoselectivity, the reaction proceeds with clean alkylation from the same face as the oxygen of the epoxide, as normally happens for palladium-catalyzed reactions of allylic substrates with soft nucleophiles. This reactivity complements the normal reactivity of allylic epoxides under standard base-catalyzed conditions. For example, 1,3-cyclohexadiene monoepoxide leads only to allylic alcohol **42** in the presence of a nucleophile of the type CH_2E_2, while the product of direct substitution with inversion of configuration **43** was obtained in basic non-catalyzed conditions (Scheme **21**).

The commonly observed 1,4-regioselectivity in palladium-catalyzed processes can be varied by an introduction of a sterically hindered group. For example, whereas the reaction of butadiene monoepoxide with ethyl acetoacetate catalyzed by palladium(0) afforded the corresponding allylic alcohol, the presence of the sterically hindered-TMS group on the double bond gave the opposite 1,2-regioselectivity and the homoallylic alcohol which was obtained in 76% yield [76]. Quite recently, allylic epoxides have been used as substrates for the allylic alkylation of chelated enolate derived from aminoacids [77]. The corresponding aminoacid containing allylic alcohols have been obtained with high *E/Z* selectivity, albeit with moderate diastereoselectivity.

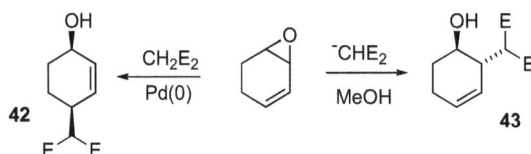

Scheme 21: Complementary reactivity of 1,3-cyclohexadiene monoepoxide with a double stabilized enolate.

The intramolecular palladium-catalyzed reaction of substrates containing both a soft carbon nucleophile and an allylic epoxide can be conveniently used to effect medium and large ring carbocyclization [78]. These reactions often proceed under neutral conditions and with an excellent regioselectivity, as reported in the total synthesis of roseophilin [79], and punctaporonin B [80].

On the other hand, only a few example describe the addition of non-stabilized metal enolates to (π-allyl)palladium complexes derived from vinyl epoxides. Trost observed that cyclopentanone could participate in this type of allylation, albeit with a modest yield (29%) of the corresponding product [75], and Tsuji described without details the same reaction from cyclohexanone silylenol ether [81]. More recently, Malacria *et al.* reported a palladium(0)-catalyzed addition of the lithium enolate of ethyl isobutyrate with isoprene monoepoxide to give allylic alcohol **44** in the first step in the total synthesis of *epi*-illudol in 90% yield (Scheme **22**) [82].

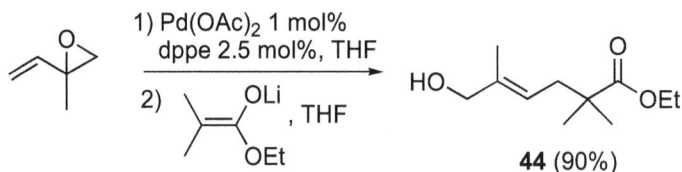

Scheme 22: Palladium-catalyzed addition of non-stabilized metal enolates to isoprene monoepoxide.

Subsequently, the same authors examined the behaviour of other vinyl epoxides featuring a novel access to synthetically useful 6-hydroxyhex-4-enoate moieties [83].

The asymmetric reactions of enolates with allylic epoxides have received little consideration. In an isolated example, Trost and Jiang described a striking asymmetric addition of ethyl acetoacetate to isoprene mono-epoxide catalyzed by 1 mol-% of [Pd$_2$(dba)$_3$]·CHCl$_3$ and 3 mol% of chiral ligand (*S,S*)-**L**^{1} (Scheme **23**) [84]. As achiral ligands give 1,4-adducts (S$_N$2' pathway), these reactions demonstrate the unique ability of the chiral ligand to control regio- and enantioselectivity, favouring the formation of the desired branched 1,2-product **45** (S$_N$2 pathway). The addition of 1 mol-% of TBAT (tetra-*n*-butylammonium triphenyldifluorosilicate) increased the rate of interconversion of diastereoisomeric π-allylpalladium complexes and gave higher 1,2-regioselectivities.

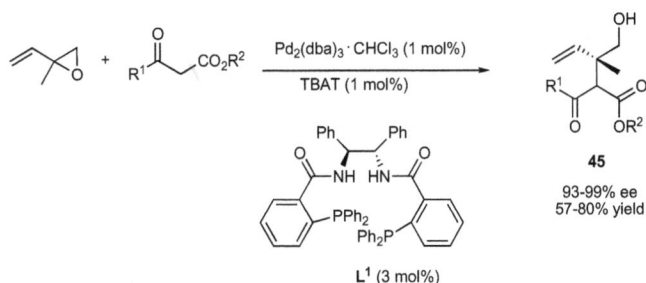

Scheme 23: Regio- and enantioselective reactions of β-ketoesters with isoprene monoepoxide.

The desired 1,2-adducts can be obtained in good yields with a diverse array of β-keto esters, and nitromethane also provided the corresponding adduct with a 51% yield and 97% *ee*. It should be noted that this asymmetric reaction creates a chiral quaternary centre with three of the groups being quite different functional groups, thus furnishing a very useful chiral building block.

2.4. Miscellaneous Ring Openings with Carbon Nucleophiles

2.4.1. Alkynylation

In a seminal work, Stork and Isobe described the reaction of the lithium salt of the 2-ethoxyethyl ether of 1-octyn-3-ol with cyclopentadiene monoepoxide to give the crude *trans* cyclopentenol **46a** (45% yield) which was then used for the synthesis of prostaglandins (Scheme **24**, eq. a) [85]. Later, it was found that this kind of *trans* 1,2-opening of cyclopentadiene monoepoxide requires the presence of a chelating group on the side chain, otherwise *cis*-1,4-opening occurred (**46b**, eq. b, Scheme **24**) [86]. The same authors found that the use of alkynyl alanes, generated *in situ* by treatment of lithium acetylide with Me$_2$AlCl, gave *cis* 1,2-opening, a mode of reaction rarely observed with carbanions (**46c**, eq. c, Scheme **24**).

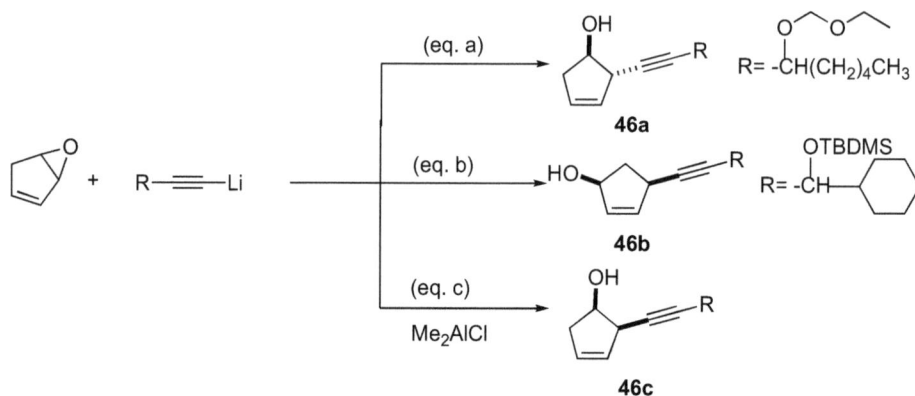

Scheme 24: Reaction of cyclopentadiene monoepoxide with metal acetylides.

Regioselective S$_N$2 alkynylation of aliphatic allylic epoxides has also been accomplished by the use of titanium acetylides generated *in situ* by treatment of the corresponding lithium acetylides with ClTi(*i*-OPr)$_3$ [87]. On the other hand, regioselective S$_N$2' alkylation of vinyl epoxides is more difficult to achieve. In fact, the conjugated alkylation of allylic epoxides is most frequently accomplished by the use of organocopper reagent (*vide supra*), but alkynyl groups are hard to transfer by means of copper, making this approach difficult to accomplish. Recently, a complete regiocontrol in the ring opening of aliphatic vinyl epoxides with ethoxyacetylene has been described by Somfai and Restorp [88]. Considering that the regiochemical outcome of the reactions of metal acetylides with vinyl epoxides was influenced by the electronic nature of the alkyne substituent, they found a particular regiodivergent behaviour using ethoxyacetylene as the alkyne: the combination of the lithium acetylide derived from ethoxyacetylene and BF$_3$·Et$_2$O gave S$_N$2 displacement (eq. a, Scheme **25**), whereas alkynyl alanes gave an S$_N$2' adduct as the sole product (eq. b) [89].

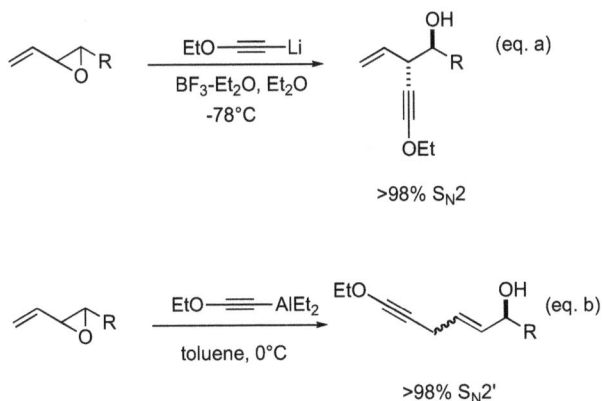

Scheme 25: Regioselective and regiodivergent opening of allylic epoxides with ethoxyacetylene.

2.4.2. Radical additions

Based on the known propensity of Ti(III) species to generate selectively free radicals from epoxides by initial C-O homolysis [90], the reduction of vinyloxiranes with Cp$_2$TiCl was examined by Barrero and co-workers [91]. This kind of reduction afforded homocoupling products with moderate-to-good control of regio- and diastereoselectivity of the carbon-carbon bond formation process. The protocol was used for a novel synthesis of the analogue **47** of natural onoceranes albeit as a mixture of stereoisomers (Scheme **26**).

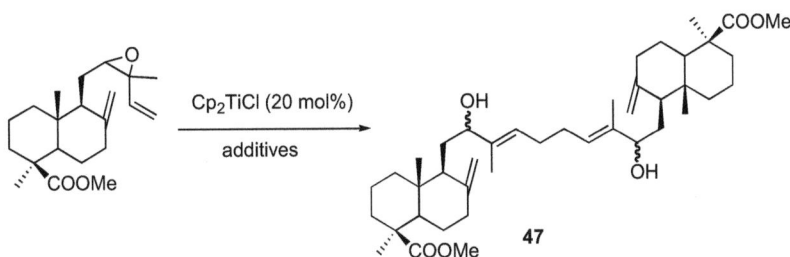

Scheme 26: Cp$_2$TiCl-mediated homocoupling of a vinyl epoxide.

A more powerful and synthetically versatile triethylborane-mediated radical addition of xanthates to vinyl epoxides has recently been developed by Zard *et al.* [92]. The method is based on the fact that xanthates are highly radicophilic and easily form a radical R· which is able to add regioselectively on the double bond of a vinyl oxirane with the formation of an alkoxy radical. The latter is rapidly intercepted by triethylborane, giving a borinate and an ethyl radical, which propagates the chain reaction by reacting with the starting xanthate. The aqueous work-up hydrolyzes the borinate intermediate, giving a ca. 80/20 mixture of *E/Z* allylic alcohols (Scheme **27**). The reaction occurs in mild experimental conditions and it is amenable also to the formation of quaternary stereocentres.

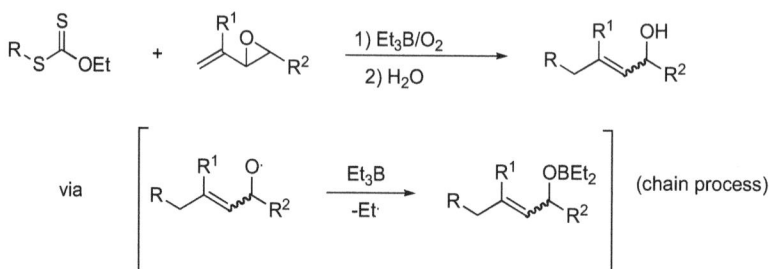

Scheme 27: Triethylborane-mediated radical addition of a xanthate to a vinyl epoxide.

2.4.3. Friedel-Crafts Type Reactions

Despite the abundance of studies on Friedel-Crafts alkylation, vinyloxiranes have rarely been used as the electrophilic partner. For example, vinyloxiranes have been used in an intramolecular process to control a 7-*endo* selective Friedel-Crafts-type cyclization promoted by $BF_3 \cdot Et_2O$ in CH_2Cl_2 at low temperature (Scheme **28**) [93]. An α,β–unsaturated ester group was found to be the best activator of the C-O bond of the epoxide adjacent to it.

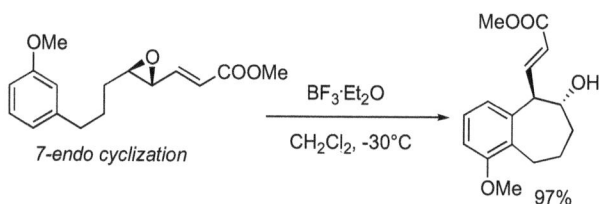

Scheme 28: 7-*Endo* selective Friedel-Crafts type cyclization of vinyloxiranes linked to an ester group.

The same authors reported later in 2009, in a full paper, that the *endo*-selective cyclization was also applicable to a six-membered cyclization [94]. Interestingly, the positional selectivity on the benzene ring shows a remarkable difference between 6-*endo* – and 7-*endo* cyclization. Thus, 6-*endo* cyclization of **48** provides *via* a direct route (Scheme **29**, paths a and b) C6'-cyclized product **49** and C2'-cyclized product **50** in a 65:35 ratio.

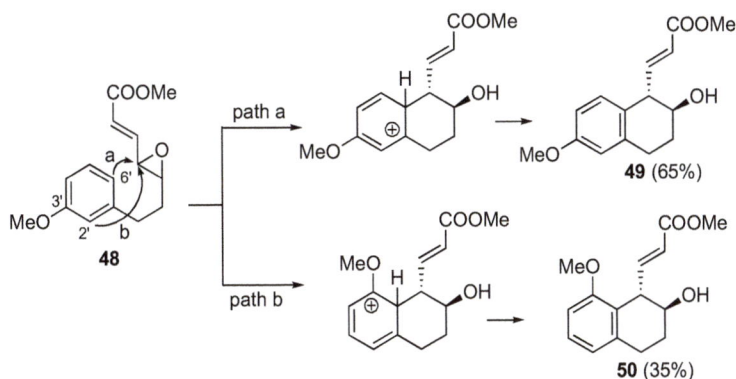

Scheme 29: Regiochemical isomers **49** and **50** by 6-*endo* cyclization of **48**.

On the other hand, 7-*endo* cyclization of **51**, proceeds regioselectively to afford only C6'-cyclized product **52** in excellent yield. In this case, in fact, spiro-benzenium ion **C**, generated from *ipso*-cyclization of **51** (Scheme **30**, path d) undergoes a skeletal rearrangement, induced by the methoxy group at the C3'-position, to afford ion **D**, that is subsequentially converted into seven-membered carbocycle **52**. This product may also be obtained by path c in a more direct way (Scheme **30**).

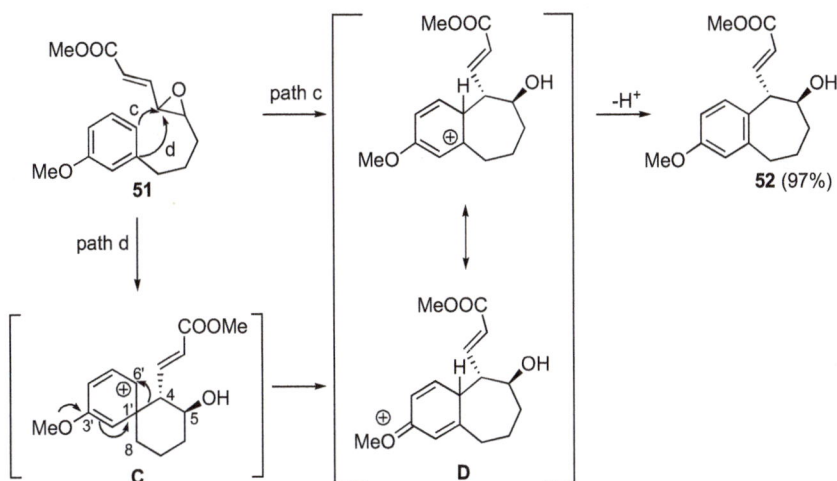

Scheme 30: Regioselective 7-*endo* cyclization of **51**.

In 2007, the same authors describe an efficient method for the construction of hydro-2-benzazepine skeleton, based on a TMSOTf-promoted Friedel-Crafts-type cyclization of vinyloxirane **53** having a nitrogen atom at the benzylic position

[95]. Also in this case, the intramolecular Friedel-Crafts reaction proceedes stereospecifically in a highly selective 7-*endo* mode, similarly to carbocyclization of **51**, due to the vinyl group linked to ester group COOMe, to afford hydro-2-benzazepine **54** in high yield. (Scheme **31**).

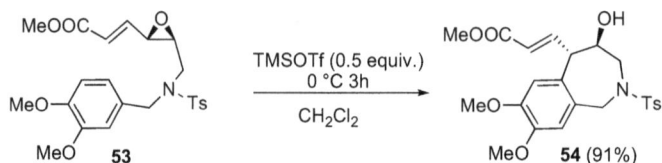

Scheme 31: 7-*endo* cyclization of vinyloxirane **53** to hydro-2-benzazepine **54**.

The previously reported results prompted the authors to perform the more difficult 8-*endo* Friedel-Craft cyclization of appropriate vinyloxiranes **55a,b** based on the installation of $Co_2(CO)_6$-complexed acetylene in the linker between the two reacting sites: the vinyloxirane as the electrophile and the benzene ring as the nucleophile [96]. The cyclization of **55a,b** having two methoxy groups in the benzene ring proceed smoothly, using $BF_3 \cdot Et_2O$ in CH_2Cl_2 at -30 °C, in an 8-*endo* mode to afford **56a,b** in excellent yields (Scheme **32**).

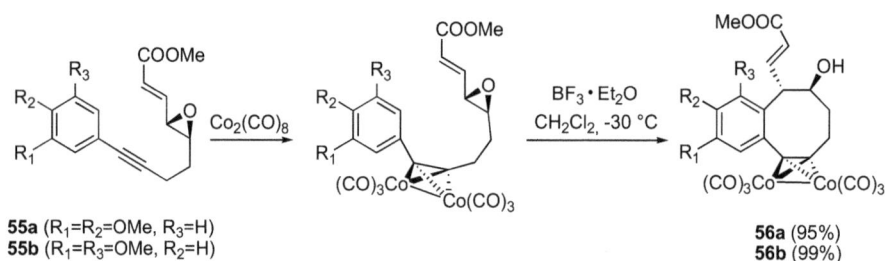

55a ($R_1=R_2=OMe$, $R_3=H$)
55b ($R_1=R_3=OMe$, $R_2=H$)

56a (95%)
56b (99%)

Scheme 32: Selective 8-*endo* Friedel-Crafts cyclization of vinyloxiranes with $Co_2(CO)_6$-complexed acetylene.

In the very limited examples regarding intermolecular processes, strong Lewis acids were invariably used to promote the reaction [97]. The drastic reaction conditions used in this work allowed direct ring opening of the oxirane ring by the Lewis acid promoter, and caused the resulting alcohols to undergo a further Friedel-Crafts alkylation, giving diarylated products.

A mild and stereoselective Friedel-Crafts alkylation of phenol derivatives with cyclic and aliphatic vinyloxiranes has recently been reported by Pineschi *et al.*

[98]. For example, the reaction of 1,3-cyclohexadiene monoepoxide with tris(3,5-dimethoxyphenyl)borate occurred at –78°C in CH_2Cl_2 to give, with complete regioselectivity and *anti*-stereoselectivity hydroxyphenol **57** with a 65% yield after chromatographic purification (Scheme **33**).

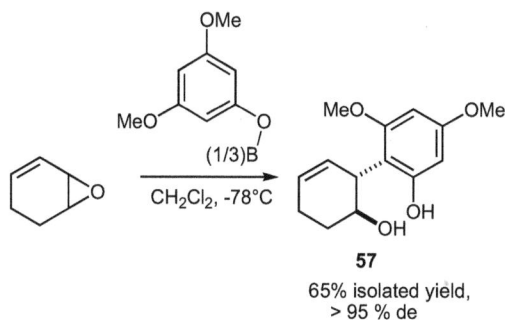

57
65% isolated yield,
> 95 % de

Scheme 33: Regio- and stereoselective Friedel-Crafts alkylation of aryl borates with vinyloxiranes.

3. OXYGEN NUCLEOPHILES

3.1. Ring-Opening Reaction with Alcohols

3.1.1. Intermolecular Ring-Opening

Since the oxirane ring of vinyl epoxides can be contemporarily considered a small-ring heterocycle and an allylic leaving group, the ring-opening reactions of these substrates with alcohols are usually performed through two main approaches:

i)   the addition of the nucleophile can be achieved in the presence of protic or Lewis acids, able to promote the cleavage of the oxirane ring by activation of the epoxide oxygen

ii)   the ring-opening reaction can occur in the presence of catalytic amounts of transition metal complexes, which can activate the electrophilic system toward allylic substitution by their coordination to the unsaturation of the substrate [99].

With regard to the former approach, Posner and Rogers described an alumina-promoted ring-opening of cyclic vinyl epoxides with allylic alcohol [100], while

Boaz reported the addition of alcohols and water to butadiene monoepoxide in the presence of catalytic amount of sulphuric acid [101]. Although these examples demonstrate that vinyl epoxides can be successfully opened by the use of acid reaction conditions, it should be noted that these are limited to cyclic compounds [100] or butadiene monoepoxide [101], respectively, and require a large excess of the alcohol, reducing the synthetic utility of these reactions.

More recently, Mioskowski and co-workers reported the ring-opening of vinyl epoxides with primary and secondary alcohols in the presence of catalytic amounts of several Brønsted and Lewis acids of which, BF$_3$'Et$_2$O was found to give the best results [102]. It is worth mentioning that the same authors took advantage of these reaction conditions synthesize (-)-muricatacin, an acetogenin having an activity as an antitumoral agent (Scheme **34**) [103].

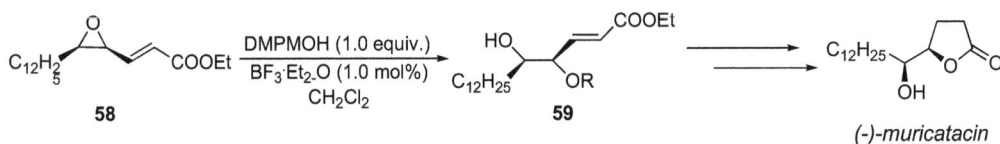

Scheme **34:** Synthesis of (-)-muricatacin *via* stereoselective ring-opening with alcohols.

The key step of this synthesis is the reaction of γ,δ-epoxy-α,β-unsaturated ester **58** with an equimolar amount of 3,4-dimethoxybenzyl alcohol (DMPMOH) in the presence of BF$_3$'Et$_2$O as a catalysts, giving compound **59** with an acceptable yield, complete regioselectivity at the allylic position and a high *anti*-diastereoselectivity.

A significant advance towards an asymmetric ring-opening of vinylepoxides with oxygen nucleophiles was reported by Jacobsen and co-workers [104]. Butadiene monoepoxide can be efficiently resolved with water in the presence of catalytic amounts of (salen)Co(III)(OAc) to give the corresponding 1,2-diol and terminal epoxide with a high yield and a high enantiomeric enrichment (Scheme **35**). Even if butadiene monoepoxide was resolved with a lower selectivity (K_{rel}= 30) with respect to alkyl-substituted epoxides (for propylene oxide K_{rel} > 400), it should be noted that highly enantiomerically enriched epoxide or diol could still be obtained in a useful yield by simply adjusting the amount of water, of the catalyst or by using a small amount of THF as the reaction solvent [1:1 (*v/v*) ratio relative to racemic starting epoxide] [105].

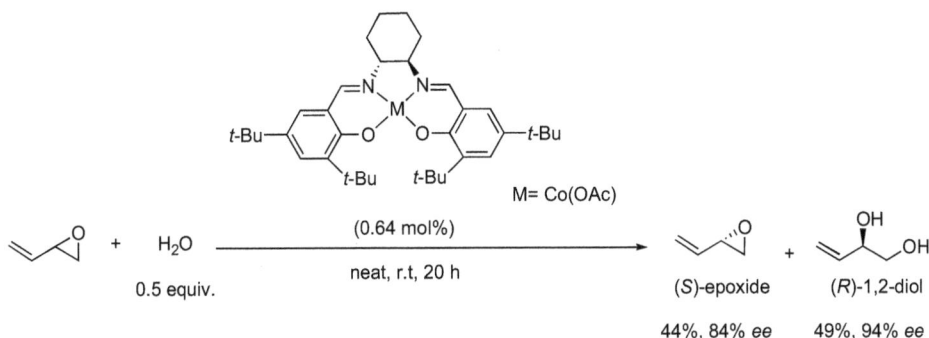

Scheme 35: Kinetic resolution of butadiene monoepoxide with water.

Although the use of vinyl epoxides as allylic electrophiles in palladium-catalyzed allylic alkylation has been considered since the early eighties [74, 75, 106], examples of ring-openings using oxygen nucleophiles are quite limited, mainly because alcohols are generally poor nucleophiles in this kind of palladium-catalyzed reaction [107]. In this connection, Trost and co-workers reported that vinyl epoxides can be opened by the use of stannyl alkoxides in the presence of catalytic amounts of Pd$_2$(dba)$_3$·CHCl$_3$ and triphenylphospine [108]. The ring-opening reaction occurred at 0°C and the ring-opened products were obtained with a high regioselectivity at the allylic position and a remarkable *syn*-stereoselectivity, in particular when cyclic vinyl epoxides were used (Scheme **36**).

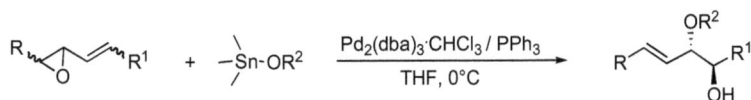

Scheme 36: Tin mediated Pd-catalyzed alkylation of vinyl epoxides.

Using an analogous approach, Trost and co-workers more recently described the Pd(0)-catalyzed dynamic kinetic resolution (DKR) of racemic vinyl epoxides [109]. Butadiene and isoprene monoepoxide reacted at room temperature with several different types of alcohols with catalysis by Pd$_2$(dba)$_3$· CHCl$_3$, ligands of type **L**, and Et$_3$B as a co-catalyst, giving the corresponding optically active β-hydroxy allyl-ethers in moderate to very good yields with a high enantiomeric enrichments (up to 98% *ee*) (Scheme **37**).

For a positive outcome of this process, it is necessary that the diastereoisomeric interconversion of the π-allyl palladium complexes should be faster than the reaction of the same complexes with the nucleophile, thus favouring a dynamic equilibrium (Scheme **38**). It is also worth mentioning the role of the co-catalyst, which activates both the nucleophile and the electrophile, and is thus able to control the regioselectivity of the reaction driving the attack of the alcohol by an S_N2 reaction variant, although Pd(0)-mediated ring-openings generally favour the formation of 1,4-products.

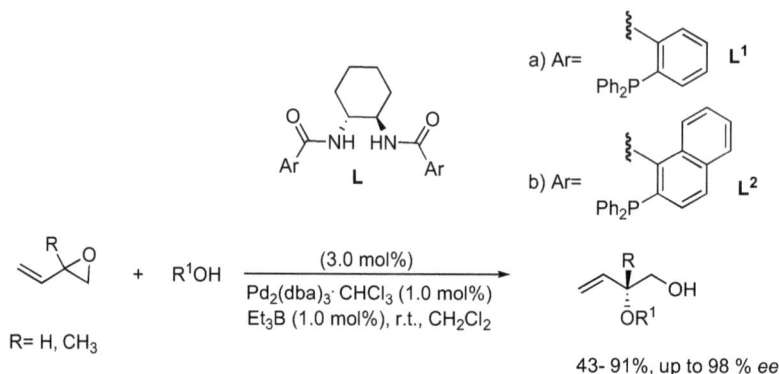

Scheme 37: Dynamic kinetic resolution of racemic terminal epoxides.

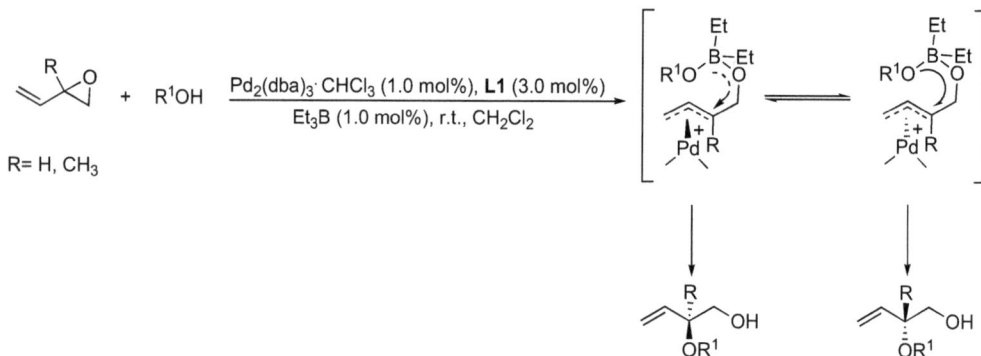

Scheme 38: Proposed mechanism for DKR.

The concept of this two-components catalyst system was subsequently exploited by Miyashita and co-workers for the synthesis of γ,δ-vicinal diols by the ring-opening reaction of α, β-unsaturated-γ,δ-epoxy esters [110]. In particular, the reaction of these substrates with butylboronic acid in the presence of a catalytic amount of Pd(PPh$_3$)$_4$ afforded the corresponding cyclic boronates, which can readily be converted to the relative γ,δ-vicinal diols by treatment with H_2O_2 in MeOH (Scheme **39**). As this substitution reaction proceeds through two S_N2

pathways, which allows formation of ring-opened products with net retention of the configuration at the cleaved centre (*d.r* up to 99/1).

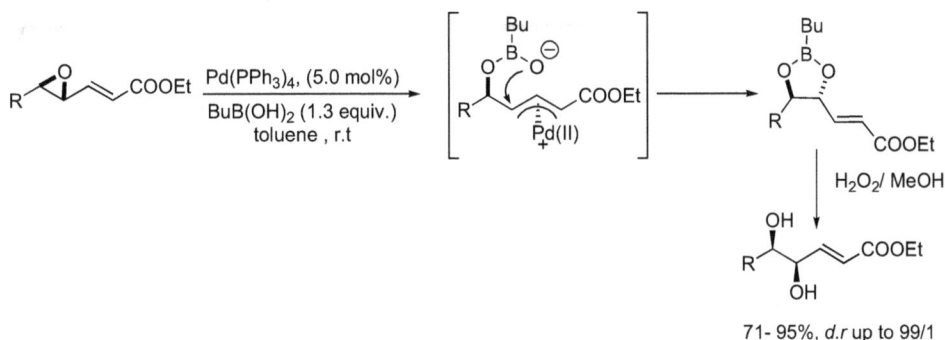

71- 95%, *d.r* up to 99/1

Scheme 39: Pd-catalyzed epoxide ring-opening with double inversion of configuration.

These types of epoxides have been successfully cleaved also with alcohols, using an analogous approach. In particular, α,β-unsaturated-γ,δ-epoxy esters reacted at 0 °C in the Pd(0)-catalyzed reaction with alkyl borates to give the corresponding alkoxy alcohols after a double inversion of configuration [111]. Although the reaction proceeded satisfactorily also using borates, the combination of pinacol, boric anhydride (B_2O_3) and the alcohol proved to be the most effective for the Pd-catalyzed substitution, giving the formation of the desired ring-opened products in good yields and with high stereoselectivities, also using base-sensitive alcohols such as 2-bromoethanol or 2-nitroethanol (Scheme **40**).

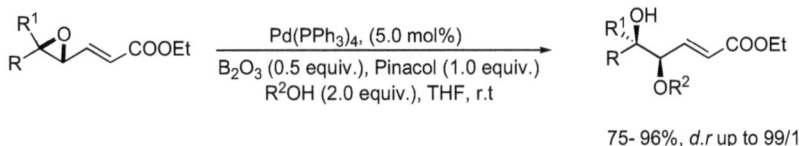

75- 96%, *d.r* up to 99/1

Scheme 40: Pd(0)-catalyzed alkoxy substitution reaction of α,β-unsaturated-γ,δ-epoxy esters.

Lautens reported that vinyl epoxides can also be opened with oxygen nucleophiles in the presence of rhodium-based catalysts [112]. Both cyclic and acyclic vinyl epoxides reacted readily with a wide range of primary and secondary alcohols catalyzed by $[Rh(CO)Cl]_2$ to give the corresponding β-hydroxy allyl ethers with a complete regioselectivity at the allylic position and with a high *anti*-stereoselectivity (Scheme **41**).

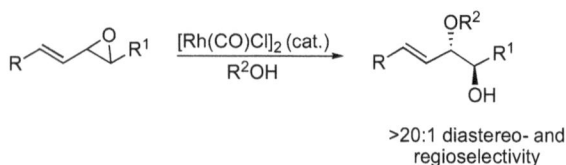

>20:1 diastereo- and
regioselectivity

Scheme 41: Rhodium-catalyzed alcoholysis of vinyl epoxides.

While the use of other rhodium salts or the presence of an added ligand did not show any reactivity, it is worth mentioning that the reaction carried out using vinyl epoxides possessing a terminal double bond produced a mixture of *1,2-* and *1,4-addition products.* This result coupled with the total absence of reactivity of styrene and cyclohexene oxide under these reaction conditions seem to indicate that rhodium is probably not acting as a Lewis acid, but the formation of ring-opened products might proceed through a π-allyl or enyl rhodium intermediate [113].

In addition to these examples of acid- or metal-catalyzed processes, Crotti and co-workers reported the preparation of 2-unsaturated-*O*-glycosides, a valuable class of synthetic intermediates, *via* the ring-opening reaction of glycal-derived vinyl epoxides **12α**, **12α-Me**, **12β**, **12β-Tr** and **12β-Me** [40-42, 114]. These substrates, which were generated *in situ* by treatment of the corresponding *trans* hydroxy mesylate with *t*-BuOK in anhydrous benzene, reacted with MeOH, primary, secondary, tertiary alcohols and partially protected monosaccharides (R^2OH in Scheme **42**) through a completely stereoselective 1,4-addition process

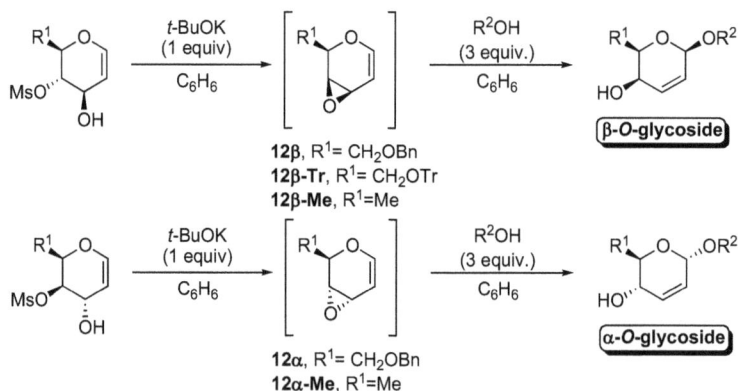

Scheme 42: Glycosylation of alcohols (R^2OH) by the *in situ*-formed glycal-derived vinyl epoxides **12α**, **12α-Me**, **12β**, **12β-Tr** and **12β-Me**.

affording the corresponding α-*O*-glycosides from α epoxides and β–*O*-glycosides from β epoxides, exclusively, in a new, uncatalyzed, substrate-dependent, stereospecific *O*-glycosylation process.

The strict correspondence between the configuration of the obtained glycosides and that of the starting vinyl epoxide has been rationalized by coordination between the oxirane oxygen and the nucleophile *via* a hydrogen bond which favours the stereoselective nucleophilic attack of the alcohol from the same side as the epoxide oxygen. An example of this is shown in Scheme **43** only for epoxides **12α** and **12β**.

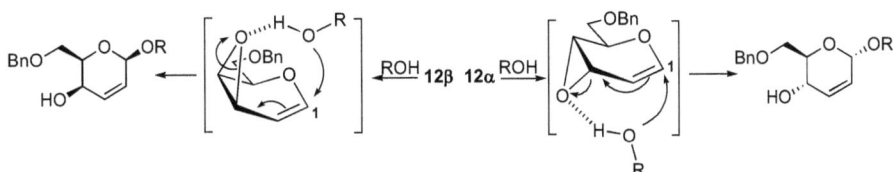

Scheme 43: Plausible mechanism for the glycosylation of alcohols (ROH) by epoxides **12α** and **12β**.

The new *O*-glycosylation process found with epoxide **12β** was adapted in a reiterative version for the complete stereoselective synthesis of β-linked 2,3-unsaturated-1,6-oligosaccharides of type **60** (Scheme **44**) [115].

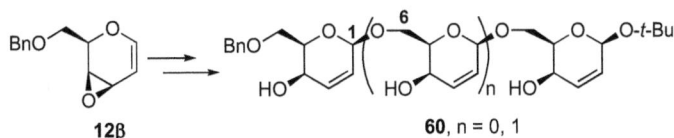

Scheme 44: Synthesis of 1,6-oligosaccharides from epoxide **12β**.

The opening reactions of vinyl epoxides *carba*-**12α** and *carba*-**12β** with MeOH and AcOH, as model *O*-nucleophiles, necessitate a basic or acid catalysis and the regiochemical outcome strictly depends on the reaction conditions and the epoxide [52, 116]. Under basic conditions (0.2 N MeONa/MeOH or AcONa/DMF-H_2O) only the corresponding *anti-1,2-addition product* is obtained from both epoxides. Under acid conditions [MeOH or AcOH (3-6 equiv) in 0.01 N TsOH in CH_2Cl_2], whereas epoxide *carba*-**12β** is once more completely 1,2-regio- and *anti*-stereoselective, epoxide *carba*-**12α** is not regio- and stereoselective affording a

43:47:10 mixture of corresponding *anti-1,2-*, *syn-1,4-* and *anti-1,4-addition products* with MeOH and a 38:30:32 mixture of *anti-1,2-*, *syn-1,2-* and *syn-1,4-addition products* with AcOH (Scheme **45**). The results have indicated that epoxide *carba-12β* has a pronounced tendency toward *anti-*1,2-addition, whereas epoxide *carba-12α* shows interesting levels of *syn-* and/or *anti-*1,4-addition processes (32-57%) to the point to be considered a potentially useful candidate for the construction of *O*-linked carba-oligosaccharides. The results were rationalized on the basis of the conformer population inside the two epoxides and stereoelectronic factors associated with the opening process of the three-membered rings which indicated, in particular, for epoxide *carba-12α* a reduced reactivity at allylic oxirane carbon [52].

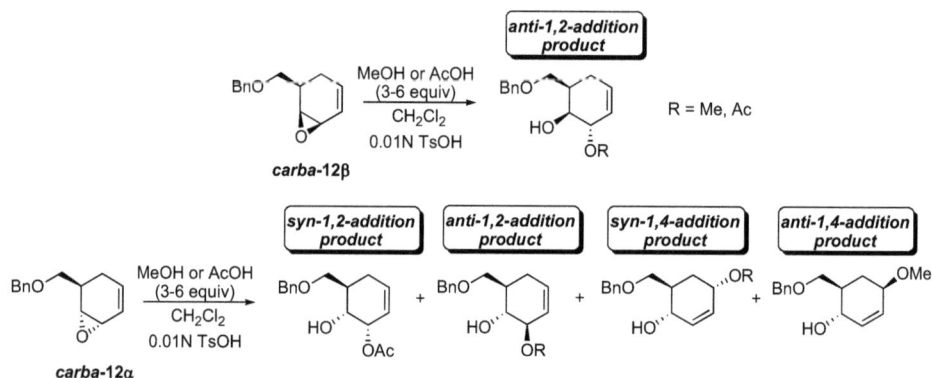

Scheme 45: Regio and stereoselectivity of the addition reaction of *O*-nucleophiles to epoxides *carba-12α* and carba **12β** under acid conditions.

Vinyl epoxides such as **61** (R^1 = Ph, aryl, 2-furyl, *n*-butyl, *i*-butyl), readily available by sulfonium ylide epoxidation of the corresponding aldehyde R^1-CHO were envisaged as alternative approach to β-hydroxy-α-methylene lactones, as **63**, which are compounds having a large array of biological functions [117]. The direct lactonization of epoxides **61** (as a 95:5 to 71:39 mixture of *trans* and *cis* diastereoisomers) carried out under acid conditions (5% aqueous H_2SO_4) afforded almost corresponding mixtures of related *trans* and *cis* lactones **63** in poor (13-24%) to satisfactory yield (62-66%). The cyclization occurs by the initial formation of the corresponding diol **62** with complete inversion of configuration on the allylic oxirane carbon, followed by diol cyclization into lactones with retention of configuration at the homoallylic diol carbon (Scheme **46**). The presence of the phenyl group in the starting epoxide was found to be responsible of an epimerization process at the oxirane benzylic carbon in the formation of the initial diol with subsequent reduced *trans/cis* diastereoselective lactonization [117].

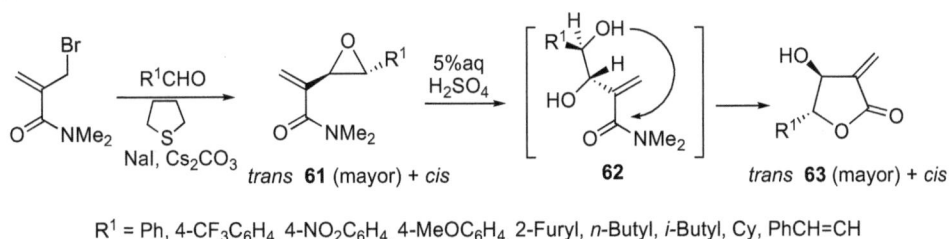

R¹ = Ph, 4-CF₃C₆H₄, 4-NO₂C₆H₄, 4-MeOC₆H₄, 2-Furyl, *n*-Butyl, *i*-Butyl, Cy, PhCH=CH

Scheme 46: Lactonization of vinyl epoxides of type **61** under acid conditions.

3.1.2. Intramolecular Ring-Opening

The preparation of tetrahydropyran and oxepane systems represents an attractive goal in organic synthesis. In this respect, the intramolecular ring-opening of hydroxy epoxides has been successfully used for the synthesis of several natural products [118, 119]. The main problem in these processes is how to drive the reaction regioselectively in an *endo* or *exo* mode of attack. Although this type of cyclization proceeds in an *exo*-mode, in accordance with Baldwin's rules [120], the presence of a vinyl group next to the epoxide, activating the cleavage of the C-O bond adjacent to it, is able to direct the attack at the allylic position by means of *endo* ring closure. One of the most detailed studies with regard to the regioselective synthesis of tetrahydropyran systems by the use of vinyl epoxides was reported by Nicolaou and co-workers [121]. Although the reaction of the epoxy alcohol **64a** with catalytic amounts of camphorsulfonic acid (CSA) afforded the expected tetrahydrofuran system **66a**, in accordance with Baldwin's rules (entry 1, Table **1**), substrate **64b**, bearing a vinyl moiety adjacent to the oxiranic ring, favoured an *endo* attack of the hydroxyl moiety giving compound **65b** as the main product (entry 2, Table **1**). The best results in terms of regioselectivity were obtained using substrates **64c** and **64d**. In these cases, the presence of more electron-rich double bonds directed the attack exclusively to the allylic position, giving corresponding products in high yields with complete regioselectivities and stereoselectivities (entries 3 and 4, Table **1**). Interestingly, *cis* isomers of epoxides of type **64** lead, under the same reaction conditions described in Table **1**, to obtain a mixture of the two possible regioisomers, even if substrates bearing electron-rich double bonds are used. As the regioselectivity of this process is reasonably ruled by the ability of the system to stabilize the electron-deficient orbital at the allylic position, the stereochemistry of *cis* epoxides may be responsible for the difficulties of these systems to assume planar arrangements, which are necessary for maximum stabilization in the transition state.

Table 1: Acid-catalyzed cyclization of *trans* epoxy alcohols

Entry	Substrate	Products (ratio)	Yield (%)
1	**64a**: R= CH$_2$CH$_2$CO$_2$Me	**65a:66a** (0:100)	94
2	**64b**: R= CH=CHCO$_2$Me	**65b:66b** (60:40)	96
3	**64c**: R= CH=CH$_2$	**65c:66c** (100:0)	95
4	**64d**: R= CH=CBr$_2$	**65d:66d** (100:0)	90

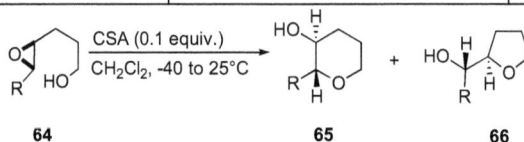

Unlike the results obtained for the synthesis of tetrahydropyran systems (Table **1**), the cyclization of the homologous epoxy alcohol **67** under the same acidic reaction conditions does not occur with complete regioselectivity (Table **2**) [122]. However, the vinyl moiety in **67b-d** directs the attack of the hydroxyl functionality *via* a 7-*endo* cyclization, giving the corresponding oxepane systems **68** in good yields and high diastereoselectivities (entries 2-4, Table **2**). On the other hand, the classical *exo* attack of the internal nucleophile give rise to tetrahydropyran derivatives **69** when saturated epoxy alcohols are used (entry 1, Table **2**).

Table 2: Acid-catalyzed cyclization of *trans* epoxy alcohols leading to oxepanes and tetrahydropyranes

Entry	Substrate	Products (ratio)	Yield (%)
1	**67a**: R= CH$_2$CH$_2$CO$_2$Me	**68a:69a** (0:100)	70
2	**67b**: R= CH=CH$_2$	**68b:69b** (82:18)	75
3	**67c**: R= Z-CH=CHCl	**68c:69c** (60:40)	70
4	**67d**: R= E-CH=CHCl	**68d:69d** (92:8)	75

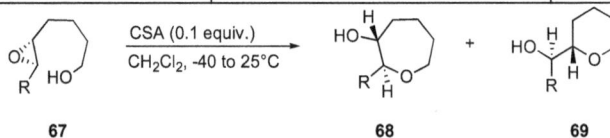

In a related study, Nakata and co-workers found that the introduction of a styryl group next to the epoxide represents a valid method to enhance the *endo* mode of attack (Scheme **47**) [123]. In this case, the additional stabilization provided by the phenyl group permits the obtainment of the desired tetrahydropyrans with complete regioselectivity also using *cis*-epoxy alcohols, whereas the application of Nicolaou's procedure resulted in a mixture of *endo*- and *exo*-cyclized compounds (see Table **1**).

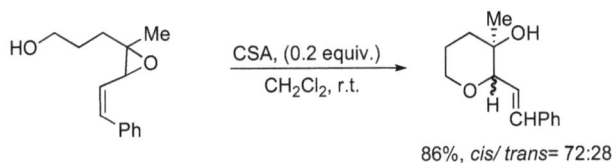

86%, *cis/trans*= 72:28

Scheme 47: Cyclization of hydroxy styryl epoxides.

As already described for the *intermolecular* ring-opening of vinyl epoxides (see above), the use of a palladium-catalyzed reaction represents a rational approach for the cyclization of vinyl epoxy alcohols. In this connection, Trost reported the synthesis of tetrahydrofuran and tetrahydropyran systems by an *intramolecular* ring-opening reaction of vinyl epoxides in the presence of $Pd_2(dba)_3 \cdot CHCl_3$ and PPh_3 [124]. Although this cyclization reaction proceeded with remarkable regioselectivity at the allylic position of the substrates, high levels of diastereoselectivity were observed only in the preparation of tetrahydrofuran systems **70** (Scheme **48**).

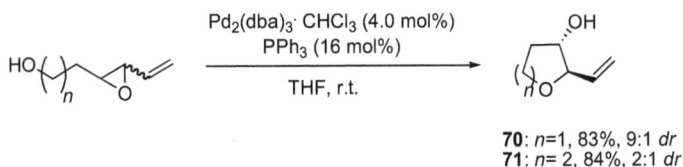

70: *n*=1, 83%, 9:1 *dr*
71: *n*= 2, 84%, 2:1 *dr*

Scheme 48: Pd-catalyzed cyclization of hydroxy vinyl epoxides.

More recently, Hirama and co-workers reported the synthesis of both *cis*- and *trans*-2,3-disubstituted tetrahydropyrans starting from silyl ethers instead of the corresponding free alcohol [125]. Ammonium alkoxides, which can be generated from the corresponding silyl ethers by TBAF, are known to be efficient nucleophiles in palladium-catalyzed allylic etherification [126]. Accordingly, *t*-

butyldiphenylsilyl (TBPS) ethers of type **72** and **73** were treated with TBAF in dichloromethane and then with a catalytic amount of Pd(PPh$_3$)$_4$ in one pot to give the corresponding ring closure compounds in good yields and high diastereoselectivities (Scheme **49**). Interestingly, this reaction exhibits an important solvent effect: for example, when THF was used as solvent, ring-opened products were obtained with good stereoselectivities, but long reaction times were required. However, when CH$_2$Cl$_2$ was used as solvent, the reaction occurred quickly, to give the desired tetrahydropyrans in high yields and stereoselections.

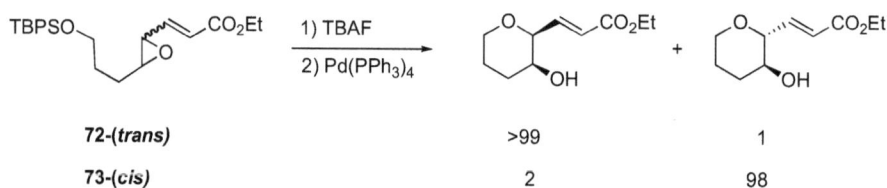

72-(trans)	>99	1
73-(cis)	2	98

Scheme 49: Palladium-catalyzed cyclization to tetrahydropyrans.

In addition to these examples of Pd-catalyzed reactions, Ha and co-workers more recently reported that rhodium complexes can also be succesfully used for the *intramolecular* ring-opening of vinyl epoxides with alcohols [127]. A variety of hydroxy vinylepoxides reacted in THF in the presence of catalytic amounts of [Rh(CO)$_2$Cl]$_2$, (used by Lautens for the *intermolecular* version of this reaction [112]), to give the corresponding tetrahydrofuran and tetrahydropyran systems in very good yields (Scheme **50**).

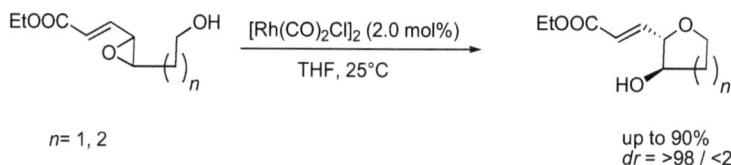

Scheme 50: [Rh(CO)$_2$Cl]$_2$-catalyzed ring closure of vinyl epoxides.

Although *trans*-vinyl epoxides reacted with high regio- and stereoselectivity, it is worth mentioning that *cis*-isomers were found to be totally inert under the same reaction conditions. Based on these results, the regio- and stereoselective outcome of this process might be explained by considering a first double coordination of rhodium with both vinyl moiety and epoxide oxygen (**i**) (Fig. **1**). This step is crucial for the oxidative addition of rhodium and the formation of the enyl or π-allyl intermediate, which ultimately affords the *trans*-cyclized products **ii**. While this process is possible for *trans*-vinyl epoxides ($R^1 = H$), for *cis*-epoxides ($R^1 =$ alkyl group) the steric repulsion between R^1 and the vinyl moiety prevents the double coordination **iii** and, in this way, the formation of an enyl or π-allyl intermediate, necessary for obtaining ring closure products.

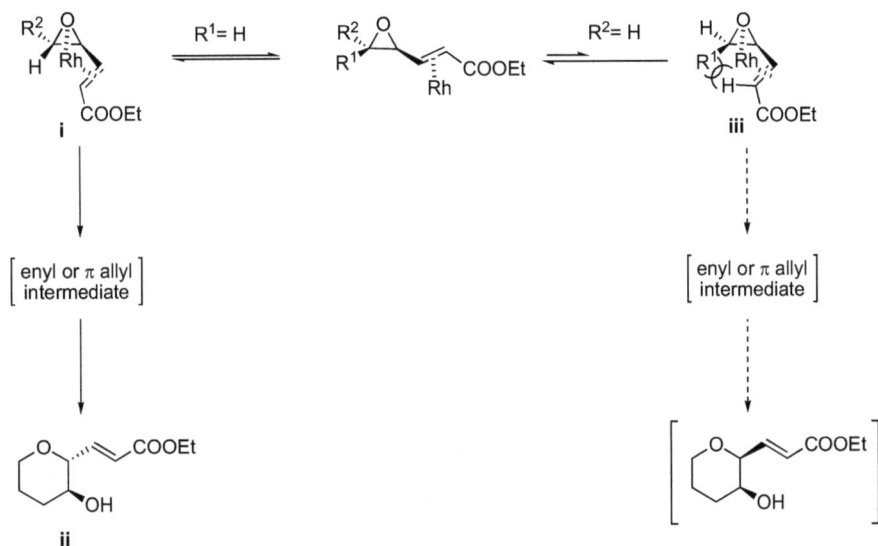

Figure 1: Plausible mechanism for the [Rh(CO)$_2$Cl]$_2$-catalyzed ring closure of vinyl epoxides.

3.2. Ring Opening with Phenols

The use of phenols in the ring-opening reaction of epoxides is rather difficult, mainly because these class of oxygen nucleophiles behave as poor nucleophilic reagents in substitution processes. Nevertheless, the use of strongly acidic or basic or metal-catalyzed conditions represent well-known methods for the ring opening of epoxides with phenols, which usually occurs in an *anti*-stereoselective manner. With regard to the use of vinyl epoxides, it should be noted that examples of their ring opening by the use of phenols are very limited and usually involve the use of transition metal catalysts [111, 112]. For example, Pineschi and co-workers reported that vinyl epoxides can be successfully opened by the use of aromatic borates [128]. Cyclic epoxides reacted at room temperature in THF with several aromatic borates to give the corresponding aryloxy alcohols with complete regioselectivity and high *syn*-stereoselectivity. The choice of the solvent was crucial in these reactions [129]. Thus, cyclohexadiene monoepoxide reacted with tris (3,5-dimethylphenyl) borate in THF to give the corresponding aryloxy alcohol exclusively with a high degree of *syn*-stereoselectivity, whereas in CH_2Cl_2 at – 78°C a mixture of C- and O-alkylated products was obtained, both with a high *anti*-stereoselectivity (Scheme **51**).

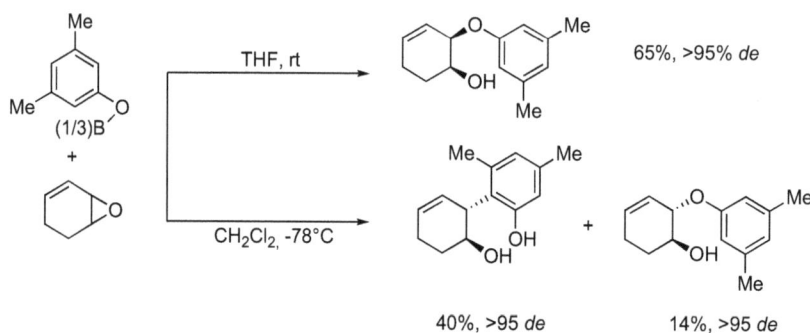

Scheme 51: Ring-opening of vinyl epoxides with aromatic borates.

These results can be explained by considering that this ring-opening reaction might occur by two possible mechanisms as a consequence of the different solvating ability of the solvents. When the reaction was carried out in the more solvating THF, the aromatic borate was able, after its coordination to the epoxide oxygen, to deliver the *internal nucleophile* with retention of configuration and with a prevalence of the "harder" character of the oxygen of the phenol derivative. On the other hand, the lesser solvating ability of the incipient allylic cationic intermediate by the use of CH_2Cl_2 did not allow the intramolecular attack of the aryl moiety. In this case, the normal *intermolecular* ring-opening with inversion of configuration was observed with a predominating attack of the nucleophilic arene at its "softer" site (the *ortho* ring carbon of the phenol derivative).

4. NITROGEN NUCLEOPHILES

The regioselectivity of the ring opening of vinyl epoxides with N-nucleophiles (amines and azide as the most common examples) can be controlled by using appropriate reaction conditions. The *1,2-addition product* (S_N2 product) with attack of the nucleophile on the oxirane C(3) allylic carbon, usually with inversion of configuration, commonly occurs with a Lewis acid catalyst, whereas *1,4-addition products* (S_N2' product) are favoured by Pd(0)-catalyzed opening processes. However, in many cases, Pd(0)-catalysis is used in order to favour the 1,2-addition process.

4.1. 1,2-Addition of Nitrogen Nucleophiles

Seminal studies of the 1,2-addition of N-nucleophiles to vinyl epoxides were carried out by Posner in the aminolysis of cyclopentadiene monoepoxide and cyclohexadiene monoepoxide by *n*-butylamine-doped alumina. The addition reaction occurred regioselectively at the allylic C(3) oxirane carbon in a completely *anti* stereoselective process, with the exclusive formation of *trans* β-amino alcohols **74** and **75**, respectively (Scheme **52**) [100].

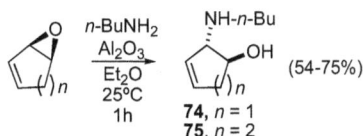

Scheme 52: Reaction of vinyl epoxides with *n*-butylamine doped alumina.

More recently, vinyl epoxides were found by Somfai to open regioselectively at the allylic C(3) oxirane carbon by ammonia and primary amines such as cyclohexylamine and benzylamine at 130°C for three days in the presence of TsOH. 2,3-Disubstituted oxiranes are the best substrates for this reaction with the *trans* epoxides (**76**) affording *anti* β-amino alcohols (**77**), while *cis* epoxides (**78**) are transformed into the corresponding *syn* derivatives (**79**) in a completely stereoselective fashion (Scheme **53**) [130]. Only in the case of epoxide **80**, due to the presence of a benzylic oxirane carbon, a 4:1 mixture of β amino alcohol **81** and regioisomeric **82** was obtained. With sterically hindered substrates [presence of a tertiary C(3) carbon], aminolysis is completely suppressed.

A practical application of this regio- and stereoselective protocol (TsOH, DMSO) is shown in Scheme **54**. Thus, benzylamine reacts with vinyl epoxide **83** to give intermediate **84** which undergoes concomitant intramolecular displacement of the mesylate by the amino group affording piperidine **85**, the key synthetic intermediate on the route to (+)-1-deoxynojirimycin (**86**), an azasugar with antiviral and antidiabetic activity (Scheme **54**) [131].

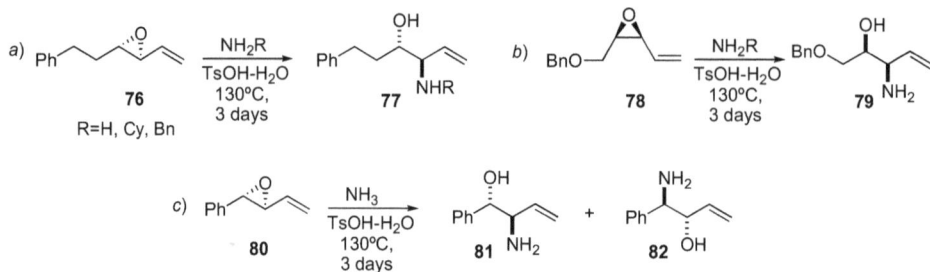

Scheme 53: Aminolysis of vinyl epoxides by ammonia and amines.

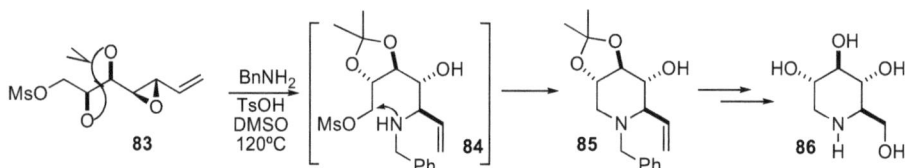

Scheme 54: Enantioselective synthesis of (+)-1-deoxynojirimycin.

An efficient microwave-assisted protocol decidedly improved [shorter reaction times (8 min), higher yields and milder reaction conditions (30W)] the aminolysis reaction with NH_4OH of di- and trisubstituted vinyl oxiranes [132].

The propensity of vinyl epoxides to undergo regio- and stereoselective ring opening at the allylic C(3) oxirane carbon under Lewis acid-catalyzed or uncatalyzed reaction conditions, was advantageously used for the regioselective insertion of an azido or amino group at C(3) of pyrimidine nucleosides. In this framework, epoxide **87** showed a poor C(2)/C(3) regiocontrol, whereas the corresponding reaction of epoxide **88** bearing a "conjugated" double bond achieved a completely regioselective addition, with the formation of the corresponding C(3) derivatives, the *trans* azido alcohol **89** and the *trans* N-benzylamino alcohol **90**. No trace of the corresponding regioisomeric C(2) addition product was observed (Scheme **55**) [133]. Only in the case of the addition

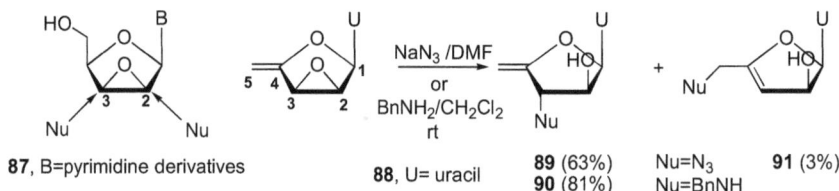

Scheme 55: Reaction of epoxide **87** with NaN_3 or $BnNH_2$.

of N_3^-, was a small amount of C(5)-adduct (azido alcohol **91**) obtained as a minor product, as a consequence of the usual [3,3]sigmatropic rearrangement or due to the soft nature of the azide [134-136].

The possibility to realize a vicinal hydroxyamination of vinyl epoxides with N-nucleophiles in the presence of the Pd(0) catalyst was accurately checked by Trost. However, because of the propensity of Pd(0)-catalyzed addition reactions to lead to the corresponding *1,4-addition products* (*vide infra*), it was thought that, in order to have a complementary regiochemistry (1,2-addition process), a tethering of the N-nucleophile to the oxygen of the leaving group was necessary. In this framework, the reactions of *p*-tolensulphonyl isocyanate with monoepoxides of cyclohexadiene and 6-phenyl-1,3-hexadiene were examined (Scheme **56**) [137].

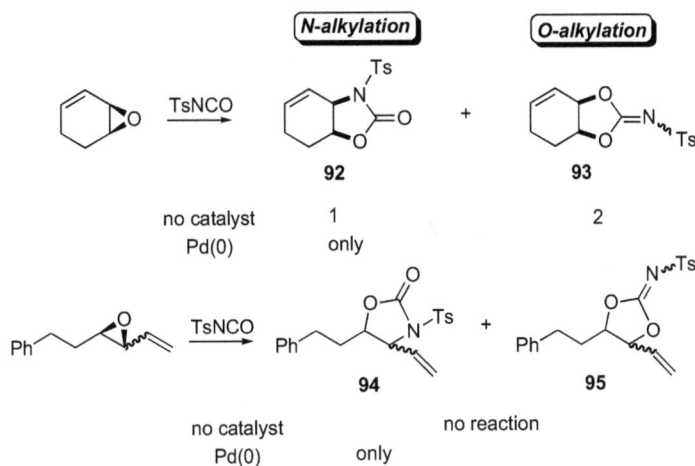

Scheme 56: Pd(0)-catalyzed vicinal *syn* hydroxyamination of vinyl epoxides.

The thermo reaction led in the first case to a 1:2 mixture of *N*-alkylated (**92**) and *O*-alkylated products (**93**), whereas the reaction with the aliphatic epoxide was unsuccessfull. In contrast the reaction of both vinyl epoxides in the presence of Pd$_2$(dba)$_3$·CHCl$_3$ (1-3 mol%) and triisopropyl phosphite in THF at room temperature were completely regioselective affording only the corresponding N-*p*-toluenesulphonyl-2-oxazolidones **92** and **94** (*N*-alkylated products) resulting from *syn* stereoselective attack of the nitrogen portion of the isocyanate (Scheme **56**). Under these conditions, no trace of the corresponding *O*-alkylated product was observed. The reaction may be interpreted, as shown in Scheme **57**, as leading to the corresponding *O*- (imino carbonates) and *N*-alkylated (2-oxazolidones) internal addition product through zwitterion **96** and the nucleophile-tethered intermediate π-palladium complex **97**. However, it has been determined that products of *O*-alkylation rearrange to the corresponding *N*-alkylation products in the presence of the Pd(0) catalyst. Thus it is possible that the kinetic products of cyclization in the Pd(0) opening of vinyl epoxides with isocyanates are the imino carbonates, which subsequently rearrange to the thermodynamically more stable 2-oxazolidones [137].

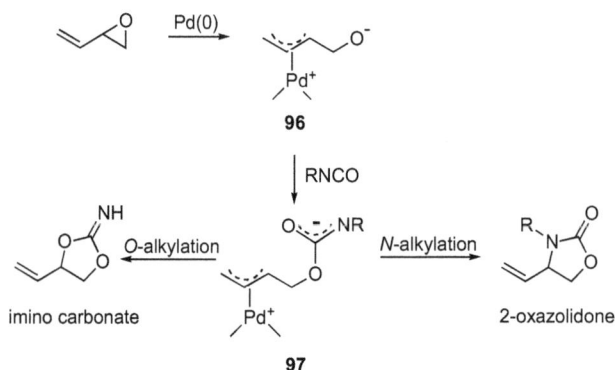

Scheme 57: Rationalization of the Pd(0)-catalyzed vicinal *syn* hydroxyamination.

The choice of isocyanate is important in these reactions. For example, phenyl and *p*-anisyl isocyanate reacted equally well. However, in the case of benzyl isocyanate the reaction fails presumably because of its failure due to intercept the zwitterion, due to its poor electrophilicity. Practically, the palladium catalyzed opening of vinyl epoxides with isocyanates proceeds with retention of configuration and considering the availability of vinyl epoxides from allylic alcohol or 1,3-dienes corresponds to a *syn* hydroxyamination of an olefin.

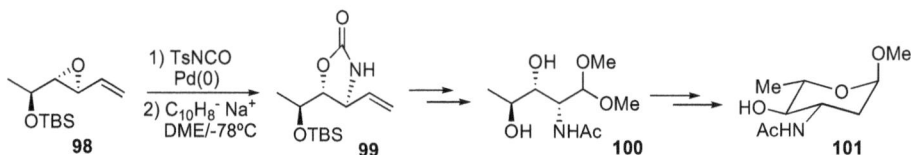

Scheme 58: Synthesis of (-)-*N*-acetyl-*O*-methylacosamine.

An example of the potential of this approach is shown in Scheme **58** which involves the synthesis of the (-)-*N*-acetyl-*O*-methyl glycoside of acosamine (**101**) starting from the enantiomerically pure vinyl epoxide **98**. The key step is the conversion of epoxide **98** to the 2-oxazolidone **99** which is subsequently converted to the *syn N*-acetyl-β-amino alcohol **100** (Scheme 58) [137].

A new strategy towards glycosidase inhibitors, represented by (±)-valienamine (**107**) evolved from the application of the Pd(0)-catalyzed protocol to the reaction of vinyl epoxide **102** with isocyanates. However, treatment of epoxide **102** under previously used conditions of Pd$_2$(dba)$_3$·CHCl$_3$ and triisopropyl phosphite gave no reaction. A number of variations involving changes of ligands and palladium gave no reaction, rearranged products or complex reaction mixtures. Addition of camphorsulfonic acid to promote ionization and stabilize the intermediate **103** by protonation led to the desired capture with ring closure at O rather than N giving the imino carbonate **105** in modest yield. Assuming that the *N*-protonated **103** was the cyclizing species, a Lewis acid that might preferentially coordinate at O rather than at N was thought to be a way of getting ring closure at N. The reaction conditions constituted by the presence of Pd(OAc)$_2$, the bidentate ligand **108** and trimethyltin acetate were successful and produced the desired oxazolidinone **104** (54%) accompanied only by a reduced amount of the corresponding imino carbonate **105** (19%). Reduction of **104** with DIBAL, removal of the silyl (HF) and tosyl group (Na/NH$_3$) provided (±)-valienamine (**107**) (Scheme 59) [138].

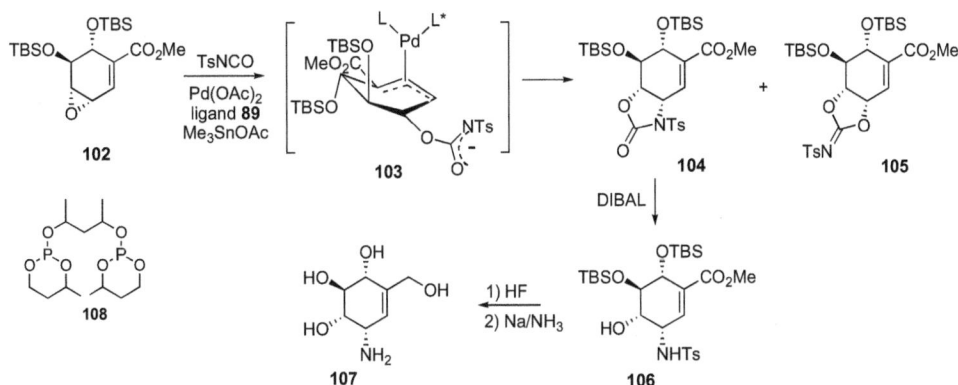

Scheme 59: Total synthesis of (±)-valienamine.

An effective, highly enanantioselective route to vinyloxazolidine derivatives was achieved in the cycloaddition of vinyl epoxides and heterocumulenes, especially carbodiimides, in the presence of Pd$_2$(dba)$_3$·CHCl$_3$ and (*S*)-TolBINAP. Under these conditions, the reaction of butadiene monoepoxide and isoprene monoepoxide with symmetrical carbodiimides **109** afforded the corresponding 4-vinyl-1,3-oxazolidin-2-imines **110** usually in high yield (92-95%) and optical purities (84-94% ee) (Scheme **60**) [139].

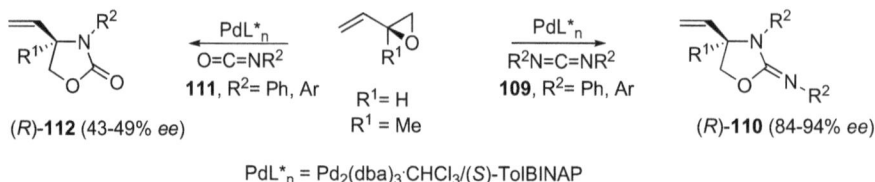

$PdL^*_n = Pd_2(dba)_3 \cdot CHCl_3/(S)\text{-TolBINAP}$

Scheme 60: Pd(0)-catalyzed asymmetric cycloaddition of vinyloxiranes with heterocumulenes.

When isocyanates **111** are used in the reaction with the same vinyloxiranes, employing a procedure identical to that used for carbodiimides, corresponding enantioenriched 4-vinyl-1,3-oxazolidin-2-ones **112** are obtained. However, the *ee* of these products is appreciabily lower (43-49%) than those derived from carbodiimides.

The high degree of asymmetric induction observed in the cycloaddition reaction of vinyl epoxides and carbodiimides is believed to proceed *via* π-allyl palladium complexes **113** and **114**, as shown in Scheme **61**. The stereodetermination step in the reaction depends on the intramolecular attack of the nitrogen nucleophile on C(3) of **113** and **114**, suggesting that the rate of interconversion between these two intermediates is much faster than the nucleophilic attack of the nitrogen nucleophile. Thus, one of the two intermediate complexes reacts faster than the other and consequently gives the major enantiomer. When (*S*)-TolBINAP is used as the chiral ligand, the intermediate **114** reacts at a greater rate affording the (*R*)-enantiomer **115** as the major stereoisomer [139].

In the reaction with vinyl epoxides, carbodiimides provide greater steric interaction between the substituent on the nucleophilic nitrogen atom and the *p*-tolyl portion of the chiral phosphine ligand, compared with isocyanates. Consequently, higher enantiomeric excesses were realized with carbodiimides than with isocyanates.

When unsymmetrical alkylarylcarbodiimides **116** were used in the reaction with 2-vinyl oxiranes, two products were always obtained, one with the alkyl group attached to the heterocyclic nitrogen (**117**, major product) and the other with the aryl group attached to the nitrogen of the oxazolidine ring (**118**, minor product) (Scheme **62**). The prevalence of **117** is due to the fact that the nucleophilicity of anionic nitrogen-containing alkyl groups is higher than that in which an aryl group is the substituent. When carbodiimides containing quite bulky alkyl substituents such as cyclohexyl and *t*-butyl are used, remarkable enantiomeric excesses (>99%) are obtained. Evidently, the bulky alkyl substituent of unsymmetrical carbodiimides may influence the steric interaction during the enantiodetermination step, thus accounting for the high enantiomeric excesses [140].

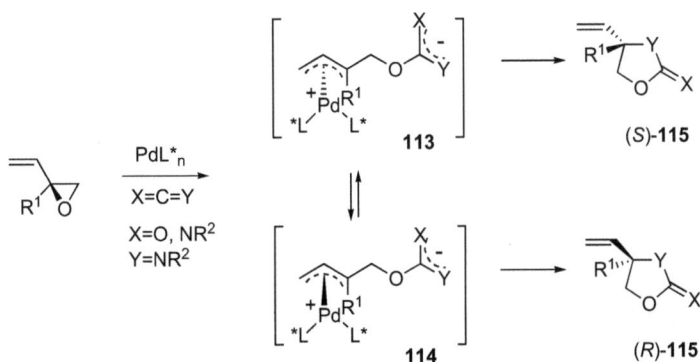

Scheme 61: Reaction pathways for the cycloaddition of heterocumulenes to vinyl oxiranes.

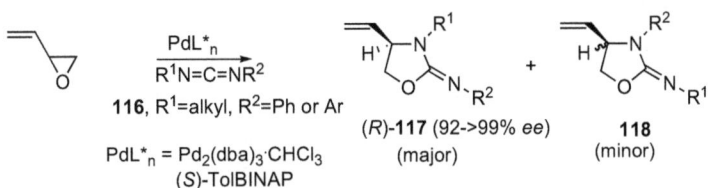

Scheme 62: Reaction of vinyloxiranes with unsymmetrical carbodiimides.

By appropriate application of the vicinal *syn* or *anti* hydroxyamination protocol, the two possible stereoisomers of types **A1** (*anti*) and **A2** (*syn*) can easily be prepared starting from a generic vinyl epoxide such as **119** (Scheme **63**). Moreover, an amino alcohol such as **A1** can be transformed into the corresponding vinyl aziridine **120** having the same configuration as the C-NH$_2$ bond of **A1**. Acid-catalyzed hydrolysis of vinyl aziridine **120** or S$_N$i rearrangement of the corresponding *N*-acetyl aziridine **121**, followed by hydrolysis of the intermediate oxazoline **122**, determines the synthesis of the two amino alcohols **B1** (*anti*) and **B2** (*syn*), regioisomers of **A1** and **A2**. Considering that a vinyl epoxide like **119** can conveniently be obtained in enantiomerically pure form from the corresponding allyl alcohol [141], a combination of all the procedures described above leads to all eight possible isomers (**A1,A2,B1,B2** + enantiomers) of a given β-amino alcohol (Scheme **63**) [142].

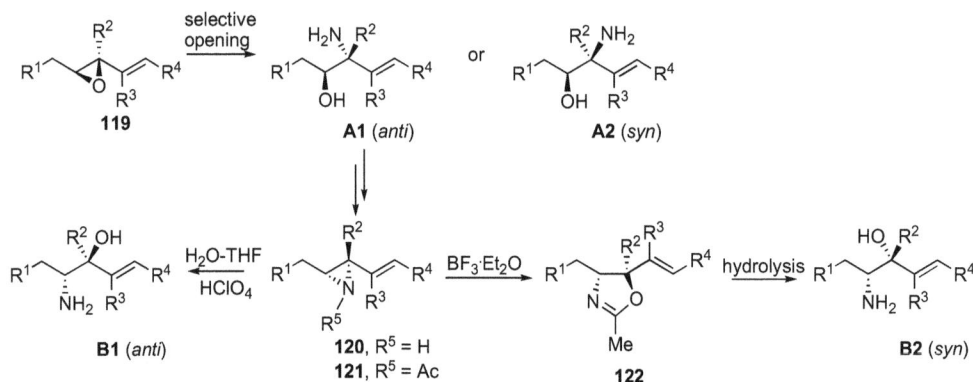

Scheme 63: Regio- and stereodivergent route to all isomers of vicinal aminoalcohols.

The concept of palladium-catalyzed asymmetric allylic alkylation (AAA) was extended to vinyl epoxides and explored in the context of a synthesis of vinylglycinol hydrochloride (**125**), a useful chiral building block, starting from racemic butadiene monoepoxide and phthalimide as the nitrogen source [143]. In a preliminary racemic version, the reaction of butadiene monoepoxide with phthalimide in the presence of a catalyst formed *in situ* from π-allylpalladium chloride dimer and PPh$_3$ as the ligand in THF afforded a 4:1 ratio of **123** (*1,2-addition product*) and **124** (*1,4-addition product*) (Scheme **64**).

When the reaction was repeated with chiral ligand **L^1**, a significantly improved regioselectivity (**123**:**124** = 16:1) and a reasonable *ee* (77%) were obtained. The modified ligand **L^2**, in which the conformational freedom of the amide carbonyl is restricted with respect to **L^1**, dramatically enhanced the reaction in terms of regioselectivity (**123**/**124**=75:1), enantioselectivity (98%), and yield (99%). All the results obtained indicated that the regioselectivity was significantly increased by the presence of the chiral ligands. Furthermore, the regioselectivity paralleled the enantioselectivity: the chiral ligand has a role not only in determining the enantioselectivity but also in directing the nucleophilic attack to the more hindered allylic oxirane carbon. The results were rationalized in accordance with the model developed by the research group, based on the minimization of interactions between the approaching nucleophile and the chiral ligand which directs the nucleophilic attack. Standard removal of the phthalimide moiety from **123** by reaction with hydrazine gave the desired vinylglycinol **125**. Of all the routes to chiral vinylglycinol described in literature, this is the shortest starting from simple commercially available compounds. The facile synthesis of (*S*)-vigabatrin, an anti-epileptic drug, and (*S,S*)-ethambuthol, a tubercolostatic drug, from chiral vinylglycinol demonstrated the power of the palladium-catalyzed approach described above (Scheme **64**) [143].

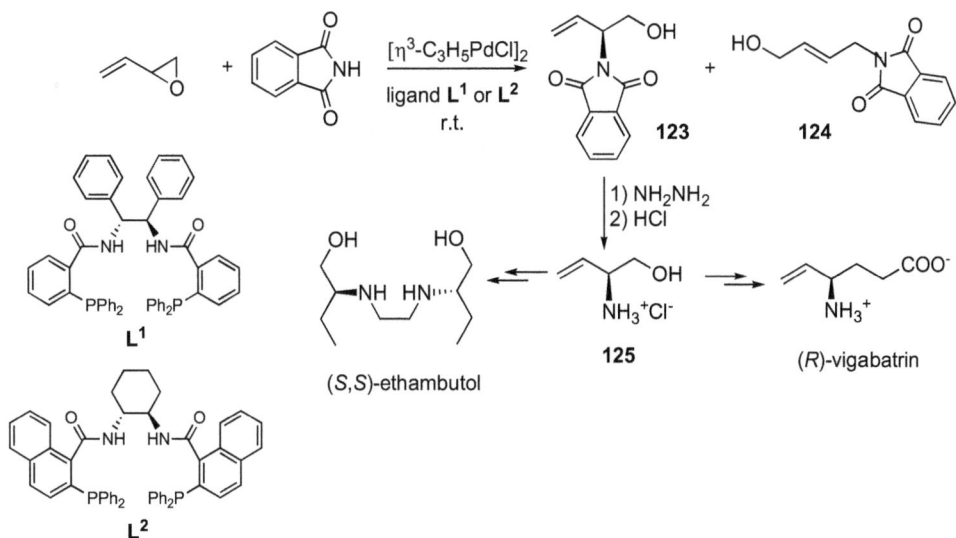

Scheme 64: Pd(0)-catalyzed addition of phthalimide to butadiene monoepoxide.

The extension of this protocol to isoprene monoepoxide demonstrated the ability of the ligands to control the regioselectivity. Actually, while the reaction of isoprene monoepoxide with phthalimide using an achiral ligand led to the *1,4-addition product* **126** (as an *E/Z* 7:1 mixture of olefin isomers), the corresponding reaction using the chiral ligand **L^1** gave a complete reversal of the regioselectivity, and produced the vicinal alkylation product **127** (83% *ee*) with only trace amounts of regioisomeric **126** (Scheme **65**). With the naphthalene-based ligand **L^2**, compound **127** was obtained in 62% isolated yield (10% of **126** present) and >90% *ee* [143].

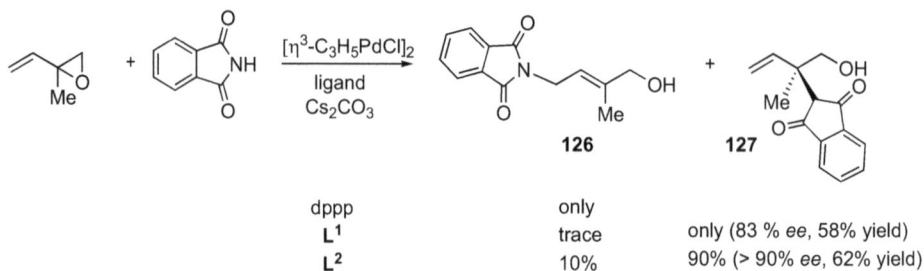

dppp	only	only (83 % *ee*, 58% yield)
L¹	trace	
L²	10%	90% (> 90% *ee*, 62% yield)

Scheme 65: Pd(0)-catalyzed addition of phthalimide to isoprene monoepoxide.

Polyhydroxylated indolizidine and pyrrolizidine alkaloids, as (-)-swainsonine **132** (Scheme **66**) and (+)-1,7-diepiaustraline **137** and (-)-7-epiaustraline **138** (Scheme **67**) are well-known potent glycosidase inhibitors [144-146].

The synthetic strategy for the synthesis of **132** involves the ring closing metathesis of diene **129b**, followed by dihydroxylation of alkene **131b**. The required diene **129b** is readily obtained from acid-catalyzed (TsOH) aminolysis with allyl amine of vinyl epoxide **128**, prepared in an enantiomerically pure form by SAE of the corresponding allyl alcohol **133** (Scheme **66**) [144].

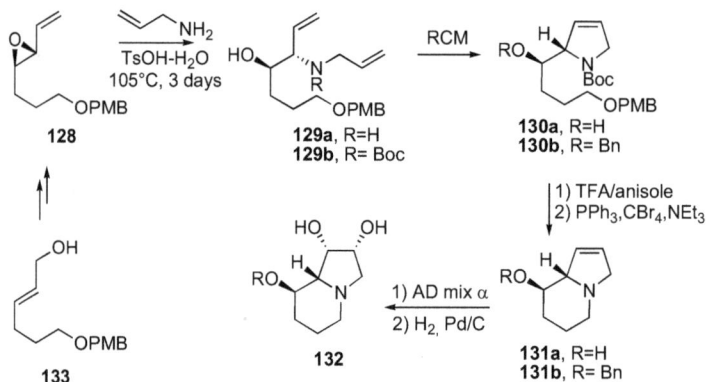

Scheme 66: Enantioselective synthesis of (-)-swainsonine.

Starting from the homologue of **128**, the synthesis of a potential glycosidase inhibitor, based on a novel 1*H*-pyrrolo[1,2-*a*]azepine structure was achieved [145].

For the synthesis of **137**, the synthetic protocol is more complex, but substantially similar to the one previously shown in Scheme **66**. In this case, the necessary starting diene **136** is prepared by LiOTf-promoted aminolysis of chiral vinyl epoxide **134** by (*S*)-allyl amine **135** (Scheme **67**) [146]. The enantiomerically pure amine **135** was prepared by Pd(0)-catalyzed enantioselective ring opening of butadiene monoepoxide by phthalimide followed by hydrolysis of the phthalimido derivative [143].

For the synthesis of the epimeric **138**, the necessary inversion of configuration at C(7) was performed through the regioselective opening of the corresponding cyclic sulphate by cesium benzoate (Scheme **67**) [146].

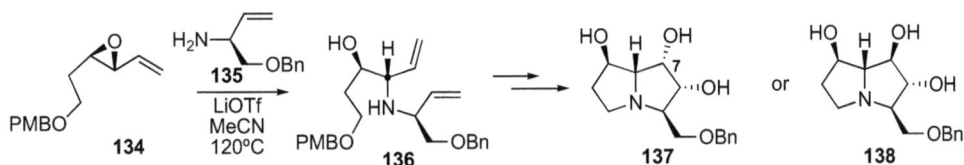

Scheme 67: Enantioselective synthesis of (+)-1,7-diepiaustraline (**137**) and (-)-7-epiaustraline (**138**).

In a completely different approach to (-)-7-epialexine (**143**), the regioselective aminolysis (NH₄OH) of vinyl epoxide **141** furnished formamide **142**, necessary for the construction of the natural product (Scheme **68**). Epoxide **141** was obtained by base-induced rearrangement of epoxy alcohol **140**, prepared by SAE of divinylcarbinol **139** [147].

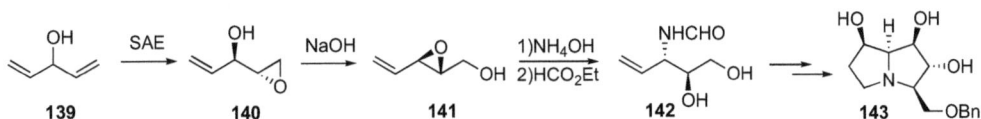

Scheme 68: Enantioselective synthesis of (-)-7-epialexine.

For the construction of the right-hand pyrrolidine core (**145**) with a quaternary centre of kaitocephalin (**146**), an antagonist of AMPA/KA and NMDA glutamate receptors, a Pd(0)-catalyzed stereoselective cyclization of oxyranylacrylate **144a** and **144b**, was envisaged (Scheme **69**). By using a catalytic amount (5 mol %) of Pd(PPh₃)₄ when the reaction was carried out with **144a** in boiling THF, the cyclized compound **145a** was obtained with 76% *ee* and a 68% yield. With **144b**, the corresponding cyclic compound **145b** was obtained with 91% *ee* and a 86% yield. The absolute configuration of **145a** and **145b** were determined to be *S* by their correlation with authentic sample, thus indicating that the Pd(0)-catalyzed cyclization took place *via* double inversion of the stereochemistry at the quaternary centre [148].

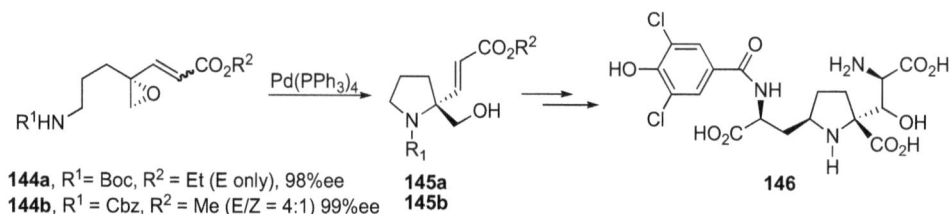

144a, R^1= Boc, R^2 = Et (E only), 98%ee   **145a**
144b, R^1 = Cbz, R^2 = Me (E/Z = 4:1) 99%ee **145b**

Scheme 69: Stereocontrolled construction of the pyrrolidine core of kaitochephalin.

In the construction of the pyrrolidine subunit of polyoxypeptin A, the Pd(0)-catalyzed cyclization of oxyranyl carbamate **147** proceeds through a double inversion from π-allyl palladium complex **A**, affording the desired cyclic compound **148** as the major isomer (90%) (Scheme 70). The isomeric compound **149**, present in a lower amount (10%) was thought to derive from π-allyl palladium complex **B**, formed by an intermolecular S$_N$2 type attack of Pd(0) on π-allyl palladium complex **A** [149].

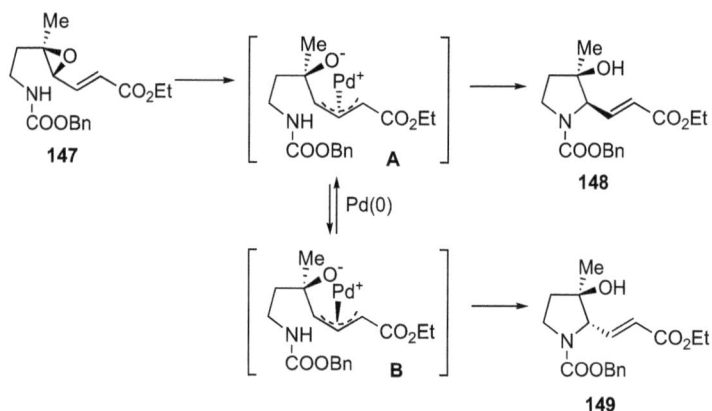

Scheme 70: Pd(0)-catalyzed cyclization of oxiranyl carbamate **147**.

A variety of amino alcohols, amino acids and quaternary amino acids, including acyclic and cyclic congeners have been prepared by a methodology which involves the palladium-catalyzed stereospecific azide substitution reaction of α,β-unsaturated γ,δ-epoxy esters with a double inversion of configuration, as the key step. This was all based on the assumption that the azide substitution in these unsaturated systems and reaction conditions might occur by a π-allyl palladium species as observed with other nucleophiles. A series of substituted *trans-* (**150**, R^2=H) and *cis*-4,5-epoxy-(2*E*)-hexenoates (**152**) were treated with TMSN$_3$ in the presence of Pd(PPh$_3$)$_4$. The azide substitution occurred at the γ position, affording the corresponding *syn* azido alcohols **151** (90% yield) and *anti* azido alcohols **153** with a complete regio- (only *1,2-addition product*) and stereoselectivity (Scheme **71**). The reaction was successful also with the trisubstituted ester **150** (R^2= Me). In all cases, the azide substitution was shown to occur stereoselectively, namely, with double inversion of configuration, regardless of the configuration and substitution pattern of the epoxide substrate [150].

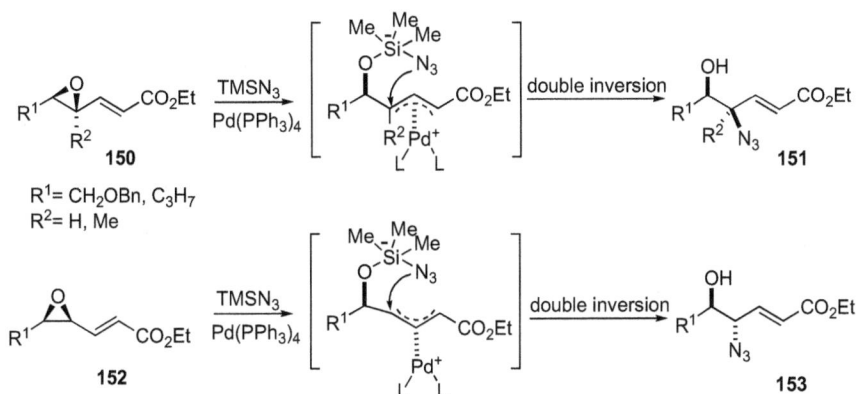

R^1= CH$_2$OBn, C$_3$H$_7$
R^2= H, Me

Scheme 71: Pd(0)-catalyzed stereospecific azidolysis of α,β-unsaturated epoxy esters.

The transfer of the same protocol [TMSN$_3$, Pd(OAc)$_2$, PPh$_3$] to corresponding cyclic systems, such as methyl 3,4-epoxy-1-cyclohexenecarboxylate (**154**), was unsatisfactory and ketone **156** was unexpectedly obtained as the major product along with a smaller amount of the desired product **155** (37%) (Scheme **72**).

Examination in detail of the effect of the phosphine ligand indicated that the reaction was dramatically improved only when P(2-furyl)$_3$ was used as the ligand. In this case, a 96:4 mixture of the desired product **155** and ketone **156** was obtained. The application of the optimized conditions to other cyclic substrates like **157** [(*E*) and (*Z*) isomers)] led to the corresponding azide substitution products **158** [150].

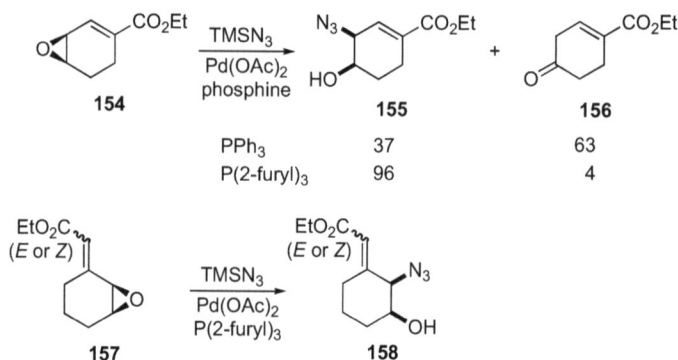

Scheme 72: Pd(0)-catalyzed stereoselective azidolysis of cyclic α,β-unsaturated epoxy esters.

To demonstrate the synthetic potential of the method, the unsaturated α,β-epoxy ester **159** was subjected to the palladium-catalyzed azide substitution to give the *syn* azido alcohol **160** exclusively (Scheme 73). Concomitant hydrogenation of the double bond and the azido group (Adams catalyst), Boc protection and oxidative cleavage of the diol moiety (RuCl$_3$/NaIO$_4$) gave (R)-N-Boc-α-methylglutamic acid γ–ethyl ester [(R)-**161**]. The corresponding enantiomer (S)-**161** could be simply obtained by applying the same two-steps sequence (hydrogenation and oxidation) to the *anti* azido alcohol **162** obtained as the only product in the azidolysis of **159** with NaN$_3$ under modified Sharpless conditions [151]. In this way, the two enantiomers (R)-**161** and (S)-**161** were efficiently and highly stereoselectively prepared starting from the same vinyl epoxide **159**, by applying two contrasting (one *syn*- and the other *anti* stereoselective) azide substitution reactions [150].

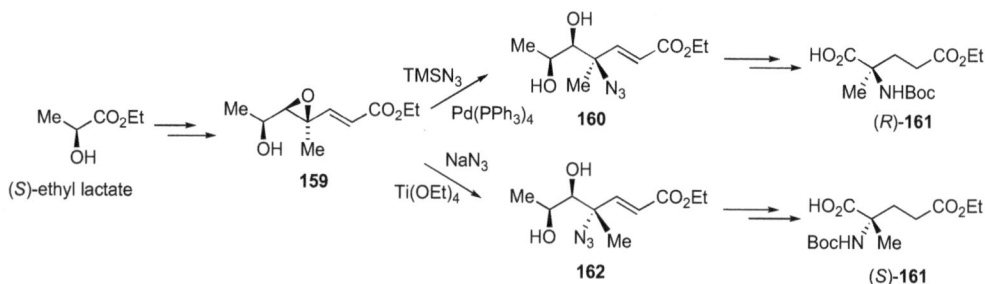

Scheme 73: Stereospecific synthesis of both enantiomers of a derivative of glutamic acid.

A simple and mild C-2 regio- (almost complete nucleophilic attack at the allylic oxirane carbon) and stereocontrolled azidolysis of vinyl epoxides of type **163** (R^1

= alkyl, Ph; R^2 = COOEt, CN, Ph, alkyl) with TMSN$_3$/BF$_3$ system has been recently reported. Only when R^1 = Ph or R^2 = alkyl, the reaction proceeds with poor or without any regioselectivity, respectively (Scheme **74**) [152].

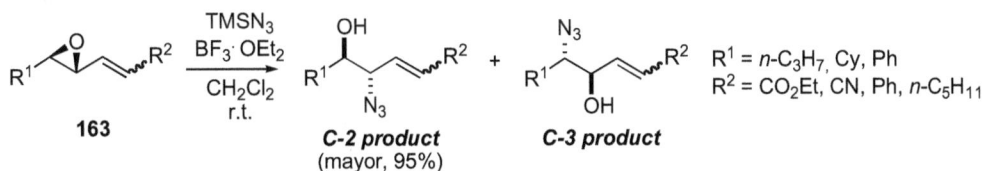

Scheme 74: Azidolysis of vinyl epoxides **163** with TMSN$_3$ in the presence of BF$_3$·OEt$_2$.

An apparent 1,2-*syn* opening process of cyclohexadiene monoepoxide with a chiral-auxiliary version of the Burgess reagent (**164**) has provided a diastereoisomeric pair of *cis*-fused cyclic sulfamidates **165a** and **165b** then transformed into a separable mixture of corresponding *trans*-amino benzoates **166a** and **166b** by treatment with sodium benzoate. Subsequently, **166a** and **166b** were converted to (-)- and (+)-balanol, respectively (Scheme **75**) [153].

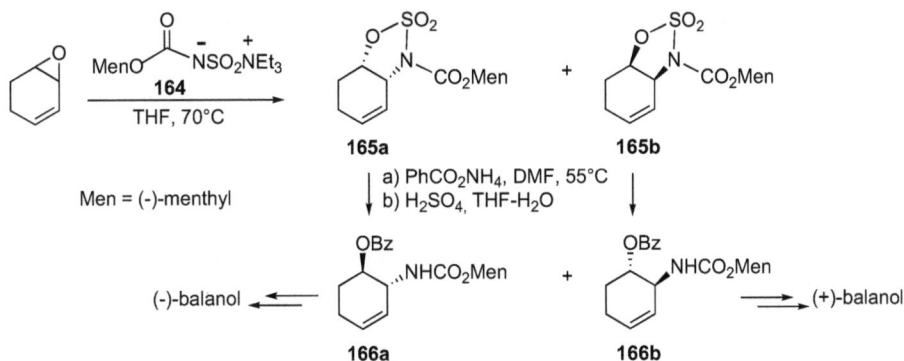

Scheme 75: Synthesis of (-)- and (+)-balanol from cyclohexadiene monoepoxide.

Contrary to expectations based on the results obtained with alcohols (see Schemes **42** and **43** and related discussion), the uncatalyzed addition reaction of epoxides **12α**, **12β**, **12α-Me** and **12β-Me** with primary and secondary aliphatic amines, necessarily used as the solvent, were completely 1,2-regio- and anti-stereoselective with the exclusive formation of the corresponding *anti-1,2-addition products*: apparently, no *1,4-addition products* were formed (Scheme **76**). Actually, appropriate control reactions carried out at different reaction times (30 s, 1.5 and 30 min) clearly indicated that the aminolysis reactions of epoxides

12α, **12β**, **12α-Me** and **12β-Me** are under thermodynamic control and the final *anti-1,2-addition product*, the only reaction product in each case obtained, is the consequence of an isomerization process by the corresponding *1,4-addition product* (apparently a mixture of α- and β-anomers, ¹H NMR), the primary (kinetic) reaction product. Unfortunately, due to their rapid isomerization and instability under any chromatographic conditions attempted, the *1,4-addition products* could not be isolated [154].

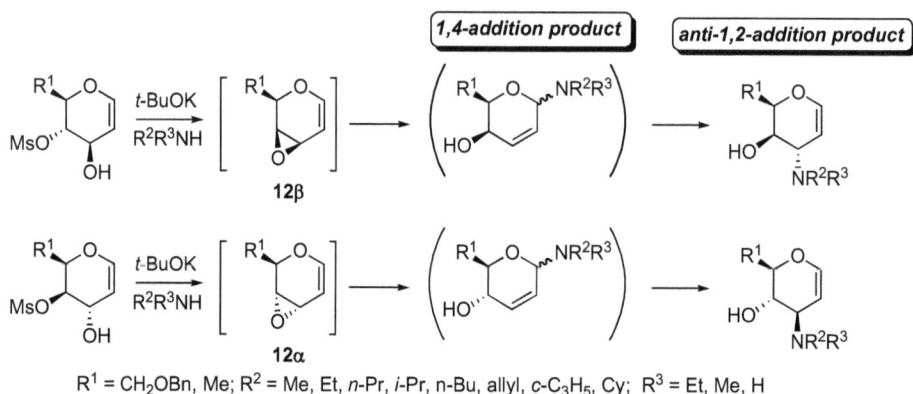

R¹ = CH₂OBn, Me; R² = Me, Et, *n*-Pr, *i*-Pr, n-Bu, allyl, *c*-C₃H₅, Cy; R³ = Et, Me, H

Scheme 76: Regio- and stereoselectivity of the addition reactions of primary and secondary aliphatic amines to glycal-derived vinyl epoxides **12α**, **12β**, **12α-Me** and **12β-Me**.

Vinyl epoxides *carba*-**12α** and *carba*-**12β** reacted with diethylamine, used as the solvent, only in the presence of a Lewis acid (LA) such as Sc(OTf)₃ (0.1-0.2 equiv) with the exclusive formation of the corresponding *anti-1,2-addition product*. Even if the regio- and stereoselectivity observed is the same, the corresponding reactivity is decidedly different: epoxide *carba*-**12β** completely reacted in 30 h, whereas for the complete conversion of epoxide *carba*-**12α** extremely long reaction times (10 days) were necessary (Scheme **77**) [52].

The azidolysis of the same epoxides carried out with NaN₃ (5 equiv)/NH₄Cl in MeOH/H₂O turned out not to be regioselective, but completely *anti* stereoselective. Actually, 55:45 and 8:2 mixtures of the corresponding *anti-1,2-* and *anti-1,4-addition product* were obtained from *carba*-**12α** and *carba*-**12β**, respectively (Scheme **78**). When the azidolysis was carried out with TMSN₃, necessarily in the presence of Sc(OTf)₃, a decidedly different result was obtained from the two epoxides: epoxide *carba*-**12β** led to a reaction mixture mostly consisting of the corresponding *anti-1,2-addition product*, whereas epoxide *carba-12α*, in accordance with its lower reactivity, led only to a *non-addition product*, the unsaturated ketone **167** (Scheme **78**). The different stereoelectronic factors present in the *trans*-diaxial opening of the oxirane ring of these epoxides were admitted responsible for the different reactivity observed in the aminolysis and in the azidolysis by $TMSN_3$ and for the larger amount of the corresponding *1,4-addition product* found in the azidolysis (NaN_3/NH_4Cl) of epoxide *carba-12α* (45%) [52].

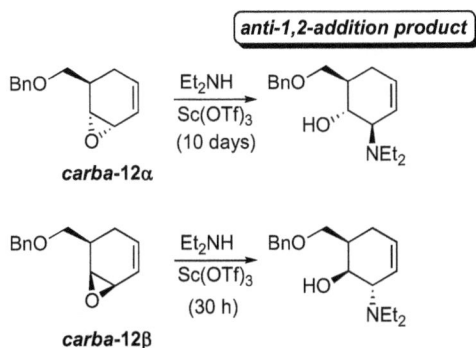

Scheme 77: Aminolysis of epoxides *carba-12α* and *carba-12β* with diethylamine.

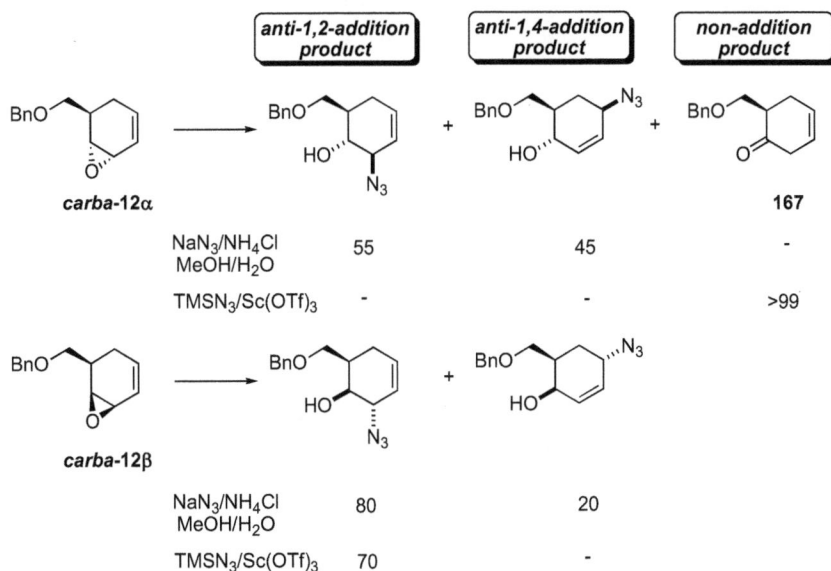

	anti-1,2-addition product	anti-1,4-addition product	non-addition product
carba-12α			**167**
NaN_3/NH_4Cl MeOH/H_2O	55	45	-
$TMSN_3/Sc(OTf)_3$	-	-	>99
carba-12β			
NaN_3/NH_4Cl MeOH/H_2O	80	20	
$TMSN_3/Sc(OTf)_3$	70	-	

Scheme 78: Azidolysis of epoxides *carba-12α* and *carba-12β* by NaN_3/NH_4Cl and $TMSN_3/Sc(OTf)_3$ protocols.

The complete absence of any *syn-1,4-* and *syn-1,2-addition product* in the azidolysis, particularly by TMSN$_3$, of epoxides ***carba*-12α** and ***carba*-12β** further indicates how different is the chemical behavior of these epoxides with respect to the corresponding, structurally strictly related, glycal-derived epoxides **12α** and **12β** (see subsequent Schemes **83** and **84**, and related discussion).

4.2. 1,4-Addition of Nitrogen Nucleophiles

By examining the reaction of the linear epoxide **168** with pyrrolidine [74] and the cyclic epoxide **169** with indole [75], Tsuji and Trost have independently shown that the reaction of these vinyl epoxides with amines, in the presence of a palladium catalyst [Pd(PPh$_3$)$_4$], led to the corresponding *1,4-addition products* [**170** from **168** and **171** from **169**], with regioselective attack of the amine on the terminal vinyl carbon (Scheme **79**). The reaction proceeds through a π-allyl complex which, as an alcoholate, behaves as a base by abstracting a proton. Subsequent nucleophilic attack affords the *1,4-addition product*.

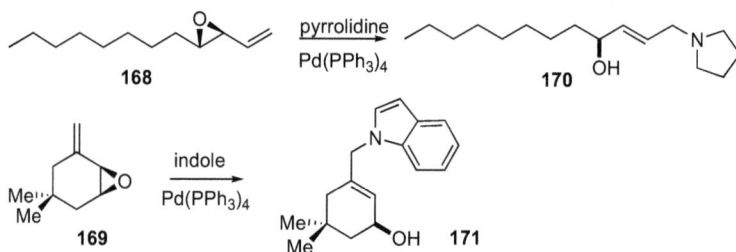

Scheme 79: Pd(0)-catalyzed 1,4-addition of amines to 1,3-diene monoepoxides.

A similar regioselective behaviour, leading to the exclusive formation of the corresponding *1,4-addition product* (1,4-azido alcohol), was found with a series of linear vinyl epoxides when azide ion was used as the nucleophile. Considering that numerous procedures are available in order to convert azides into amines, this protocol provides a new easy access to 4-amino-alcohols from vinyl epoxides [155].

When the Pd(0)-catalysis was applied to the addition reaction of 2-pyrimidinone to acyclic and cyclic vinyl epoxides, it appeared that the choice of the epoxide might play a role. Actually, exclusive 1,2-attack was achieved with butadiene monoepoxide, giving **172**, and exclusive 1,4-attack onto the cyclopentadiene monoepoxide gave the addition product **173** (Scheme **80**) [156].

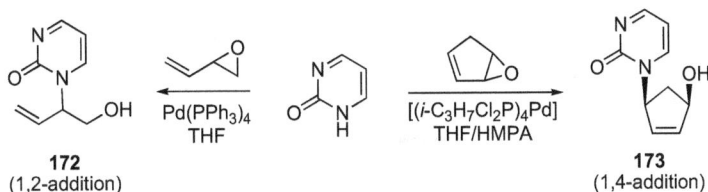

Scheme 80: Palladium-catalyzed addition of 2-pyrimidinone to vinyl epoxides.

The latter result illustrated a short and convenient synthetic pathway to the carbanucleoside natural product (±)-aristeromycin (**175**). Using [(*i*-PrO)$_3$P]$_4$Pd, generated *in situ*, with a 1:1 mixture of adenine and cyclopentadiene monoepoxide gave the *cis* 1,4-alkylated product **174**, as expected (Scheme **81**). The reaction proceeds with the clean addition of the purine base from the same face as the oxygen of the epoxide. Subsequent substitution of the generated allyl alcohol with retention of configuration and stereoselective dihydroxylation (basic potassium permanganate) of the residual double bond afforded the desired natural product. Other members of this class of compounds may be readily obtained by similar procedures [156].

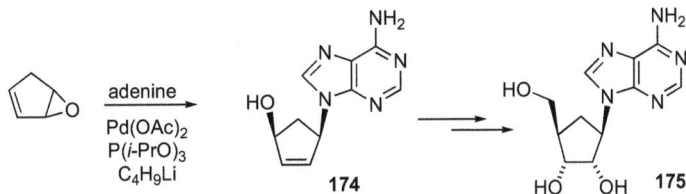

Scheme 81: A key step in the synthesis of (±)-aristeromycin.

A general method for the enantioselective preparation of 2,5-disubstituted pyrrolidine systems like **179a,b**, present in natural products of the alkaloid group, has been developed by using appropriate 4-amido-2-alken-1-ols derivatives. Pd(0)-catalyzed nucleophilic opening (1,4-addition) of *cis* vinyl epoxides **176a** [(*Z*):(*E*) isomer ratio = 90:10] by NaNHTs/NH$_2$Ts in MeCN at 40°C gave *syn* amidoalkenols **177a** with high stereo- and regioselectivity, indicating that a palladium-mediated double inversion had occurred. None of the corresponding *1,2-addition products* were obtained and no epimerization at C(5) was observed. The open compounds **177a** were subsequently cyclized to **178a** and finally elaborated to the pyrrolidine-derived (*cis* isomers) target compounds **179a**. The corresponding *anti* amidoalkenols **177b** and the related diastereoisomeric pyrrolidine derivatives (*trans* isomers) **179b** were obtained from *trans* vinyl epoxides **176b** [(Z):(E) isomer ratio, in the same range as for **176a**] *via* an analogous synthetic route (Scheme **82**) [157].

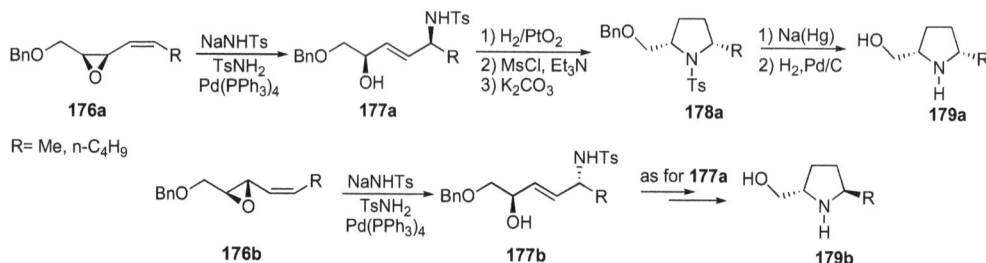

Scheme 82: Diastereoselective route to *cis*- and *trans*-2,5-disubstituted pyrrolidines.

The behaviour of glycal-derived allyl epoxides in their reaction with azide-based *N*-nucleophiles (TMSN$_3$ and TMGA) turned out to be borderline between 1,2- and 1,4-addition process. The reaction of epoxides **12α**, and **12β-Tr** and corresponding 6-deoxy derivatives **12α-Me** and **12β-Tr** with TMSN$_3$ gave reaction mixtures consisting mainly of the corresponding *cis* azido alcohol **180** and **181** (*syn-1,2-addition product*), accompanied by the corresponding glycosyl azide **182α-OTMS** and **183β-OTMS** (*1,4-addition product*), having the same configuration as the starting epoxide (Scheme **83**) [42].

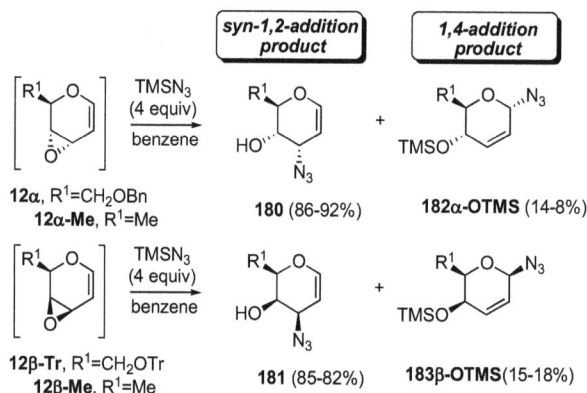

Scheme 83: Regio- and stereoselectivity of the reaction of glycal-derived allyl epoxides with TMSN$_3$.

Appropriate control experiments indicated the existence of a suprafacial [3,3]-sigmatropic rearrangement process of the glycosyl azides **182α-OTMS** and **183β-**

OTMS (the primary reaction product, formed by a coordination of the epoxide with the TMS group of the reagent) into the configurationally related *syn-1,2-addition product*, the corresponding *cis* azido alcohols **180** and **181** (the main reaction product), as shown in Scheme **84** only for β epoxides.

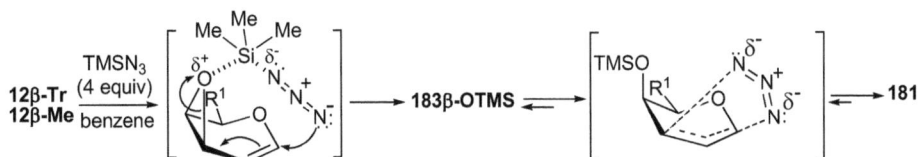

Scheme 84: Formation and suprafacial [3,3] sigmatropic rearrangement of glycosyl azides into *cis* azido alcohols.

The reaction of epoxides **12β**, **12β-Tr** and **12β-Me** with tetramethyl guanidinium azide (TMGA) are completely regioselective, with exclusive formation of the corresponding *trans* azido alcohol **184** (*anti-1,2-addition product*), by nucleophile attack at the C(3) allylic oxirane carbon [40, 42]. This is not the case with the diastereoisomeric epoxides **12α** and **12α-Me** in which the corresponding D-glucal-derived *trans* azido alcohol **185** (*anti-1,2-addition product*, 79-84%) is unexpectedly accompanied by a small, but significant, amount (almost 15-20%) of the corresponding *syn-1,2-addition product*, the *cis* azido alcohol **180** which was demonstrated deriving from an isomerization process of the corresponding α-glycosyl azide **182α** (Scheme **85**). The formation of **182α** was attributed to the

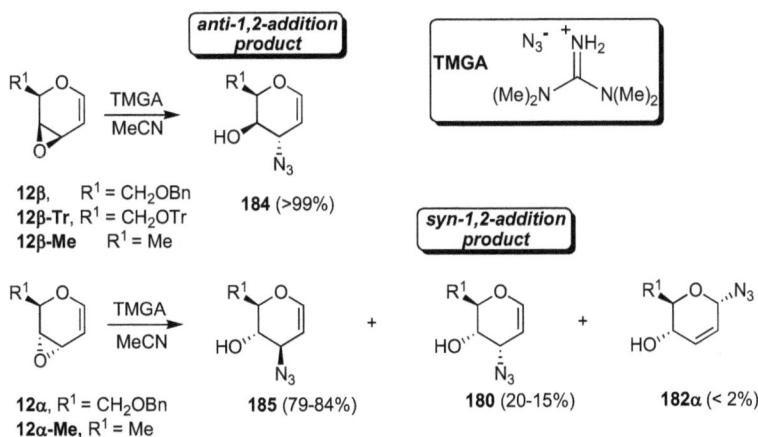

Scheme 85: Azidolysis of glycal-derived allyl epoxides by TMGA.

ability, even at a limited extent, of TMGA to coordination. When the azidolysis reaction of epoxides **12α** and **12α-Me** was repeated with the non-coordinating tetrabutylammonium azide (TBAN₃), the corresponding *trans* azido alcohol **185** was the only reaction product [40, 42].

The results obtained have indicated that 3-deoxy-3-azido glycals with a well-defined relative (*cis/trans*) and absolute configuration can be regio- and stereoselectively obtained depending on the azide-based nucleophile (TMSN₃, TMGA or TBAN₃) and the configuration (α or β) of the epoxide [42].

5. HALIDE IONS

The ring opening of vinyl epoxides with halide ions has been studied quite extensively, especially in view of the versatility of the products obtained. One of the first examples of regioselective synthesis of *O*-protected vicinal chlorohydrins under extremely mild conditions was reported by Andrews *et al.* [158]. The reaction of butadiene monoepoxide with chlorotrimethylsilane in the presence of triphenylphosphine as a nucleophilic catalyst at a low temperature (-50 °C), allows the uncommon regioselective formation of primary chloride **186** in 99 % yield (Scheme **86**).

Scheme 86: Regioselective opening of butadiene monoepoxide with TMSCl.

A 1,2-regio- and stereoselective ring opening of vinyl epoxides with a lithium cuprate halide source was employed by Taylor for the preparation of dichloroconduritol, a compound endowed with potential antiviral activity as glucosidase inhibitor [159]. When the *anti*-benzene dioxide (**187**) is subjected to reaction with dilithium tetrachlorocuprate (Li₂CuCl₄) in THF at r.t., a double ring opening occurs, giving the desired double *trans* chlorohydrin (**188**) in 65% yield (Scheme **87**).

In 1999, Llebaria and co-workers reported a formal protection of *N*-Boc vinyl epoxides *cis*-**189** and *trans*-**189** *via* stereospecific formation of the corresponding chlorohydrins (Scheme **88**) [160]. After protection of the hydroxyl functionality as silyl ether and deprotection of the *N*-Boc group under acidic conditions, followed by tosylation and hydroxyl deprotection, the intermediate new chlorohydrins underwent basic cyclization to generate the parent *N*-tosylamino substituted epoxides *cis*-**190** and *trans*-**190** which were obtained as single products in an overall retention process, resulting from a double inversion at the epoxide ring opening and ring closure reactions.

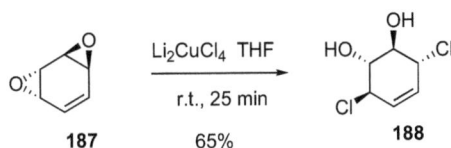

Scheme 87: Reaction of *anti*-benzene dioxide with Li$_2$CuCl$_4$.

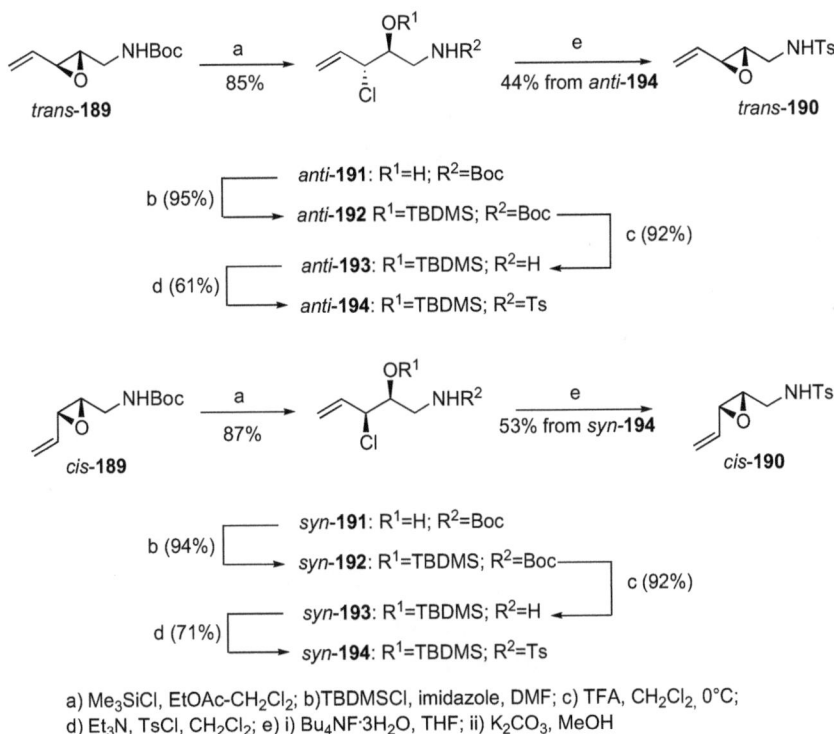

a) Me$_3$SiCl, EtOAc-CH$_2$Cl$_2$; b)TBDMSCl, imidazole, DMF; c) TFA, CH$_2$Cl$_2$, 0°C;
d) Et$_3$N, TsCl, CH$_2$Cl$_2$; e) i) Bu$_4$NF·3H$_2$O, THF; ii) K$_2$CO$_3$, MeOH

Scheme 88: Formal protection of vinyl epoxides *via* stereospecific chlorohydrin formation.

The key step in this synthetic transformation is the regio- and stereospecific ring opening of the *N*-Boc substituted oxiranes *cis*-**189** and *trans*-**189** by treatment with TMSCl in EtOAc/CH$_2$Cl$_2$ under mild conditions (promoters or catalysts are not required), giving the chlorohydrins *syn*-**191** and *anti*-**191** as single products, respectively, in high yields.

An interesting deoxygenation process with LiI/Amberlyst 15 in acetone at room temperature transforming vinyl epoxides **195** into *trans,trans*-dienes **196** has been reported by Righi (Scheme **89**) [161]. This deoxygenation of epoxides conjugated with *sp*^{2} systems, makes it possibile to use the oxirane ring as a protecting group for conjugated double bonds and thus control the geometry of the olefins obtained.

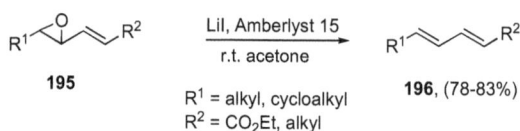

195

LiI, Amberlyst 15
r.t. acetone

R^1 = alkyl, cycloalkyl
R^2 = CO$_2$Et, alkyl

196, (78-83%)

Scheme 89: Deoxygenation process of vinyl epoxides.

Substrates with various R^1 and R^2 substituent were prepared to investigate the influence of the steric hindrance of R^1 on the oxirane ring and the electronic effects of R^2 on the double bond. When R^2 was an ester functionality, the deoxygenation reaction worked very well, affording the corresponding *trans*, *trans*-diene in high yield, independently of the steric hindrance of R^1, whereas when R^2= alkyl, a complex mixture of products was obtained.

As reported in previous works on α,β-epoxy ketone substrates [162, 163], the first step of the reaction is a regioselective nucleophilic attack of the halide on the oxirane carbon α to the *sp*^{2} carbon, and the second step is the elimination of HOI in a non-concerted process. The driving force in the elimination step is the production of a conjugated system.

When vinyl epoxides **197** were treated with LiBr/Amberlyst 15, a regio- and stereoselective ring opening occurred, affording the corresponding *anti*-α-bromo-β-hydroxy unsaturated derivatives **198** (Scheme **90**).

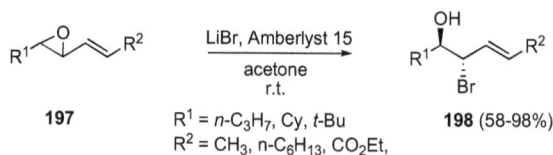

197

LiBr, Amberlyst 15
acetone
r.t.

R^1 = *n*-C$_3$H$_7$, Cy, *t*-Bu
R^2 = CH$_3$, *n*-C$_6$H$_{13}$, CO$_2$Et,

198 (58-98%)

Scheme 90: Reaction of vinyl epoxides with LiBr/Amberlyst **15**.

The regio- and stereoselectivity of the oxirane ring opening did not depend on the size of R^1. The known high reactivity of the allylic position might be responsible for the complete regioselectivity observed, particularly when an electron-withdrawing substituent on the double bond makes this position more electron-deficient.

LiBr in AcOH was used by Reymond for a very clean regio- and stereoselective conversion of the allylic epoxide **199** into allylic bromohydrin **200**, a key intermediate for the synthesis of aminocyclopentitol (**201**), an analog of α-L-fucose (Scheme **91**) [164]. Acetic acid presumably acts both as a favorable solvent for the reaction, and as a weak acid which activates the epoxide for nucleophilic bromide ion induced ring opening by nucleophilic attack. Interestingly, the acetonide protecting group was stable under these conditions.

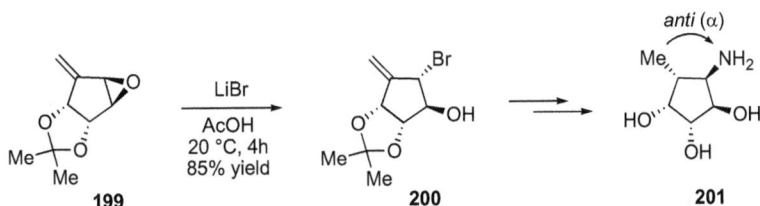

Scheme 91: Preparation of a key intermediate for the synthesis of aminocyclopentitol.

A stereoselective synthesis of silylated polyunsaturated halides **205a-c** was developed by Babudri starting from (*E, E*)-1,4-bis(trimethylsilyl)-1,3-butadiene (**202**) in which the ring opening of the α,β-epoxysilane **203** by lithium halides in the presence of AcOH is the key step (Scheme **92**) [165]. The reaction is completely regioselective, and only α ring opening has been observed, with formation of the corresponding halohydrins **204a-c**, with the halogen on the carbon bearing the trimethylsilyl group, in high yields. The reaction is also highly stereoselective and the halohydrins **204a-c** are the main diastereoisomers.

In particular, the stereoselectivity of the reaction depends on the nature of the halide ions (the stereoselectivity follows the order Cl⁻ >Br⁻ >I⁻). Due to the low reactivity of LiCl (the reactivity of lithium halides is in the order LiI>LiBr>LiCl), the reaction requires the use of a significant excess of halide. Finally, the reaction of halohydrins **204a-c** with $BF_3 \cdot Et_2O$ in CH_2Cl_2 at 0°C leads to the desired compounds **205a-c** with a (Z, E)-configuration and high isomeric purity, by an *anti* β-elimination process.

Scheme 92: Ring opening of an unsaturated α,β-epoxysilane by lithium halides/AcOH.

In order to produce bromohydrins in a regioselective manner, introduction of a π system adjacent to the epoxide was generally employed. In 1986, Corey reported a stereoselective synthesis of 5,6(*S,S*)-epoxy-15(*S*)-hydroxy-7(*E*),9(*E*),11(*Z*),19(*F*)-eicosatetraenoic acid (**208**), a possible biosynthetic precursor of lipoxins [166]. In this synthesis, the key intermediate, bromohydrin **207**, was obtained with a 93% yield by regioselective ring opening of vinyl epoxide **206** with HBr in CH$_2$Cl$_2$ at 0°C (Scheme **93**).

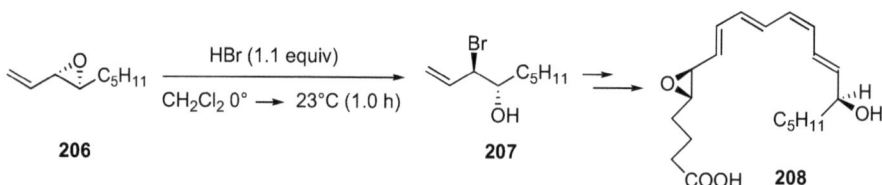

Scheme 93: Preparation of a key intermediate for the synthesis of an eicosatetraenoic acid derivative.

Subsequently, in 2001, Martín and co-workers reported a stereocontrolled synthesis of unsaturated bromohydrins **210** from unsaturated epoxides **209** using Ph$_3$P/Br$_2$ as a convenient reagent (Scheme **94**) [167].

Assuming the active species is the partially ionic adduct Ph$_3$PBr$_2$, a plausible mechanism for the epoxide ring opening would imply an electrophilic activation of the C-O bond of the epoxide by a positive phosphorus, as shown in Scheme **94**. Interestingly, chlorohydrins could also be obtained when commercially available Ph$_3$PCl$_2$ was used. The reaction is completely stereoselective for both bromine and chlorine, since the treatment of the corresponding halohydrins with TBAF or Et$_3$N yielded the starting enantiomeric epoxides.

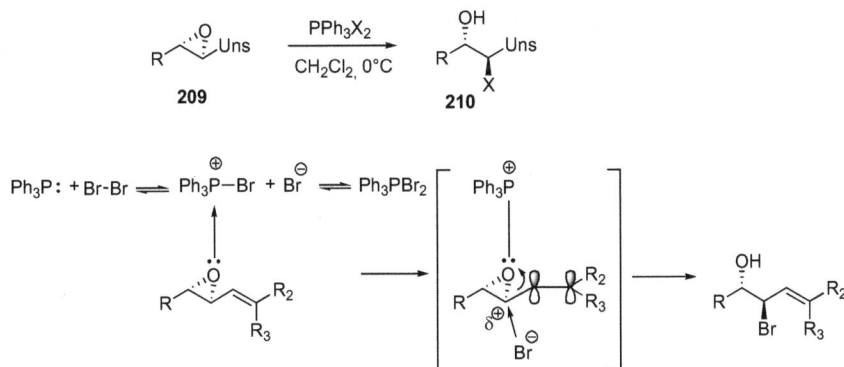

Scheme 94: Regio- and stereoselective synthesis of halohydrins from unsaturated epoxides.

An alternative method to obtain bromohydrins with high regio- and stereoselectivity, from various vinyl epoxides bearing ester groups, has been reported by Ha *et al.* using, for the first time on this substrates, MgBr$_2$ as the Lewis acid [168]. In a typical experimental procedure, vinyl epoxide **211** in CH$_3$CN was treated with MgBr$_2$ (2 equiv.) for 5h at −10 °C, to give the corresponding α-bromohydrin **212**, along with its regioisomer (β-bromohydrin) in a ratio from 15:1 to 19:1 in a nearly quantitative yield. In the presence of internal nucleophiles, such as NHCbz and hydroxyl groups at the vinyl epoxide terminus, the Lewis acid did not promote intramolecular cyclization, and only the desired ring opening occurred (Scheme **95**).

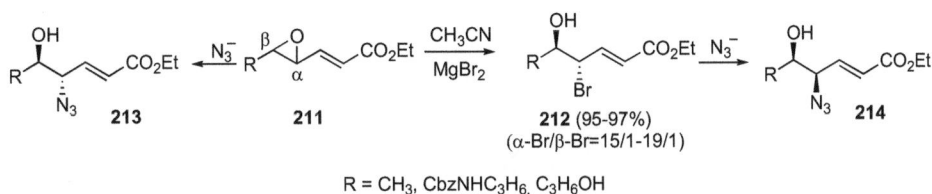

R = CH$_3$, CbzNHC$_3$H$_6$, C$_3$H$_6$OH

Scheme 95: Regio- and stereoselective synthesis of bromohydrins from vinyl epoxides.

Vinyl epoxides **211** and bromohydrins **212** were transformed into the corresponding enantiopure azido-alcohols *anti*-**213** and *syn*-**214**, respectively, which possess useful functional groups for further elaboration towards several classes of pyrrolizidine and indolizidine alkaloids (Scheme **95**).

An interesting regio- and stereoselective reaction of *gem*-difluorinated vinyl oxiranes, which are useful synthetic intermediates for difluorinated compounds, with heteronucleophiles has been developed by Kitazume [169, 170]. The reaction of epoxide **215** with MgBr$_2$·Et$_2$O in CH$_3$CN, which has a moderate donor strength, results in the formation of bromohydrin *anti*-**216** with a 76% yield of isolated pure product (Scheme **96**). In CH$_3$CN, the Lewis acid Mg^{++} ion is strong enough to activate the epoxide moiety, but it is not too strong to give rearranged products. Furthermore, in CH$_3$CN the release of the nucleophile bromide from MgBr$_2$ is efficient.

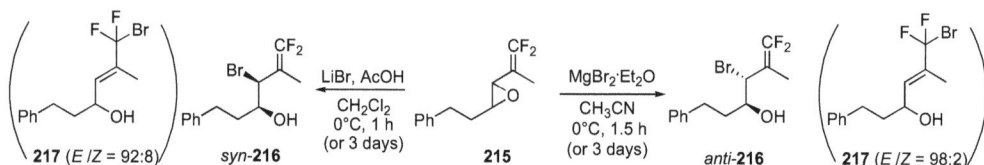

Scheme 96: Bromination of a *gem*-difluorinated vinyl oxirane.

When the mixture of **215** with MgBr$_2$·Et$_2$O was stirred for 3 days in CH$_3$CN, a further reaction occurred, giving the S$_N$2' product, the allyl alcohol **217** (89%, *E/Z* ratio = 98:2) without forming bromohydrin *anti*-**216**. This observation implies that bromohydrin *anti*-**216** is the kinetically favored product and the (*E*)-allylic alcohol **217** is the thermodynamically favored product. The (*E*)-allylic alcohol **217** is formed selectively by way of a further bromination to the vic-bromohydrin *anti*-**216** in an S$_N$2' halogenation process extremely rare on vinyl oxirane substrates.

The authors also investigated the bromination of epoxide **215** with LiBr in the presence of AcOH. When the reaction was carried out in CH$_2$Cl$_2$ with a 3:2 LiBr/AcOH mixture at 0°C for 1h, the bromohydrin *syn*-**216** was obtained in excellent yield (95%) accompanied by a small amount (5%) of allyl alcohol **217** (*E/Z* ratio = 56:44) (Scheme **96**). Interestingly, the lithium cation was crucial for this reaction, as both NaBr and KBr in the place of LiBr resulted in no reaction. As in the case with MgBr$_2$·Et$_2$O, also using LiBr/AcOH for a longer reaction time (3 days) or at higher temperature (100°C), the (*E*)-isomer of allylic alcohol **217** was produced, selectively (82 and 91%, *E/Z* ratio = 92:8 and 98:2, respectively) [169, 170].

A plausible reaction mechanism of these selective brominations is depicted in Scheme **97**. Usually, oxirane-opening halogenations take place in an S$_N$2 manner with inversion of configuration. In the case of the reaction of epoxide **215** with MgBr$_2$·Et$_2$O, the bromohydrin *anti*-**216** is probably produced through the intermediate **Int-A** by way of an S$_N$2 type reaction. However, in the case of the reaction of **215** with LiBr/AcOH, the results clearly indicated that under these conditions, the acid determines the activation of the epoxide moiety, producing a partially cationic intermediate like **Int-B**. Considering that non-fluorinated epoxides usually yield only *anti*-halohydrins in reactions using LiX/AcOH, whereas the *syn* isomer was formed selectively in this case, the results obtained were rationalized by admitting that the supposed intermediate **Int-B**, strongly stabilized by the difluoropropylene moiety, undergoes an internal delivery of the nucleophile at the same face to the starting epoxide to give a bromination process, in a retention manner, which produces the bromohydrin *syn*-**216** (Scheme **97**) [170].

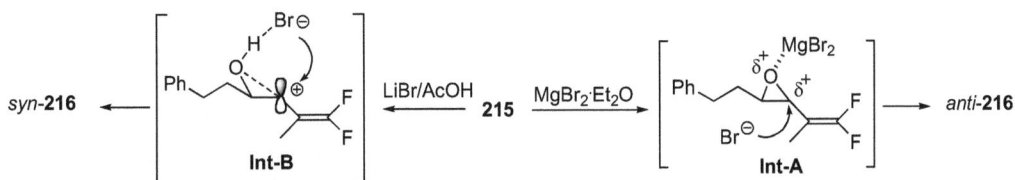

Scheme 97: Plausibile reaction mechanism of the selective bromination of epoxide **215**.

Also the selective chlorination of *gem*-difluorinated vinyl epoxide **215** was performed and the stability of **215** under Brønsted acidic conditions was examined [170].

Only a few other examples of S$_N$2' halogenation processes on vinyl epoxides have been reported. The first oxidative chlorination of a vinyl epoxide *via* a S$_N$2' addition was described by Re in 1976 [171]. In this paper isoprene monoepoxide was treated with cupric chloride in the presence of lithium chloride to give as the major product (80% yield) (*E*)–but-2-enal (**219**), *via* cupric alcoholate **218** (Scheme **98**). The α,β-unsaturated chloride **219** is the direct precursor of acetate **220**, a key intermediate in the Pommer industrial synthesis of vitamin A acetate from β-ionone.

CuCl$_2$ and LiCl have been employed as efficient reagents to convert isoprene monoepoxide into **219** by Koo in 1999 [172] and later in 2002 by Rohdich [173]. Treatment of isoprene monoepoxide with TiCl$_4$ at −80°C, made it possible to synthesize (*E*)-4-chloro-2-methylbut-2-en-1ol (**221**) (Scheme **98**) [173].

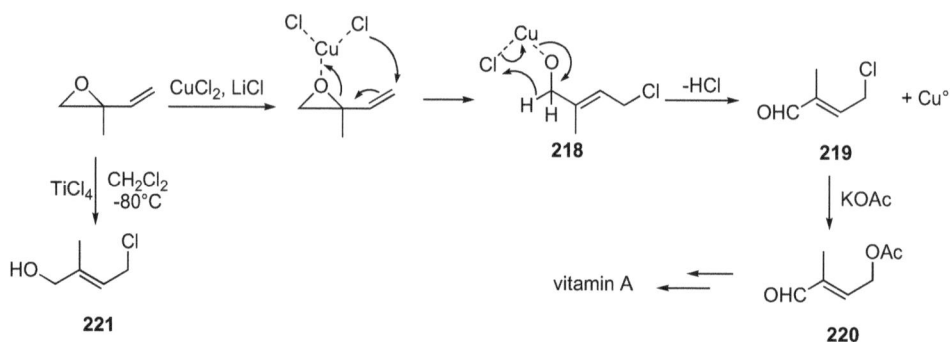

Scheme 98: S_N2'-Chlorination of isoprene monoepoxide with $CuCl_2/LiCl$ and $TiCl_4$.

An interesting transformation of the allylic oxirane group of **222** stereoselectively into the allylic halides **223** and **224** has been achieved by a proper choice of the Lewis acid activator (Scheme **99**) [174, 175]. Specifically, the treatment of a mixture of **222** and LiI (5 equiv) with Sc(OTf)$_3$ (1 equiv, hydrate) in THF (-78→-25°C) gave the corresponding (*E*)-allylic iodide **223** selectively, whereas addition of 9-bromo-9-borabicyclo[3.3.1]nonane to **222** (THF, 0°C; H$_2$O$_2$, H$_2$O, MeOH) selectively furnished the (*Z*)-allylic bromide **224** in 64% yield, presumably by intramolecular bromide delivery to the s-*cis* epoxide conformer. Allylic iodide **223** turned out to be an important intermediate in the synthesis of the fungal metabolite terpestacin that inhibits the formation of syncytia, multinucleated cell bodies that are part of the pathology of AIDS infection.

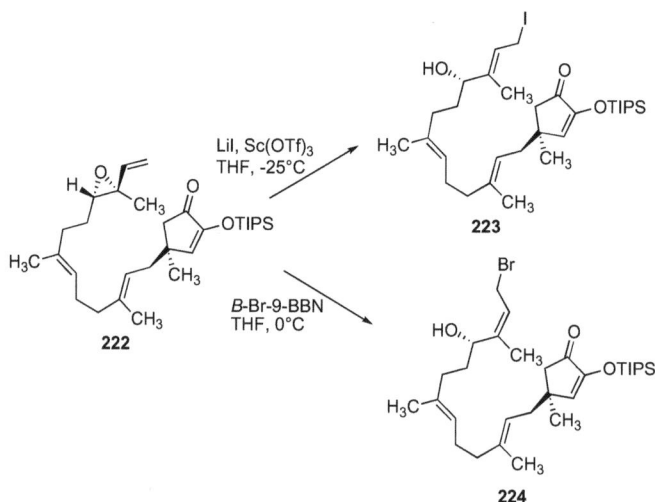

Scheme 99: (*E*)- and (*Z*)-allylic halides from a vinyl epoxide.

6. SULFUR NUCLEOPHILES

Among the many types of nucleophiles that have been used in the ring opening of vinyl epoxides, sulfur nucleophiles have so far received little attention; usually *1,2-addition products* have been observed in these reactions. In one of the first studies, Corey [176] reports that the reaction of epoxy tetraene **225** in methanol with a variety of sulfhydryl compounds (2-3 equiv) and triethylamine (3-4 equiv) at 25°C results in good yields of the product of S_N2 displacement by nucleophilic sulfur at allylic C(6) oxirane carbon forming derivatives of structure **226** (Scheme **100**).

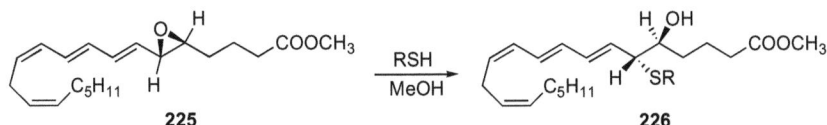

Scheme 100: Reaction of epoxy tetraene **225** with sulfhydryl compounds.

An interesting thiolysis of vinyl epoxide **227a** with PhSH, in the presence of $Yb(OTf)_3$ as a catalyst, has been reported as the key step for the synthesis of enantiomerically pure phenylthio conduritol F (**230**) (Scheme **101**) [177].

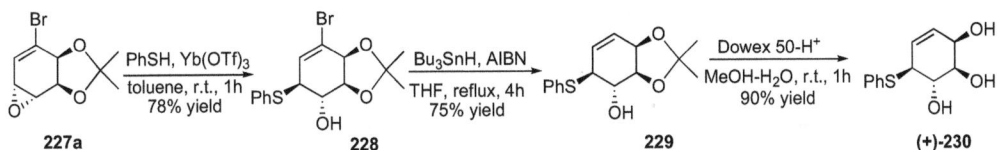

Scheme 101: Synthesis of phenylthio conduritol F.

With a view to an efficient preparation of optically pure vinyl episulfides, Bellomo and Gonzalez reported the ring opening of vinyl epoxides **227a** and **227b** by thiocyanates in CH_3CN at room temperature without any use of a catalyst to synthesize previously unknown β-hydroxy thiocyanates **231a** and **231b**. Then the hydroxy thiocyanates isolated were converted in good yields into the corresponding episulfides **232a** and **232b** by treatment with K_2CO_3 and 18-crown-6 ether in CH_2Cl_2. The direct oxirane-episulfide conversion of epoxides **227a** and **227b**, by exposing these compounds to KSCN in the presence of 18-crown-6, is feasible and the desired compounds (**232a** and **232b**, respectively) are obtained in one-pot than in a stepwise procedure, but in a slightly lower yield (Scheme **102**) [178].

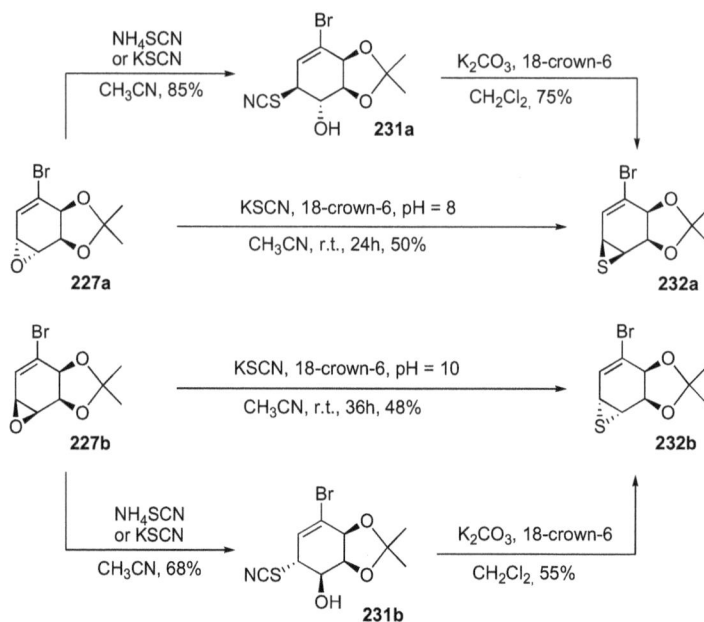

Scheme 102: Synthesis of enantiomerically pure vinyl episulfides.

A rare stereoselective 1,4-addition of dialkylaluminum thiolate to a vinyl oxirane has been reported by Takaya [179]. In this paper isoprene monoepoxide was allowed to react with 2 equiv. of diethylaluminum benzenethiolate at room temperature in benzene affording the (Z)-allylic alcohol **233** and its (E)-isomer **234** with a 92% combined yield in a ratio of 98:2. The strong stereoselectivity observed in the (Z)-allylic alcohol **233** suggests the occurrence of a cyclic transition state such as **235** (Scheme **103**).

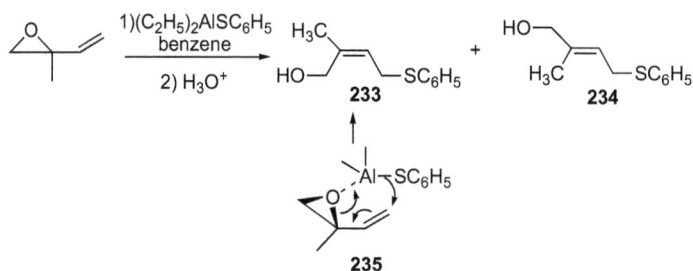

Scheme 103: 1,4-Addition of diethylaluminum benzenethiolate to isoprene monoepoxide.

A palladium-catalyzed ring opening of 1,3-cyclohexadiene monoepoxide and vinyl epoxide **236** with trimethylsilyl sulfur nucleophiles drammatically dir.ects the 1,4 regioselectivity of the process: with epoxide **236**, the allyl sulfide **237** (*1,4-addition product*) is the only reaction product, whereas in the case of 1,3-cyclohexadiene monopepoxide the corresponding *1,4-addition product*, the allyl sulfide **238** is accompanied by a certain amount of the corresponding *anti-1,2-addition product*, the *trans*-phenylthiocyclohexenol **239**. The presence of the *1,2-addition product* in the reaction of 1,3-cyclohexadiene monoepoxide is attributed to a non-palladium catalyzed reaction which cannot be completely suppressed in this case. However, this example shows the complementarity of the palladium- (1,4-addition) and non-palladium-catalyzed reactions (1,2-addition) (Scheme **104**) [180].

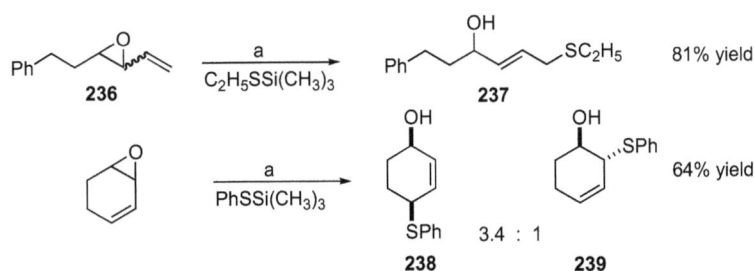

a) Pd₂(dba)₃·CHCl₃ [5 mol % Pd(0)], 30 mol% triisopropyl phosphite, THF

Scheme 104; Palladium-catalyzed ring opening of vinyl epoxides with trimethylsilyl sulfur nucleophiles.

7. MISCELLANEOUS METHODS

7.1. Insertion of Sylenes into Allylic Epoxides

An interesting synthesis of vinyl-1,2-silaoxetanes, useful to obtain valuable functionalized compounds, has been recently reported by Woerpel and coworkers [181, 182]. This process involves stereospecific metal-catalyzed silylene insertions into the carbon-oxygen bonds of vinyl epoxides, as spiro-epoxide **240**, to form reactive vinyl silaoxetane intermediates of type **241** that can undergo nucleophilic diastereoselective additions with a variety of aldehydes to give *trans*-dioxasilacyclooctanes such as **242** and **243** (Scheme **105**).

The strained allylic silanes intermediate **241** displays an enhanced nucleophilicity in the additions to unactivated aldehydes, serving as both the Lewis acid and the nucleophilic allyl moiety. The *anti*-Bredt olefins so obtained (**242** and **243**), having *trans* double bonds in bridged eight-membered rings, were oxidized diastereoselectively on the external face of the cyclic double bond with various electrophilic oxidizing agents (es. MCPBA, OsO$_4$).

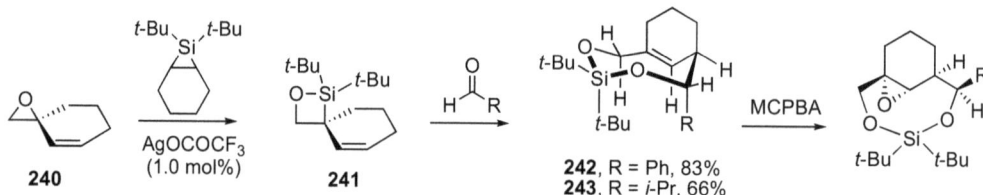

Scheme 105: *anti*-Bredt olefines from vinyl spiro-epoxide **240**.

7.2. Borylative Ring Opening

A mild ring-opening reaction of vinyl epoxides with bis(pinacolate)diboron (B$_2$pin$_2$) catalyzed by Ni(0)-Binap was reported by Pineschi *et al.* [183]. The reaction affords new functionalized allylic boron derivatives, such as **244**, which can be directly treated with aldehydes to give the corresponding *trans-threo* 1,3-diol **245** with very high diastereoselectivity and high yield (Scheme **106**).

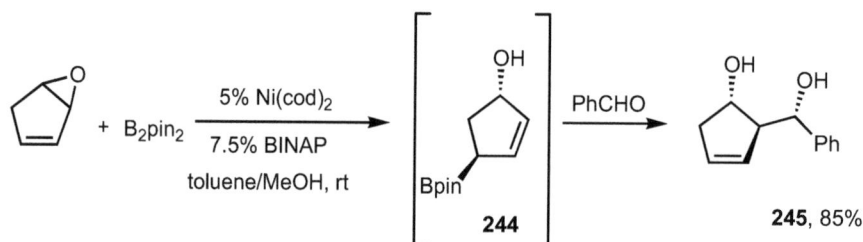

Scheme 106: Nickel-catalyzed borylative ring opening-allylation of cyclopentadiene monoepoxide.

In general, the obtained hydroxyl-containing allylboronate, such as **244**, showed to be not stable enough to be isolated. Very recently, Tortosa showed that 1,4-hydroxyboronate intermediates can be isolated after transformation into the corresponding 1,4-silyloxyboronate with common silylating agent such as triethylchlorosilane [184]. Moreover, the same author reported a S$_N$2'-regio- and *anti*-stereoselective copper-catalyzed boration of allylic epoxides to give 1,4-diols after a mild oxidation with H$_2$O$_2$ at 0 °C in the presence of KHCO$_3$ [183]. In this way it was possibile to achieve a diastereoselective synthesis of aliphatic *anti*- and *syn*-1,4-diols starting from aliphatic (*E*)- or (*Z*)-allylic epoxides (Scheme **107**).

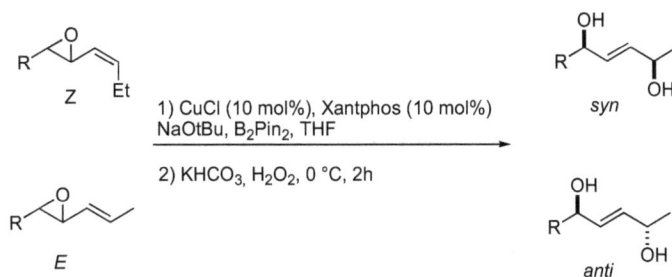

Scheme 107: Copper-catalyzed diastereoselective synthesis of *syn-* and *anti-*1,4-diols from allylic epoxides.

ACKNOWLEDGEMENTS

The authors gratefully acknowledge funding by the MIUR (PRIN 2008) (Roma) and the University of Pisa.

CONFLICT OF INTEREST

The author(s) confirm that this chapter content has no conflict of interest.

DISCLOSURE

The chapter submitted for series eBook titled **"Advances in Organic Synthesis, Volume 5"**is an update of our article published in **CURRENT ORGANIC SYNTHESIS, Volume 6, Number 3, August Issue 2009,** with additional text and references.

REFERENCES

[1]   Olofsson, B.; Somfai, P. In *Aziridines and Epoxides in Organic Synthesis*; Yudin, A. K., Ed.; Wiley-VCH: Weinheim, Germany, 2006; pp 315-347.

[2]   Batory, L. A.; McInnis, C. E.; Njardarson, J. T. Copper-catalyzed rearrangement of vinyl oxiranes. *J. Am. Chem. Soc.* **2006**, *128*, 16054-16055.

[3]   Lautens, M.; Tayama, E.; Nguyen, D. Direct vinylogous Mannich-type reactions *via* ring opening and rearrangement of vinyloxiranes. *Org.Lett.* **2004**, *6*, 345-347.

[4]   Jaime, C.; Ortuño, R. M.; Font, J. Interpretation of conjugated oxiranes behavior toward nucleophiles. *J. Org. Chem.* **1988**, *53*, 139-141.

[5]   Anderson, R. J. Additions of organocopper reagents to allylic epoxides. *J. Am. Chem. Soc.* **1970**, *92*, 4978-4979.

[6]   Rose, C. B.; Taylor, S. K. Reactions of organometallic reagents with unsaturated epoxides. II. Control of product ratios. *J. Org. Chem.* **1974**, *39*, 578-581.

[7]   Herr, R. W.; Johnson, C. R. Comparison of the reactions of methylmagnesium, methyllithium, and methylcopper reagents with 1,2-epoxybutane and 3,4-epoxy-1-butene. *J. Am. Chem. Soc.* **1970**, *92*, 4979-4981.

[8]   Bloodworth, A. J.; Curtis, R. J.; Spencer, M. D. T.; Tallant, N. A. Oxymetalation. Part 24. Preparation of cyclic peroxides by cycloperoxymercuriation of unsaturated hydroperoxides. *Tetrahedron* **1993**, *49*, 2729-2750.

[9]   Söderberg, B. C.; Austin, L. R.; Davis, C. A.; Nyström, J. E.; Vagborg, J. O. Regioselective palladium-catalyzed allylation of fulvenes. *Tetrahedron* **1994**, *50*, 61-76.

[10]   Staroscik, J.; Rickborn, B. Reaction of 1,3- and 1,4-cyclohexadiene monoepoxides with methyl organometallic reagents. *J. Am. Chem. Soc.* **1971**, *93*, 3046-3047.

[11]   Wieland, D. M.; Johnson, C. R. Additions of organometallic reagents to 3,4-epoxycyclohexene. *J. Am. Chem. Soc.* **1971**, *93*, 3047-3049.

[12]   Stork, G.; Kowalski, C. ; Garcia, G. Route to prostaglandins *via* a general synthesis of 4-hydroxycyclopentenones. *J. Am. Chem. Soc.* **1975**, *97*, 3258-3260.

[13]   Crosby, G. A.; Stephenson, R. A. Solvent mediated reactions of diethylhex-1-ynylaluminium with 3,4-epoxycyclopentene. *J. Chem. Soc. Chem. Comm.* **1975**, 287-288.

[14]   Briggs, A. J.; Walker, K. A. M. Unexpected cis-openings of cyclopentadiene monoepoxide with lithium acetylides and dialkylalkynylalanes. *J. Org. Chem.* **1990**, *55*, 2962-2963.

[15]   Abe, N.; Hanawa, H.; Maruoka, K.; Sasaki, M.; Miyashita, M. Highly efficient alkylation of epoxides with R_3Al/H_2O systems based on the double activation of epoxy oxygens. *Tetrahedron Lett.* **1999**, *40*, 5369-5372.

[16]   Eshelby, J. J.; Parsons, P. J.; Crowley, P. J. Organozinc reagents in synthesis: the facile generation of 2-(trialkysilyl)prop-2-enylzinc from 2-bromo-1-trimethylsilylprop-2-ene. *J. Chem. Soc. Perkin 1* **1996**, 191-199.

[17]   Bernard, N.; Chemla, F.; Normant, J. F. Lewis acid-mediated ring opening of propargylic epoxides: a stereospecific synthesis of 1,2-disubstituted homopropargylic alcohols. *Tetrahedron Lett.* **1998**, *39*, 6715-6718.

[18]   Alexakis, A.; Vrancken, E.; Mangeney, P. Effect of BF_3-Et_2O reagent on the base-promoted rearrangements of epoxides attached to eight-membered rings. *J. Chem. Soc, Perkin 1* **2000**, 3354-3355.

[19]   Equey, O.; Vrancken, E.; Alexakis, A. Regioselective S_N2 opening of vinylic epoxides with trialkylzincates and trialkylaluminates. *Eur. J. Org. Chem.* **2004**, 2151-2159.

[20]   El-Awa, A.; du Jourdin, X. M.; Fuchs, P. L. Asymmetric Synthesis of All Eight Seven-Carbon Dipropionate Stereotetrads. *J. Am. Chem. Soc.* **2007**, *129*, 9086-9093.

[21]   Naruta, Y.; Maruyama,, K. Highly regioselective addition of allylstannanes to vinyl epoxides by Lewis acid mediation. *Chem. Lett.* **1987**, 963-966.

[22]   Zaidlewicz, M.; Krzeminski, M. P. Syntheses with Organoboranes. XI. Allylboration of Vinylic Epoxides with Allylic Dialkylboranes. *Org. Lett.* **2000**, *2*, 3897-3899.

[23]   Xue, S.; Li, Y.; Han, K.; Yin, W.; Wang, M.; Guo, Q. Addition of Organozinc Species to Cyclic 1,3-Diene Monoepoxides. *Org. Lett.* **2002**, *4*, 905-907.

[24]   For a review, see: Marshall, J. A. S_N2' Additions of organocopper reagents to vinyloxiranes. *Chem. Rev.* **1989**, *89*, 1503-1511.

[25]   Marino, J. P.; Floyd, D. M. Regioselective reactions of organocuprates with cycloalkene epoxides. *Tetrahedron Lett.* **1979**, *20*, 675-678.

[26]   Marino, J. P.; Hatanaka, N. Stereospecific and regiospecific methodology for the synthesis of chiral molecules. *J. Org. Chem.* **1979**, *44*, 4467-4468.

[27]   Marino, J. P.; Abe, H. The Regiospecific Synthesis of Substituted Cyclohex-2-enols from α,β-Epoxycyclohexanones. *Synthesis* **1980**, 872-873.

[28]   Marino, J. P.; Jaen, J. C. Stereospecific umpolung α' substitution of ketones *via* reactions of organocuprates with enol ethers of α,β-epoxycyclohexanones. *J. Am. Chem. Soc.* **1982**, *104*, 3165-3172.

[29]   Marino, J. P.; Fernandez de la Pradilla, R.; Laborde, E. Regio- and stereoselectivity of the reaction between cyanocuprates and cyclopentene epoxides. Application to the total synthesis of prostaglandins. *J. Org. Chem.* **1987**, *52*, 4898-4913.

[30]   Lipshutz, B. H.; Woo, K.; Gross, T.; Buzard, D. J.; Tirado, R. Alkylations of Functionalized Organozinc Reagents with Allylic Epoxides *Catalyzed* by A Cyanocuprate. *Synlett* **1997**, 477-478.

[31]   Deslogchamps, P. *Stereoelectronic Effects in Organic Chemistry;* Pergamon Press: Oxford, **1983**, pp. 174-178.

[32]   Corey, E.J.; Boaz, N.W. d-Orbital stereoelectronic control of the stereochemistry of S_N2' displacements by organocuprate reagents. *Tetrahedron Lett.* **1984**, *25*, 3063-3066.

[33]   Goering, H. L.; Kantner, S. S. Alkylation of allylic derivatives. 8. Regio- and stereochemistry of alkylation of allylic carboxylates with lithium methylcyanocuprate. *J. Org. Chem.* **1984**, *49*, 422-426.

[34]   Marino, J. P.; Anna, L. J.; Fernández de la Pradilla, R.; Martínez, M. V.; Montero, C.; Viso, A. Sulfoxide-controlled S(N)2' displacements between cyanocuprates and epoxy vinyl sulfoxides. *J. Org. Chem.* **2000**, *65*, 6462-6473.

[35]   Dieter, R. K.; Huang, Y.; Guo, F. Regio- and stereoselectivity in the reactions of organometallic reagents with an electron-deficient and an electron-rich vinyloxirane: applications for sequential bis-allylic substitution reactions in the generation of vicinal stereogenic centers. *J. Org. Chem.* **2012**, *77*, 4949-4967.

[36]   Ueki, H.; Chiba, T.; Yamazaki, T.; Kitazume, T. Preparation and regioselective S_N2' reaction of novel gem-difluorinated vinyloxiranes with RLi. *J. Org. Chem.* **2004**, *69*, 7616-7627.

[37]   Smith III, A. B.; Pitram, S. M.; Gaunt, M. J.; Kozmin, S. A. Dithiane additions to vinyl epoxides: steric control over the S_N2 and S_N2' manifolds. *J. Am. Chem. Soc.* **2002**, *124*, 14516-14517.

[38]   Smith III, A. B.; Foley, M. A.; Dong, S.; Orbin, A. (+)-Rimocidin synthetic studies: construction of the C(1-27) aglycone skeleton. *J. Org. Chem.* **2009**, *74*, 5987-6001.

[39]   Di Bussolo, V.; Caselli, M.; Pineschi, M.; Crotti, P. New stereoselective beta-C-glycosidation by uncatalyzed 1,4-addition of organolithium reagents to a glycal-derived vinyl oxirane. *Org. Lett.* **2003**, *5*, 2173-2176.

[40]   Di Bussolo, V.; Caselli, M.; Romano, M. R.; Pineschi, M.; Crotti, P. Regio- and stereoselectivity of the addition of O-, S-, N-, and C-nucleophiles to the beta vinyl oxirane derived from D-glucal. *J. Org. Chem.* **2004**, *69*, 8702-8708.

[41]   Di Bussolo, V.; Caselli, M., Romano, M. R.; Pineschi, M.; Crotti, P. Stereospecific uncatalyzed alpha-O-glycosylation and alpha-C-glycosidation by means of a new D-Gulal-derived alpha vinyl oxirane. *J. Org. Chem.* **2004**, *69*, 7383-7386.

[42]   Di Bussolo, V.; Favero, L.; Romano, M. R.; Pineschi, M.; Crotti, P. Synthesis of diastereoisomeric 6-deoxy-D-allal- and 6-deoxy-D-galactal-derived allyl epoxides and examination of the regio- and stereoselectivity in nucleophilic addition reactions. Comparison with the corresponding 6-O-functionalized allyl epoxides. *Tetrahedron* **2008**, *64*, 8188-8201.

[43]   Marié, J.-C.; Courillon, C.; Malacria, M. Compared Behaviors of *trans*- and *cis*-α,β-Epoxy-γ,δ-vinyl-silanes Towards Nucleophiles and Bases: High Regioselective Ring Opening and Deprotonation. *Synlett* **2002**, 553-556.

[44]   Marié, J.-C.; Courillon, C.; Malacria, M. SN2' reactions between lithiated carbon nucleophiles and silylated vinyloxiranes. Effects of salts and solvents on the stereocontrol. *Eur. J. Org. Chem.* **2006**, 463-470.

[45]   Bertozzi, F.; Crotti, P.; Del Moro, F.; Di Bussolo, V.; Macchia, F.; Pineschi, M Copper-Catalyzed Addition of Organometallic Reagents to Vinyl Diepoxides. A Novel Entry to Oxabridged Systems and to Substituted Allylic Alcohols. *Eur. J. Org. Chem.* **2003**, 1264-1270.

[46]   For a review, see: Pineschi, M. Asymmetric Ring Opening of Epoxides and Aziridines with Carbon Nucleophiles *Eur. J. Org. Chem.* **2006**, 4979-4988.

[47]   Feringa, B. L.; Pineschi, M.; Arnold, L. A.; Imbos, R.; de Vries, A. H. M. Highly Enantioselective Catalytic Conjugate Addition and Tandem Conjugate Addition-Aldol Reactions of Organozinc Reagents. *Angew. Chem., Int. Ed.* **1997**, *36*, 2620-2623.

[48]   Bertozzi, F.; Crotti, P.; Feringa, B. L.; Macchia, F.; Pineschi, M. A Multigram, Catalytic and Enantioselective Synthesis of Optically Active 4-Methyl-2-cyclohexen-1-one: a Useful Chiral Building Block. *Synthesis* **2001**, 483-486.

[49]   Pineschi, M.; Del Moro, F.; Crotti, P.; Di Bussolo, V.; Macchia, F. Copper-Catalyzed Highly Enantioselective Synthesis of Cyclic Allylic and Homoallylic Alcohols with Dialkylzinc Reagents. *Synthesis* **2005**, 334-337.

[50]   Bertozzi, F.; Crotti, P.; Macchia, F.; Pineschi, M.; Feringa, B. L. Highly Enantioselective Regiodivergent and Catalytic Parallel Kinetic Resolution. *Angew. Chem., Int. Ed.* **2001**, *40*, 930-932.

[51]   Pineschi, M.; Del Moro, F.; Crotti, P.; Di Bussolo, V.; Macchia, F. Catalytic Regiodivergent Kinetic Resolution of Allylic Epoxides: a New Entry to Allylic and Homoallylic Alcohols with High Optical Purity. *J. Org. Chem.* **2004**, *69*, 2099-2105.

[52]   Di Bussolo, V.; Frau, I.; Checchia, L.; Favero, L.; Pineschi, M.; Uccello-Barretta, G.; Balzano, F.; Roselli, G.; Renzi, G.; Crotti, P. Synthesis of carba analogs of 6-O-(benzyl)-D-allal- and -D-galactal-derived allyl epoxides and evaluation of the regio- and stereoselective behavior in nucleophilic addition reactions. *Tetrahedron* **2011**, *67*, 4696-4709.

[53]   Equey, O.; Alexakis, A. Enantioselective opening of cyclic vinyl epoxides with organoaluminum reagents catalyzed by copper salts. *Tetrahedron: Asymmetry* **2004**, *15*, 1531-1536.

[54]   Millet, R.; Alexakis, A. Copper-Catalyzed Kinetic Resolution of 1,3-Cyclohexadiene Monoepoxide with Grignard Reagents. *Synlett* **2007**, 435-438.

[55]   Millet, R.; Alexakis, A. SimplePhos as Efficient Ligand for the Copper-Catalyzed Kinetic Resolution of Cyclic Vinyloxiranes with Grignard Reagents. *Synlett* **2008**, 1797-1800.

[56]   Bertozzi, F.; Crotti, P.; Macchia, F.; Pineschi, M.; Arnold, A.; Feringa, B. L. A New Catalytic and Enantioselective Desymmetrization of Symmetrical Methylidene Cycloalkene Oxides. *Org. Lett.* **2000**, *2*, 933-936.

[57]   Bertozzi, F.; Crotti, P.; Del Moro, F.; Feringa, B. L.; Macchia, F.; Pineschi, M. Unprecedented Catalytic Enantioselective Trapping of Arene Oxides with Dialkylzinc Reagents. *Chem. Commun.* **2001**, 2606-2607.

[58]   Matsuda, T.; Sugishita, M. Reaction of cyclooctatetraene oxide with Grignard reagents. *Bull. Chem. Soc. Jpn.* **1967**, *40*, 174-177.

[59]   Del Moro, F.; Crotti, P.; Di Bussolo, V.; Macchia, F.; Pineschi, M. Catalytic Enantioselective Desymmetrization of COT-Monoepoxide. Maximum Deviation from Coplanarity for an S_N2'-Cuprate Alkylation. *Org. Lett.* **2003**, *5*, 1971-1974.

[60]   Pineschi, M.; Del Moro, F.; Crotti, P.; Macchia, F. Simple Synthetic Transformations of Highly Enantioenriched 4-Alkyl-2,5,7-Cyclooctatrienols into Functionalized Bicyclo[4.2.0]octa-2,4-dienes and 2,6-Cyclooctadienones. *Eur. J. Org. Chem.* **2004**, 4614-4620.

[61]   Clive, D. L. J.; Wickens, P. L.; da Silva, G. V. J. Preparation of a Semisynthetic Analog of Mevinolin and Compactin. *J. Org. Chem.* **1995**, *60*, 5532-5536.

[62]   Nagumo, S.; Irie, S.; Hayashi, K.; Akita, H. A formal total synthesis of xanthorrhizol based on nucleophilic opening of vinyloxirane by arylcopper reagent *Heterocycles* **1996**, *43*, 1175-1178.

[63]   Wang, Q.; Huang, Q.; Chen, B.; Lu, L.; Wang, H.; She, X.; Pan, X. Total synthesis of (+)-machaeriol D with a key regio- and stereoselective SN2' reaction. *Angew. Chem., Int. Ed.* **2006**, *45*, 3651-3653.

[64]   Marino, J. P.; Fernandez de la Pradilla, R.; Laborde, E. Direct formation of organocopper compounds by oxidative addition of zerovalent copper to organic halides. *J. Org. Chem.* **1984**, *49*, 5280-5382.

[65]   Marino, J. P.; Tucci, F.; Comasseto, J. V. An efficient preparation of functionalized Z-vinylcuprates from terminal acetylenes and their reactions with epoxides. *Synlett* **1993**, 761-763.

[66]   Larock, R. C.; Ilkka, S. Synthesis of allylic alcohols *via* organopalladium additions to unsaturated epoxides. *Tetrahedron Lett.* **1986**, *27*, 2211-2214.

[67]   Tueting, D. R.; Echavarren, A. M.; Stille, J. K. Palladium-catalyzed coupling of organostannanes with vinyl epoxides. *Tetrahedron* **1989**, *45*, 979-992.

[68]   Castaño A. M.; Méndez, M.; Ruano, M.; Echavarren, A. M. Reaction of vinyl epoxides with palladium-switchable bisnucleophiles: synthesis of carbocycles. *J. Org. Chem.* **2001**, *66*, 589-593.

[69]   Miyaura, N.; Tanabe, Y.; Suginome, H.; Suzuki, A. Cross-coupling reactions of 1-alkenylboranes with 3,4-epoxy-1-butene catalyzed by palladium or nickel complexes. *J. Organomet. Chem.* **1982**, *233*, C13-C16

[70]   Kjellgren, J.; Aydin, J.; Wallner, O. A.; Saltanova, I. V.; Szabó, K. J. Palladium pincer complex catalyzed cross-coupling of vinyl epoxides and aziridines with organoboronic acids. *Chem. Eur. J.* **2005**, *11*, 5260-5268.

[71]   Guan, Y.; Zhang, H.; Pan, C.; Wang, J.; Huang, R.; Li, Q. Flexible synthesis of montanine-like alkaloids: revisiting the structure of montabuphine. *Org. Biomol. Chem.* **2012**, *10*, 3812-3814.

[72]   Herron, J. R.; Russo, V.; Valente, E. J.; Ball, Z. T. Catalytic Organocopper Chemistry from Organosiloxane Reagents. *Chem. Eur. J.* **2009**, *15*, 8713-8716.

[73]   Jeganmohan, M.; Bhuvaneswari, S.; Cheng, C.-H. A cooperative copper- and palladium-catalyzed three-component coupling of benzynes, allylic epoxides, and terminal alkynes. *Angew. Chem. Int. Ed.* **2009**, *48*, 391-394.

[74]   Tsuji, J.; Kataoka, H.; Kobayashi, Y. Regioselective 1,4-addition of nucleophiles to 1,3-diene monoepoxides catalyzed by palladium complex. *Tetrahedron Lett.* **1981**, *22*, 2575-2578.

[75]   Trost, B. M.; Molander, G. A. Neutral alkylations *via* palladium(0) catalysis. *J. Am. Chem. Soc.* **1981**, *103*, 5969-5972.

[76]   Tsuji, J.; Yuhara, M.; Minato, M.; Yamada, H.; Sato, F.; Kobayashi, Y. Palladium-catalyzed regioselective reactions of silyl-substituted allylic carbonates and vinyl epoxide. *Tetrahedron Lett.* **1988**, *29*, 343-346.

[77]   Thies, S.; Kazmaier, U. Vinylepoxides as versatile substrates for allylations of amino acids and peptides. *Synlett* **2010**, 137-141.

[78]   For a review, see: Trost, B. M. Cyclization *via* palladium catalyzed allylic alkylation *Angew. Chem., Int. Ed.* **1989**, *28*, 1173-1192.

[79]   Fürstner, A.; Weintritt, H. Total Synthesis of Roseophilin. *J. Am. Chem. Soc.* **1998**, *120*, 2817-2825.

[80]   Kende, A. S.; Kaldor, I.; Aslanian, R. *J. Am. Chem. Soc.* **1988**, *110*, 6265.

[81]   Tsuji, J. New synthetic reactions catalyzed by palladium complexes. *Pure & Appl. Chem.* **1989**, *58*, 869-878.

[82]   Rychlet Elliott, M.; Dhimane, A.-L.; Malacria, M. Biomimetic Diastereoselective Total Synthesis of epi-Illudol *via* a Transannular Radical Cyclizations Strategy. *J. Am. Chem. Soc.* **1997**, *119*, 3427-3428.

[83]   Rychlet Elliott, M.; Dhimane, A.-L.; Malacria, M. Regio- and stereoselective palladium(0)-catalyzed alkylation of vinyloxiranes with non-stabilized lithium ester enolates nucleophiles. A direct access to highly functionalized allylic alcohols. *Tetrahedron Lett.* **1998**, *39*, 8849-8852.

[84]   Trost, B. M.; Jiang, C. Atom economic asymmetric creation of quaternary carbon: regio- and enantioselective reactions of a vinylepoxide with a carbon nucleophile. *J. Am. Chem. Soc.* **2001**, *123*, 12907-12908.

[85]   Stork, G.; Isobe, M. General approach to178prostaglandins *via* methylenecyclopentanones. †Total synthesis of178(+-)-prostaglandin F2α. *J. Am. Chem. Soc.* **1975**, *97*, 4745-4746.

[86]   Briggs, A. J.; Walker, K. A. M. Unexpected cis-openings of cyclopentadiene monoepoxide with lithium acetylides and dialkylalkynylalanes. *J. Org. Chem.* **1990**, *55*, 2962-2964.

[87]   Krause, N.; Seebach, D. The ring opening of unsymmetrical allylic, benzylic, propargylic, and silicon-substituted epoxides by titanium acetylides. A convenient access to certain 2-substituted 3-butyn-1-ols. *Chem. Ber.* **1988**, *121*, 1315-1320.

[88]   Restorp, P.; Somfai, P. Regioselective and divergent opening of vinyl epoxides with ethoxyacetylene. *Chem. Comm.* **2004**, 2086-2087.

[89]   Restorp, P.; Somfai, P. Regioselective and divergent opening of vinyl epoxides with alkyne nucleophiles. *Eur. J. Org. Chem.* **2005**, 3946-3951.

[90]   RajanBabu, T. V.; Nugent, W. A. Selective generation of free radicals from epoxides using a transition-metal radical. A powerful new tool for organic synthesis. *J. Am. Chem. Soc.,* **1994**, *116*, 986-997.

[91]   Barrero, A. F.; Quílez del Moral J. F.; Sánchez, E. M.; Arteaga, J. F. Regio- and diastereoselective reductive coupling of vinylepoxides catalyzed by titanocene chloride. *Org. Lett.,* **2006**, *8*, 669-672.

[92]   Charrier, N.; Gravestock, D.; Zard, S. Z. Radical additions of xanthates to vinyl epoxides and related derivatives: a powerful tool for the modular creation of quaternary centers. *Angew. Chem. Int. Ed.,* **2006**, *45*, 6520-6523.

[93]   Nagumo, S.; Miyoshi, I.; Akita, H.; Kawahara, N. 7-*endo* Selective Friedel-Crafts-type cyclization of vinyloxiranes linked to an ester group. *Tetrahedron Lett.,* **2002**, *43*, 2223-2226.

[94]   Nagumo, S.; Miura, T.; Mizukami, M.; Miyoshi, I.; Imai, M.; Kawahara, N.; Akita, H. Intramolecular Friedel-Crafts-type reaction of vinyloxiranes linked to an ester group. *Tetrahedron,* **2009**, *65*, 9884-9896.

[95]   Nagumo, S.; Mizukami, M.; Wada, K.; Miura, T.; Bando, H.; Kawahara, N.; Hashimoto, Y.; Miyashita, M.; Akita, H. Novel construction of hydro-2-benzazepines based on 7-*endo* selective Friedel-Crafts-type reaction of vinyloxiranes. *Terahedron Lett.* **2007**, *48*, 8558-8561.

[96]   Nagumo, S.; Ishii, Y.; Sato, G.; Mizukami, M.; Imai, M.; Kawahara, N.; Akita, H. 8-*endo* Selective Friedel-Crafts cyclization of vinyloxiranes with Co2(CO)6-complexed acetylene. *Tetrahedron Lett.*, **2009**, *50*, 26-28.

[97]   Taylor, S. K.; Clark, D. L.; Heinz, K. L.; Schramm, S. B.; Westermann, C. D.; Barnell, K. K. Friedel-Crafts reactions of some conjugated epoxides. *J. Org. Chem.*, **1983**, *48*, 592-596.

[98]   Bertolini, F.; Di Bussolo, V.; Crotti, P.; Pineschi, M. Mild and stereoselective Friedel-Crafts alkylation of phenol derivatives with vinyloxiranes: a new access to cycloalkenobenzofurans. *Synlett*, **2007**, 3011-3015.

[99]   Crotti, P.; Pineschi, M. In: *Aziridines and Epoxides in Organic Synthesis*; Yudin, A. K., Ed.; Wiley-VCH: Weinheim, Germany, **2006**; pp 271-313.

[100]   Posner, G. H.; Rogers, D., Z. Organic reactions at alumina surfaces. Mild and selective opening of arene and related oxides by weak oxygen and nitrogen nucleophile. *J. Am. Chem. Soc.*, **1977**, *99*, 8214-8218.

[101]   Boaz, N. W. The stereochemistry of solvolysis of an acyclic allylic epoxide.*Tetrahedron; Asymm.*, **1995**, *6*, 15-16.

[102]   Prestat, G.; Baylon, C.; Heck, M. P.; Mioskowski, C. Lewis acid-catalyzed regiospecific opening of vinyl epoxides by alcohols. *Tetrahedron*, **2000**, *41*, 3829-3831.

[103]   Baylon, C.; Prestat, G.; Heck, M. P.; Mioskowski, C. Synthesis of (-)-(4*R*,5*R*)-muricatacin using a regio- and stereospecific ring-opening of a vinyl epoxide. *Tetrahedron*, **2000**, *41*, 3833-3835.

[104]   Tokunaga, M.; Larrow, J. F. Kakiuchi, F; Jacobsen, E. N. Asymmetric catalysis with water: efficient kinetic resolution of terminal epoxides by means of catalytic hydrolysis. *Science*, **1997**, *277*, 936-938.

[105]   Schaus, S. E.; Brandes, B. D.; Larrow, J. F.; Tokunaga, M; Hansen, K. B.; Gould, A. E.; Furrow, M. E.; Jacobsen, E. N. Highly selective hydrolytic kinetic resolution of terminal epoxides catalyzed by chiral (salen)Co^{III} complexes. Practical synthesis of enantioenriched terminal epoxides and 1,2-diols. *J. Am. Chem. Soc.,* **2002**, *124*, 1307-1315.

[106]   Suzuki, M.; Oda, Y.; Noyori, R. Palladium(0) catalyzed reaction of 1,3-diene epoxides. A useful method for the site-specific oxygenation of 1,3-dienes. *J. Am. Chem. Soc.*, **1979**, *101*, 1623-1625.

[107]   Santon, S. A.; Felman, S. W.; Parkhurst, C. S.; Godleski, S. A. Alkoxides as nucleophiles in (π-allyl)palladium chemistry. Synthetic and mechanistic studies. *J. Am. Chem. Soc.*, **1983**, *105*, 1964-1969.

[108]   Trost, B. M,; Tenaglia, A. Tin mediated catalyzed regiocontrolled alkylation of vinyl epoxides. *Tetrahedron Lett.*, **1988**, *29*, 2931-2934.

[109]   Trost, B. M.; McEachern, E. J.; Toste, F. D. A two-component catalyst system for asymmetric allylic alkylations with alcohol pronucleophiles. *J. Am. Chem. Soc.*, **1998**, *120*, 12702-12703.

[110]   Hirai, A.; Yu, X.-Q.; Tonooka, T.; Miyashita, M. Palladium-Catalyzed Stereospecific Epoxide-Opening Reaction of γ,δ-Epoxy-α,β-Unsaturated Esters with an Alkylboronic Acid Leading to γ,δ-Vicinal Diols with Double Inversion of the Configuration. *Chem. Commun.*, **2003**, 2482-2483.

[111] Yu, X.-Q.; Yoshimura, F.; Ito, F.; Sasaki, M.; Hirai, A.; Tanino, K.; Miyashita, M. Palladium-Catalyzed Stereospecific Substitution of α,β-Unsaturated γ,δ-Epoxy Esters by Alcohols with Double Inversion of Configuration: Synthesis of 4-Alkoxy-5-hydroxy-2-pentenoates. *Angew. Chem. Int. Ed.*, **2008**, *47*, 750-754.

[112] Fagnou, K.; Lautens, M. Rhodium-catalyzed ring opening of vinyl epoxides with alcohols and aromatic amines. *Org. Lett.*, **2000**, *2*, 2319-2321.

[113] Evans, P. A.; Nelson, J. D. Conservation of Absolute Configuration in the Acyclic Rhodium-Catalyzed Allylic Alkylation Reaction: Evidence for an Enyl (σ + π) Organorhodium Intermediate. *J. Am. Chem. Soc.*, **1998**, *120*, 5581-5582.

[114] Di Bussolo, V.; Caselli, M.; Pineschi, M.; Crotti, P. New stereoselective β-glycosylation *via* vinyl oxrane derived from D-glucal. *Org. Lett.*, **2002**, *4*, 3695-3698.

[115] Di Bussolo, V.; Checchia, L.; Romano, M. R.; Pineschi, M.; Crotti, P. Stereoselective synthesis of 2,3-unsaturated 1,6-oligosaccharides by means of a glycal-derived allyl epoxide and *N*-nosyl aziridine. *Org. Lett.*, **2008**, *10*, 2493-2496.

[116] Frau, I.; Di Bussolo, V.; Favero, L.,; Pineschi, M.; Crotti, P. Stereodivergent synthesis of diastereoisomeric carba analogs of glycal-derived vinyl epoxides: a new access to carbasugars. *Chirality*, **2011**, *23*, 820-826.

[117] Davoust, M.; Cantagrel, F.; Metzner, P.; Brière, J.-F. A stereodivergent synthesis of β-hydroxy-α-methylene lactones *via* vinyl epoxides. *Org. Biomol. Chem.*, **2008**, *6*, 1981-1993.

[118] Nicolaou, K. C.; Rutjes, F. P. J. T.; Theodorakis, E. A.; Tiebes, J.; Sato, M.; Untersteller, E. Total synthesis of brevetoxin B. 2. Completion. *J. Am. Chem. Soc.*, **1995**, *117*, 1173-1174.

[119] Sasaki, M.; Inoue, M.; Takamatsu, K.; Tachibana, K. Stereocontrolled synthesis of the JKLM ring fragment of ciguatoxin. *J. Org. Chem.*, **1999**, *64*, 9399-9415.

[120] Baldwin, E. J. Rules for ring closure. *J. Chem. Soc. Chem. Commun.*, **1976**, 734-736.

[121] Nicolaou, K. C.; Prasad, C. V. C.; Somers, P. K.; Hwang, C.-K. Activation of 6-*endo* over 5-*exo* hydroxy epoxide openings. Stereoselective and ring selective synthesis of tetrahydrofuran and tetrahydropyran systems. *J. Am. Chem. Soc.*, **1989**, *111*, 5330-5334.

[122] Nicolaou, K. C.; Prasad, C. V. C.; Somers, P. K.; Hwang, C.-K. Activation of 7-*endo* over 6-*exo* epoxide openings. Synthesis of oxepane and tetrahydropyran systems. *J. Am. Chem. Soc.*, **1989**, *111*, 5335-5340.

[123] Matsukura, H.; Morimoto, M.; Koshino, H.; Nakata, T. Stereoselective synthesis of tetrahydropyran and oxepane systems by the *endo*-cyclization of hydroxy styrylepoxides. *Tetrahedron Lett.*, **1997**, *38*, 5545-5548.

[124] Trost, B. M,; Tenaglia, A. Palladium-catalyzed chemoselective cyclization to cyclic ethers. *Tetrahedron Lett.*, **1988**, *29*, 2927-2930.

[125] Suzuki, T.; Sato, O.; Hirama, M. Palladium catalyzed stereospecific cyclization of hydroxy epoxides. Stereocontrolled synthesis of cis- and trans-2-alkenyl-3-hydroxytetrahydropyrans. *Tetrahedron Lett.*, **1990**, *31*, 4747-4750.

[126] Sakaitani, M.; Ohfune, Y. Syntheses and reactions of silyl carbamates. 2. A new mode of cyclic carbamate formation from tert-butyldimethylsilyl carbamate. *J. Am. Chem. Soc.*, **1990**, *112*, 1150-1158.

[127] Ha, J. D.; Shin, E. Y.; Kang, S. K.; Ahn, J. H.; Choi, J-K. Studies of rhodium-catalyzed ring opening of vinyl epoxides. *Tetrahedron Lett.*, **2004**, *45*, 4193-4195.

[128] Pineschi, M.; Bertolini, F.; Haak, R. M.; Crotti, P.; Macchia, F. Mild metal-free syn-stereoselective ring opening of activated epoxides and aziridines with aryl borates. *Chem. Commun.*, **2005**, 1426-1428.

[129] Bertolini, F.; Di Bussolo, V.; Crotti, P.; Pineschi, M. Mild and stereoselective Friedel-Crafts alkylation of phenol derivatives with vinyloxiranes: a new access to cycloalkenobenzofurans. *Synlett*, **2007**, 3011-3015.

[130] Lindström, U. M.; Franckowiak, R.; Pinault, N.; Somfai, P. A stereospecific synthesis of vicinal amino alcohols by aminolysis of vinylepoxides. *Tetrahedron Lett.*, **1997**, *38*, 2027-2030.

[131] Lindström, U. M.; Somfai, P. Asymmetric synthesis of (+)-1-deoxynojirimycin. *Tetrahedron Lett.*, **1998**, *39*, 7173-7176.

[132] Lindström, U. M.; Olofsson, B.; Somfai, P. Microwave-assisted aminolysis of vinylepoxides. *Tetrahedron Lett.*, **1999**, *40*, 9273-9276.

[133] Takasu, H.; Tsuji, Y.; Sajiki, H.; Hirota, K. 3'-Selective modification of a 4', 5'-didehydro-5'-deoxy-2',3'-epoxyuridine using nucleophiles. *Tetrahedron*, **2005**, *61*, 8499-8504.

[134] Heyns, K.; Hohlweg, R. [3.3]-Sigmatropic rearrangements of glycals and pseudoglycals. *Chem. Ber.*, **1978**, *111*, 1632-1645.

[135] Paulsen, H.; Heiker, F. R. Cyclite reaction. VI. Synthesis of 7-aminovaleinamine and other branched conduritol derivatives by [3,3]   sigmatropic rearrangement. *Carbohydr. Res.*, **1982**, *102*, 83-98.

[136] Takasu, H.; Tsuji, Y.; Sajiki, H.; Hirota, K. Rearrangement of allylic azide and phenylthio groups of 3'-azido- or 3'-phenylthio-4',5'-didehydro-5'-deoxyarabinofuranosyluridines. *Tetrahedron*, **2005**, *61*, 11027-11031.

[137] Trost, B. M.; Sudhakar A. R. A cis hydroxyamination equivalent: application to the synthesis of (-)-acosamine. *J. Am. Chem. Soc.*, **1987**, *109*, 3792-3794.

[138] Trost, B. M. ; Chupak, L. S.; Lübbers, T. Total synthesis of (±)- and (+)-valienamine *via* a strategy derived from new palladium-calalyzed reactions. *J. Am. Chem. Soc.*, **1998**, *120*, 1732-1740.

[139] Larksarp, C.; Alper, H. Palladium(0)-catalyzed asymmetric cycloaddition of vinyloxiranes with heterocumulenes using chiral phosphine ligands: an effective route to highly enantioselective vinyloxazolidine derivatives. *J. Am. Chem. Soc.*, **1997**, *119*, 3709-3715.

[140] Larksarp, C.; Alper, H. Highly enantioselective synthesis of 1,3-oxazolidin-2-imine derivatives by asymmetric cycloaddition reactions of vinyloxiranes with unsymmetrical carbodiimides calalyzed by palladium(0) complexes. *J. Org. Chem.*, **1998**, *63*, 6229-6233.

[141] Díez-Martin, D.; Kotecha, N. R.; Ley, S. L.; Mantegani, S.; Menéndez, J. C.; Organ, H. M.; White, A. D.; Banks, J. B. Total synthesis of the ionophore antibiotic CP-61,405 (routiennocin). *Tetrahedron*, **1992**, *48*, 7899-7938.

[142] Olofsson, B.; Somfai, P. A regio- and stereodivergent route to all isomers of *vic*-amino alcohols. *J. Org. Chem.*, **2002**, *67*, 8574-8583.

[143] Trost, B. M.; Bunt, R. C.; Lemoine, R. C.; Calkins, T. L. Dynamic kinetic asymmetric transformation of diene monoepoxides: a practical asymmetric synthesis of vinylglycinol, vigabatrin, and ethambutol. *J. Am. Chem. Soc.*, **2000**, *122*, 5968-5976.

[144] Lindsay, K. B.; Pyne, S. G. Asymmetric synthesis of (-)-swainsonine, (+)-1,2-di-*epi*-swainsonine, and (+)-1,2,8-tri-*epi*-swainsonine. *J.Org. Chem.*, **2002**, *67*, 7774-7780. See also: Lindsay, K. B.; Tang, M.; Pyne, S. G. Diastereoselective synthesis of polyfunctional-pyrrolidines *via* vinyl epoxide aminolysis/ring-closing metathesis: synthesis of chiral 2,5-dyhydropyrroles and (1*R*,2*S*,7*R*,7a*R*)-1,2,7-trihydroxypyrrolizidine. *Synlett*, **2002**, 731-734.

[145] Lindsay, K. B.; Pyne, S. G. Synthesis of (1*R*,2*S*,9*S*,9a*R*)-octahydro-1*H*-pyrrolo-[1,2-*a*]azepine-1,2,9-triol: a potential glycosidase inhibitor. *Tetrahedron,* **2004**, *60*, 4173-4176.

[146] Tang, M.; Pyne, S. G. Asymmetric synthesis of (-)-7- epiaustraline and (+)-1,7-diepiaustraline. *J. Org. Chem.,* **2003**, *68*, 7818-7824.

[147] Romero, A.; Wong, C.-H. Chemo-enzymatic total synthesis of 3-epiaustraline, australine, and 7-epialexine. *J. Org. Chem.,* **2000**, *65*, 8264-8268.

[148] Takahashi, K.; Haraguchi, N.; Ishihara, J.; Hatakeyama, S. Synthetic studies directed toward kaitocephalin: a highly stereocontrolled route to the right-hand pyrrolidine core. *Synlett,* **2008**, 671-674

[149] Noguchi, Y.; Uchiro, H.; Yamada, T.; Kobayashi, S. Synthetic study of polyoxypeptin: stereoselective synthesis of (2*S*, 3*R*)-3-hydroxy-3-methylproline. *Tetrahedron Lett.,* **2001**, *42*, 5253-5256.

[150] Miyashita, M.; Mizutani, T.; Tadano, G.; Iwata, Y.; Miyazawa, M.; Tanino, K. Pd-Catalized stereospecific azide substitution of α,β-unsaturated γ,δ-epoxy esters with double inversion of configuration. *Angew. Chem. Int. Ed.,* **2005**, *44*, 5094-5097.

[151] Caron, M; Sharpless, K. B. Titanium isopropoxide-mediated nucleophilic openings of 2,3-epoxy alcohols. A mild procedure for regioselective ring-opening. *J. Org. Chem.,* **1985**, *50*, 1557-1560.

[152] Righi, G.; Salvati Manni, L.; Bovicelli, P.; Pelagalli, R. A new, simple, and mild azidolysis of vinylepoxides. *Tetrahedron Lett.* **2011**, *52*, 3895-3896.

[153] Gilmet, J.; Sullivan, B.; Hudlicky, T. Formal total symthesis of (-)- and (+)-balanol: two complementary enantiodivergent routes from vinyloxiranes and vinylaziridines. *Tetrahedron,* **2009**, *65*, 212-220.

[154] Di Bussolo, V.; Checchia, L.; Romano, M. R.; Favero, L.; Pineschi, M., Crotti, P. Aminolysis of glycal-derived allyl epoxides and activated aziridines. Effects of the absence of coordination processes on the regio- and stereoselectivity. *Tetrahedron,* **2010**, *66*, 689-697.

[155] Tenaglia, A.; Waegell, B. Palladium-catalyzed reaction of 1,3-diene monoepoxides with sodium azide. 1,4-Azidohydroxylation of conjugated dienes. *Tetrahedron Lett.,* **1988**, *29*, 4851-4854.

[156] Trost, B. M.; Kuo, G.-H.; Benneche, T. A transition- metal-controlled synthesis of (±)-aristeromycin and (±)-2',3'-*diepi*-aristeromycin. An unusual directive effect in hydroxylations. *J. Am. Chem. Soc.,* **1988**, *110*, 621-622.

[157] Pettersson-Fasth, H.; Riesinger, S. W.; Bäckvall, J.-E. An enantioselective route to *cis*- and *trans*-2-(hydroxymethyl)-5-alkylpyrrolidines. *J. Org. Chem.,* **1995**, *60*, 6091-6098.

[158] Andrews, G. C.; Crawford, T. C.; Contillo, L. G. Jr. Nucleophilic catalysis in the insertion of silicon halides into oxiranes: a synthesis of *O*-protected vicinal halohydrins. *Tetrahedron Lett.* **1981**, *22*, 3803-3806.

[159] Guo, Z.-X.; Haines, A.H.; K.; Taylor, R.J.K. Reaction of dilithium tetrachlorocuprate and dilithium tetrabromonickelate with unsaturated epoxides: the preparation of novel analogs of the antiviral agent, bromoconduritol. *Synlett* **1993**, 607-608.

[160] Daviu, N.; Delgado, A.; Llebaria, A. Formal protection of vinyl epoxides *via* stereospecific chlorohydrin formation. *Synlett* **1999**, 1243-1244.

[161] Antonioletti, R.; Bovicelli, P.; Fazzolari, E.; Righi, G. Stereo- and regiocontrolled transformations of vinyloxiranes with metal halides. *Tetrahedron Lett.* **2000**, *41*, 9315-9318.

[162] Righi, G.; Bovicelli, P.; Sperandio, A. An easy deoxygenation of conjugated epoxides *Tetrahedron* **2000**, *56*, 1733-1737.

[163] Righi, G.; Bovicelli, P.; Sperandio, A. A mild preparation of α-halo-α,β-enones from cyclic enones. *Tetrahedron Lett.* **1999**, *40*, 5889-5892.

[164] Blaser, A.; Reymond, J.-L. Stereoselective Synthesis of an Aminocyclopentitol Analog of α-*l*-Fucose *via* an Allylic Bromohydrin.183 *Synlett* **2000**, 817-819.

[165] Babudri, F.; Fiandanese, V.; Marchese, G.; Punzi, A. A stereoselective synthesis of silylated polyunsaturated halides from alpha,beta-epoxysilanes. *Tetrahedron* **2001**, *57*, 549-554.

[166] Corey, E.J.; Mehrotra, M.M. A stereoselective and practical synthesis of 5,6(S,S)-epoxy-15(S)-hydroxy-7(E),9(E),11(Z),13(E)-eicosatetraenoic acid, possible precursor of the lipoxins. *Tetrahedron Lett.* **1986**, *27*, 5173-5176.

[167] Díaz, D.; Martín, T.; Martín, S.V. Stereocontrolled Synthesis of Unsaturated Halohydrins from Unsaturated Epoxides. *J. Org. Chem.* **2001**, *66*, 7231-7233.

[168] Ha, J.D.; Kim, S.Y.; Lee, S.J.; Kang, S. K.; Ahn, J. H.; Kim, S.S.; Choi, J.-K. Regio- and stereoselective ring opening of vinyl epoxides with MgBr$_2$. *Tetrahedron Lett.* **2004**, *45*, 5969-5972.

[169] Ueki, H.; Kitazume, T. Highly regio- and stereocontrolled brominations of gem-difluorinated vinyloxiranes. *Tetrahedron Lett.* **2005**, *46*, 5439-5442.

[170] Ueki, H.; Kitazume, T. Regio- and Stereoselective Reactions of gem-Difluorinated Vinyloxiranes with Heteronucleophiles. *J. Org. Chem.* **2005**, *70*, 9354-9363.

[171] Eletti-Bianchi, G.;Centini, F.; Re, L. Regio- and Stereoselective Reactions of gem-Difluorinated Vinyloxiranes with Heteronucleophiles. *J. Org. Chem.* **1976**, *41*, 1648-1650.

[172] Choi H.; Ji, M.; Park, M.; Yun, I.-K.; Oh, S.-S.; Baik, W.; Koo, S. Diallylic Sulfides as Key Structures for Carotenoid Syntheses. *J. Org. Chem.* **1999**, *64*, 8051-8053.

[173] Hecht, S.; Amslinger, S.; Jauch, J.; Kis, K.; Trentinaglia, V.; Adam, P.; Eisenreich, W.; Bacher, A.; Rohdich, F. Studies on the non-mevalonate isoprenoid biosynthetic pathway. Simple methods for preparation of isotope-labeled (E)-1-hydroxy-2-methylbut-2-enyl 4-diphosphate. *Tetrahedron Lett.* **2002**, *43*, 8929-8933.

[174] Myers, A.G.; Siu, M.; Ren, F. Enantioselective Synthesis of (-)-Terpestacin and (-)-Fusaproliferin: Clarification of Optical Rotational Measurements and Absolute Configurational Assignments Establishes a Homochiral Structural Series. *J. Am. Chem. Soc.* **2002**, *124*, 4230-4232.

[175] Myers, A.G.; Siu, M. Lewis acid mediated control of allylic epoxide opening in carbocyclization and halide addition pathways. *Tetrahedron* **2002**, *58*, 6397-6404.

[176] Corey, E.J.; Clark, D.A.; Goto, G.; Marfat, A.; Mioskowski, C. Samuelsson, B.; Hammarstroem, S. Stereospecific total synthesis of a "slow reacting substance" of anaphylaxis, leukotriene C-1. *J. Am. Chem. Soc.* **1980**, *102*, 1436-1439.

[177] Bellomo, A.; Gonzalez, D. Catalytic thiolysis of chemoenzymatically derived vinylepoxides. Efficient synthesis of homochiral phenylthioconduritol F. *Tetrahedron: Asymmetry* **2006**, *17*, 474-478.

[178] Bellomo, A.; Gonzalez, D. 183Diasterodivergent synthesis of optically pure vinyl episulphides and b-hydroxy thiocyanates from a bacterial metabolite. *Tetrahedron Lett.* **2007**, *48*, 3047-3051.

[179] Yasuda, A.; Takahashi, M.; Takaya, H. Stereoselective 1,4-addition of dialkylaluminum benzenethiolate to vinyl oxiranes. *Tetrahedron Lett.* **1981**, *22*, 2413-2416.

[180]   Trost, B. M.; Scanlan, T.S. Synthesis of allyl sulfides *via* a palladium-mediated allylation. *Tetrahedron Lett.* **1986**, *27*, 4141-4144.

[181]   Prévost M.; Woerpel, K.A. Insertions of Silylenes into Vinyl Epoxides: Diastereoselective Synthesis of Functionalized, Optically Active trans-Dioxasilacyclooctenes. *J. Am. Chem. Soc.* **2009**, *131*, 14182-14183.

[182]   Prévost M.; Ziller, J. W.; Woerpel, K.A. Strained organosilacyclic compounds: synthesis of anti-Bredt olefins and *trans*-dioxasilacyclooctenes. *Dalton Trans.* **2010**, *39*, 9275-9281.

[183]   Crotti, S.; Bertolini, F.; Macchia, F.; Pineschi, M. Nickel-Catalyzed Borylative Ring Opening of Vinyl Epoxides and Aziridines *Org. Lett.* **2009**, *11*, 3762-3765.

[184]   Tortosa, M. Synthesis of *syn* and *anti* 1,4-Diols by Copper-Catalyzed Boration of Allylic Epoxides *Angew. Chem. Int. Ed.* **2011**, *50*, 3950-3953.

Send Orders of Reprints at bspsaif@emirates.net.ae

CHAPTER 4

Conjugated Dendrimers with Poly(Phenylenevinylene) and Poly(Phenyleneethynylene) Scaffolds

Joaquín C. García-Martínez[1], Enrique Díez-Barra[2] and Julián Rodríguez-López[2,*]

[1]*Facultad de Farmacia, Universidad de Castilla-La Mancha, Avda. de los Estudiantes s/n, 02071-Albacete, Spain and* [2]*Facultad de Ciencias y Tecnologías Químicas, Universidad de Castilla-La Mancha, Avda. Camilo José Cela, 10, 13071-Ciudad Real, Spain*

Abstract: Fully conjugated dendrimers are interesting materials as active chromophores for a range of optoelectronic applications. The rigid and well-defined framework of these materials enables structure-property relationships to be more easily understood. In this paper, we summarize the main strategies for the synthesis of poly(phenylenevinylene)- and poly(phenyleneethynylene)-based dendrimers as well as approaches to hybrid dendritic structures that combine both scaffolds. Along with the synthetic features, some physical and chemical properties are also discussed.

Keywords: Dendrimers, poly(phenylenevinylene), poly(phenyleneethynylene), organic synthesis, chromophores, luminescence, stilbenes, phenylacetylenes, π-conjugation, macromolecules, organic electronics, wittig-Horner reaction, hybrid dendritic structures, light harvesting, optoelectronic devices, organic light-emitting diodes, energy transfer, electron transfer, orthogonal synthesis, branched molecules.

1. INTRODUCTION

Organic conjugated polymers with good photoluminescent responses have proved to be promising candidates for photonic, electronic and optoelectronic applications [1]. The organic compounds typically used have some common structural characteristics and a few specific modifications make them suitable for a specific application. As a result, organic synthesis represents a powerful tool to introduce the appropriate chemical functionalization to tune the final properties of these materials [2].

*Address correspondence to Julián Rodríguez-López: Facultad de Ciencias y Tecnologías Químicas, Universidad de Castilla-La Mancha, Avda. Camilo José Cela, 10, 13071-Ciudad Real, Spain; Tel: +34-926295300 ext. 3462; Fax: +34-926295318; E-mail: julian.rodriguez@uclm.es

These systems usually have a fully conjugated backbone with alternate double and single bonds (or triple and single) along a chain. Numerous technological devices have been developed that incorporate this type of material. The poly(*p*-phenylenevinylene) (PPVs) and the polythiophene (PTs) families are among the most useful examples in the preparation of PLEDs (polymer light-emitting diodes). These compounds glow in response to electrical currents and they have been used in the manufacture of luminescent displays such as those found in clocks, VCRs and DVD players, microwave ovens, cell phones, laptop computers and dashboards. In addition, PTs have been employed as magnetic field transistors with applications in anti-theft devices at stores and supermarkets. PPVs have been the most widely used materials in this field because of their multiple applications, despite their moderate stability and the sensitive process needed to obtain good-quality devices. PPVs, together with polyfluorenes, dominate the field of LEDs and similar applications [3], whereas poly(*p*-phenyleneethynylenes) (PPEs) are increasingly important for use as chemical sensors and molecular wires, showing a spectacular oligomeric chemistry [4].

Nevertheless, the synthesis of linear polymers implies a wide distribution of molecular weights and poor control of the structure, both factors that lead to difficulties in the development of clear correlations between structures and photochemical responses. The boom in dendrimer chemistry has provided both new methodologies and molecular architectures, which allow the controlled synthesis of nanosized globular molecules with well-defined branches [5]. For example, the synthesis of dendrimers is a fundamental tool in the bottom-up approach to nanomaterials. Synthetic methodologies, involving convergent and divergent methods, are well established and allow total control of the critical molecular design parameters such as size, shape and functionality present in the core, the branches and/or the periphery. Not only are the synthetic approaches to dendrimers well-known, but also their unusual chemical, physical and biological properties and the wide range of potential applications are increasingly understood [6]. In addition, the responses of dendrimers can be tuned through the appropriate choice of functional groups and connectivity of the branches.

These macromolecules are currently attracting the interest of a large number of scientists in the design of photo- and electroluminescent materials. In the last years, several families of dendrimers containing different chromophoric and luminophoric units have been synthesized [7]. Due to the analogy between these materials and chromophoric arrays discovered in photosynthetic natural centers, these new materials have also been used in the preparation of artificial photosynthetic antennae and for charge transport [8].

Luminescent dendrimers fall into two classes, those in which the dendrimer is fully conjugated and those where the active chromophores are connected by insulating dendrons. The advantage of the former dendrimer type is that the conjugation provides a rigid framework for the dendrimer, which enables structure-property relationships to be more easily understood. Nevertheless, dendrimers with π-extended structures are associated with growth and stability problems. Additionally, the non-globular, almost planar shapes of these systems often bring about π-stacking of molecules with loss of the luminescent properties. The possibility of overcoming of these drawbacks and gaining a better understanding of the optical and electrochemical properties of these materials justify the study of these compounds.

The latter class of dendrimers has flexible branches and photo- and/or electroactive units in the different regions of the structure and these units can be bonded in a covalent or non-covalent way. Dendrimers of this type have been modified, mainly in the periphery, with units such as azobenzene [9], dansyl sulfonate [10], porphyrins [11], oligo(p-phenylenevinylenes) [12], PPVs [13] and pyrenes [14], in the search for new photochemical and photophysical properties. However, it should be noted that structure-property relationships are more difficult to determine and predict in materials with flexible branches.

The present review gives a general outline of some of the synthetic methodologies that have been investigated for the preparation of π-conjugated dendritic architectures, and is an update of a previously published article [15]. Coverage of the huge number of reported structures is beyond the scope of this review and, as a result, we will concentrate the discussion on dendrimers with poly(phenylenevinylene) and poly(phenyleneethynylene) scaffolds and focus particular emphasis on those materials prepared in our research group. Firstly, we will discuss the main strategies developed for the synthesis of poly(phenylenevinylene)-based dendrons and the bonding of such units to different cores; secondly, dendritic architectures based on poly(phenyleneethynylene) will be covered; and finally we will describe our approach to dendrimers with a hybrid scaffold (bearing both double and triple bonds). In addition to a discussion of synthetic routes, some physical and chemical properties of the structures will also be pointed out.

2.   DENDRITIC   ARCHITECTURES   BASED   ON   POLY (PHENYLENEVINYLENE)

2.1. Synthetic Approaches

Poly(phenylenevinylene)-based dendrimers (PPVdend) have become one of the most studied families of conjugated dendrimers. These compounds are also called stilbenoid dendrimers [16] and they usually have the optoelectronic component at the core, although there have been a few reports that describe materials with the active moiety at the surface. The chemistry of PPVdend systems is relatively new, although the first report on the synthesis of stilbenoid derivatives was published in 1969. The reaction of mesitylene with different aromatic imines in a basic medium yields what can be considered as the first generation of a PPVdend (Scheme **1**) [17].

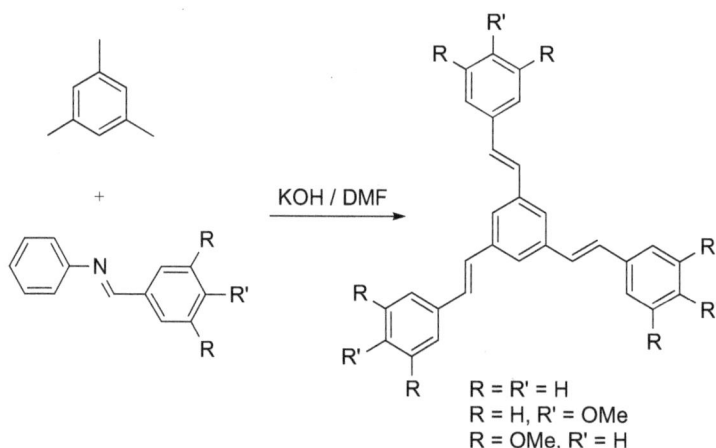

R = R' = H
R = H, R' = OMe
R = OMe, R' = H

Scheme 1: First 1,3,5-tris(styryl)benzene derivatives reported by Weber and colleagues.

However, in 1997 the concept of dendritic molecules bearing phenylenevinylene linkages was first considered [18]. Two main strategies have been reported for the construction of stilbenyl-based dendrimers: firstly, the components can be connected by forming the double bond using Wittig-type chemistry and, secondly, the vinyl unit can be added directly to the next aryl unit using palladium-catalyzed chemistry [5]. Both of these processes have been successfully applied in convergent syntheses. Recently, an alternative approach to stilbenyl dendrons has been reported in which the basic dendron framework is built up with sulfur linkages. In this case, oxidation of the sulfides to sulfones followed by a series of Ramburg–Backlund reactions is used to introduce the vinylene moieties late in the synthesis [19].

The normal route used to make optoelectronic dendrimers consists of the separate preparation of the core and the dendrons, which are then linked together in the final reaction step. This method gives rise to reproducible syntheses of mono(disperse) materials.

Yu and colleagues were the first to describe a simple orthogonal approach to PPVdend, combining the Heck and Horner–Wadsworth–Emmons (HWE) reactions [18]. They designed two AB$_2$-type building monomers (**1** and **2**, Scheme **2**), one of which underwent the HWE reaction and the other the Heck coupling reaction to produce homogeneous dendrons up to the fourth-generation with a single functionality, which is either a bromo or formyl group, at the focal point.

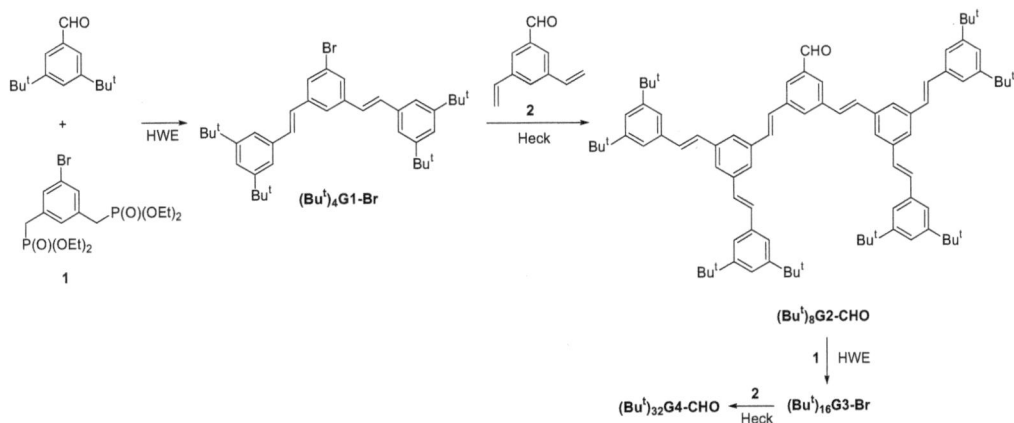

Scheme 2: Synthesis of PPV dendrons by Yu and colleagues.

The yields of the reactions diminished for higher generations because of the steric hindrance at the focal point of larger molecules. Therefore, for the preparation of a fourth-generation dendrimer it was necessary to incorporate a spacer in the third-generation dendron in order to generate a molecule in which the aldehyde functional group had a lower level of steric hindrance. In spite of this change, and due to solubility problems, the yield of this coupling process with 1,3,5-tris(diethoxyphosphorylmethyl)benzene (**3**) was as low as 9% (Scheme **3**).

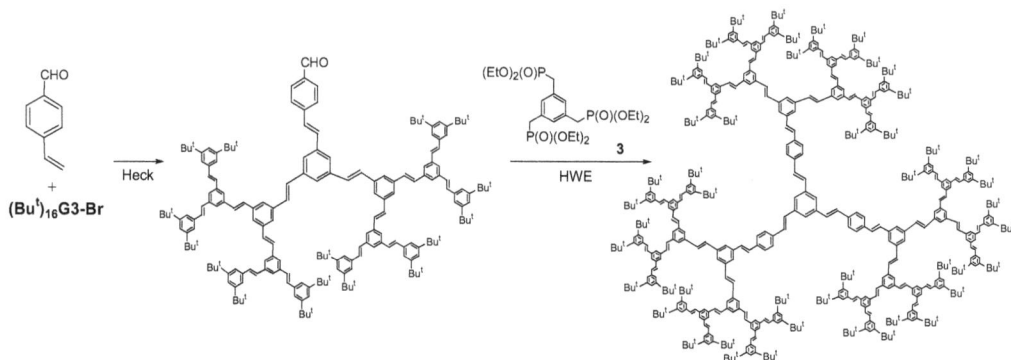

Scheme 3: Fourth-generation dendrimer prepared by Yu and colleagues.

These molecules were almost colorless and exhibited virtually identical UV-vis spectra, implying that the major absorbing unit was stilbene in all cases. These materials fluoresced in the near UV or blue region of the spectrum.

Later, Meier and colleagues described a methodology based exclusively on the HWE reaction. They used diphosphonate **4**, with protected aldehyde functionality, as the building block. The reaction of **4** with appropriate dendritic aldehydes followed by deprotection gave the higher dendrons. Once again, the last reaction step in the formation of the dendrimers consisted of a threefold HWE reaction between triphosphonate **3** and the appropriate dendrons (Scheme **4**) [20]. In this case, the yields also decreased as the steric hindrance increased, and the process reached its limits with the fifth generation dendrimer.

This kind of macromolecule has been widely studied in terms of structure and properties. Full conjugation in *trans*-stilbene can be achieved in the completely planar conformation. However, the potential energy for a slight torsion around the single bond is low in the ground state of *trans*-stilbene. The stilbenoid dendrimers are expected to adopt a rather flat shape in solution. Indeed, molecular modelling suggests a nearly planar conformation for the first and the second generation

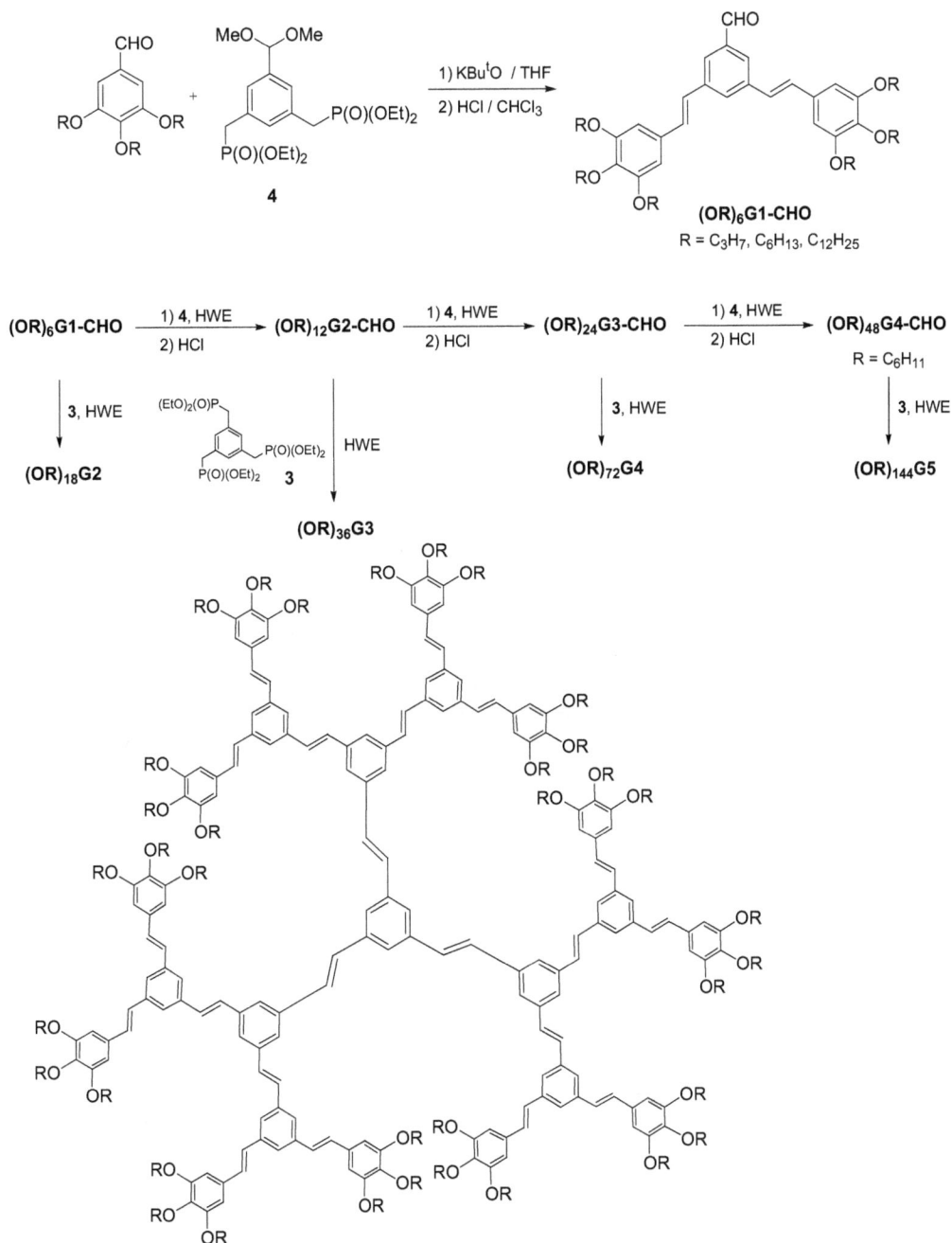

Scheme 4: Synthesis of PPV dendrons and dendrimers by Meier and colleagues.

[20a]. Thus, the first two generations, **(OR)₉G1** and **(OR)₁₈G2**, have a relatively compact disklike shape, and it is not until the third generation that steric

interactions cause a more significant distortion from planarity, forcing an extension into the "third dimension" and leading to a more cylindrical shape. A small-angle scattering study clearly confirmed this situation [21].

PPVdend are not capable of backfolding; therefore the molecular density increases from the core to the periphery of the systems. As a consequence cavities exist close to the core. These cavities are large enough to host guest molecules [22]. Toluene, for example, is so firmly included in the higher PPVdend that it flies in the vacuum together with the molecular ions $M^{+\cdot}$ in MALDI-TOF mass spectrometry [20a].

The long alkyl groups at the periphery not only enhance solubility but also induce liquid crystalline behavior [20a]. Only the first and the second generation materials form thermotropic liquid crystals (columnar discotic phases), provided that hexyloxy or dodecyloxy side chains are attached to the periphery; the propoxy chain is too short to permit such intermolecular stacking. The higher generations were oils or they formed glassy or solid states, which directly transformed to isotropic phases on heating. The molecular mobility in liquid crystalline phases of stilbenoid mesogens has been correlated with photochemical and photophysical properties, *e.g.* molecular motion allows photoreactions to proceed even when the molecules are photostable in crystalline phases [23]. The driving force for the formation of columnar mesophases in stilbenoid dendrimers has been proposed to be microsegregation of the rigid conjugated scaffold and the flexible aliphatic chains, a process that occurs due to the preorganization of these molecular units in the mesogen [16, 20a, 24]. The conjugated nature of the stilbenyl moieties also means that surface groups are required to facilitate purification and processing of the materials.

Burn and colleagues also reported a simple two-step iterative procedure in which the Heck and Wittig reactions were alternated for the preparation of PPV dendrons; however, only the Heck step led to growth of the dendron (Scheme **5**) [25].

The aldehyde-focused dendrons could be successfully coupled in a single step to different phosphonate derivatives for the formation of a new family of dendrimers

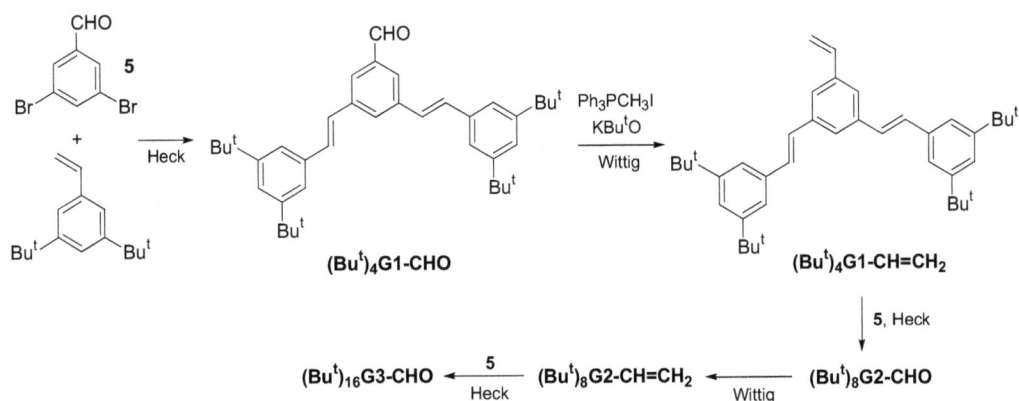

Scheme 5: Iterative synthesis of dendrons by Burn and colleagues.

that contained luminescent chromophores [26]. All of the resulting dendrimers were luminescent, with emission observed from the core. Appropriate selection of the core enabled the preparation of dendrimers that gave blue (distyrylbenzene), yellow-green (distyrylanthracene), and red (*meso*-tetraphenylporphyrin) light emissions. Excitation of the dendrons led to energy transfer from the branches to the luminescent cores. An important characteristic of these materials for use in organic LEDs [27] is that all the dendrimers, regardless of generation or core, could be spin-coated from solution to form optical-quality thin films [26, 28]. It was demonstrated that dendrimer generation is a powerful tool for controlling the intermolecular interactions that govern OLED performance. The effect of generation was first studied on three generations of dendrimers with distyrylbenzene chromophores as cores (the first generation in the series is shown in Fig. **1**) [29]. These materials were incorporated into LED structures consisting of a single dendrimer layer between ITO and calcium contacts, and it was found that the device efficiency was greater for the higher generations than for the first generation. It was also found that while the emission spectrum of the first generation was quite broad because of the excimer emission, the spectra of the higher generations narrowed as the dendrons protected the emissive chromophores of the dendrimers and, furthermore, reduced the intermolecular interactions that lead to excimer emission.

The same authors also prepared dendrons that contained phosphonate focal points and coupled these to cores that contained aldehyde functionalities. In this way,

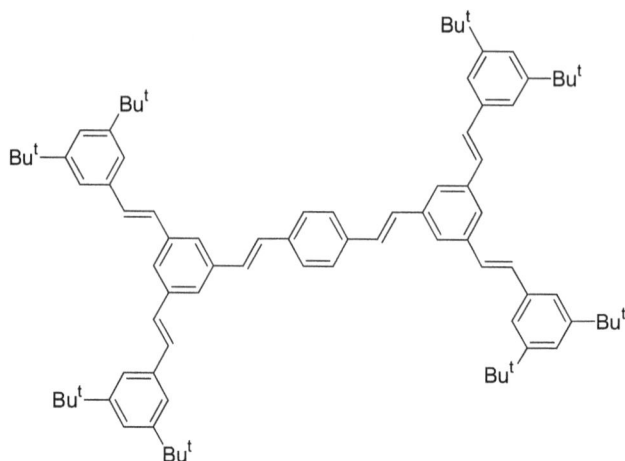

Figure 1: First generation PPVdend with a distyrylbenzene chromophore as the core.

four generations of dendrimers with fluorescent tris(distyrylbenzene)amine chromophores were prepared in order to investigate the precise reason for the improved device performance (the second generation is shown in Scheme **6**) [30]. It was found that charge was directly injected into the tris(distyrylbenzene)amine chromophores at the center of the dendrimers and that the mobility of the injected holes decreased by a factor of 100 on going from the "zeroth-" to third-generation dendrimer [31]. In other words, as the hopping distance between the chromophores increased with generation, the mobility of the holes decreased. The change in mobility was accompanied by a concomitant increase in the light-emitting device efficiency, providing the first direct evidence that generation could control charge mobility and give improved device performance.

At the same time, our research group reported a convergent synthesis of new PPVdends bearing a 1,3,5-tris(phenylenevinylene)benzene core [32]. Two strategies were employed for the preparation of the first generation of these compounds (Scheme **7**). The first route started with the palladium-catalyzed coupling of three molecules of *para*-substituted styrene derivatives with 1,3,5-tribromobenzene. We found that when *p*-methylstyrene was used, the triple Heck reaction successfully led to **(Me)₃G1** as the major product. However, a tedious and difficult isolation and purification procedure was required, and **(Me)₃G1** was obtained in only 39% yield. When *p*-methoxystyrene was used, only the undesired

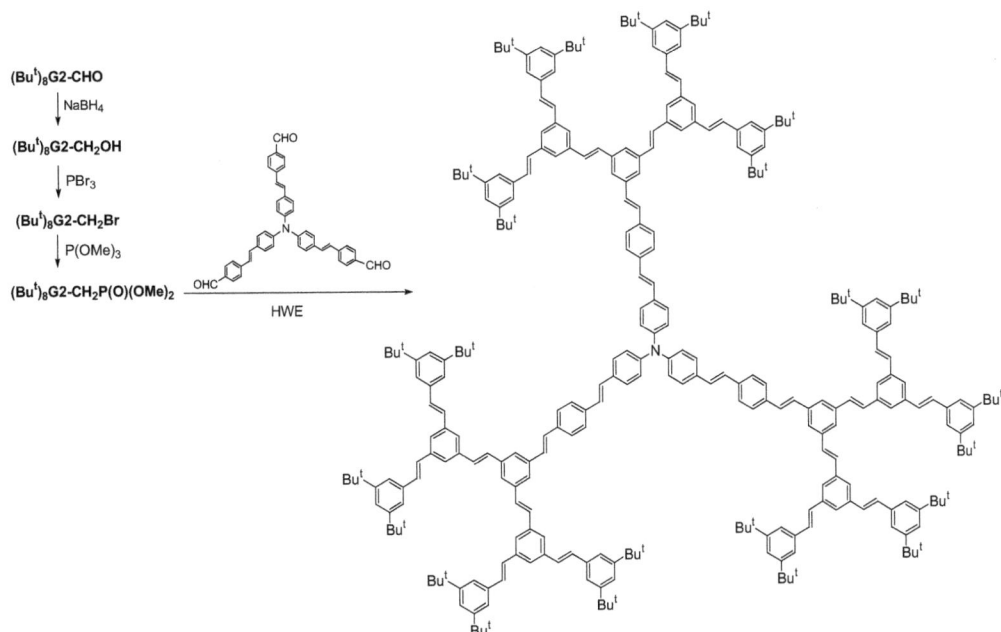

Scheme 6: Synthesis of a second-generation PPVdend with a tris(distyrylbenzene)amine core.

disubstituted product could be isolated (in 23% yield). Neither of the purified compounds contained any double bonds with the Z-configuration. Overall, the Heck reaction proved to be unsatisfactory in terms of obtaining first-generation dendrimers with different end groups in good yields (reaction times were also extremely long). The second method involved the formation of the required E-configured double bonds by a HWE reaction. The readily available 3-fold triphosphonate **3** was used as a basic unit to form the core. When this compound was coupled with a set of *para*-substituted benzaldehydes, using KBu^{t}O as the base and THF as the solvent, only the all-*trans* isomers **(R)$_3$G1** were observed within the limits of NMR detection.

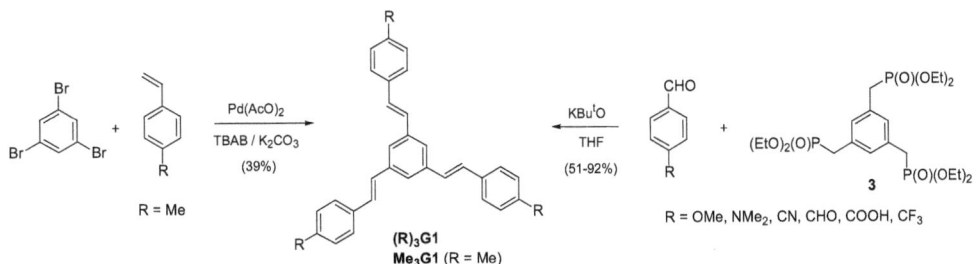

Scheme 7: Approaches for the preparation of first-generation PPVden.

This reaction afforded cleaner crude products, which were correspondingly easier to purify and gave better yields than those obtained in the Heck reaction. These advantages, together with the abundant number of commercially available aldehydes, led us to focus our efforts on the synthesis of dendrimers using exclusively the HWE methodology.

The reaction of phosphonate **6** with the appropriate *p*-substituted benzaldehyde gave first generation alcohols **(R)₂G1-CH₂OH** in good overall yields after *in situ* hydrolysis of the ester group with NaOH/H₂O (Scheme **8**). These alcohols are also precursors of the next generation dendrons, which were synthesized by a two-step iterative procedure in which the dendron was doubled in size in each iteration. The synthesis was based around the oxidation of the alcohol group with either PCC or MnO₂ to give the corresponding aldehyde-containing dendrons. This step was followed by coupling with the diphosphonate **6** (Scheme **8**, route a). However, the oxidation reaction failed when the side chains in the periphery were dialkylamino groups. We were therefore obliged to prepare this kind of aldehyde by applying the methodology previously described by Meier *et al.* [20], which makes use of compound **4** as a core (Scheme **8**, route b). Diphosphonates **4** and **6**

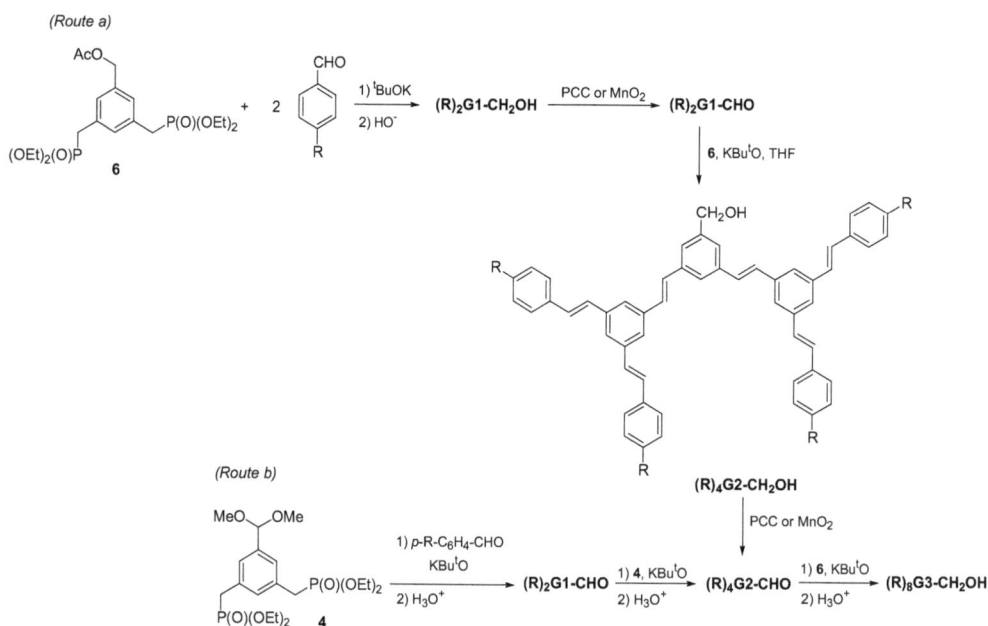

Scheme 8: Different routes based on the HWE reaction for the synthesis of PPV dendrons.

served as the extension and branching units for the next higher generation of dendron. The preparation of third generation dendrons required longer reaction times; however, the molecular peaks corresponding to the desired product and the carboxylic acid derived from the oxidation of starting material were always detected with different relative intensities by mass spectrometry. These results can be explained by considering the higher level of steric hindrance around the reactive site.

Aldehyde-focused compounds can also be regarded as starting materials for the corresponding dendrimers. For example, the 3-fold reaction of **(OC$_{12}$H$_{25}$)$_2$G1-CHO** with the core 1,3,5-tris(diethoxyphosphorylmethyl)benzene (**3**) gave the compound **(OC$_{12}$H$_{25}$)$_6$G2** in excellent yield (Scheme **9**). This versatile synthetic strategy allowed us to prepare a variety of peripherally modified dendritic architectures (R = alkyl, alkyloxy, trifluoromethyl, dialkylamine, cyano, aldehyde, nitro, *etc.*), which can be bonded to different cores (*vide infra*, section 2.3).

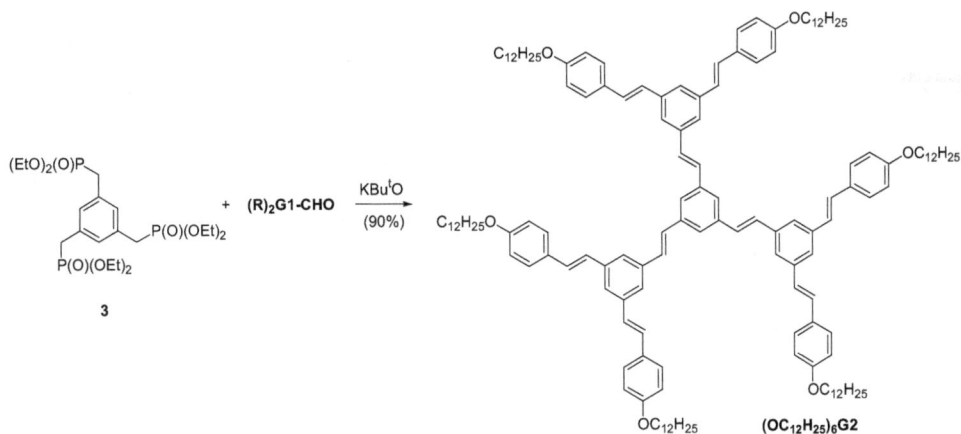

Scheme 9: Second-generation PPVdend with dodecyloxy peripheral groups.

The optical properties of these structures were also evaluated. Owing to the *meta* arrangement through which the stilbene units are linked, the observed absorption spectra consisted of a simple superposition of the absorptions due to the independent constituent chromophores. Only small changes were observed in the onset of absorption in the UV/Vis spectra between the different generations. As one would expect, the molar extinction coefficients increased with increasing generation. The very high intensity of absorption could be deduced from the large extinction coefficients of these molecules at their absorption maxima. These results are comparable to those obtained by Yu and colleagues for unfunctionalized compounds (with *tert*-butyl groups at the periphery) [18]. The introduction of different groups has a negligible influence on the extinction coefficients. On the other hand, this peripheral substitution resulted in a red shift of the absorption band with increasing electron-withdrawing or electron-donating strength in comparison to unfunctionalized dendrimers [18]. However, the less effective conjugation due to *meta* substitution meant that all compounds were absolutely transparent above 400 nm, with the exception of the dialkylamino derivatives.

The materials all emitted strong blue light under UV irradiation, showing typical bands for stilbenoid compounds. The intensity of the fluorescence from the higher generation structures was substantially lower than that from their corresponding lower generation counterparts. Similar behavior has previously been observed for other structurally rigid dendrons [33] and was accounted for by the many modes used to dissipate the excitation energy in larger dendrons, a property that enhances the nonradiative decay. The through-space interaction between the fluorescent units became more significant with increasing molecular size, providing additional fluorescence quenching pathways. Although contributions from more than one fluorescent species cannot be excluded, we assume that in the emission spectra only the lowest energy $\pi \rightarrow \pi^*$ transition was fluorescent. The maximum emission wavelength of these compounds strongly depends on the peripheral substitution. Desirable wavelength emission could be obtained by appropriate choice of the peripheral group of the dendritic compound, even in solid thin films.

A number of research groups around the world have also focused on the creation of new families of PPVdend with different optoelectronically active components and a diverse array of strategies have been used for their synthesis, all of which are based on slight modifications of the aforementioned methodologies. For example, ferrocenyl [34], diphenylquinoline [35], phenothiazine [36], triazine [37], hydrophilic oligo(ethylene oxide) [38] and triphenylamine [39] units have been incorporated at the periphery and distyrylbenzene [37a], anthracene [37b, 40] and distyrylstilbene [40] groups as the core. Guldi and Martín proposed an alternative synthesis, based on the HWE reaction, in which the monomer employed was a diphosphonate functionalized with a cyano group. Activation of the focal point was achieved by reduction of the cyano group to an aldehyde group. This approach was used to prepare up to the second generation of PPV dendrons with dibutylaniline or dodecyloxynaphthalene peripheral groups as antennae/electron donors, which were covalently attached to a C_{60} core [41]. The PPVdend systems containing donors (acceptors) at the core and acceptors (donors) at the edge of peripheral groups have also been prepared and these showed very large first hyperpolarizability and good thermal stability, making them attractive candidates for nonlinear optical materials [42].

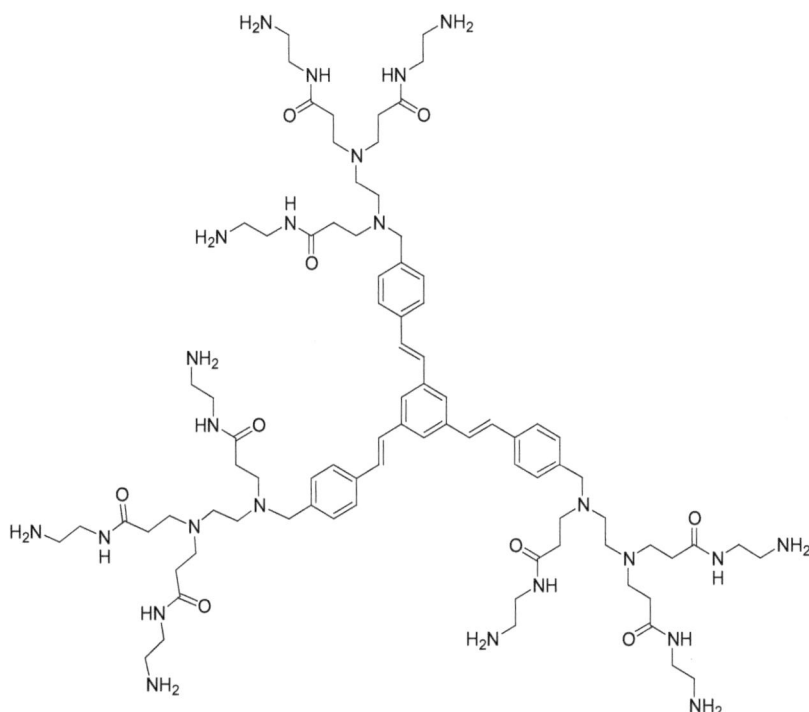

Figure 2: Structure of the hybrid PPV-PAMAM dendrimer named Transgeden®.

Recently, we have reported a novel hybrid dendrimer (Transgeden®) that combines a conjugated rigid PPV core with flexible polyamidoamine (PAMAM) branches at the surface (Fig. **2**) [43]. The potential of this material as a nonviral gene delivery system was also examined, and it was observed that dendriplexes formed by Transgeden® and small interfering ribonucleic acids (siRNAs) can be incorporated into >90% of neuronal cells without any toxicity up to a dendrimer concentration of 3 μM. This result suggests that this novel dendrimer is a simple and efficient alternative to viral vectors and warrants further investigation as a promising nonviral gene delivery carrier.

2.2. Control of the Peripheral Functionality

The convergent approach appears ideally suited for the preparation of dendrimers in which control over both the number and the location of end functionalities is achieved. This is because the number of reactions that are performed with each generation remains the same in the convergent synthesis. This methodology has therefore been widely used by different authors for the synthesis of highly asymmetrical dendrimers containing a predetermined and well-defined number and arrangement of functional groups at their periphery [44]. Thayumanavan and colleagues were the first to build dendrons and dendrimers in which all the monomer units are different [45]. They prepared benzyl ether-based dendritic structures and the synthesis of dendrimers based on melamine with a multifunctional periphery has also been achieved [46]. Diversity in functional groups has also been achieved through divergent approaches [47]; however, their inherent nature does not allow the generation of structures in which every peripheral unit is different.

The methodology developed in our research group made it possible to control the functionality introduced at the periphery. The appropriate choice of starting monomer enabled the synthesis of dendritic molecules bearing electron-donating and electron-withdrawing groups as well as long-chain substituents, with the latter aimed at improving the solubility of these systems.

Initially, we conceived two different synthetic strategies to prepare the first-generation PPVdend with two different end units: the bonding of one of the branches to the core followed by the other two (identical) branches or, alternatively, the coupling of two identical branches to the core and the subsequent linking of the third branch in the final reaction step. The initial idea was to use a combination of Heck and HWE reactions. In this way, the use of phosphonates **1** and **7** and following Scheme **10** allowed us to prepare the compounds $(R)_2(R')G1$, where R and R' can be either electron-withdrawing or electron-donating groups.

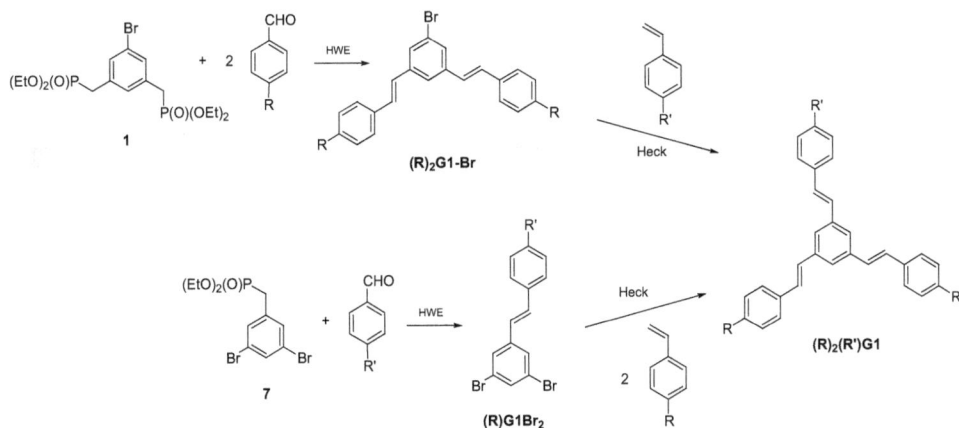

Scheme 10: Approaches to the synthesis of first-generation PPVdend with two different peripheral units.

However, for the Heck reaction we encountered all of the drawbacks detailed above for the preparation of threefold symmetrical dendrimers (R = R', see section 2.1). Once again, the exclusive use of the HWE reaction allowed access to the desired compounds in a versatile fashion according to Scheme **11**. Thus, treatment of alcohols **(R)$_2$G1-CH$_2$OH** (obtained as mentioned in Scheme **8**, section 2.1) with CBr$_4$/PPh$_3$ in acetonitrile gave the corresponding benzyl bromides, which could be easily transformed into the phosphonates **(R)$_2$G1-CH$_2$P(O)(OEt)$_2$**. The route was successfully completed by reaction with the appropriate aldehydes under a new HWE reaction.

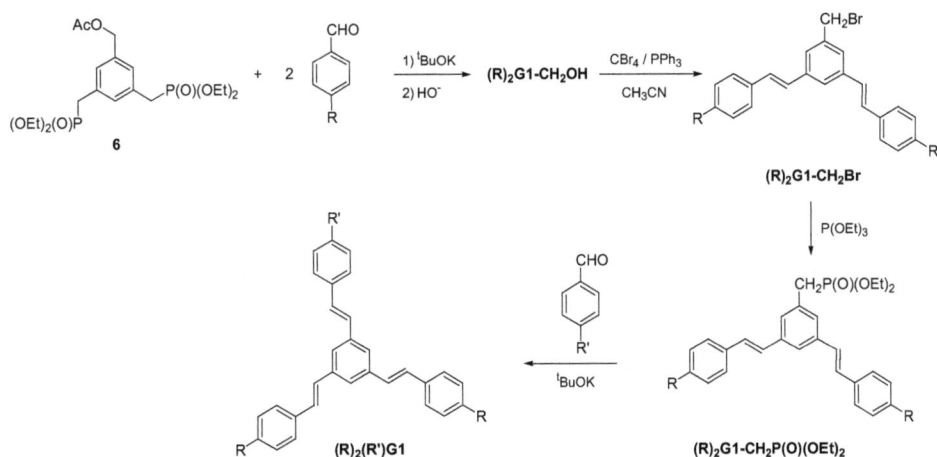

Scheme 11: Preparation of first-generation PPVdend with two different peripheral units by exclusive use of the HWE reaction.

The second-generation phosphonates **(R)₄G2-CH₂P(O)(OEt)₂** were analogously synthesized from the corresponding alcohols. The last reaction step in the formation of the second-generation dendrimers was their coupling with the appropriate dendritic aldehyde. The HWE reaction of aldehydes **(R)₄G2-CHO** (see Scheme **8**, section 2.1) with the first-generation phosphonates also yielded

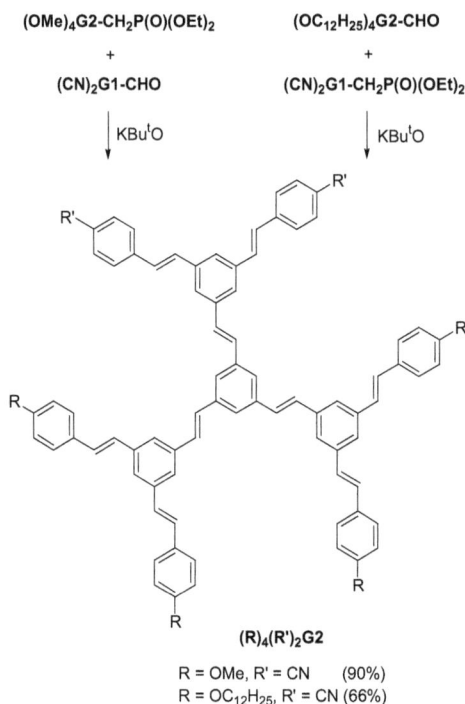

Scheme 12: Second-generation PPVdend containing both π-donor and π-acceptor end groups.

the target compounds. In this way, dendrimers **(MeO)₄(CN)₂G2** and **(OC₁₂H₂₅)₄(CN)₂G2** could be obtained in good yields (Scheme **12**). These new PPVdend are comparable to other dendritic systems that contain π-donor and π-acceptor groups and have been claimed as interesting and attractive targets. These materials should possess amphoteric redox properties under electrochemical control, and may engage in inter- and/or intramolecular charge-transfer interactions [48]. As far as the optical properties of these materials are concerned, the familiar crossing between absorption and emission was far from both maxima, *i.e.*, large Stokes shifts occurred. The magnitude of the Stokes shifts decreased from the first to the second generation, indicating that likely charge transfer processes are only effective when π-donor and π-acceptor groups are directly connected by the 1,3-phenylenevinylene system, *i.e.*, in the first-generation compounds.

We also developed an efficient orthogonal and convergent-growth sequence methodology based on the HWE reaction for the synthesis of new PPV dendritic structures with fully differentiated surface functional groups. This was achieved by the stepwise incorporation of functionalities in a controlled manner onto an AB$_2$-type monomer unit, diphosphonate **8**. In this way, a large variety of dendritic structures were prepared in good yields, including dendrons and dendrimers up to the third generation with eight different functional groups at the periphery (Scheme **13**) [49].

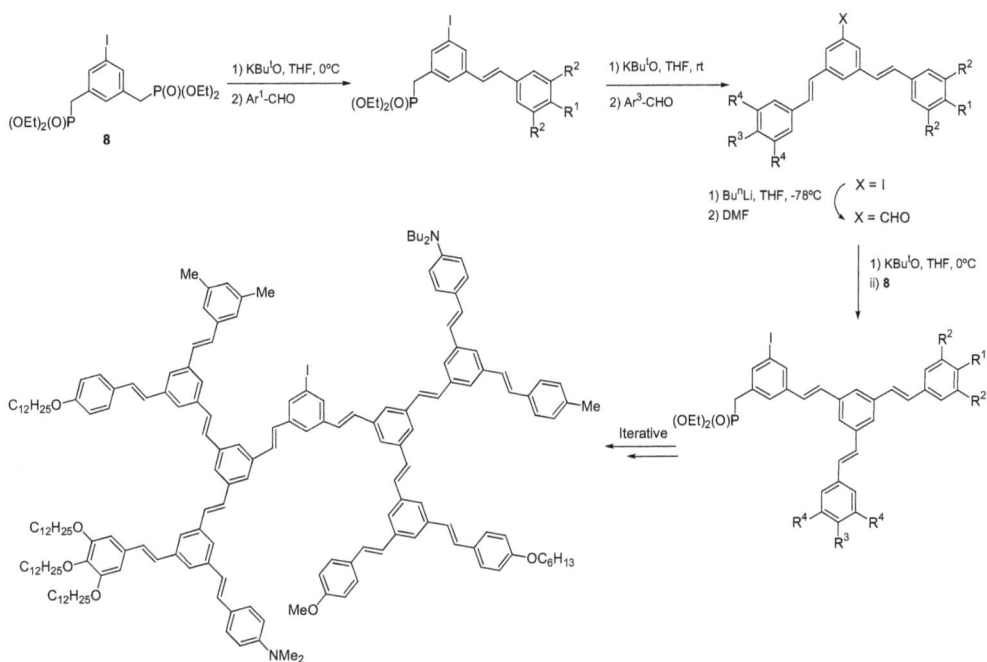

Scheme 13: Convergent methodology to access PPV dendrons in which all the monomer units are different.

This iterative strategy is extremely general, permitting not only the use of a wide variety of functional groups but also excellent control over their placement. In principle, these reaction sequences can be further progressed through multiple cycles, although in practice one would expect that higher molecules would be difficult to prepare because of the well-documented progressive reduction in the reactivity of the focal point in successive generations [50].

The groups selected at the periphery provided good control of solubility. Indeed, all of the new molecules prepared are highly soluble in THF as well as in chlorinated solvents such as dichloromethane and chloroform, thus allowing purification by conventional silica gel chromatography. A suitable choice of starting materials could provide a wide variety of dendritic architectures with different peripheral moieties, although peripheral pendant groups should be chosen carefully in order to impart solubility.

This approach does have a clear drawback: the desired monosubstituted product is obtained in statistical yield together with the corresponding undesired disubstituted byproduct. However, all of the monophosphonates could be easily separated by column chromatography (silica gel) in good yields because of their very different polarity, thus avoiding the need for protection-deprotection steps that are much more time-consuming. This methodology can therefore be used as an accelerated route to a wide range of asymmetrically conjugated dendritic architectures. The use of this methodology will allow the synthesis of numerous PPVdend systems with specific substitution patterns.

2.3. Modification of the Core

It can be clearly inferred that one of the main advantages of the convergent methodology for the synthesis of dendrimers is the possibility of tuning their physical properties by the appropriate choice of either the peripheral groups in the initial stage or the core in the last step. The functionality at the focal point of the dendrons plays a crucial role as it provides the means of creating the core. Besides the examples described above – in which a benzene unit is placed at the core – our research group has also prepared a wide range of new families of PPVdend with different cores, such as distyrylbenzene [49], polyene [51], β-lactams [52] (Fig. 3), binaphthyl [53] and C_{60} [54], and studied their photophysical and/or electrochemical properties.

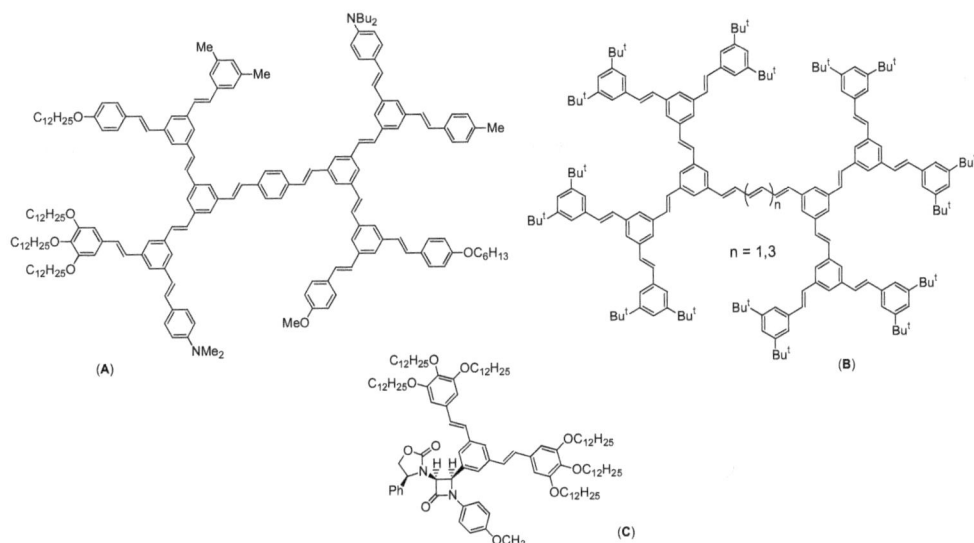

Figure 3: Some PPVdend with different cores: distyrylbenzene (A), polyene (B) and β-lactam (C).

The preparation of dendrimers containing enantiopure binaphthyl cores was achieved in a convergent manner by HWE reaction of the (*R*)-1,1'-binaphthyl derivative **9** and the appropriate aldehyde dendrons in KBu^{t}O/THF (the second-generation series is shown in Scheme **14**). Different electroactive units were incorporated in the peripheral positions of the dendrons in order to tune both the optical and electrochemical behavior of these systems. All of the molecules described were new, chiral, blue-emitting phenylenevinylene-based dendrimers that were highly fluorescent. Once again, the second-generation compounds showed a lower quantum yield than the corresponding first-generation dendrimers. The redox properties of the dendrimers were determined by cyclic voltammetry, which demonstrated the influence of the functional groups at the peripheral positions of the dendrimer on the redox behavior of these systems. A most interesting observation is that the first two irreversible oxidation potentials, which are observed after one single scan, disappear after multiple scans, while a new chemically reversible oxidation wave ($\Delta E \cong 100$ mV) appears between the previously observed waves. This behavior clearly indicates that the dendrimers are electrochemically unstable and, after the first oxidation, give rise to a new electroactive species that exhibits electrochemically reversible behavior. This lack of stability of the radical-cations could be a problem for their further use in the preparation of organic LEDs.

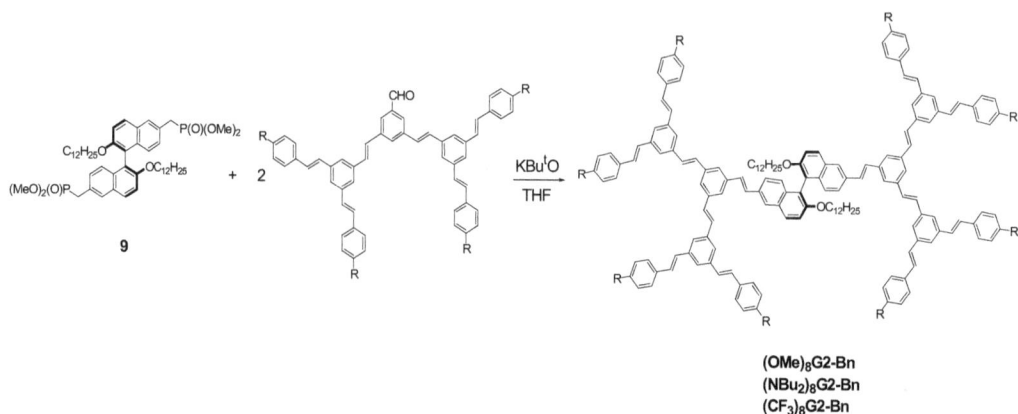

Scheme 14: Formation of second-generation PPVdend containing an enantiopure binaphthyl core.

C_{60}-based dendrimers with ferrocene units as donors located on the periphery are particularly attractive targets. The functionalization of C_{60} was accomplished by means of the 1,3-dipolar cycloaddition of the corresponding azomethine ylides [55], generated *in situ* from the appropriate aldehydes and *N*-methylglycine (sarcosine), as shown in Scheme **15** for the second-generation compound. Cyclic voltammetry studies indicated an electrochemically determined HOMO–LUMO gap as low as 0.99 eV for the second-generation fullerodendrimer. The UV/Vis spectra showed similar λ_{max} values for both first and second generation fullerodendrimers, suggesting that these compounds behave as an assembly of two and four vinylferrocene moieties, respectively. The appearance of a broad band between 450 and 500 nm, which is enhanced in more polar solvents, suggests the existence of weak ground-state interactions between the vinylferrocene moieties and the fullerene sphere. Photophysical investigations showed an efficient photoinduced electron-transfer process even in toluene, with higher quantum yields as the dendrimer generation is increased. The lifetimes of the charge-separated state vary from tens of nanoseconds in nonpolar solvents to hundreds of nanoseconds in polar solvents, with a higher value for the second-generation fullerodendrimer due to the longer distance between the ferrocene and the C_{60} cage.

A novel family of liquid-crystalline fullerene–ferrocene dyads was designed by assembling a first- or second-generation ferrocene-based dendrimer, a second-generation mesomorphic poly(arylester) dendrimer carrying cyanobiphenyl units

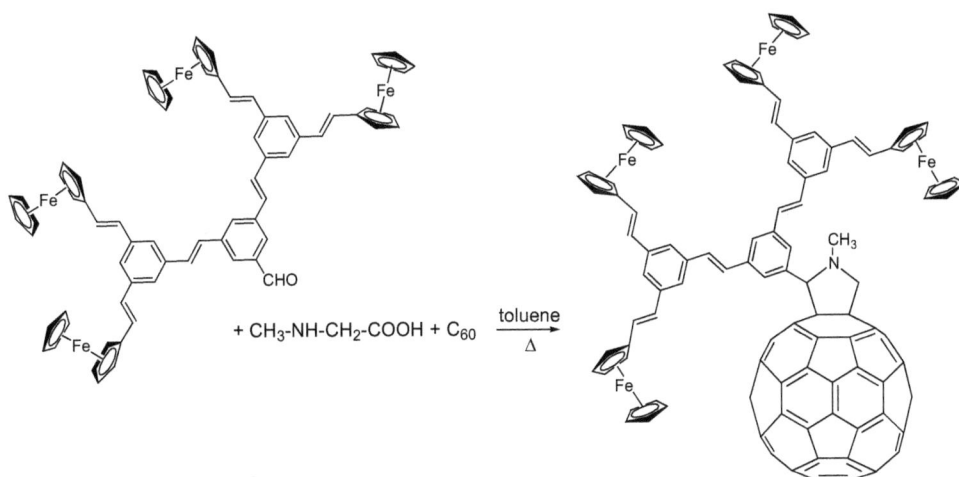

Scheme 15: Synthesis of the second-generation fulleroferrocene dendrimer.

and fullerene, see Fig. **4** [56]. The materials displayed smectic A phases and organized into bilayer structures within the smectic layers. The supramolecular organization was governed by steric constraints. The electrochemical behavior was consistent with the redox-activity of the building-blocks. Steady-state emission spectra showed that the fluorescence was totally quenched, suggesting the existence of an efficient electron transfer process from the ferrocene dendrimer to C_{60}. The use of ferrocene-based dendrimers as electron donor moieties is therefore appealing to elaborate supramolecular switches based on fullerene and ferrocene.

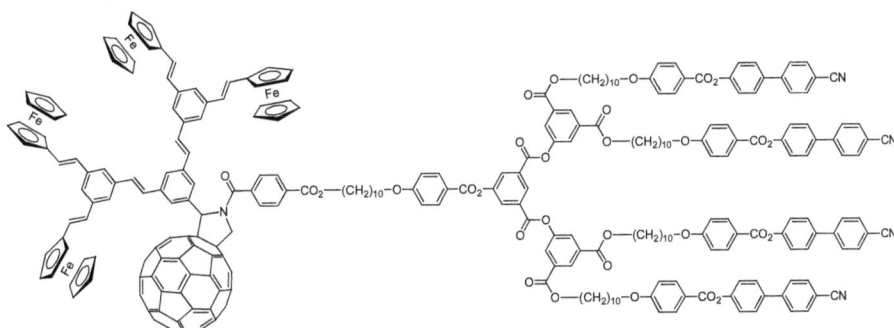

Figure 4: Second-generation liquid-crystalline fullerene–ferrocene dyad.

More recently, new cores have enlarged our catalogue of PPVdend. Thus, we have been successful in preparing lanthanide(III)-cored dendrimers [57], dendritic PPV pyrimidines [58] and dendritic 2,6-di(pyridin-2-yl)pyrimidines (Fig. **5**) [59].

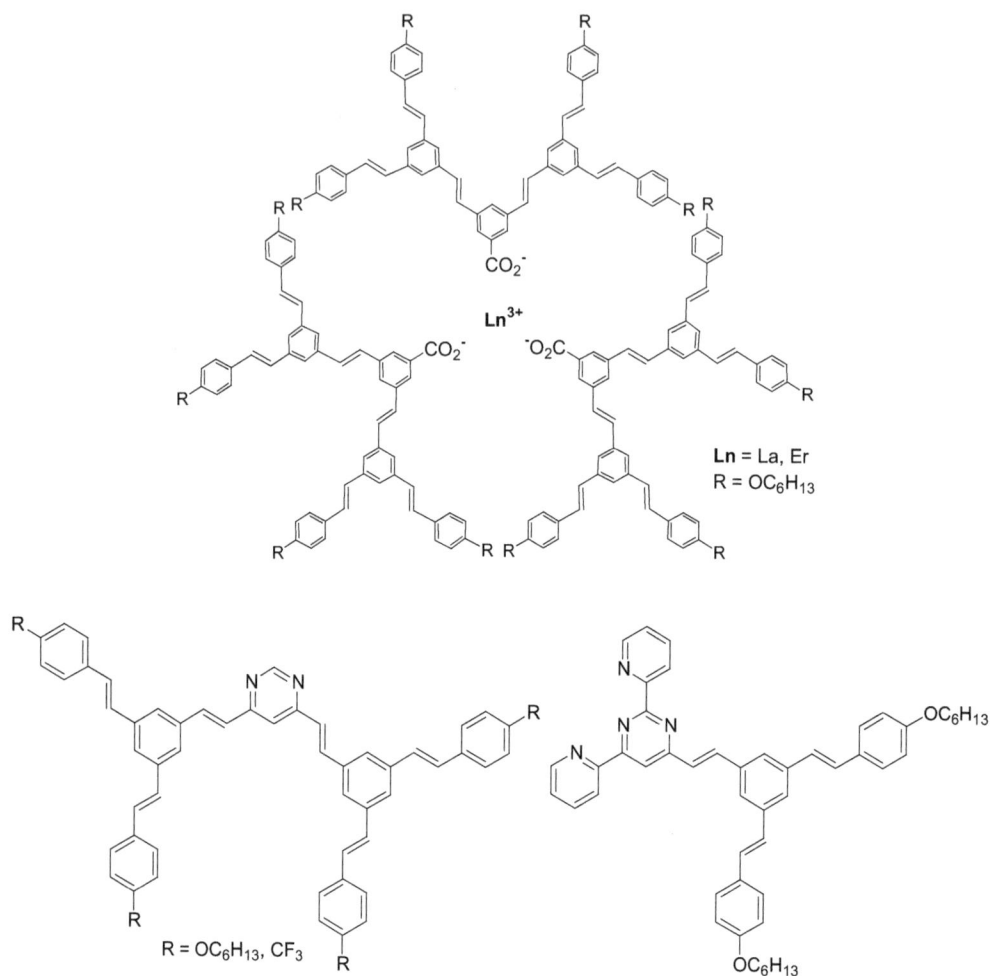

Figure 5: PPVdend prepared recently in our research group.

In this context, we have also functionalized the dithieno[3,2-*b*:2',3'-*d*]phosphole scaffold at the 2- and 6-position with phenylenevinylene groups through a HWE protocol [60]. Studies towards the general accessibility have provided dithienophosphole materials with significantly reduced band gaps covering the full optical spectrum, depending on the electronic nature of the terminal groups used [60b]. Extension of this protocol towards first generation PPVdend has provided a series of dendrimers with electron-withdrawing, -neutral, as well as -donating terminal groups (Scheme **16**). The nature of the end groups used has been found to have significant impact on the self-organization, but more importantly, also on the optical and energy-transfer properties of the materials. Due to the rigid, planar, extended π-system of the phenyl-terminated dendrimer (R = H), this compound have the ability to self-assemble into large and rigid one-dimensional microfibers in THF by π-π stacking, whereas the bulky end groups in dendrimers with R = CF₃, NPh₂ do not seem to allow for extended structures. In terms of optical properties, the three dendrimers show enhanced fluorescence emission from the core ranging between 40% to almost 60% intensity increase, again depending on the end-groups employed. Thus, the intriguing photophysical properties of the dithienophosphole system can be further improved by embedding this unit as core in π-conjugated dendrimers.

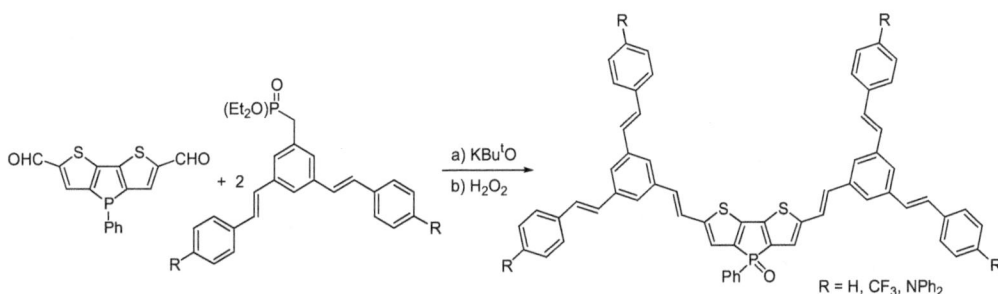

Scheme 16: Preparation of dendritic dithieno[3,2-*b*:2',3'-*d*]phospholes.

The possibility of engineering the self-assembly behavior of dendritic phosphole-cored derivatives was also envisaged, with the aim of obtaining dynamic and well ordered supramolecular structures, such as gels and/or liquid crystals [61]. A series of dendritic phospholes with PPV (Fig. **6**) as well as Fréchet type units was found to exhibit the ability to gel organic solvents at different transition temperatures. The spectroscopic investigation revealed that the use of different solvents and/or temperatures allowed modulating the emissive properties of the gel phases. Moreover, in the solid state these molecules were also capable of self-assembly into highly luminescent and room temperature liquid crystals. By means of polarizing optical microscopy (POM) and X-ray diffraction (XRD) the self-organization was determined to be columnar hexagonal. These compounds represent an interesting example of molecules that combine liquid crystal behavior and gelation properties. Moreover, on the basis of the strong photoluminescence observed in the different states (solid, solution, and gel), the electrochromic properties of compound depicted in (Fig. **6**) were representatively investigated highlighting the switchability of the color emission of this phosphole-cored dendrimer *via* applying a potential.

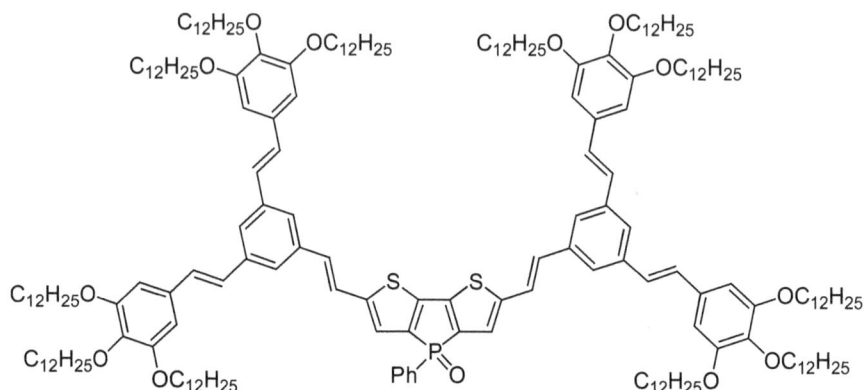

Figure 6: Dendritic phosphole with PPV units that combines liquid crystal behavior and gelation properties.

3. DENDRITIC ARCHITECTURES BASED ON POLY (PHENYLENEETHYNYLENE)

Elegant syntheses of poly(phenyleneethynylene)-based dendrimers, PPEdend (also known as phenylacetylene dendrimers), have been widely reported. In spite of the structural variations, the main method used for the preparation of these materials is based on Sonogashira chemistry and a convergent strategy. It is important to note that the conjugated and linear nature of the phenylacetylene moiety means that in low generation systems the dendrimers can adopt a relatively planar form in the solid state. However, higher generation dendrimers cannot adopt a planar arrangement in the solid state due to steric interactions [62]. Moore was the first to study extensively this kind of compound [63]. The poor intrinsic solubility of these compounds means that the poly(phenyleneethynylene) dendrons require solubilizing end groups. The 4-*tert*-butylphenyl peripheral units initially incorporated proved useful as solubilizing groups only up to the third generation dendron; however, the use of 3,5-di-*tert*-butylphenyl peripheral groups provided sufficient solubility to access fourth-generation dendrons (Scheme **17**) [64].

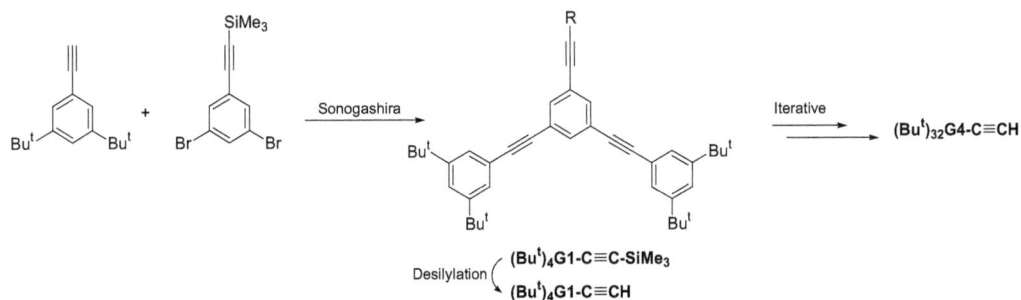

Scheme 17: Convergent synthesis of PPE dendrons developed by Moore and colleagues.

The only significant side reaction involved the oxidative dimerization of the desilylated dendrons to give the corresponding symmetrical diacetylenes. Surprisingly, the amount of these diacetylene byproducts increased with each generation and resulted in a dramatic reduction in dendron yield.

The dendrimers were obtained by Pd(0)-catalyzed cross-coupling between the appropriate dendron with acetylene at its focal point and tribromo- or triiodobenzene. As one would expect, yields are lower as the generation increases and diminish rapidly on using 1,3,5-tribromobenzene. The fifth-generation dendrimer could not be prepared in this manner as only the diacetylene byproduct of (mono)dendron oxidative dimerization was formed. However, the fifth-generation material was obtained in 37% yield when 1,3,5-triiodobenzene was used as the core (Scheme **18**). Molecular models indicated that this compound has a globular shape with a diameter of roughly 55 Å. Aryl iodides allow the coupling reaction to be performed at lower temperatures than the corresponding bromides and, as a consequence, the cross-coupling becomes competitive with the self-coupling reaction. As the molecular weight of the dendrons increases, it becomes necessary to use larger quantities of the catalyst. In this case, it is essential to run the reaction at low temperatures (35–40 °C) for longer periods of time (> 48 h).

Functionalization on the focal point also proved important for the reactivity of the compound. Indeed, the synthesis was later optimized by simply reversing the functionalities on the monomer while using a dialkyltriazene precursor for the focal iodo functionality (Fig. **7**) [65]. Yields were improved significantly and remained good even for the higher generations. The oxidative dimerization side reaction was essentially eliminated on using this approach. In addition, this reverse approach enabled the synthesis of multigram quantities of PPEdend even beyond the fifth generation, and allowed these compounds to be synthesized through a solid-supported technique.

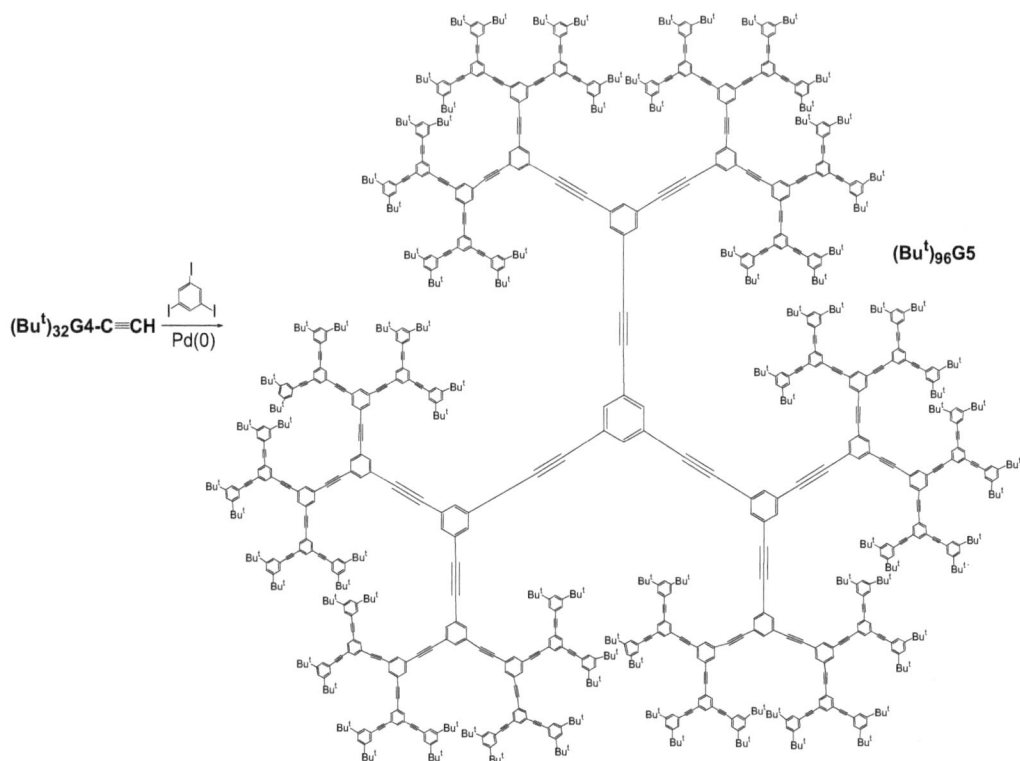

Scheme 18: Fifth-generation PPEdend described by Moore.

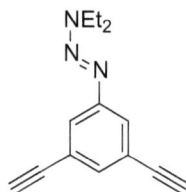

Figure 7: Monomer with a dialkyltriazene group at the focal point.

Generations higher than fourth are difficult to obtain because of De Gennes dense packing, a steric limit to regular growth [66]. One way to counteract this steric hindrance is the combination of convergent and linear growth approaches. For example, Moore and colleagues employed the concept of dendritic spacer elongation for the construction of geometrically expanded, large-size phenylacetylene macromolecules [33c, 67]. The methodology required a series of monomers and cores that gradually increased in size. The enlargement of the monomers and cores was accomplished by repetitive catenation of *para*-linked phenylacetylene structural units. In this way, the compound shown in Fig. **8** was obtained and is 12.5 nm in diameter. The approach involving the avoidance of steric inhibition to dendrimer construction through the use of increasingly longer spacer moieties gave rise to the acronym *SYNDROME* (*SYN*thesis of *D*endrimers by *R*epetition *O*f *M*onomer *E*nlargement).

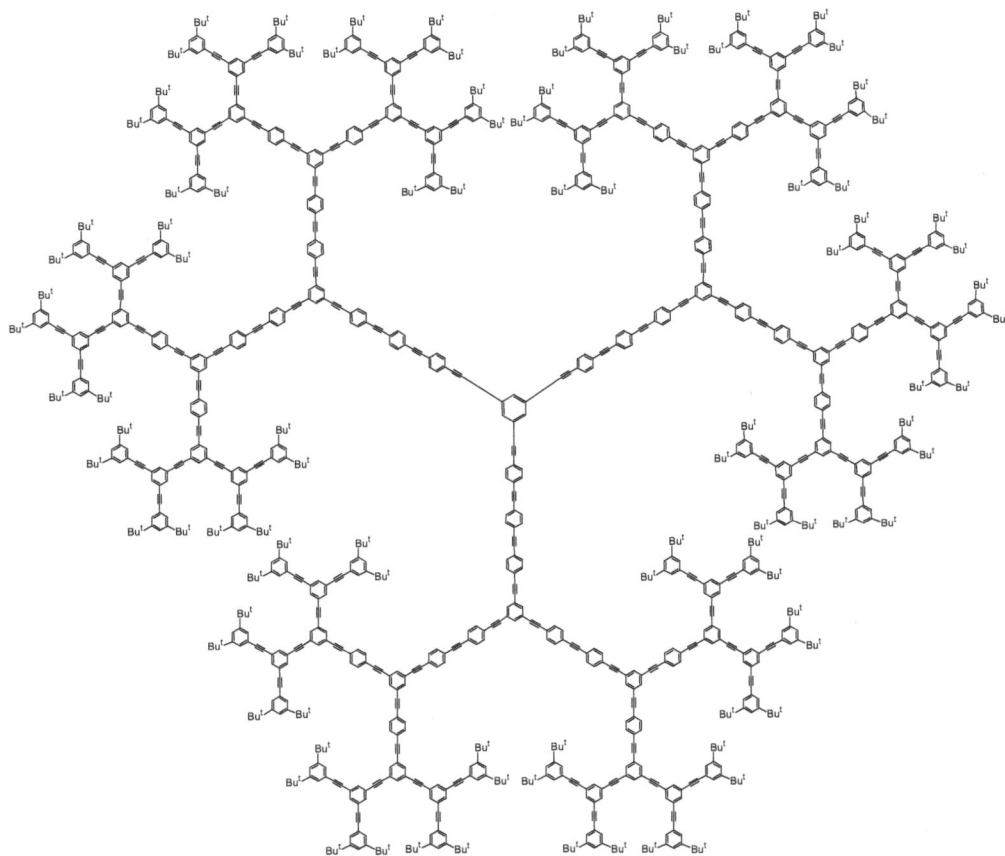

Figure 8: Dendrimer with a diameter of 12.5 nm described by Moore.

However, the strategy that allowed rapid access to high generation dendrimers was double exponential growth. This approach made use of a doubly protected monomer and combined both a convergent and divergent methodology (Scheme **19**) [68].

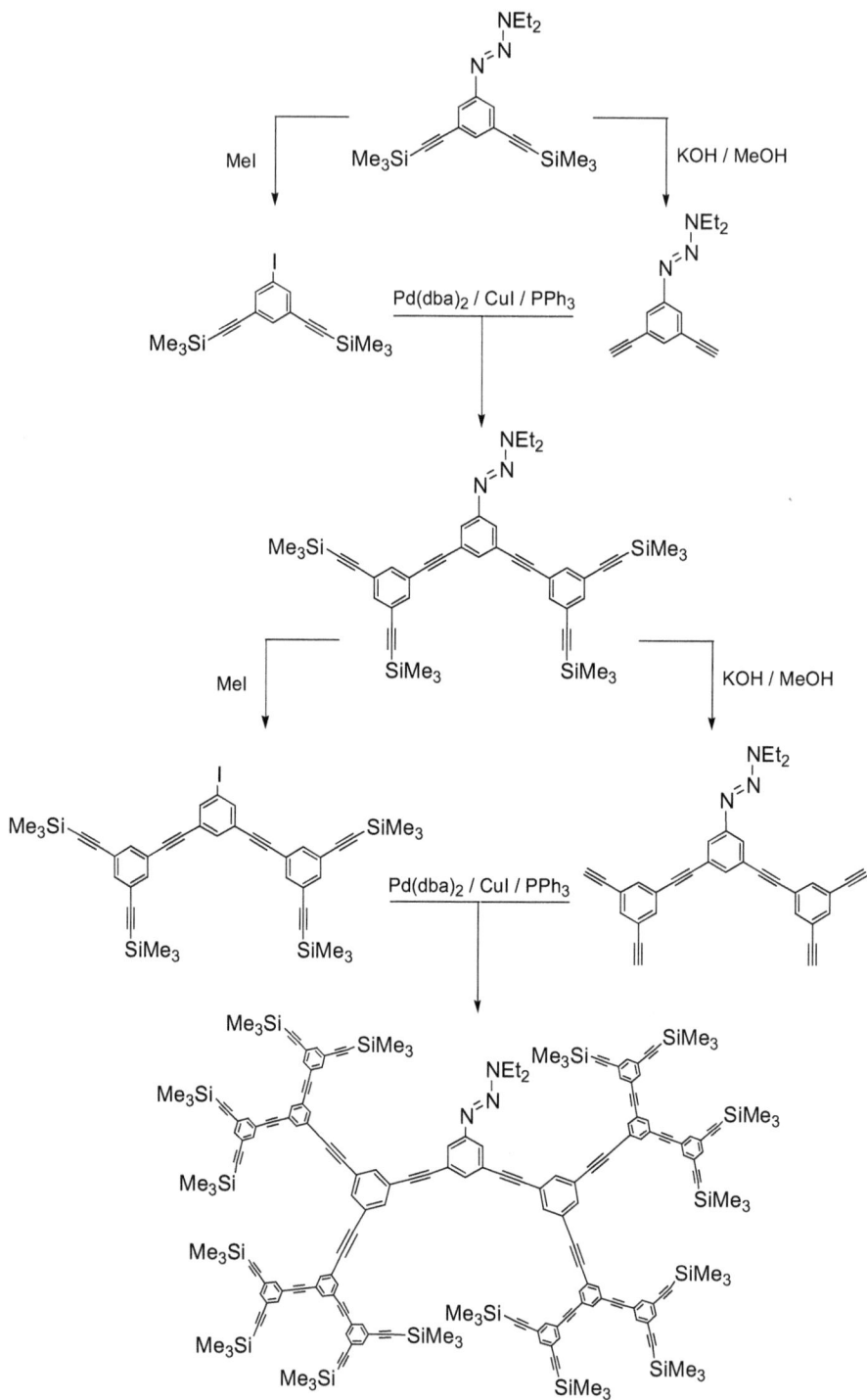

Scheme 19: Double exponential growth to access PPEdend.

The notable features of these phenylacetylene dendritic structures are the rigid repeat units that lead to 'shape-persistent' and 'dimension-persistent' dendrimers [69]. These factors can play an important role in the development of molecular frameworks that require rigorous control of functional group juxtaposition and the conjugated segments within the structure, which impart interesting photophysical properties [70]. One of the main aims of the work on phenylacetylene-based dendrimers has been to develop materials for the study of energy transfer. The treelike structure of a dendritic wedge naturally suggests itself as a molecular antenna suitable for the transfer of energy or electrons from the multitopic surface to the dendron focus. This process would involve excitation of the dendrimer primarily at a chromophore on the surface and the energy would then be transferred to a chromophore with a smaller HOMO-LUMO energy gap at the core [71].

The research of Moore and co-workers has been again particularly elegant. They reported a series of dendritic wedges with fluorescent perylene chromophores at the focal point [33c, 62, 72]. When the dendrons were irradiated at 310 nm (absorption band of the dendritic branching), the perylene chromophore emitted at 484 and 518 nm with quantum yields up to 98%. This indicates that singlet-singlet energy transfer occurs in the wedge. The light-harvesting ability increased with increasing generation, while the efficiency of the energy transfer decreased. The compound shown in Scheme **20** was particularly interesting as it contained an inbuilt energy gradient obtained by using phenylacetylene spacers that increased in size by one repeat unit proceeding from the periphery to the core. This design means that the levels of the localized electronic states decrease smoothly in energy from the exterior to the interior, creating a directional energy flow. This dendron was compared with analogous systems without an energy gradient, and its rate of energy transfer was two orders of magnitude faster. Carbonyl-focused PPEdend systems have also been studied in order to examine the basic process of electronic energy transfer through the phenylacetylene network by removing effects due to through-space interactions and size-dependent nonradiative relaxation processes [73]. Efficient energy migration followed by emission with high quantum yield was also confirmed in a series of dendrimers bearing a core stilbene moiety [74]. When the PPE conjugated units were attached to the BINOL structure, the resulting multiarmed dendritic molecules showed greatly amplified fluorescence responses [75]. Thus, the light harvesting effect of dendrimers could be used to greatly increase the sensitivity of the BINOL-based fluorescent sensors. Energy transfer was also observed in carbazole and silole-cored phenylacetylene dendrimers. These compounds were tested as chemical sensors for explosives [76].

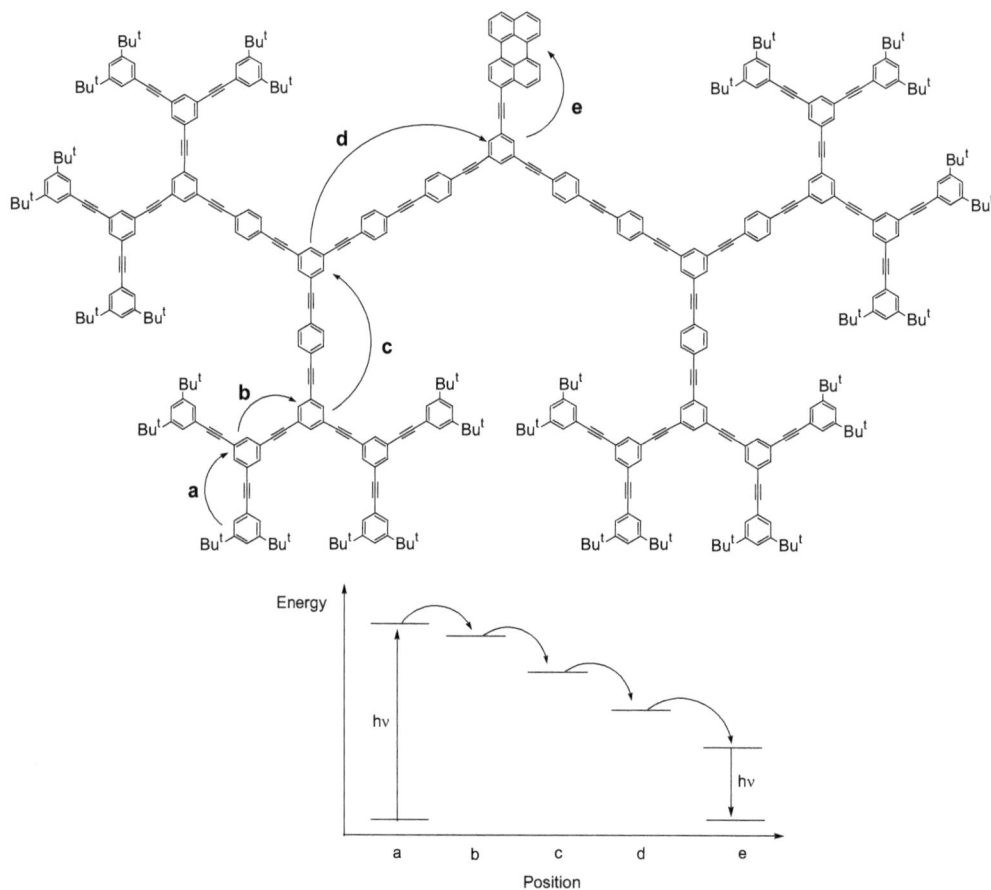

Scheme 20: Light harvesting process in PPE dendritic architectures.

In spite of the extensive synthesis and photophysical studies on dendrimers containing acetylene units, there have been few reports of their use in optoelectronic devices. The first reported OLEDs with a light-emitting dendrimer layer contained materials with structures based on the compound shown in Fig. **9**, which has diphenylacetylene moieties in the dendrons and a 9,10-

di(phenylethynyl)anthracene core [77]. Although the OLEDs emitted light, the emission was broad, which is indicative of excimer emission, and efficiency data were not reported, suggesting that these systems did not work particularly well. The excimer emission arises from the planarity of the phenylacetylene-containing dendrimers, a structural feature that allows the emissive chromophores to interact strongly in the solid state. The incorporation of hole-transport diphenylamine groups in place of the *tert*-butyl groups on the surface of the dendrimer [77] or the inclusion of the electron-transporting oxadiazole units [78] did not significantly improve the efficiency of the devices.

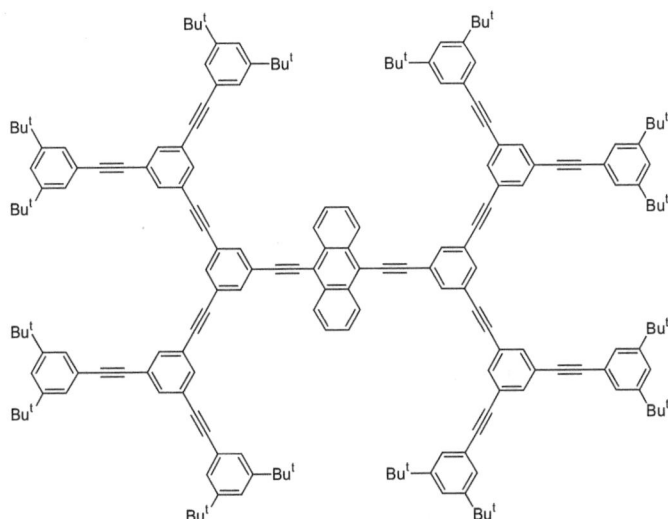

Figure 9: The first PPEdend used for the fabrication of an OLED.

Applications of PPEdends are not limited to electroluminescence. Replacement of the terminal *tert*-butyl groups by carboxylic groups gives rise to dendrimers that are soluble in water [79] and columnar liquid crystals have been prepared through substitution by polyether chains [80].

Other research groups have prepared this kind of dendrimer with porphyrins at the periphery [81] or hexasubstituted benzenes [82] and azobenzene [83] as cores. Peng *et al.* were the first to describe a new family of asymmetric conjugated dendrimers incorporating a 1,2-disubstituted benzene as a monomer (Scheme **21**) [84]. The fourth-generation dendrimer showed a broad absorption in the UV and visible range and possessed an intrinsic energy gradient from the outside branches to the core (due to the extended conjugation beyond its repeating units). This material showed efficient energy transfer from higher band-gap branches to the longest conjugated segment and naturally suggests itself as a potentially efficient light harvesting material. On the other hand, fullerene derivatives possessing similar asymmetric phenyleneethynylene-based dendrons have been synthesized. The photophysical investigations performed in CH_2Cl_2 have revealed an ultrafast dendron $\rightarrow C_{60}$ energy transfer in all these systems [85].

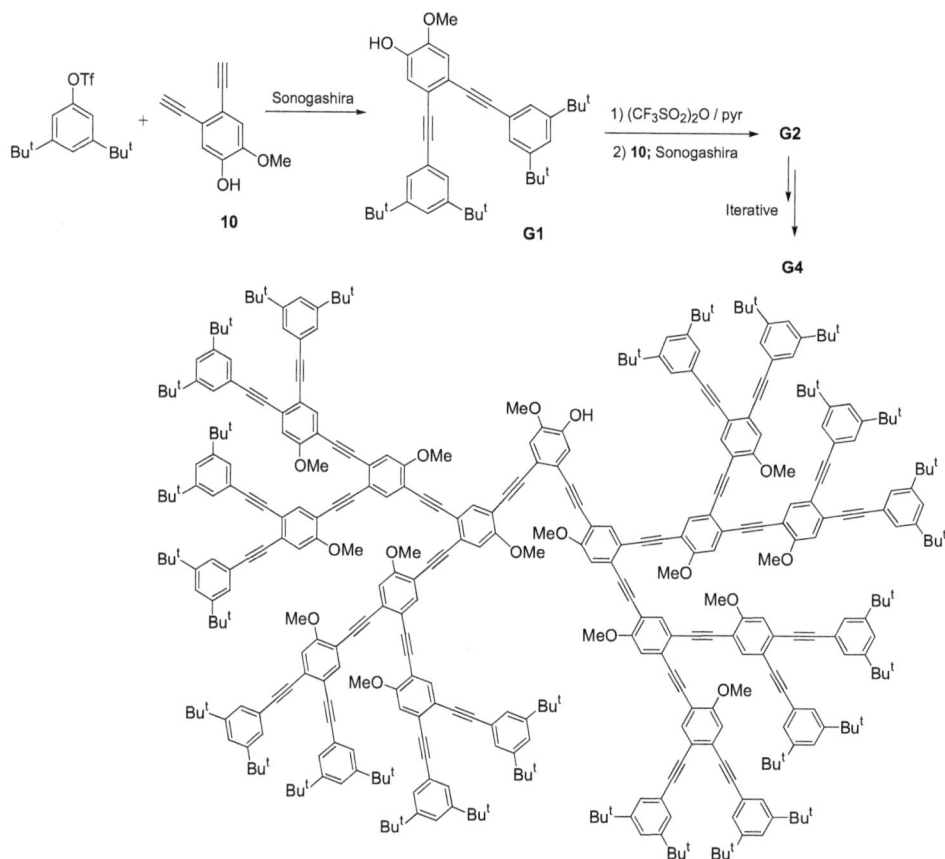

Scheme 21: Asymmetric PPEdend reported by Peng *et al.*

PPEdend have also found applications in catalysis. For example, 2,9-dimethyl-1,10-phenanthroline (neocuproine) has been attached to dendritic structures through its 5-position in order to study the copper-binding capability. The complexes were used to catalyze the Cu^+-promoted substitution of 4-iodoanisole to give 1,4-dimethoxybenzene with potentially easy recovery of the catalyst [86].

First-, second- and third-generation phenyleneethynylene monodendrons with a terminal acetylene at the focal point were readily polymerized using a Rh catalyst to yield the corresponding polydendrons, which had good membrane-forming ability and were available for oxygen-permselective membranes [87].

4. DENDRIMERS CONTAINING BOTH PHENYLENEVINYLENE AND PHENYLENE-ETHYNYLENE MOIETIES

As described throughout this review, several studies have been published to date concerning the synthesis and properties of conjugated light-emitting dendrimers based on either the phenylenevinylene or phenyleneethynylene structure. However, to the best of our knowledge there was no precedent for the synthesis of dendrimers whose branches incorporate both of these moieties at the same time, although it has recently been demonstrated that the presence of both double and triple bonds in a conjugated structure had a strong influence on the resulting optoelectronic properties. For example, poly(*p*-phenylenevinylene)s (PPVs) with tolane-bis-benzyl moieties [88], conjugated oligomers and polymers with double and triple bonds [89], and dithiafulvene-acetylene hybrid chromophores [90] have been studied as PLEDs, fluorophores, and materials with electronic properties of interest, respectively.

In this context, we developed an efficient orthogonal and convergent synthesis for new dendritic architectures containing both phenyleneethynylene and phenylenevinylene in an alternating manner within the dendritic arms [91]. These materials were effectively synthesized up to the fourth generation using two routes that allowed specific control over the location of double and triple bonds in the interior of the dendrimers. Our plan was again based on the combination of two coupling methods; the Sonogashira and HWE reactions. These reactions were used in an alternating order to build up each successive layer (Scheme 22).

The key to the overall strategy is that building blocks 5, 8, and 10 possess mutually complementary functional groups at the terminal positions while avoiding the need for the introduction of additional functional group protection/deprotection steps.

8 + 2 OHC—⬡—OC$_{12}$H$_{25}$ $\xrightarrow[93\%]{\text{HWE}}$ (OC$_{12}$H$_{25}$)$_2$G1-I

10 + 2 (OC$_{12}$H$_{25}$)$_2$G1-I $\xrightarrow[64\%]{\text{Sonogashira}}$ (OC$_{12}$H$_{25}$)$_4$G2-CHO

8 + 2 (OC$_{12}$H$_{25}$)$_4$G2-CHO $\xrightarrow[51\%]{\text{HWE}}$ (OC$_{12}$H$_{25}$)$_8$G3-I

11 + 3 (OC$_{12}$H$_{25}$)$_2$G1-I $\xrightarrow[54\%]{\text{Sonogashira}}$ (OC$_{12}$H$_{25}$)$_6$G2

3 + 3 (OC$_{12}$H$_{25}$)$_4$G2-CHO $\xrightarrow[71\%]{\text{HWE}}$ (OC$_{12}$H$_{25}$)$_{12}$G3

5 + 2 ⬡—C$_5$H$_{11}$ $\xrightarrow[97\%]{\text{Sonogashira}}$ (C$_5$H$_{11}$)$_2$G'1-CHO

8 + 2 (C$_5$H$_{11}$)$_2$G'1-CHO $\xrightarrow[91\%]{\text{HWE}}$ (C$_5$H$_{11}$)$_4$G'2-I

10 + 2 (C$_5$H$_{11}$)$_4$G'2-I $\xrightarrow[41\%]{\text{Sonogashira}}$ (C$_5$H$_{11}$)$_8$G'3-CHO

8 + 2 (C$_5$H$_{11}$)$_8$G'3-CHO $\xrightarrow[32\%]{\text{HWE}}$ (C$_5$H$_{11}$)$_{16}$G'4-I

3 + 3 (C$_5$H$_{11}$)$_2$G'1-CHO $\xrightarrow[86\%]{\text{HWE}}$ (C$_5$H$_{11}$)$_6$G'2

11 + 3 (C$_5$H$_{11}$)$_4$G'2-I $\xrightarrow[38\%]{\text{Sonogashira}}$ (C$_5$H$_{11}$)$_{12}$G'3

3 + 3 (C$_5$H$_{11}$)$_8$G'3-CHO $\xrightarrow[65\%]{\text{HWE}}$

(C$_5$H$_{11}$)$_{24}$G'4
R = C$_5$H$_{11}$

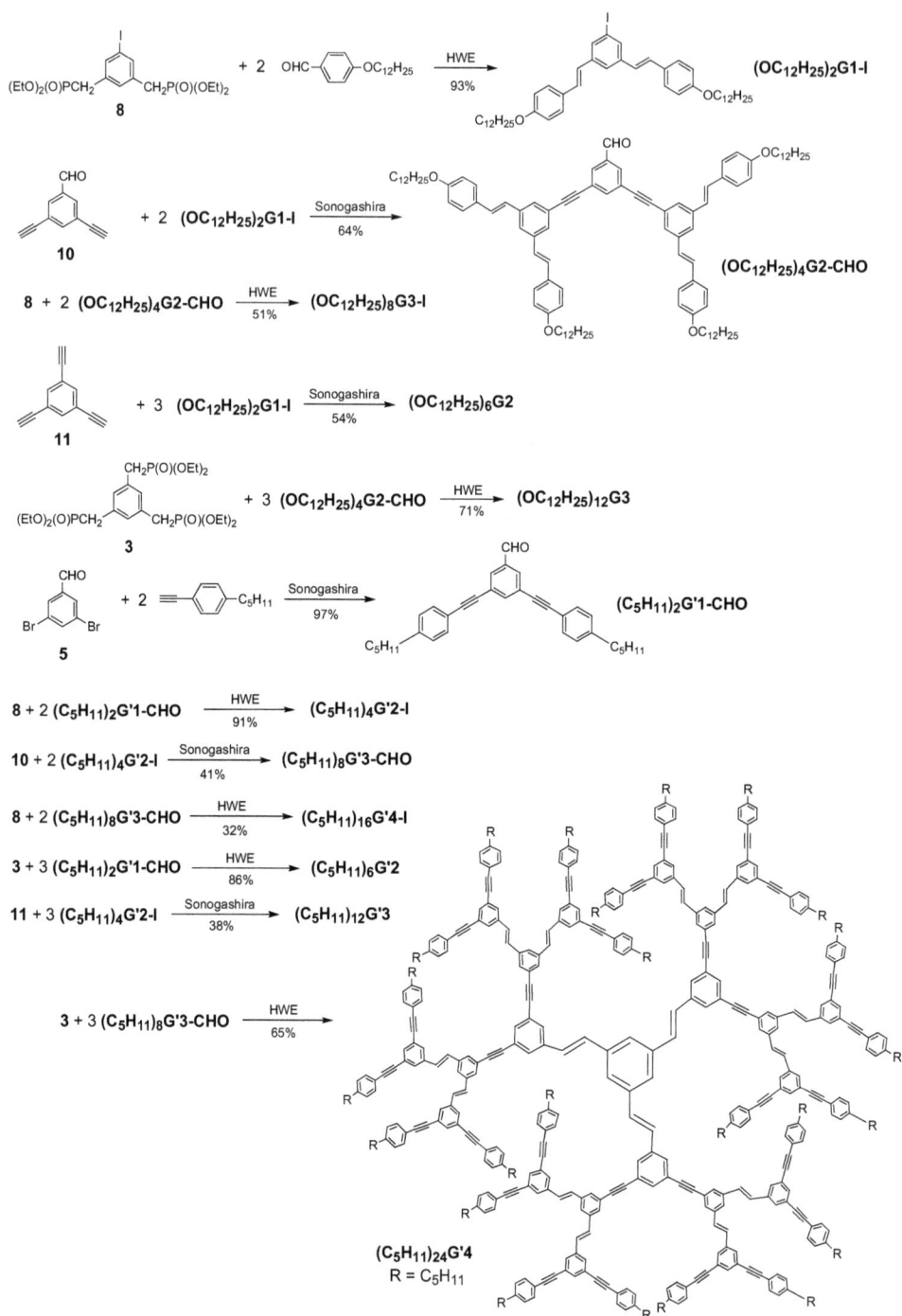

Scheme 22: Synthesis of hybrid dendrons and dendrimers containing both phenylenevinylene and phenyleneethynylene moieties.

The peripheral groups selected in this work provided good control of the solubility. All of the new compounds were highly soluble in THF as well as in chlorinated solvents such as dichloromethane and chloroform, thus allowing purification by conventional silica gel chromatography. The appropriate choice of starting materials could provide a wide variety of dendritic architectures with different peripheral moieties, although the use of peripheral pendant groups other than alkyl would be expected to afford molecules with decreased solubility.

This methodology could be used for rapid access to a wide range of conjugated dendrimers. An example of this potential was the synthesis of the second-generation dendrimers shown in Scheme **23**, starting from 1,3,5-tris(2-ethynylphenyl)benzene as the core [91, 92].

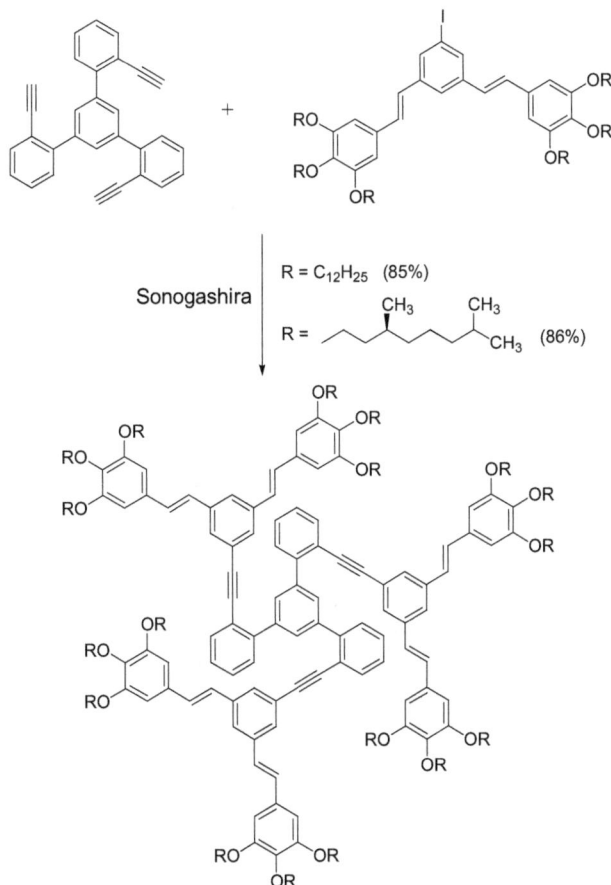

Scheme 23: Hybrid dendrimers with 1,3,5-tris(2-ethynylphenyl)benzene as the core.

In a similar way to their counterparts with homogeneous scaffolds (only double or triple bonds), the UV-vis absorption spectra consisted essentially of a simple superposition of the absorptions due to the different chromophores, *i.e.* stilbene and tolane moieties. These molecules were almost colorless and were absolutely transparent above 400 nm. All of the compounds prepared were fluorescent and emitted blue light when irradiated at the absorption maxima. The intensity of the fluorescence from the higher generation dendrimers was substantially lower than that from the lower generations at the same molar concentration – behavior that is comparable to that of PPVdend (see section 2.2) and similar to that noted by Moore and colleagues in a phenylacetylene dendrimer series [33c], as well as in monodendrons based on 9-phenylcarbazole [33a]. On the other hand, $(C_5H_{11})_4G'2$-I showed a low fluorescence quantum yield and this finding can be attributed to effective quenching by the iodine atom. The heavy atom effect was lower for $(OC_{12}H_{25})_8G3$-I and $(C_5H_{11})_{16}G'4$-I because the iodine atoms are located far away from the peripheral fluorescent units. The fluorescence of dendrons bearing formyl groups at the focal point is also strongly quenched. The higher quantum yield observed for $(OC_{12}H_{25})_6G2$ when compared with $(C_5H_{11})_6G'2$ could be a consequence of the different peripheral functionalization, although the bigger dendrimers, $(OC_{12}H_{25})_{12}G3$ and $(C_5H_{11})_{12}G'3$, showed similar values. A complete explanation for these results would require further investigations. The magnitudes of the Stokes shifts decreased as the generation number increased, a situation that was also encountered for the phenylenevinylene series (see section 2.2) [32a].

CONCLUSIONS

This article has provided a general overview of the main strategies for the synthesis of PPVdend, PPEdend and hybrid dendritic structures that combine both scaffolds, as well as highlighting some physical and chemical properties of these structures. Specifically, PPVdend systems are typically prepared through convergent approaches that involve either Wittig-type chemistry, palladium-catalyzed chemistry, or a combination of the two. In general, the former strategy is more suitable because of the easy treatment of the crude reaction mixtures, high control of stereochemistry and the versatility in the final functionalization that, as a last resort, determines the physical and chemical properties of materials for applications. The preparation of PPEdend mainly relies on cross-coupling chemistry and variations in the starting material make it possible to overcome steric hindrance and allow growth toward larger generation dendrimers. Finally, the synthesis of hybrid dendrimers is usually carried out by an orthogonal combination of HWE and Sonogashira reactions.

In all cases, the *meta* arrangement, through which the stilbene and/or tolane units are linked, results in π-conjugated dendrimers that behave as an ensemble of almost independent chromophores. These materials emit strong blue light under UV irradiation and the intensity of the fluorescence decreases as the generation increases; probably because the through-space interaction between the fluorescent units becomes more significant on increasing molecular size, providing additional fluorescence quenching pathways. The final properties of these compounds strongly depend on the peripheral substitution and/or the nature of the core. Thus, materials with potential applications as diverse as OLEDs, light-harvesting antennae or solar cells (photoinduced charge-separation) have been produced.

Nowadays, the synthetic strategies to prepare fully conjugated dendrimers are highly developed and only slight modifications from those shown here might appear. Therefore, the current high level of interest from the scientific community in this area is focused on addressing the needs of the industry, and on overcoming implementation difficulties of materials in real devices. From a molecular standpoint, the development of materials with good charge transportation properties – without a significant quenching of the fluorescence in solid state – as well as phosphorescent compounds – especially short wavelength emitters – are still significant challenges. Furthermore, these materials can serve as models to address various scientific questions.

ACKNOWLEDGEMENTS

Financial support from the Ministerio de Ciencia e Innovación (Spain)/FEDER (European Union) is gratefully acknowledged — project BFU2011-30161-C02-02. The authors wish to express their gratitude to the present and former members of our research group and to our collaborators. Without their exceptional efforts this body of work would not have been possible.

CONFLICT OF INTEREST

The author confirms that this chapter content has no conflict of interest.

DISCLOSURE

The chapter submitted for series eBook entitled "**Advances in Organic Synthesis, Volume 5**" is an update of our article published in **CURRENT ORGANIC SYNTHESIS, Volume 5, Number 3, 2008,** with additional text and references.

REFERENCES

[1]   (a) Organic Electronics and Optoelectronics; Special Issue: Forrest, S. R.; Thompson, M. E., Eds., Chem. Rev. **2007**, 107(4). (b) Dai, L. Intelligent Macromolecules for Smart Devices: From Materials Synthesis to Device Applications; Springer-Verlag: London, **2004**. (c) Dai, L.; Winkler, B.; Dong, L.; Tong, L.; Mau, A. W. H. Conjugated polymers for light-emitting applications. Adv. Mater., **2001**, 13, 915-925. (d) Tour, J. M. Molecular electronics. Synthesis and testing of components. Acc. Chem. Res., **2000**, 33, 791-804. (e) McQuade, D. T.; Pullen, A. E.; Swager, T. M. Conjugated polymer-based chemical sensors. Chem. Rev., **2000**, 100, 2537-2574. (f) Sheats, J. R.; Barbara, P. F. Molecular materials in electronic and optoelectronic devices. Acc. Chem. Res., **1999**, 32, 191-192. (g) Electronic Materials: The Oligomer Approach; Müllen, K.; Wegner, G., Eds.; Wiley-VCH: Weinheim, **1998**. (h) Swager, T. M. The molecular wire approach to sensory signal amplification. Acc. Chem. Res., **1998**, 31, 201-207. (i) Handbook of Conducting Polymers, 2nd ed.; Skotheim, T. A., Elsenbaumer, R. L., Reynolds, J., Eds.; Marcel Dekker: New York, **1997**. (j) Tour, J. M. Conjugated macromolecules of precise length and constitution. Organic synthesis for the construction of nanoarchitectures. Chem. Rev., **1996**, 96, 537-554.

[2]   Conjugated Polymer Synthesis; Chujo, Y., Ed.; Wiley-VCH: Weinheim, **2010**.

[3]   (a) Murray, M. M.; Holmes, A. B. Poly(arylene vinylene)s–synthesis and applications in semiconductor devices, In: Semiconducting Polymers, Chemistry, Physics and Engineering; Hadziioannou, G.; van Hutten, P. F., Eds.; Wiley-VCH: Weinheim, **2005**. (b) Scherf, U. Oligo- and polyarylenes, oligo- and polyarylenevinylenes. Top. Curr. Chem., **1999**, 201, 163-222. (c) Kraft, A.; Grimsdale, A. C.; Holmes, A. B. Electroluminescent conjugated polymers – seeing polymers in a new light. Angew. Chem. Int. Ed., **1998**, 37, 402-428. (d) Hide, F.; Díaz-García, M. A.; Schwartz, B. J.; Heeger, A. J. New developments in the photonic applications of conjugated polymers. Acc. Chem. Res., **1997**, 30, 430-436.

[4]   (a) Poly(arylene ethynylene)s; Special Issue: Weder, C., Ed., Adv. Polym. Sci., **2005**, 177. (b) Bunz, U. H. F. Poly(aryleneethynylene)s: syntheses, properties, structures, and applications. Chem. Rev., **2000**, 100, 1605-1644. (c) Yang, J.-S.; Swager, T. J. Fluorescent porous polymer films as TNT chemosensors: electronic and structural effects. J. Am. Chem. Soc., **1998**, 120, 11864-11873. (d) Bumm, L. A.; Arnold, J. J.; Cygan, M. T.;

Dunbar, T. D.; Burgin, T. P.; Jones, II, L.; Allara, D. L.; Tour, J. M.; Weiss, P. S. Are single molecular wires conducting?. Science, **1996**, 271, 1705-1707. (e) Zhou, Q.; Swager, T. M. Fluorescent chemosensors based on energy migration in conjugated polymers: the molecular wire approach to increased sensitivity. J. Am. Chem. Soc., **1995**, 117, 12593-12602.

[5]   (a) Designing Dendrimers; Campagna, S.; Ceroni, P.; Puntoriero, F., Eds.; Wiley: Hoboken, NJ, **2011**. (b) Hourani, R.; Kakkar, A. Advances in the elegance of chemistry in designing dendrimers. Macromol. Rapid Commun., **2010**, 31, 947-974. (c) Vögtle, F.; Richard, G.; Werner N. Dendrimer Chemistry: Concepts, Syntheses, Properties, Applications; Wiley-VCH: Weinheim, **2009**.

[6]   (a) Majoral, J.-P., Ed., New J. Chem., **2012**, 36 (2). (b) Duarte, A.; Pu, K.-Y.; Liu, B.; Bazan, G. C. Recent advances in conjugated polyelectrolytes for emerging optoelectronic applications. Chem. Mater., **2011**, 23, 501-515. (c) Cameron, D. J. A.; Shaver, M. P. Aliphatic polyester polymer stars: synthesis, properties and applications in biomedicine and nanotechnology. Chem. Soc. Rev., **2011**, 40, 1761-1776. (d) Dendrimers: Towards Catalytic, Material and Biomedical Uses; Caminade, A.-M.; Turrin, C.-D.; Laurent, R.; Ouali, A.; Delavaux-Nicot, B., Eds.; Wiley: Chichester, **2011**. (e) Röglin, L.; Lempens, E. H. M.; Meijer, E. W. A synthetic "tour de force": well-defined multivalent and multimodal dendritic structures for biomedical applications. Angew. Chem. Int. Ed., **2011**, 50, 102-112. (f) Astruc, D.; Boisselier, E.; Ornelas, C. Dendrimers designed for functions: from physical, photophysical and supramolecular properties to applications in sensing, catalysis, molecular electronics, photonics and nanomedicine. Chem. Rev., **2010**, 110, 1857-1959. (g) Paleos, C. M.; Tsiourvas, D.; Sideratou, Z.; Tziveleka, L.-A. Drug delivery using multifunctional dendrimers and hyperbranched polymers. Expert Opin. Drug Del., **2010**, 7, 1387-1398. (h) Dendrimer-Based Nanomedicine; Majoros, I.; Baker, J. R., Eds.; Pan Stanford Publishing Pte. Ltd.: Singapore, **2008**. (i) Majoral, J.-P., Ed., New J. Chem. **2007**, 31 (7). (j) Boas, U.; Christensen, J. B.; Heegaard, P. M. H. Dendrimers in Medicine and Biotechnology: New Molecular Tools; RSC Publishing: Cambridge, **2006**.

[7]   See, for example: (a) Bergamini, G.; Marchi, E.; Ceroni, P. Metal ion complexes of cyclam-cored dendrimers for molecular photonics. Coord. Chem. Rev., **2011**, 255, 2458-2468. (b) Balzani, V.; Bergamini, G.; Ceroni, P.; Vögtle, F. Electronic spectroscopy of metal complexes with dendritic ligands. Coord. Chem. Rev., **2007**, 251, 525-535. (c) Ceroni, P.; Bergamini, G.; Marchioni, F.; Balzani, V. Luminescence as a tool to investigate dendrimer properties. Prog. Polym. Sci., **2005**, 30, 453-473. (d) De Schryver, F. C.; Vosch, T.; Cotlet, M.; Van der Auweraer, M.; Müllen, K.; Hofkens, J. Energy dissipation in multichromophoric single dendrimers. Acc. Chem. Res., **2005**, 38, 514-522. (e) Nierengarten, J.-F.; Armaroli, N.; Accorsi, G.; Rio, Y.; Eckert, J. F. [60]Fullerene: a versatile photoactive core for dendrimer chemistry. Chem. Eur. J., **2003**, 9, 36-41. (f) Balzani, V.; Ceroni P.; Maestri, M.; Saudan, C.; Vicinelli, V. Luminescent dendrimers. Recent advances. Top. Curr. Chem., **2003**, 228, 159-191. (g) Yokoyama, S.; Otomo, A.; Nakahama, T.; Okuno, Y.; Mashiko, S. Dendrimers for optoelectronics applications. Top. Curr. Chem., **2003**, 228, 197-220.

[8]   (a) Andrews D. L. Light harvesting in dendrimer materials: designer photophysics and electrodynamics. J. Mat. Research, **2012**, 27, 627-638. (b) Fukuzumi, S.; Ohkubo, K. Assemblies of artificial photosynthetic reaction centres. J. Mater. Chem., **2012**, 22, 4575-4587. (c) Balzani, V.; Bergamini, G.; Ceroni, P.; Marchi, E. Designing light harvesting

antennas by luminescent dendrimers. New J. Chem., **2011**, 35, 1944-1954. (d) Köse, M. E.; Long, H.; Kim, K.; Graf, P.; Ginley, D. Charge transport simulations in conjugated dendrimers. J. Phys. Chem. A, **2010**, 114, 4388-4393. (e) Wang, J.-L.; Yan, J.; Tang, Z.-M-; Xiao, Q.; Ma, Y.; Pei, J. Gradient shape-persistent π-conjugated dendrimers for light-harvesting: synthesis, photophysical properties, and energy funneling. J. Am. Chem. Soc., **2008**, 130, 9952-9962. (f) Choi, M.-S.; Yamazaki, T.; Yamazaki, I.; Aida, T. Bioinspired molecular design of light-harvesting multiporphyrin arrays. Angew. Chem. Int. Ed., **2004**, 43, 150-158. (g) Kawa, M. Antenna effects of aromatic dendrons and their luminescence applications. Top. Curr. Chem., **2003**, 228, 193-204. (h) Balzani, V.; Juris, A. Photochemistry and photophysics of Ru(II)–polypyridine complexes in the Bologna group. From early studies to recent developments. Coord. Chem. Rev., **2001**, 211, 97-115. (i) Adronov, A.; Fréchet, J. M. J. Light-harvesting dendrimers. Chem. Commun., **2000**, 1701-1710.

[9]   (a) Tsuda, K.; Dol, G. C.; Gensch, T.; Hofkens, J.; Latterini, L.; Weener, J. W.; Meijer, E. W.; De Schryver, F. C. Fluorescence from azobenzene functionalized poly(propylene imine) dendrimers in self-assembled supramolecular structures. J. Am. Chem. Soc,. **2000**, 122, 3445-3452. (b) Archut, A.; Azzellini, G. C.; Balzani, V.; De Cola, L.; Vögtle, F. Toward photoswitchable dendritic hosts. Interaction between azobenzene-functionalized dendrimers and eosin. J. Am. Chem. Soc., **1998**, 120, 12187-12191. (c) Archut, A.; Vögtle, F ; De Cola, L.; Azzellini, G. C.; Balzani, V.; Ramanujam, P. S.; Berg, R. H. Azobenzene-functionalized cascade molecules: photoswitchable supramolecular systems. Chem. Eur. J., **1998**, 4, 699-706.

[10]   (a) Vicinelli, V.; Ceroni, P.; Maestri, M.; Balzani, V.; Gorka, M.; Vögtle, F. Luminescent lanthanide ions hosted in a fluorescent polylysin dendrimer. Antenna-like sensitization of visible and near-infrared emission. J. Am. Chem. Soc., **2002**, 124, 6461-6468. (b) Vögtle, F.; Gestermann, S.; Kauffmann, C.; Ceroni, P.; Vicinelli, V.; De Cola, L.; Balzani, V. Poly(propylene amine) dendrimers with peripheral dansyl units: protonation, absorption spectra, photophysical properties, intradendrimer quenching, and sensitization processes. J. Am. Chem. Soc., **1999**, 121, 12161-12166.

[11]   Yeow, E. K. L.; Ghiggino, K. P.; Reek, J.; N. H.; Crossley, M. J.; Bosman, A. W.; Schenning, A. P. H. J.; Meijer, E. W. The dynamics of electronic energy transfer in novel multiporphyrin functionalized dendrimers: a time-resolved fluorescence anisotropy study. J. Phys. Chem. B, **2000**, 104, 2596-2606.

[12]   (a) Precup-Blaga, F. S.; García-Martínez, J. C.; Schenning, A. P. H. J.; Meijer, E. W. Highly emissive supramolecular oligo(p-phenylene vinylene) dendrimers. J. Am. Chem. Soc., **2003**, 125, 12953-12960. (b) Schenning, A. P. H. J.; Jonkheijm, P.; Hofkens, J.; De Feyter, S.; Asavei, T.; Cotlet, M.; De Schryver, F. C.; Meijer, E. W. Formation and manipulation of supramolecular structures of oligo(p-phenylenevinylene) terminated poly(propylene imine) dendrimers. Chem. Commun., **2002**, 1264-1265. (c) Meskers, S. C. J.; Bender, M.; Hübner, J.; Romanovskii, Y. V.; Oestreich, M.; Schenning, A. P. H. J.; Meijer, E. W.; Bässler, H. Interchromophoric coupling in oligo(p-phenylenevinylene)-substituted poly(propyleneimine) dendrimers. J. Phys. Chem. A, **2001**, 105, 10220-10229. (d) Schenning, A. P. H. J.; Peeters, E.; Meijer, E. W. Energy transfer in supramolecular assemblies of oligo(p-phenylene vinylene)s terminated poly(propylene imine) dendrimers. J. Am. Chem. Soc., **2000**, 122, 4489-4495.

[13] Martín-Zarco, M.; Toribio, S.; García-Martínez, J. C.; Rodríguez-López, J. Polyamido amine dendrimers functionalized with poly(phenylenevinylene) dendrons at their periphery. J. Polym. Sci., Part A: Polym. Chem., **2009**, 47, 6409-6419.

[14] Baker, L. A.; Crooks, R. M. Photophysical properties of pyrene-functionalized poly(propylene imine) dendrimers. Macromolecules, **2000**, 33, 9034-9039.

[15] García-Martínez, J. C.; Díez-Barra, E.; Rodríguez-López, J. Conjugated dendrimers with poly(phenylenevinylene) and poly(phenyleneethynylene) scaffolds. Curr. Org. Synth., **2008**, 5, 267-290.

[16] Meier, H.; Lehmann, M. In: Encyclopedia of Nanoscience and Nanotechnology; Nalwa, H. S., Ed.; American Scientific Publishers: Stevensons Ranch, **2005**; Vol. 10, pp. 95-106.

[17] Siegrist, A. E.; Liechti, P.; Meyer, H. R.; Weber, K. Anil-synthese. 3. Mitteilung [1] über die darstellung von styryl-derivaten aus methyl-substituierten carbocyclischen aromaten. Helv. Chim. Acta, **1969**, 52, 2521-2554.

[18] Deb, S. K.; Maddux, T. M.; Yu, L. A simple orthogonal approach to poly(phenylenevinylene) dendrimers. J. Am. Chem. Soc. **1997**, 119, 9079-9080.

[19] Chow, H.-F.; Ng, M.-K.; Leung, C.-W.; Wang, G.-X. Dendrimer interior functional group conversion and dendrimer metamorphosis – new approaches to the synthesis of oligo(dibenzyl sulfone) and oligo(phenylenevinylene) dendrimers. J. Am. Chem. Soc., **2004**, 126, 12907-12915.

[20] (a) Meier, H.; Lehmann, M.; Kolb, U. Stilbenoid dendrimers. Chem. Eur. J., **2000**, 6, 2462-2469. (b) Meier, H.; Lehmann, M. Stilbenoid dendrimers. Angew. Chem. Int. Ed., **1998**, 37, 643-645.

[21] Rosenfeldt, S.; Karpuk, E.; Lehmann, M.; Meier, H.; Lindner, P.; Harnau, L.; Ballauff, M. The solution structure of stilbenoid dendrimers: a small-angle scattering study. ChemPhysChem, **2006**, 7, 2097-2104.

[22] Meier, H.; Karpouk, E.; Lehmann, M.; Schollmeyer, D.; Enkelmann, V. Guest-host systems of 1,3,5-tristyrylbenzenes. Z. Naturforsch., **2003**, 58b, 775-781.

[23] Lehmann, M.; Fischbach, I.; Spiess, H. W.; Meier, H. Photochemistry and mobility of stilbenoid dendrimers in their neat phases. J. Am. Chem. Soc., **2004**, 126, 772-784.

[24] Tschierske, C. Micro-segregation, molecular shape and molecular topology – partners for the design of liquid crystalline materials with complex mesophase morphologies. J. Mater. Chem., **2001**, 11, 2647-2671.

[25] Pillow, J. N. G.; Burn, P. L.; Samuel, I. D. W.; Halim, M. Synthetic routes to phenylene vinylene dendrimers. Synth. Met., **1999**, 102, 1468-1469.

[26] Pillow, J. N. G.; Halim, M.; Lupton, J. M.; Burn, P. L.; Samuel, I. D. W. A facile iterative procedure for the preparation of dendrimers containing luminescent cores and stilbene dendrons. Macromolecules, **1999**, 32, 5985-5993.

[27] (a) Lo, S.-C.; Burn, P. L. Development of dendrimers: macromolecules for use in organic light-emitting diodes and solar cells. Chem. Rev., **2007**, 107, 1097-1116. (b) Burn. P. L.; Lo, S.-C.; Samuel, I. D. W. The development of light-emitting dendrimers for displays. Adv. Mater., **2007**, 19, 1675-1688.

[28] Halim, M.; Samuel, I. D. W.; Pillow, J. N. G.; Burn, P. L. Conjugated dendrimers for LEDs: control of colour. Synth. Met., **1999**, 102, 1113-1114.

[29] (a) Pillow, J. N. G.; Samuel, I. D. W.; Burn, P. L. Conjugated dendrimers for light-emitting diodes: effect of generation. Adv. Mater. **1999**, 11, 371-374. (b) Halim, M.; Pillow, J. N.

G.; Samuel, I. D. W.; Burn, P. L. The effect of dendrimer generation on LED efficiency. Synth. Met., **1999**, 102, 922-923.

[30] Pålsson, L.-O.; Beavington, R.; Frampton, M. J.; Lupton, J. M.; Magennis, S. W.; Markham, J. P. J.; Pillow, J. N. G.; Burn, P. L.; Samuel, I. D. W. Synthesis and excited state spectroscopy of tris(distyrylbenzenyl)amine-cored electroluminescent dendrimers. Macromolecules, **2002**, 35, 7891-7901.

[31] (a) Lupton, J. M.; Samuel, I. D. W.; Beavington, R.; Burn, P. L.; Bässler, H. Control of charge transport and intermolecular interaction in organic light-emitting diodes by dendrimer generation. Adv. Mater., **2001**, 13, 258-261. (b) Lupton, J. M.; Samuel, I. D. W.; Beavington, R.; Framptom, M. J.; Burn, P. L.; Bässler, H. Control of mobility in molecular organic semiconductors by dendrimer generation. Phys. Rev. B, **2001**, 63, 155206.

[32] (a) Díez-Barra, E.; García-Martínez, J. C.; Merino, S.; del Rey, R.; Rodríguez-López, J.; Sánchez-Verdú, P. Tejeda, J. Synthesis, characterization, and optical response of dipolar and non-dipolar poly(phenylenevinylene) dendrimers. J. Org. Chem., **2001**, 66, 5664-5670. (b) Díez-Barra, E.; García-Martínez, J. C.; Rodríguez-López, J. A Horner-Wadsworth-Emmons approach to dipolar and non-dipolar poly(phenylenevinylene)-dendrimers. Tetrahedron Lett., **1999**, 40, 8181-8184.

[33] (a) Zhu, Z.; Moore, J. S. Synthesis and characterization of monodendrons based on 9-phenylcarbazole. J. Org. Chem., **2000**, 65, 116-123. (b) Devadoss, C.; Bharathi, P.; Moore, J. S. Photoinduced electron transfer in dendritic macromolecules. 1. Intermolecular electron transfer. Macromolecules, **1998**, 31, 8091-8099. (c) Devadoss, C.; Bharathi, P.; Moore, J. S. Energy transfer in dendritic macromolecules: molecular size effects and the role of an energy gradient. J. Am. Chem. Soc., **1996**, 118, 9635-9644.

[34] (a) Palomero, J.; Mata, J. A.; González, F.; Peris, E. Facile synthesis of first generation ferrocene dendrimers by a convergent approach using ditopic conjugated dendrons. New J. Chem., **2002**, 26, 291-297. (b) Peruga, A.; Mata, J. A.; Sainz, D.; Peris, E. Facile synthesis of bidimensional ferrocenyl-based branched oligomers by palladium-catalyzed coupling reactions. J. Organomet. Chem., **2001**, 637, 191-197.

[35] Kwon, T. W.; Alam, M. M.; Jenekhe, S. A. n-Type conjugated dendrimers: convergent synthesis, photophysics, electroluminescence, and use as electron-transport materials for light-emitting diodes. Chem. Mater., **2004**, 16, 4657-4666.

[36] Zhang, X.-H.; Choi, S.-H.; Choi, D. H.; Ahn, K.-H. Synthesis and photophysical properties of phenothiazine-labeled conjugated dendrimers. Tetrahedron Lett., **2005**, 46, 5273-5276.

[37] (a) Kim, C. K.; Song, E. S.; Kim, H. J.; Park, C.; Kim, Y. C.; Kim, J. K.; Yu, J. W.; Kim, C. Synthesis and luminescence characteristics of conjugated dendrimers with 2,4,6-triaryl-1,3,5-triazine periphery. J. Polym. Sci., Part A: Polym. Chem., **2006**, 44, 254-263. (b) Kim, C. K.; Song, E. S.; Kim, H. J.; Park, C.; Kim, J. K.; Kim, Y. C.; Yu, J. W.; Kim, C. Conjugated dendrimers with triazine peripheries and a distyrylanthracene core. J. Polym. Sci., Part A: Polym. Chem., **2006**, 44, 5855-5862.

[38] (a) Chang, D. W.; Dai, L. Luminescent amphiphilic dendrimers with oligo(p-phenylene vinylene) core branches and oligo(ethylene oxide) terminal chains: syntheses and stimuli-responsive properties. J. Mater. Chem., **2007**, 17, 364-371. (b) Ding, L.; Chang, D.; Dai, L.; Ji, T.; Li, S.; Lu, J.; Tao, Y.; Delozier, D.; Connell, J. Luminescent dendrons with oligo(phenylenevinylene) core branches and oligo(ethylene oxide) terminal chains. Macromolecules, **2005**, 38, 9389-9392.

[39]   Mizusaki, M.; Yamada, Y.; Obara, S.; Tada, K. Synthesis of π-conjugated two generation dendrimer composed of p-phenylenevinylene dendron and triphenylamine surface group. Chemistry Lett., **2012**, 41, 516-517.

[40]   Sengupta, S.; Sadhukhan, S. K.; Singh, R. S.; Pal, N. Synthesis of dendritic stilbenoid compounds: Heck reactions for the periphery and the core. Tetrahedron Lett., **2002**, 43, 1117-1121.

[41]   (a) Guldi, D. M.; Swartz, A.; Luo, C.; Gómez, R.; Segura, J. L.; Martín, N. Rigid dendritic donor–acceptor ensembles: control over energy and electron transduction. J. Am. Chem. Soc., **2002**, 124, 10875-10886. (b) Segura, J. L.; Gómez, R.; Martín, N.; Guldi, D. M. Synthesis of photo- and electroactive stilbenoid dendrimers carrying dibutylamino peripheral groups. Org. Lett., **2001**, 3, 2645-2648.

[42]   (a) Cho, B. R.; Chajara, K.; Oh, H. J.; Son, K. H.; Jeon, S.-J. Synthesis and nonlinear optical properties of 1,3,5-methoxy-2,4,6-tris(styryl)benzene derivatives. Org. Lett., **2002**, 4, 1703-1706. (b) Cho, B. R.; Park, S. B.; Lee, S. J.; Son, K. H.; Lee, S. H.; Lee, M.-J.; Yoo, J.; Lee, Y. K.; Lee, G. J.; Kang, T. I.; Cho, M.; Jeon, S.-J. 1,3,5-Tricyano-2,4,6-tris(vinyl)benzene derivatives with large second-order nonlinear optical properties. J. Am. Chem. Soc., **2001**, 123, 6421-6422.

[43]   Rodrigo, A. C.; Rivilla, I.; Pérez-Martínez, F. C.; Monteagudo, S.; Ocaña, V.; Guerra, J.; García-Martínez, J. C.; Merino, S.; Sánchez-Verdú, P.; Ceña, V.; Rodríguez-López, J. Efficient, non-toxic hybrid PPV-PAMAM dendrimer as gene carrier for neuronal cells. Biomacromolecules, **2011**, 12, 1205-1213.

[44]   (a) Ganesh, R. N.; Shraberg, J.; Sheridan, P. G.; Thayumanavan, S. Synthesis of difunctionalized dendrimers: an approach to main-chain poly(dendrimers). Tetrahedron Lett., **2002**, 43, 7217-7220. (b) Zhang, W.; Nowlan, D. T., III; Thomson, L. M.; Lackowski, W. M.; Simanek, E. E. Orthogonal, convergent syntheses of dendrimers based on melamine with one or two unique surface sites for manipulation. J. Am. Chem. Soc., **2001**, 123, 8914-8922. (c) Bo, Z.; Schäfer, A.; Franke, P.; Schlüter, A. D. A facile synthetic route to a third-generation dendrimer with generation-specific functional aryl bromides. Org. Lett., **2000**, 2, 1645-1648. (d) Wooley, K. L.; Hawker, C. J.; Fréchet, J. M. J. Polymers with controlled molecular architecture: control of surface functionality in the synthesis of dendritic hyperbranched macromolecules using the convergent approach. J. Chem. Soc., Perkin Trans. 1, **1991**, 1059-1076. (e) Hawker, C. J.; Fréchet, J. M. J. Control of surface functionality in the synthesis of dendritic macromolecules using the convergent-growth approach. Macromolecules, **1990**, 23, 4726-4729.

[45]   (a) Ambade, A. V.; Chen, Y.; Thayumanavan, S. Controlled functional group presentations in dendrimers as a tool to probe the hyperbranched architecture. New J. Chem., **2007**, 31, 1052-1063. (b) Sivanandan, K.; Sandanaraj, B. S.; Thayumanavan, S. Sequences in dendrons and dendrimers. J. Org. Chem., **2004**, 69, 2937-2944. (c) Vutukuri, D.; Bharathi, P.; Yu, Z.; Rajasekaran, K.; Tran, M.-H.; Thayumanavan, S. A mild deprotection strategy for allyl-protecting groups and its implications in sequence specific dendrimer synthesis. J. Org. Chem., **2003**, 68, 1146-1149. (d) Vutukuri, D. R.; Sivanandan, K.; Thayumanavan, S. Synthesis of dendrimers with multifunctional periphery using an ABB' monomer. Chem. Commun., **2003**, 796-797. (e) Sivanandan, K.; Vutukuri, D.; Thayumanavan, S. Functional group diversity in dendrimers. Org. Lett., **2002**, 4, 3751-3753.

[46]   Steffensen, M. B.; Simanek, E. E. Synthesis and manipulation of orthogonally protected dendrimers: building blocks for library synthesis. Angew. Chem., Int. Ed., **2004**, 43, 5178-5180.

[47]   (a) Hollink, M.; Simanek, E. E. A divergent route to diversity in macromolecules. Org. Lett., **2006**, 8, 2293-2295. (b) Newkome, G. R.; Kim, H. J.; Moorefield, C. N.; Maddi, H.; Yoo, K.-S. Syntheses of new 1 → (2 + 1) C-branched monomers for the construction of multifunctional dendrimers. Macromolecules, **2003**, 36, 4345-4354. (c) Newkome, G. R.; Yoo, K. S.; Hwang, S.-H.; Moorefield, C. N. Metallodendrimers: homo- and heterogeneous tier construction by bis(2,2':6',2''-terpyridinyl)Ru(II) complex connectivity. Tetrahedron, **2003**, 59, 3955-3964. (d) Woller, E. K.; Walter, E. D.; Morgan, J. R.; Singel, D. J.; Cloninger, M. J. Altering the strength of lectin binding interactions and controlling the amount of lectin clustering using mannose/hydroxyl-functionalized dendrimers. J. Am. Chem. Soc., **2003**, 125, 8820-8826. (e) Newkome, G. R.; Weis, C. D.; Moorefield, C. N.; Baker, G. R.; Childs, B. J.; Epperson, J. Isocyanate-based dendritic building blocks: combinatorial tier construction and macromolecular-property modification. Angew. Chem. Int. Ed., **1998**, 37, 307-310. (f) Larré, C.; Donnadieu, B.; Caminade, A.-M.; Majoral, J.-P. Phosphorus-containing dendrimers: chemoselective functionalization of internal layers. J. Am. Chem. Soc., **1998**, 120, 4029-4030. (g) Lartigue, M.-L.; Slany, M.; Caminade, A.-M.; Majoral, J.-P. Phosphorus-containing dendrimers: synthesis of macromolecules with multiple tri- and tetrafunctionalization. Chem. Eur. J., **1996**, 2, 1417-1426.

[48]   (a) Bryce, M. R.; de Miguel, P.; Devonport, W. Redox-switchable polyester dendrimers incorporating both π-donor (tetrathiafulvalene) and π-acceptor (anthraquinone) groups. Chem. Commun., **1998**, 2565-2566. (b) Wooley, K. L.; Hawker, C. J.; Fréchet, J. M. J. Unsymmetrical three-dimensional macromolecules: preparation and characterization of strongly dipolar dendritic macromolecules. J. Am. Chem. Soc., **1993**, 115, 11496-11505.

[49]   Tolosa, J.; Romero-Nieto, C.; Díez-Barra, E.; Sánchez-Verdú, P.; Rodríguez-López, J. Control of surface functionality in poly(phenylenevinylene) dendritic architectures. J. Org. Chem., **2007**, 72, 3847-3852.

[50]   Grayson, S. M.; Fréchet, J. M. J. Convergent dendrons and dendrimers: from synthesis to applications. Chem. Rev., **2001**, 101, 3819-3868.

[51]   Cano-Marín, A. R.; Díez-Barra, E.; Rodríguez-López, J. Synthesis, characterization and optical response of polyene-cored stilbenoid dendrimers. Tetrahedron, **2005**, 61, 395-400.

[52]   Díez-Barra, E.; García-Martínez, J. C.; Rodríguez-López, J. Synthesis of 4-dendronized β-lactams. Synlett, **2003**, 1587-1590.

[53]   (a) Díez-Barra, E.; García-Martínez, J. C.; del Rey, R.; Rodríguez-López, J.; Giacalone, F.; Segura, J. L.; Martín, N. Synthesis and photoluminiscent properties of 1,1'-binaphthyl-based chiral phenylenevinylene dendrimers. J. Org. Chem., **2003**, 68, 3178-3183. (b) Díez-Barra, E.; García-Martínez, J. C.; Rodríguez-López, J.; Gómez, R.; Segura, J. L.; Martín, N. Synthesis of new 1,1'-binaphthyl-based chiral phenylenevinylene dendrimers. Org. Lett., **2000**, 2, 3651-3653.

[54]   (a) Pérez, L.; García-Martínez, J. C.; Díez-Barra, E.; Atienzar, P.; García, H.; Rodríguez-López, J.; Langa, F. Electron Transfer in non Polar Solvents in Fullerodendrimers with Peripheral Ferrocene Units. Chem. Eur. J., **2006**, 12, 5149-5157. (b) Langa, F.; Gómez-Escalonilla, M. J.; Díez-Barra, E.; García-Martínez, J. C.; de la Hoz, A.; Rodríguez-López, J.; González-Cortés, A.; López-Arza, V. Synthesis, electrochemistry and photophysical properties of phenylenevinylene fullerodendrimers. Tetrahedron Lett., **2001**, 42, 3435-3438.

[55]   Prato, M.; Maggini, M. Fulleropyrrolidines: a family of full-fledged fullerene derivatives. Acc. Chem. Res., **1998**, 31, 519-526.

[56]   Campidelli, S.; Pérez, L.; Rodríguez-López, J.; Barberá, J.; Langa, F.; Deschenaux, R. Dendritic liquid-crystalline fullerene-ferrocene dyads. Tetrahedron, **2006**, 62, 2115-2122.

[57]   García-Martínez, J. C.; Atienza, C.; de la Peña, M.; Rodrigo, A. C.; Tejeda, J.; Rodríguez-López, J. A MALDI-TOF MS study of lanthanide(III)-cored poly(phenylenevinylene) dendrimers. J. Mass Spectrom., **2009**, 44, 613-620.

[58]   Achelle, S.; Nouira, I.; Pfaffinger, B.; Ramondenc, Y.; Plé, N.; Rodríguez-López, J. V-Shaped 4,6-bis(arylvinyl)pyrimidine oligomers: synthesis and optical properties. J. Org. Chem., **2009**, 74, 3711-3717.

[59]   Hadad, C.; Achelle, S.; García-Martínez, J. C.; Rodríguez-López, J. 4-Arylvinyl-2,6-di(pyridin-2-yl)pyrimidines: synthesis and optical properties. J. Org. Chem., **2011**, 76, 3837-3845.

[60]   (a) Romero-Nieto, C.; Kamada, K.; Cramb, D.T.; Merino, S.; Rodríguez-López, J.; Baumgartner, T. Synthesis and photophysical properties of donor-acceptor dithienophospholes. Eur. J. Org. Chem., **2010**, 5225-5231. (b) Romero-Nieto, C.; Merino, S.; Rodríguez-López, J.; Baumgartner, T. Dendrimeric OPV-extended dithieno[3,2-b:2',3'-d]phospholes – synthesis, self-organization and optical properties. Chem. Eur. J., **2009**, 15, 4135-4145.

[61]   Romero-Nieto, C.; Marcos, M.; Merino, S.; Barberá, J.; Baumgartner, T.; Rodríguez-López, J. Room temperature multifunctional organophosphorus gels and liquid crystals. Adv. Funct. Mater., **2011**, 21, 4088–4099.

[62]   Xu, Z.; Moore, J. S. Design and synthesis of a convergent and directional molecular antenna. Acta Polym., **1994**, 45, 83-87.

[63]   Moore, J. S. Shape-persistent molecular architectures of nanoscale dimension. Acc. Chem. Res., **1997**, 30, 402-413.

[64]   (a) Xu, Z.; Kahr, M.; Walker, K. L.; Wilkins, C. L.; Moore, J. S. Phenylacetylene dendrimers by the divergent, convergent, and double-stage convergent methods. J. Am. Chem. Soc., **1994**, 116, 4537-4550. (b) Xu, Z.; Moore, J. S. Synthesis and characterization of a high molecular weight stiff dendrimer. Angew. Chem., Int. Ed. Engl., **1993**, 32, 246-248. (c) Moore, J. S.; Xu, Z. Synthesis of rigid dendritic macromolecules: enlarging the repeat unit size as a function of generation, permitting growth to continue. Macromolecules, **1991**, 24, 5893-5894.

[65]   Bharathi, P.; Patel, U.; Kawaguchi, T.; Pesak, D. J.; Moore, J. S. Improvements in the synthesis of phenylacetylene monodendrons including a solid-phase convergent method. Macromolecules, **1995**, 28, 5955-5963.

[66]   De Gennes, P. G.; Hervet, H. Statistics of « starburst » polymers. J. Phys. Lett., **1983**, 44, 351-360.

[67]   Xu, Z.; Moore, J. S. Rapid construction of large-size phenylacetylene dendrimers up to 12.5 nanometers in molecular diameter. Angew. Chem., Int. Ed. Engl., **1993**, 32, 1354-1357.

[68]   Kawaguchi, T.; Walker, K. L.; Wilkins, C. L.; Moore, J. S. Double exponential dendrimer growth. J. Am. Chem. Soc., **1995**, 117, 2159-2165.

[69]   Gorman, C. B.; Smith, J. C. Effect of repeat unit flexibility on dendrimer conformation as studied by atomistic molecular dynamics simulations. Polymer, **2000**, 41, 675-683.

[70]   (a) Yamaguchi, Y.; Ochi, T.; Miyamura, S.; Tanaka, T.; Kobayashi, S.; Wakamiya, T.; Matsubara, Y.; Yoshida, Z.-I. Rigid molecular architectures that comprise a 1,3,5-trisubstituted benzene core and three oligoaryleneethynylene arms: light-emitting characteristics and π conjugation between the arms. J. Am. Chem. Soc., **2006**, 128, 4504-

4505. (b) Itoh, T.; Maemura, T.; Ohtsuka, Y.; Ikari, Y.; Wildt, H.; Hirai, K.; Tomioka, H. Synthesis of phenylacetylene dendrimers having sterically congested diazo units and characterization of their photoproducts. Eur. J. Org. Chem., **2004**, 2991-3003. (c) Rodríguez, J. G.; Esquivias, J.; Lafuente, A.; Díaz, C. Synthesis of nanostructures based on 1,4- and 1,3,5-ethynylphenyl subunits with π-extended conjugation. Carbon dendron units. J. Org. Chem., **2003**, 68, 8120-8128.

[71]   (a) Swallen, S. F.; Shi, Z.-Y.; Tan, W.; Xu, Z.; Moore, J. S.; Kopelman, R. Exciton localization hierarchy and directed energy transfer in conjugated linear aromatic chains and dendrimeric supermolecules. J. Lumin., **1998**, 76-77, 193-196. (b) Kopelman, R.; Shortreed, M.; Shi, Z.-Y.; Tan, W.; Xu, Z.; Moore, J. S.; Bar-Haim, A.; Klafter, J. Spectroscopic evidence for excitonic localization in fractal antenna supermolecules. Phys. Rev. Lett., **1997**, 78, 1239-1242.

[72]   Shortreed, M. R.; Swallen, S. F.; Shi, Z. Y.; Tan, W.; Xu, Z.; Devadoss, C.; Moore, J. S.; Kopelman, R. Directed energy transfer funnels in dendrimeric antenna supermolecules. J. Phys. Chem. B, **1997**, 101, 6318-6322.

[73]   Ahn, T. S.; Thompson, A. L.; Bharathi, P.; Müller, A.; Bardeen, C. J. Light-harvesting in carbonyl-terminated phenylacetylene dendrimers: the role of delocalized excited states and the scaling of light-harvesting efficiency with dendrimer size. J. Phys. Chem. B, **2006**, 110, 19810-19819.

[74]   Nishimura, Y.; Kamada, M.; Ikegami, M.; Nagahata, R.; Arai, T. The relaxation dynamics of the excited state of stilbene dendrimers substituted with phenylacetylene groups. J. Photochem. Photobiol. A: Chem., **2006**, 178, 150-155.

[75]   Pu, L. Enantioselective fluorescent sensors: a tale of BINOL. Acc. Chem. Res., **2012**, 45, 150-163.

[76]   (a) Clulow, A. J.; Burn, P. L.; Meredith, P.; Shaw, P. E. Fluorescent carbazole dendrimers for the detection of nitroaliphatic taggants and accelerants. J. Mater. Chem., **2012**, 22, 12507-12516. (b) Li, S.; Leng, T.; Zhong, H.; Wang, C.; Shen, Y. Silole-core phenylacetylene dendrimers and their application in detecting picric acid. J. Heterocyclic Chem., **2012**, 49, 64-70.

[77]   Wang, P.-W.; Liu, Y.- J.; Devadoss, C.; Bharathi, P.; Moore, J. S. Electroluminescent diodes from a single component emitting layer of dendritic macromolecules. Adv. Mater., **1996**, 8, 237-241.

[78]   Cha, S. W.; Choi, S.-H.; Kim, K.; Jin, J.-I. Synthesis and luminescence properties of four-armed conjugated structures containing 1,3,4-oxadiazole moieties. J. Mater. Chem., **2003**, 13, 1900-1904.

[79]   Pesak, D. J.; Moore, J. S.; Wheat, T. E. Synthesis and characterization of water-soluble dendritic macromolecules with a stiff, hydrocarbon interior. Macromolecules, **1997**, 30, 6467-6482.

[80]   Pesak, D. J.; Moore, J. S. Columnar liquid crystals from shape-persistent dendritic molecules. Angew. Chem. Int. Ed. Engl., **1997**, 36, 1636-1639.

[81]   Mongin, O.; Papamicaël, C.; Hoyler, N.; Gossauer, A. Modular synthesis of benzene-centered porphyrin trimers and a dendritic porphyrin hexamer. J. Org. Chem., **1998**, 63, 5568-5580.

[82]   Kayser, B.; Altman, J.; Beck, W. Benzene-bridged hexaalkynylphenylalanines and first-generation dendrimers thereof. Chem. Eur. J., **1999**, 5, 754-758.

[83]   Liao, L.-X.; Stellacci, F.; McGrath, D. V. Photoswitchable flexible and shape-persistent dendrimers: comparison of the interplay between a photochromic azobenzene core and dendrimer structure. J. Am. Chem. Soc., **2004**, 126, 2181-2185.

[84]   Peng, Z.; Pan, Y.; Xu, B.; Zhang, J. Synthesis and optical properties of novel unsymmetrical conjugated dendrimers. J. Am. Chem. Soc., **2000**, 122, 6619-6623.

[85]   Clifford, J. N.; Gégout, A.; Zhang, S.; Pereira de Freitas, R.; Urbani, M.; Holler, M.; Ceroni. P; Nierengarten, J.-F.; Armaroli, N. Fullerene derivatives substituted with differently branched phenyleneethynylene dendrons: synthesis, electronic and excited state properties. Eur. J. Org. Chem., **2007**, 5899-5908.

[86]   Lüning, U.; Eggert, J. P. W.; Hagemann, K. A second-generation dendrimer with six 2,9-dimethyl-1,10-phenanthroline units as ligand for copper-catalyzed reactions. Eur. J. Org. Chem., **2006**, 2747-2752.

[87]   (a) Kaneko, T.; Yamamoto, K.; Asano, M.; Teraguchi M.; Aoki, T. Synthesis of poly(phenylacetylene)-based polydendrons consisting of a phenyleneethynylene repeating unit, and oxygen/nitrogen permeation behavior of their membranes. J. Membr. Sci., **2006**, 278, 365-372. (b) Aoki, T.; Kaneko, T. New macromolecular architectures for permselective membranes – gas permselective membranes from dendrimers and enantioselectively permeable membranes from one-handed helical polymers –. Polym. J., **2005**, 37, 717-735. (c) Kaneko, T.; Asano, M.; Yamamoto, K.; Aoki, T. Polymerization of phenylacetylene-based monodendrons and structure of the corresponding polydendrons. Polym. J., **2001**, 33, 879-890. (d) Kaneko, T.; Horie, T.; Asano, M.; Matsumoto, S.; Yamamoto, K.; Aoki, T.; Oikawa, E. Polydendrons and polymacrocycles with phenyleneethynylene repeating unit. Polym. Adv. Technol., **2000**, 11, 685-691. (e) Kaneko, T.; Horie, T.; Asano, M.; Aoki, T.; Oikawa, E. Polydendron: polymerization of dendritic phenylacetylene monomers. Macromolecules, **1997**, 30, 3118-3121.

[88]   Becker, H.; Spreitzer, H.; Kreuder, W.; Kluge, E.; Schenk, H.; Parker, I.; Cao, Y. Soluble PPVs with enhanced performance – a mechanistic approach. Adv. Mater., **2000**, 12, 42-48.

[89]   (a) Mongin, O.; Porrès, L.; Moreaux, L.; Mertz, J.; Blanchard-Desce, M. Synthesis and photophysical properties of new conjugated fluorophores designed for two-photon-excited fluorescence. Org. Lett., **2002**, 4, 719-722. (b) Schenning, A. P. H. J.; Tsipis, A. C.; Meskers, S. C. J.; Beljonne, D.; Meijer, E. W.; Brédas, J. L. Electronic structure and optical properties of mixed phenylene vinylene/phenylene ethynylene conjugated oligomers. Chem. Mater., **2002**, 14, 1362-1368. (c) Egbe, D. A. M.; Roll, C. P.; Birckner, E.; Grummt, U.-W.; Stockmann, R.; Klemm, E. Side chain effects in hybrid PPV/PPE polymers. Macromolecules, **2002**, 35, 3825-3837. (d) Hwang, G. T.; Son, H. S.; Ku, J. K.; Kim, B. H. Novel fluorophores: efficient synthesis and photophysical study. Org. Lett., **2001**, 3, 2469-2471. (e) Brizius, G.; Pschirer, N. G.; Steffen, W.; Stitzer, K.; zur Loye, H.-C.; Bunz, U. H. F. Alkyne metathesis with simple catalyst systems: efficient synthesis of conjugated polymers containing vinyl groups in main or side chain. J. Am. Chem. Soc., **2000**, 122, 12435-12440.

[90]   Nielsen, M. B.; Moonen, N. N. P.; Boudon, C.; Gisselbrecht, J.-P.; Seiler, P.; Gross, M.; Diederich, F. Tetrathiafulvalene–acetylene scaffolding: new π-electron systems for advanced materials. Chem. Commun., **2001**, 1848-1849.

[91]   Díez-Barra, E.; García-Martínez, J. C.; Rodríguez-López, J. Synthesis of novel cross-conjugated dendritic fluorophores containing both phenylenevinylene and phenyleneethynylene moieties. J. Org. Chem., **2003**, 68, 832-838.

[92]   García-Martínez, J. C.; Díez-Barra, E.; Rodríguez-López, J. Study of the aggregation behavior of a π-conjugated dendrimer with a twisted core. Tetrahedron Lett., **2012**, 53, 2752-2755.

Send Orders of Reprints at bspsaif@emirates.net.ae

CHAPTER 5

α-Activated Cross Conjugated Cycloalkenone Systems in Organic Synthesis

Yen-Ku Wu[1], Tai Wei Ly[2] and Kak-Shan Shia[3,*]

[1]Department of Chemistry, National Tsing Hua University, Hsinchu, 30013, R.O.C. Taiwan; [2]Axikin Pharmaceuticals, Inc., 10835 Road to the Cure, Suite 250, San Diego, California 92121, USA and [3]Institute of Biotechnology and Pharmaceutical Research, National Health Research Institutes, Miaoli County, 35053, R.O.C. Taiwan

Abstract: Under catalysis with an appropriate Lewis acid, the intermolecular Diels-Alder cycloaddition and intramolecular polyene cyclization of the α-activated cross conjugated cycloalkenone systems, possessing particularly increased dienophilicity/electrophilicity in the cross conjugated double bond, proceeded constantly with a high degree of stereochemical control, leading to a variety of synthetically useful molecules, which may serve as advanced intermediates towards many structurally challenging natural products, especially *cis*-clerodane diterpenoids. Mechanistically, an additional conjugated double bond incorporated into the cycloalkenone core of the titled systems might contribute synergistically to the classical secondary orbital effects, thus significantly enhancing the *endo*-to-ketone addition. In addition, a tandem multiple σ-bond migration process, an enzymatic pathway prevalent in nature, was proposed to rationalize the formation of structurally unusual polyene-cyclization products occurring under standard chemical conditions. On the other hand, these highly polarized cross-conjugated compounds was found to be a superior Michael acceptor for Grignard reagents. In that regard, the introduction of ω-unsaturation to the titled substrates *via* 1,4-addition could serve as a springboard for a diverse array of annulation processes.

Keywords: Diels-Alder, cross conjugated, endo selectivity, secondary orbital interaction, facial selectivity, dienophile, cycloaddition, polyene cyclization, Michael reaction, Lewis acid, autoxidative cascade, cationic cascade, Conia-ene reaction, lithium naphthalenide, cis-clerodane diterpenoid, hydride shift, catalysis, palladium(II)-mediated oxidative cyclization, annulation, electron-withdrawing group.

1. INTRODUCTION

Owing to the nature of their muti-functionality, α-activated cross conjugated cycloalkenones (Fig. **1**) play a key role in many synthetically useful transformations. From a synthetic standpoint, the α-activating group not only

---

*Address correspondence to Kak-Shan Shia: Institute of Biotechnology and Pharmaceutical Research, National Health Research Institutes, Miaoli County, 35053, R.O.C. Taiwan; Tel: 886-37-246-166; Fax: 886-37-586-456; E-mail: ksshia@nhri.org.tw

Atta-ur-Rahman (Ed)

significantly polarized the alkene toward cycloaddition or Michael addition, but also served as a versatile handle for succeeding chemical conversions. Reaction designs based on features of these substrates were particularly fruitful in the past decade [1]. Therefore, the following account aims to provide an overview of the use of the titled systems in the Diels-Alder reaction, the polyene cyclization as well as the sequential Michael/annulation processes with emphasis on their recent advances in organic synthesis.

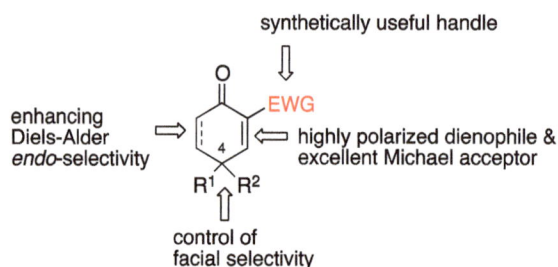

Figure 1: Features of cross conjugated cycloalkenone systems.

2. DIELS-ALDER REACTIONS

The Diels-Alder cycloaddition reaction was and continues to be of extreme utility in synthetic organic chemistry [2]. Its capacity to construct, in one step, fused polycyclic skeletons in a highly regio- and stereoselective manner is unparalleled. It was quickly delineated however, that, under solely thermal activation, the cycloaddition of dienes to cyclic dienophiles, such as cyclohex-2-enone, is a rather poor process, often resulting in low yields of products [3]. As such, subsequent efforts in the past three decades have been devoted to the enhancement of the dienophilicity of various cycloalkenone systems, leading to the development of two general solutions. As illustrated in Scheme **1**, one approach

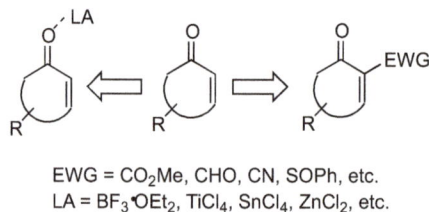

EWG = CO$_2$Me, CHO, CN, SOPh, etc.
LA = BF$_3$·OEt$_2$, TiCl$_4$, SnCl$_4$, ZnCl$_2$, etc.

Scheme 1: Enhancement of dienophilicity of cycloalkenone systems.

employs Lewis acid catalysis while the other involves the introduction of an additional electron-withdrawing group to the dienophilic carbon-carbon double bond to further activate the dienophile. The latter strategy proved to be particularly useful and has been successfully applied to facilitate the total synthesis of various polycyclic natural products as exhibited in section 2.6.

2.1. Historical Perspective and Early Studies

Thermal Diels-Alder reaction of enone **1** (Fig. **2**) with simple acyclic dienes was first examined in Liu's laboratories [4, 5]. In these instances, dimeric compounds **2** and **3** were obtained as the major products while the desired Diels-Alder adduct was only isolated in an insignificant quantity if at all [5]. This was the case even under extreme conditions (sealed tube, 250 °C). The poor dienophilicity of **1** was further confirmed by performing the reactions in the presence of boron trifluoride etherate or aluminum chloride and systematically varying reaction temperature and solvent. Subsequently, Wenkert and co-workers successfully generated the Diels-Alder adduct between **1** and 1,3-butadiene in synthetically useful yield by pre-forming the complex between **1** and aluminum chloride [6].

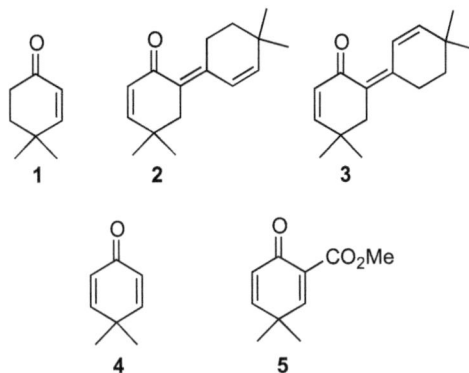

Figure 2: Structures of compounds **1-5**.

Based on the aforementioned observation, it was concluded that the Diels-Alder reaction of **1** would proceed only if its dienophilicity could be enhanced while reducing the propensity for the aldol type side reaction. Towards this end, it was postulated that the introduction of a second double bond into the ring of enone **1** would accomplish such a goal. As such, dienone **4** was generated and its Diels-Alder chemistry was examined [5]. When compound **4** was treated with an excess of various dienes in the presence of boron trifluoride, the projected cycloadduct exemplified by **6** was indeed obtained, albeit in modest yield and under sub-optimal conditions (Scheme **2**), lending credence to the above supposition. Mechanistically, most of adducts thus obtained obeyed the Diels-Alder addition rules with the execption of the reaction involving isoprene from which adduct **6a** was formed exclusively in violation of the *para* rule.

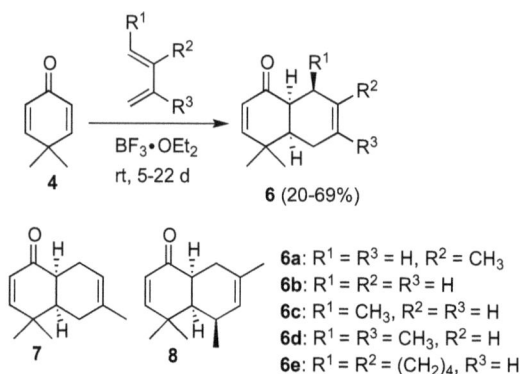

Scheme 2: Diels-Alder reaction of dienophile **4**.

Normally, addition of the 2-substituted dienes should result in adduct **7** as a result of the preferred electronic effects in accordance with the *para* rule [3]. On the other hand, if sufficient steric repulsion is encountered, the anti-*para* addition pathway leading exclusively to an unusual but conceivable product **6a** (Fig. **3**, **A1** ($R^1 = R^3 =$ H, $R^2 =$ Me)) may be preferred. Comparison of the various dienes is interesting in this context. *trans*-Piperylene was expected not to suffer any steric barrier and thus the *endo*-to-ketone addition adduct **6c**, following the *ortho* rule, was formed exclusively. The addition of *trans*-2-methyl-1,3-pentadiene however, would be problematic since the steric factor would act in opposition to the combined directing influence of the *ortho* and *para* rules. The preference for steric control would afford enone **8** whereas electronic control would result in the formation of **6d**. Experimentally, enone **6d** was the sole product obtained, indicating that electronic effects, including a secondary orbital effect originating from the presence of the neighboring carbon-carbon double bond, predominated over steric effects in the course of the reaction. As illustrated in Fig. **3**, the aforementioned results disclosed that *endo*-to-ketone addition *via* transition state **A1** might hold a higher priority than *exo*-to-ketone addition *via* transition state **B1**.

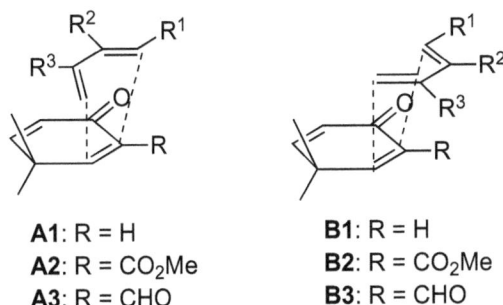

A1: R = H
A2: R = CO$_2$Me
A3: R = CHO

B1: R = H
B2: R = CO$_2$Me
B3: R = CHO

Figure 3: Proposed Diels-Alder transition states **A1-3** and **B1-3**.

2.2. α-Carbalkoxy/Formyl/Keto Cycloalkenones as Dienophiles

In theory, one possible method to increase the influence of the electronic effect of **4** might be the use of a stronger and more electron-withdrawing Lewis acid as catalyst to induce greater polarization of the dienophilic double bond and promote *endo*-to-ketone addition. Indeed, as shown in Scheme **2**, this inference was viable and verified by Wenkert *et al.* [7], where product **6a** was produced more effectively in the presence of aluminum chloride as compared to boron trifluoride under similar reaction conditions. Subsequent development into the use of dienophiles as exemplified by **4** involved the introduction of an additional electron-withdrawing substituent onto the dienophilic double bond (*e.g.* dienone ester **5**) to increase both the polarization and strength of the *para*-directing effect. The choice of the carbomethoxyl moiety as an activating group stems from the belief that it could impart several advantages onto the resulting Diels-Alder adducts with respect to chemical stability and malleability. According to the Alder rule, *endo* addition might occur on the more dienophilic double bond with a secondaryl orbital overlap either with the ketone or ester carbonyl group as depicted in transition states **A2** and **B2** (Fig. 3). It was highly conceivable that the unsubstituted enone unit of dienophile **5** is much less reactive and thus would be intact during Diels-Alder addition. In practice, as expected, *endo*-to-ketone addition occurred exclusively for various dienes examined (Scheme 3) [5]. However, when 2-methyl-1,3-pentadiene was employed, products **9e** (*endo*-to-ketone) and **10** (*endo*-to-ester, R^1 = R^3 = CH$_3$, R^2 = H) were obtained in a 1:1 ratio, implying that 2-substituted diene might cause considerable steric repulsion in approach **A2**, and thus part of the addition process was switched to **B2** (Fig. 3).

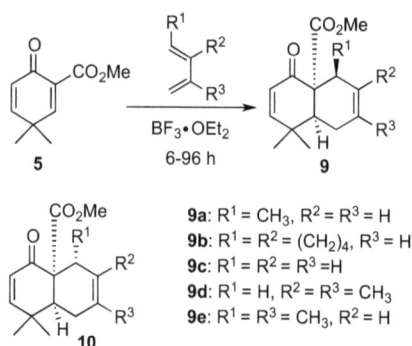

Scheme 3: Diels-Alder reaction of dienophile **5**.

Further demonstration of steric encumberances driving the product of the Diels-Alder reaction of dienone **5** was observed in its reaction with isoprene (Scheme **4**) [5]. In all cases employing various Lewis acids, a mixture of *para* addition product **9g** was observed admixed with the anti-*para* product **9f** as a minor product or, in the case of AlCl$_3$, SbCl$_5$, and SnCl$_4$, as the major product. Interestingly, this steric encumberance to a clean reaction result was completely eliminated by removing the non-participating olefin in the dienophile. Specifically, in the Diels-Alder cycloaddition reaction between enone **11** and isoprene (Scheme **5**) [8], the predicted *para*-addition product **12**, assumed to be *endo*-to-ester addition to avoid steric hindrance, was achieved exclusively irrespective of catalyst or temperature applied.

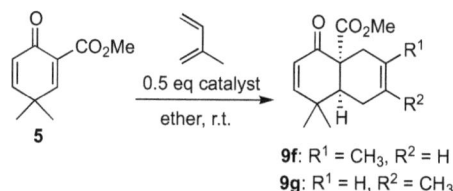

Lewis acid	time	ratio (**9f:9g**)
BF$_3$·OEt$_2$	72 h	70:30
TiCl$_4$	14 h	52:48
ZnCl$_2$	48 h	52:48
FeCl$_3$	0.3 h	50:50
AlCl$_3$	4 h	47:53
SbCl$_5$	4 h	34:66
SnCl$_4$	44 h	18:82

Scheme 4: Screening various Lewis acid catalysts for dienophile **5**.

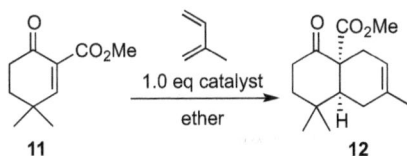

Scheme 5: Diels-Alder reaction of dienophile **11**.

Dienone **13**, bearing a formyl instead of an ester functionality as an activating group, was also prepared and tested as a dienophile (Scheme **6**) [9]. As expected, Diels-Alder reactions occurred solely at the more active double bond *via* transition state **A3** (R^1 = Me, R^2 = R^3 = H) to produce adduct **14** (*endo*-to-ketone) as a single compound. This result in good agreement with the preceding discussion and readily understood by the *ortho* rule with favorable electronic effects.

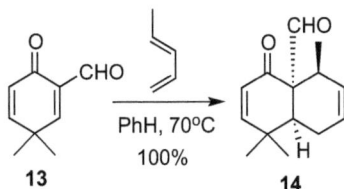

Scheme 6: Diels-Alder reaction of dienophile **13**.

Up to this point, one remaining issue to investigate in the Diels-Alder reaction of α-activated cyclohexenones involved the question of facial selectivity. Towards this end, the symmetrically substituted C-4 in dienones **5** and **11** were individually replaced with asymmetrical substitutions to give methyl esters **15** and **16**. These facially asymmetrical dienophiles were, as expected, highly amenable towards the Diels-Alder cycloaddition reaction and induced a high degree of facial selectivity (Scheme **7**) [10]. Under zinc chloride catalysis, the addition of *trans*-piperylene individually to dienophiles **15** and **16** yielded the *endo*-to-ketone cycloadducts **17** and **18** respectively in virtually quantitative yields. Remarkably, these cycloadducts also revealed an overwhelming preference for the C-4 ester face approach. This unusual high selectivity was totally unexpected particularly in light of the relative van der Waals sizes estimated for carbomethoxy (n = 12.1) and methyl group (n = 8.5), and was tentatively rationalized by the electrostatic π-factors of the carbonyl, rendering the ester face more electron-poor characteristic and addition of electron-rich dienes from this side is therefore favored [11].

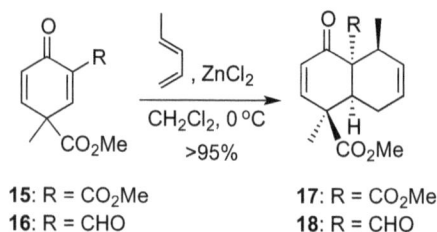

15: R = CO₂Me   **17**: R = CO₂Me
16: R = CHO   **18**: R = CHO

Scheme 7: Diels-Alder reaction of C-4 ester dienophiles **15** and **16**.

Along these lines, the application of α-activated cyclopentenone systems in Diels-Alder reactions were also investigated [12]. Using dienophile **19** as a typical example (Scheme **8**), under catalysis with bidentate Lewis acids (*e.g.* SnCl₄), addition of isoprene gave rise to the *para*-addition adduct **20a** exclusively; however, in the presence of a monodentate catalyst (*e.g.* BF₃•OEt₂), a significant proportion of anti-*para* product **20b** was obtained in addition to **20a**. Compared to the corresponding cyclohexenone **11**, addition of dienes to dienophile **19** occurred with a higher degree of *exo-endo* scrambling, presumably attributable to the steric impediment of gem-dimethyl group.

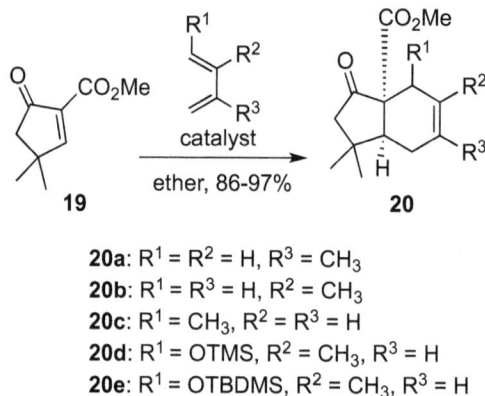

20a: R¹ = R² = H, R³ = CH₃
20b: R¹ = R³ = H, R² = CH₃
20c: R¹ = CH₃, R² = R³ = H
20d: R¹ = OTMS, R² = CH₃, R³ = H
20e: R¹ = OTBDMS, R² = CH₃, R³ = H

Scheme 8: Diels-Alder reaction of α-ester cyclopentenone **19**.

More recently, Schotes and Mezzetti demostrated that the asymmetric [4+2] cycloaddition between cyclic unsaturated β-keto esters and substituted 2,3-dienes was effectively catalyzed by dicationic ruthenium/PNNP complex furnishing tetrahydro-1-indanone derivatives in good yields and enantioselectivities (Scheme **9**) [13, 14]. The steric bulk of alkyl group on the ester portion was essential for the *endo-exo* selectivity of cycloaddition as exemplified in the synthetic application toward estrone precursors. On the other hand, Ohfusa and Nishida examined the Diels-Alder reactivity of alkylidene β-keto esters with a Rawal-type diene [15] using chiral Cr/salen complex, wherein poor enantio- and *endo-exo* selectivities were observed in most cases [16].

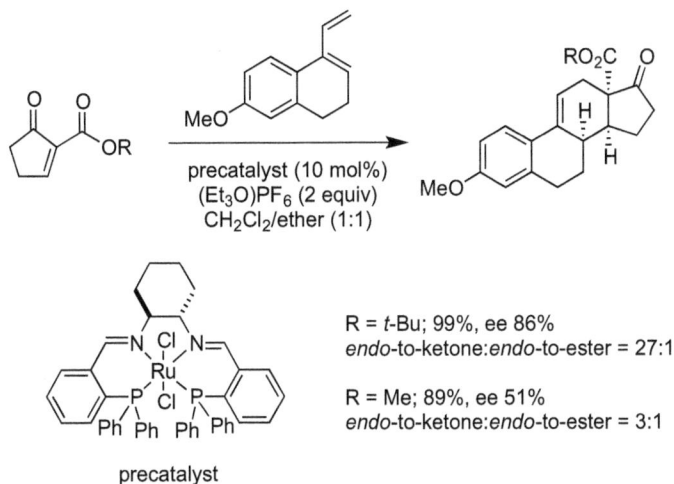

R = *t*-Bu; 99%, ee 86%
endo-to-ketone:*endo*-to-ester = 27:1

R = Me; 89%, ee 51%
endo-to-ketone:*endo*-to-ester = 3:1

precatalyst

Scheme 9: Asymmetric Diels-Alder reaction using ruthenium salen complex.

For comparison purposes, enone ester **21** without a geminal dimethyl at the γ-carbon was also designed and examined as the dienophile in the Diels-Alder reaction. Experimentally, enone **21** was found to be easily enolizable and underwent polymerization rapidly, thus necessitating its immediate consumption after preparation [17]. Under thermal condtions, addition of various dienes to **21** gave only a trace amount of the desired Diels-Alder adducts, presumably due to the liability mentioned above. However, under catalysis with stannic chloride, the expected adducts **22** could be obtained in moderate to good yields at low temperatures (Scheme **10**). As indicated, the major isomers were found to be *endo*-to-ester addition products when an additional C5-C6 double bond was absent in the six-membered ring systems.

Similarly, when the same cycloalkenone system was activated by α-keto carbonyl as the dienophile (*e.g.* **23**) as reported by Snider (Scheme **11**) [18], the corresponding adducts **24** were also obtained at a low level of ketone-*endo* selectivity.

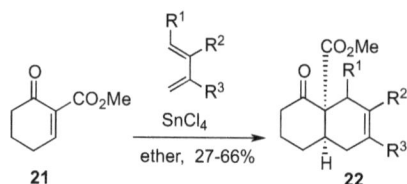

22a: R^1 = H, R^2 = R^3 = CH_3
22b: R^1 = R^2 = H, R^3 = OTMS
22c: R^1 = R^2 = H, R^3 = CH_3
22d: R^1 = CH_3, R^2 = R^3 = H
22e: R^1 = R^3 = CH_3, R^2 = H
22f: R^1 = CH_3, R^2 = H, R^3 = OTMS
22g: R^1 = CH_3, R^2 = H, R^3 = OPO(OEt)$_2$

Scheme 10: Diels-Alder reaction of α-ester cyclohexenone **21**.

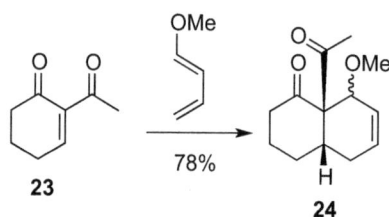

Scheme 11: Diels-Alder reaction of α-keto cyclohexenone **23**.

Dienophiles with large ring systems were further designed and examined in Liu's laboratories. As demonstrated by 2-carbalkoxy-2-cycloheptenones **25** under catalysis with $ZnCl_2$, the resulting Diels-Alder adducts were obtained with low degrees of *endo-exo* selectivity without fail as shown in Scheme **12** [19]. Intriguingly, an unexpectedly high level of face selectivity was observed in cycloheptenones bearing a C-7 methyl group (R^4 = Me). However, what controls this high degree of face selectivity is still not fully understood.

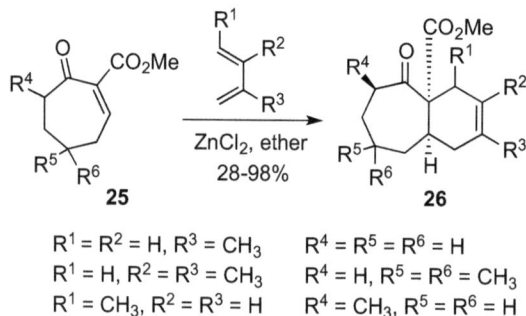

R^1 = R^2 = H, R^3 = CH_3 R^4 = R^5 = R^6 = H
R^1 = H, R^2 = R^3 = CH_3 R^4 = H, R^5 = R^6 = CH_3
R^1 = CH_3, R^2 = R^3 = H R^4 = CH_3, R^5 = R^6 = H

Scheme 12: Diels-Alder reaction of α-ester cycloheptenone **25**.

As for eight-membered ring dienophile **27**, the addition took place following *ortho-* and *para-*rule with a major preference for the *endo-*to-ester addition (Scheme **13**) [20]. In general, the larger-ring dienophiles mentioned above are chemically more stable than the corresponding six-membered ring analogues (*e.g.* **21**), but the reaction rate is somewhat slower.

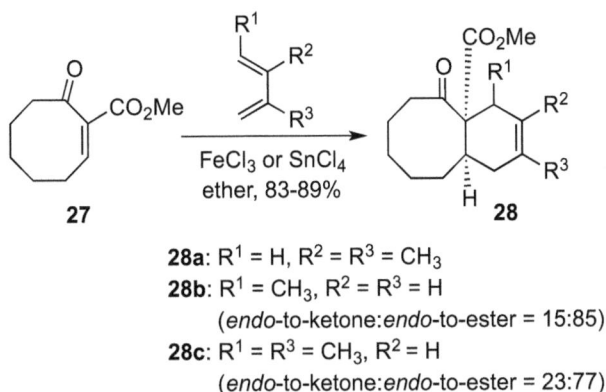

28a: R^1 = H, R^2 = R^3 = CH_3
28b: R^1 = CH_3, R^2 = R^3 = H
    (*endo*-to-ketone:*endo*-to-ester = 15:85)
28c: R^1 = R^3 = CH_3, R^2 = H
    (*endo*-to-ketone:*endo*-to-ester = 23:77)

Scheme 13: Diels-Alder reaction of α-ester cyclooctenone **27**.

The facially differentiated enone ester **29** has been applied to study the effects of steric influence on the Diels-Alder reaction (Scheme **14**) [21]. Under catalysis with $ZnCl_2$, only products *via* addition of dienes to the *Si* face were obtained. Regiochemically, the adducts were completely governed by the *ortho* and *para* rules, and stereochemically, *endo*-to-ester addition was dramatically favored. This excellent stereoselectivity (**30:31** > 9:1) appeared to originate from a combination of both favorable electronic and steric factors *via* ester-*endo* transition state **C1** over the keto-*endo* transition state **D1** during the course of the reaction (Fig. **4**).

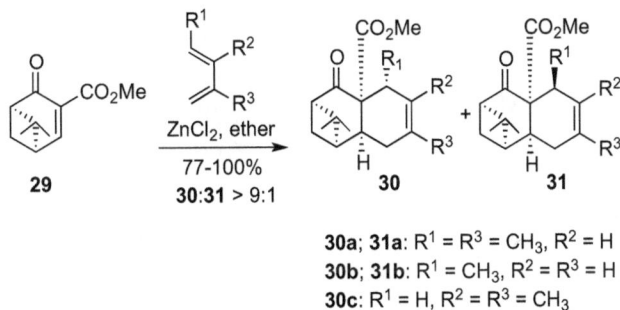

30a; 31a: R^1 = R^3 = CH_3, R^2 = H
30b; 31b: R^1 = CH_3, R^2 = R^3 = H
30c: R^1 = H, R^2 = R^3 = CH_3

Scheme 14: Diastereoselective Diels-Alder reaction of chiral dienophile **29**.

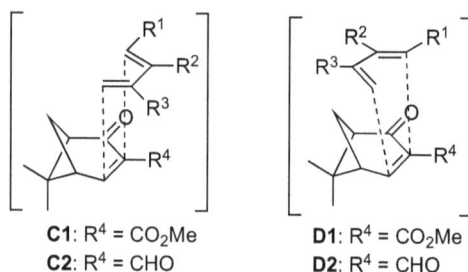

C1: R⁴ = CO₂Me
C2: R⁴ = CHO

D1: R⁴ = CO₂Me
D2: R⁴ = CHO

Figure 4: Proposed Diels-Alder transition states **C1,2** and **D1,2**.

In contrast to enone ester **29**, studies on the stereoselective outcome of the corresponding enone aldehyde **32** in Diels-Alder reaction were also carried out [22]. The results revealed that the yields of adduct **33** were extremely catalyst dependent, and better yields were obtained only when mild Lewis acids such as zinc chloride was employed (Scheme **15**). Apparently, relative to the moderate ester-*endo* selectivity with **29**, the formyl functionality appears to be more electron-withdrawing and less sterically demanding, resulting in a profound enhacement of the aldehyde-*endo* selectivity as illustrated with the proposed transition state **C2** in Fig. **4**.

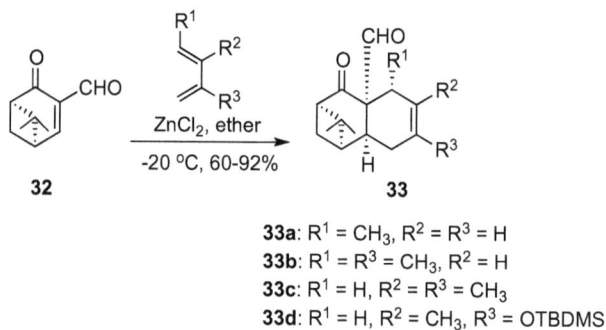

32

ZnCl₂, ether
-20 °C, 60-92%

33

33a: R¹ = CH₃, R² = R³ = H
33b: R¹ = R³ = CH₃, R² = H
33c: R¹ = H, R² = R³ = CH₃
33d: R¹ = H, R² = CH₃, R³ = OTBDMS

Scheme 15: Diastereoselective Diels-Alder reaction of chiral dienophile **32**.

Enantioselective Diels-Alder reaction of carbomethoxy-*para*-benzoquinone **34** with a variety of dienes, under catalysis with chiral lanthanide complexes [Sm(pybox)](OTf)₃ or [Gd(pybox)](OTf)₃, has recently been described by Evans *et al.* (Scheme **16**) [23]. In all cases examined, the exclusive formation of the *endo*–to-ketone adduct **35** was observed with excellent yield and enantioselectivity, again indicating that the C5-C6 double bond unit might play a significant role in affording the extra secondary orbital effects.

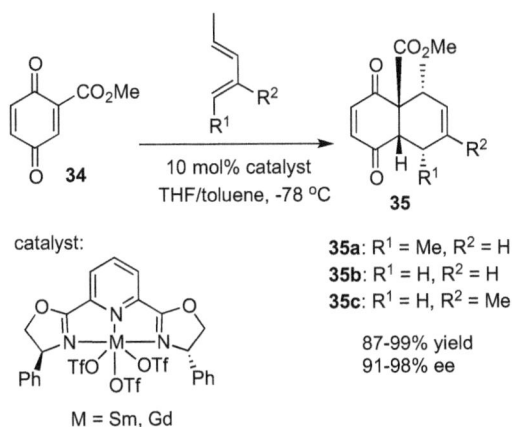

Scheme 16: Enantioselective Diels-Alder reaction of ester-substituted 1,4-benzoquinone **34**.

2.3. α-Cyano Cycloalkenones as Dienophiles

Cycloalkenone systems with a cyano functionality as the α-activating group were found to be synthetically useful and versatile by Liu's laboratories. As exemplified by Scheme **17** [24], the regiochemistry of the obtained adducts follows, in general, the *ortho-* and *para*-rules whereas the stereochemistry follows mainly the *endo*-to-ketone addition route. As usual, there were some exceptions occurring with the addition of isoprene where the *anti-para* product **37f** was formed predominantly. As well, addition of 2-methyl-1,3-pentadiene afforded products **37g** (*endo*-to ketone) and **38** (*endo*-to-nitrile) in almost an equal amount. These results are consistent with those observed for the ester dienophile **5** in reaction with isoprene and 2-methyl-1,3-pentadiene, respectively, and thus similar rationales are proposed (*vide supra*).

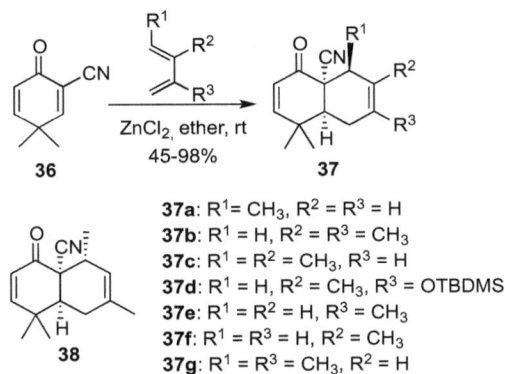

37a: R^1= CH$_3$, R^2 = R^3 = H
37b: R^1 = H, R^2 = R^3 = CH$_3$
37c: R^1 = R^2 = CH$_3$, R^3 = H
37d: R^1 = H, R^2 = CH$_3$, R^3 = OTBDMS
37e: R^1 = R^2 = H, R^3 = CH$_3$
37f: R^1 = R^3 = H, R^2 = CH$_3$
37g: R^1 = R^3 = CH$_3$, R^2 = H

Scheme 17: Diels-Alder reaction of α-cyano cyclohexenone **36**.

Also encouraging is the finding that unlike the corresponding 2-carbalkoxy-2-cyclohexenone **21**, 2-cyano-2-cycloalkenones **39**, ranging from five to eight membered ring, are rather stable and could be distilled, chromatographed and stored at 0 °C over a long period of time without significant decomposition. As illustrated in Scheme **18** [25], under catalysis with zinc choride, Diels-Alder reactions of dienophile **39** occurred readily at room temperature with a variety of dienes to give adducts in high yields. In general, the regioselective outcomes were found to strictly follow the *ortho* and *para* rules, and where applicable, *endo*-to-ketone addition products were obtained without fail. Interestingly, it is noteworthy that, in the case of 2-cyano-8-methyl-2-cyclooctenone, a complete face selectivity from the side opposite to the R^4 methyl group (*e.g.* **40f** and **40g**) was observed.

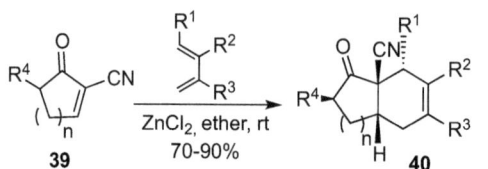

40a: n = 1, R^1 = R^2 = R^4 = H, R^3 = CH$_3$
40b: n = 2, R^1 = R^2 = R^4 = H, R^3 = CH$_3$
40c: n = 2, R^1 = CH$_3$, R^2 = R^3 = R^4 = H
40d: n = 2, R^1 = R^4 = H, R^2 = R^3 = CH$_3$
40e: n = 3, R^1 = R^2 = R^4 = H, R^3 = CH$_3$
40f: n = 4, R^1 = R^4 = CH$_3$, R^2 = R^3 = H
40g: n = 4, R^1 = R^4 = CH$_3$, R^2 = H, R^3 = OTBDMS

Scheme 18: Diels-Alder reaction of α-cyano cycloalkenone **39**.

As mentioned above, the cyano group present in each of the adducts was found to be quite useful for further synthetic elaboration. As demonstrated in Scheme **19** [25], the cyano moiety of adduct **41** could be readily replaced with an appropriate electrophile in one step *via* lithium naphthalenide-induced reductive alkylation to afford ketone **42** with retention of the complete *cis* configuration on the ring junction, which is difficult to achieve by any existing methods.

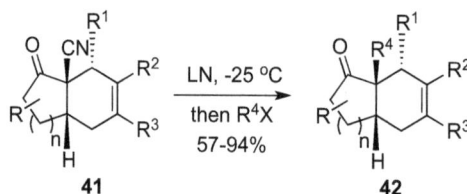

Scheme 19: LN-mediated reductive alkylation of α-cyano cycloalkanone **41**.

2.4.   α-Phenylthio/Phenylselenenyl/*p*Tolylsulphinyl   Cycloalkenones   as Dienophiles

Dienophile **43**, activated with a sulfur-containing moiety, was first studied by Knapp *et al.* [26]. This phenylthio group was found to have a positive effect on accelerating the Diels-Alder reaction as compared to the parent cyclopentenone (Scheme **20**). The adducts **44** thus obtained could undergo a facile oxidative elimination to afford dihydro-1-indanone skeletons **45** and **46** with a remarkable preference for the former, presumably owing to the fact that a strong dipole repulsion between the carbon-oxygen and sulfur-oxygen could be avoided in the course of *syn* elimination toward the adjacent angular hydrogen.

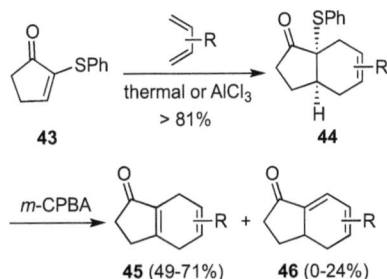

Scheme 20: Diels-Alder reaction of dienophile **43** and subsequent sulfoxide elimination.

A more versatile S-containing dienophile was extensively studied by Carretero *et al.* [27]. As illustrated in Scheme **21**, cycloaddition of dienophile **47** with Dane's diene proceeded with excellent π-facial and *endo*-to-ketone selectivity to give adduct **48** exclusively. The absolute configuration for enantiomer **48** has been assigned based on the proposed chelated model (Fig. **5**). Accordingly, the approach of the diene from the less hindered face of the chelate **I**, a complex arising from the association of the catalyst with both oxygens of the β-ketosulfoxide, is assumed to govern the formation of the observed product.

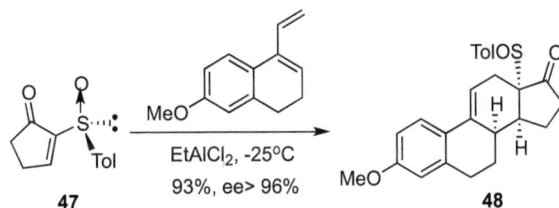

Scheme 21: Diastereoselective Diels-Alder reaction of chiral sulfoxide **47**.

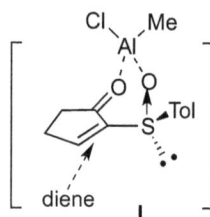

Figure 5: Proposed Diels-Alder transition state I.

In principle, the corresponding α-phenylseleno alkenones could also serve as effective dienophiles for Diels-Alder reaction [28], providing facile access to α-phenylseleno ketones, which might be useful in the preparation of various α,β-unsaturated cyclic ketones or have potential to serve as anti-cancer and anti-inflammatory agents [29]. Initial attempts to carry out Diels-Alder reactions with these dienophiles under thermal conditions were fruitless. Since the selenium species is naturally an electron-donating group, this electron-rich feature was highly suspected to be responsible for the experienced failure, but might be counteracted by introducing a Lewis acid. In fact, Diels-Alder reactions of dienophiles **49** with dienes underwent smoothly under catalysis with a variety of Lewis acids, and among them, stannic chloride was found to be the most effective one. In addition, this methodology also proved to be an effective benzoannulation process for preparation of benzoketones **51** (Scheme **22**).

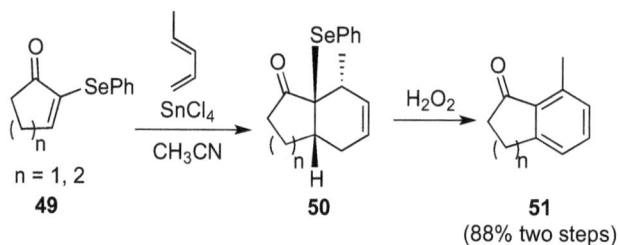

Scheme 22: Diels-Alder reaction of dienophile **49** and subsequent oxidative aromatization.

2.5. α-Halo/Diethoxyphosphinyl Cycloalkenones as Dienophiles

In contrast to the poor dienophilicity of the parent cycloalkenones, 2-bromo-2-cyclopentenone **52** and 2-bromo-2-cyclohexenone **53** were found to possess desirable Diels-Alder reactivity. Under stannic chloride catalysis [30], these dienophiles were shown to undergo cycloaddition reaction readily with various dienes to give the corresponding bromo adducts in good yields following the general Diels-Alder rules (Scheme **23**).

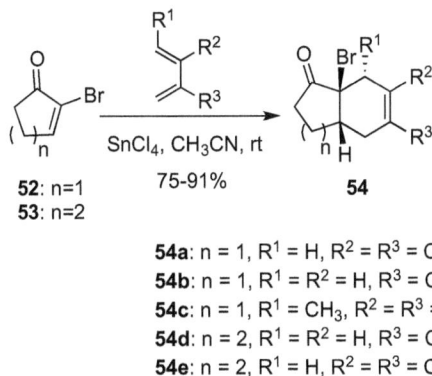

52: n=1
53: n=2

SnCl$_4$, CH$_3$CN, rt

75-91%

54

54a: n = 1, R^1 = H, R^2 = R^3 = CH$_3$
54b: n = 1, R^1 = R^2 = H, R^3 = CH$_3$
54c: n = 1, R^1 = CH$_3$, R^2 = R^3 = H
54d: n = 2, R^1 = R^2 = H, R^3 = CH$_3$
54e: n = 2, R^1 = H, R^2 = R^3 = CH$_3$

Scheme 23: Diels-Alder reaction of α-bromo cycloalkenones **52** and **53**.

Adducts **54** thus obtained were easily converted to the corresponding dienones **55** and **56** under basic treatment at room temperature (Scheme **24**) [30]. The formation of doubly cisoid fully conjugated dienone **55** was found to be predominant, presumably due to the preference for the *trans*-elimination with existence of methylene protons adjacent to bromine. These structurally unusual compounds, possessing three electrophilic sites and dual Diels-Alder reactivity, are potentially useful intermediates for further synthetic elaboration.

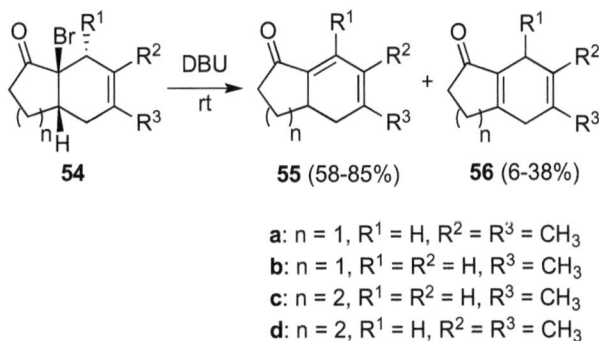

54   **55** (58-85%)   **56** (6-38%)

a: n = 1, R^1 = H, R^2 = R^3 = CH$_3$
b: n = 1, R^1 = R^2 = H, R^3 = CH$_3$
c: n = 2, R^1 = R^2 = H, R^3 = CH$_3$
d: n = 2, R^1 = H, R^2 = R^3 = CH$_3$

Scheme 24: Generation of isomeric dienes **55** and **56** from α-bromo ketone **54**.

Recently, Danishefsky's laboratories revisited the Diels-Alder chemistry of α-halo cycloalkenones with a range of dienes, where the cycloadducts were obtained in good yields in the presence of 10 mol% aluminum methyl dichloride (Scheme **25**)

[31]. Notably, the LN-mediated reductive alkylation of the resulting α-halo ketones provided the *trans*-fused bicyclic products bearing angular functional group, which was complementary to Liu's protocol in preparing related *cis*-fused systems (Scheme **19** and **27** *vs.* Scheme **25**) [32].

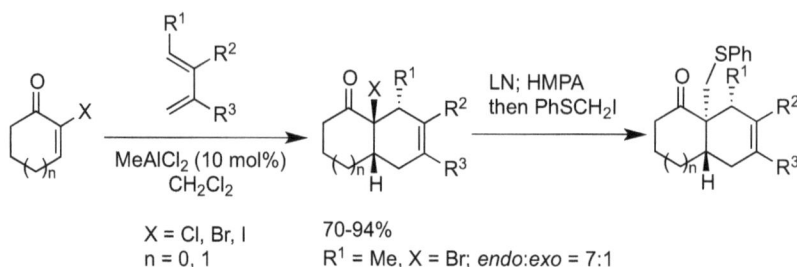

X = Cl, Br, I   70-94%
n = 0, 1   R^1 = Me, X = Br; *endo:exo* = 7:1

Scheme 25: Sequential Diels-Alder/reductive alkylation process.

The use of the phosphonate moiety as an activating group for the Diels-Alder reaction has also been investigated by Liu's group [33]. Under Lewis acid catalysis, dienophiles **57** were found to undergo facile Diels-Alder reaction to provide novel access to angularly phosphonated polycycles (Scheme **26**). Among the Lewis acids examined, including AlCl$_3$, ZnCl$_2$, SnCl$_4$, BF$_3$. OEt$_2$, TiCl$_4$, and FeCl$_3$, stannic chloride was found to be particularly effective in terms of reaction rate and yields of cycloadducts **58**. The experimental results revealed that most of the reactions, where applicable, follow the *ortho-* and *para*-rule. In terms of stereoselectivity, the addition reaction consistently follow the *cis* principle and *endo*-addition to the ketone carbonyl, suggesting that the ketone carbonyl plays a more important role in directing the *endo*-addition than the phosphonate group.

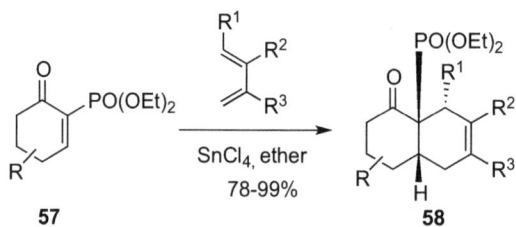

58a: R^1 = CH$_3$, R^2 = R^3 = H
58b: R^1 = H, R^2 = R^3 = CH$_3$
58c: R^1 = R^3 = CH$_3$, R^2 = H
58d: R^1 = R^2 = CH$_3$, R^3 = H

Scheme 26: Diels-Alder reaction of enone phosphonate **57**.

In addition to providing direct access to various phosphonate-containing polycyclic compounds, the aforementioned Diels-Alder approach is also synthetically useful in accessing angularly substituted *cis*-decalin systems. As demonstrated in Scheme **27** [33], adducts thus obtained could undergo direct reductive alkylation under mild reaction conditions to afford the desired products with complete retention of the *cis*-ring juncture.

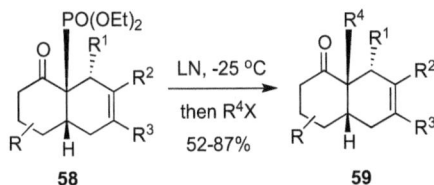

Scheme 27: LN-mediated reductive alkylation of β-keto phosphonate **58**.

2.6. Synthetic Applications to Formal/Total Syntheses of Natural Products

Forskolin (**64**), a naturally occurring polycyclic compound of biological interest, was the target of total synthesis by E.J. Corey and K.C. Nicolaou *via* key intermediate **63** for the former and **62b** for the latter. Liu and co-workers subsequently completed a formal synthesis of **64** by developing a Diels-Alder approach to access both aforementioned key intermediates. Towards this end, compound **5** was utilized as a dienophile to rapidly establish the required stereochemical arrangements for further synthetic elaboration (Scheme **28**) [34]. Adduct **60** thus formed was transformed into ketal diene **62a** in a eight-step sequence, involving an unusual radical dehydrogenation of **61c** under reduction with Zn metal in the presence of NaI. This was followed by sequentially acidic hydrolysis, selective oxidation with $(Ph_3P)_3RuCl_2$, oxidative-cyclization with Jones reagent and photooxygenation to give the targeted endoperoxide **63**. Similarly, intermediate **61** was applied to provide compound **62b** [35].

The synthetic utility of enone ester **29** as a dienophile was further validated in a total synthesis of (+)-qinghaosu (**68**) by Liu *et al.* [36] (Scheme **29**). Compound **29** could undergo Diels-Alder reaction with isoprene with complete facial, regio- and stereoselectivity to provide adduct **30c** possessing virtually all the carbon units arranged in the correct sense for target molecule **68**. Adduct **30c** was then transformed into the protected intermediate **65** in four steps, which in turn was converted into benzoate **66** in a sequence of eleven steps, involving the installment of a required methyl group by a series of stereochemical-control operations. Compound **66** thus obtained was further elaborated into isomeric mixture **67** under standard reaction conditions, which was subjected to photooxygenation followed by treatment with trifluoroacetic acid, triggering a process of multi-step ring closure, to accomplish the total synthesis of (+)-qinghaosu (**68**).

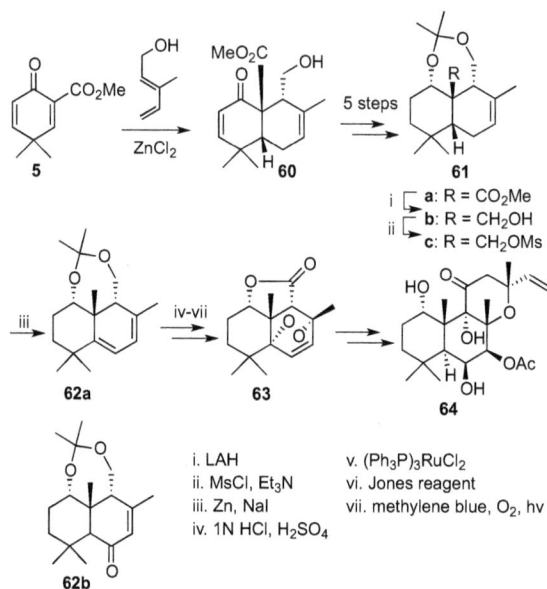

Scheme 28: Formal synthesis of forskolin (**64**) *via* a Diels-Alder approach.

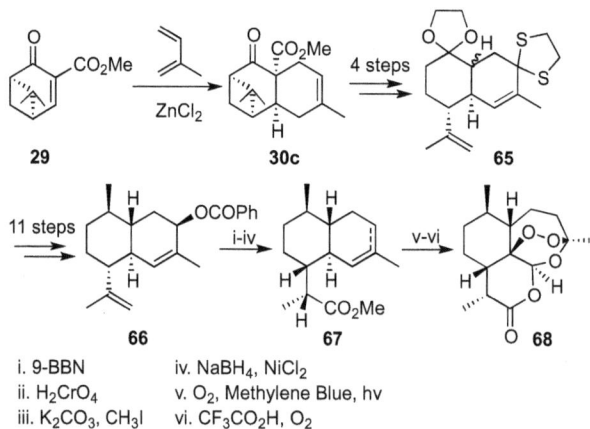

Scheme 29: Asymmetric total synthesis of (+)-qinghaosu (**68**) *via* a Diels-Alder approach.

The first total synthesis of 2-oxo-5α,8α-13,14,15,16-tetranorclerod-3-en-12-oic acid (**72**) in racemic form was accomplished by Liu *et al.* [37], again utilizing a Diels-Alder reaction of a α-activated cross conjugated cyclohexenone system. The synthetic design called for a Diels-Alder reaction of dienone **69** with *trans*-piperylene as a key operation, the product of which was transformed efficiently in nine steps to afford target **72** (Scheme **30**). This Diels-Alder strategy effectively provided a general synthetic scheme towards a plethora of naturally occurring *cis*-clerodane diterpenoids since it allowed for the simultaneous generation of four contiguous stereogenic centers in the correct fashion.

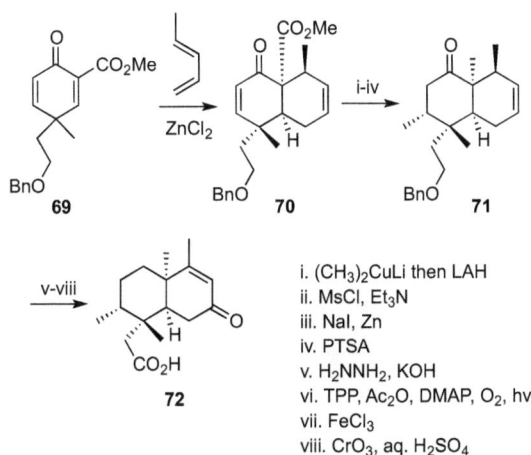

i. (CH$_3$)$_2$CuLi then LAH
ii. MsCl, Et$_3$N
iii. NaI, Zn
iv. PTSA
v. H$_2$NNH$_2$, KOH
vi. TPP, Ac$_2$O, DMAP, O$_2$, hv
vii. FeCl$_3$
viii. CrO$_3$, aq. H$_2$SO$_4$

Scheme 30: Total synthesis of 2-oxo-5,8-13,14,15,16-tetranor-clerod-3-en-12-oic acid (**72**) *via* a Diels-Alder approach.

To demonstrate this, *cis*-clerodane diterpenoid 6β-acetoxy-2-oxokolavenool (**77**) was also synthesized in a similar Diels-Alder approach using **73** as a dienophile (Scheme **31**) [38]. Dienone **73** was allowed to react with *trans*-piperylene in the presence of zinc chloride to provide the expected cycloadduct **74** predominantly *via* the addition from less hindered C-4 methyl face. After serving its purpose as an activating group for the Diels-Alder reaction, the nitrile moiety in adduct **74** was reductively removed by treatment with lithium naphthalenide followed by alkylation with methyl iodide to give enone **75**. As expected, the reductive alkylation proceeded in a complete stereoselctive manner from the convex face of the **74**, leading to the introduction of the methyl group with retention of the *cis* configuration. Compound **75** thus obtained was transformed into diketone **76** by a two-step sequence, involving a stereoselective 1,4-addition process and Wacker reaction. Finally, target **77** was achieved straightforwardly by further elaboration of **76** in three steps. Obviously, this improved synthetic strategy is more efficient than the original approach using **69** as a dienophile, in which four steps are required for converting the ester moiety to the methyl group (**70**→**71**) rather than one (**74**→**75**) [39].

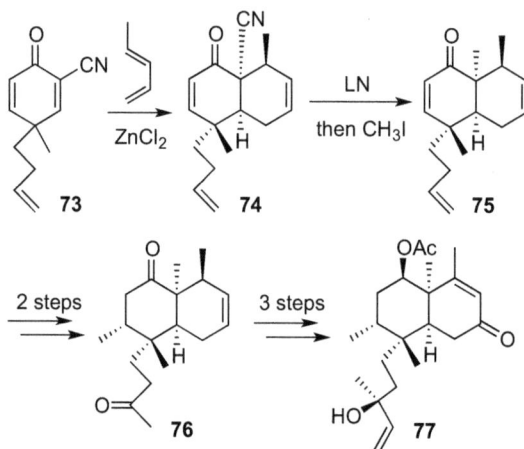

Scheme 31: Total Synthesis of 6-Acetoxy-2-oxokolavenool (**77**) *via* a Diels-Alder approach.

The total synthesis of structurally complex teucvin (**82**) and its 12-epimer **83** were first accomplished by Liu *et al.* in 2003 [40]. As illustrated in Scheme **32**, under zinc chloride catalysis, cycloaddition of dienophile **78** with *trans*-2,4-pentadien-1-ol occurred in a completely face-selective fashion in favor of the carbomethoxy-containing face, which was highly consistent with the previous observation that the direct placement of an ester group at C-4 on various cyclohexenone dienophiles could induce a profound face selectivity during the addition of dienes (see Scheme 7). In addition, this key operation also proceeded with remarkable stereo- and regioselectivity, giving rise to adduct **79** appended with an angular hemiacetal moiety. Compound **79** was, in five steps, converted to keto acid **80** which was treated with *p*-TSA to introduce an α,β-unsaturated γ-lactone ring in **81** *via* enol lactone ring formation with a concomitant isomerization of the double bond α to the carbonyl. Finally, lactone **81** underwent a four-step synthetic sequence to attain two separable 19-*nor*-clerodanoid natural products **82** and **83** in racemic form. In addition to the synthetic efforts described above, the utility of α-activated cross conjugated cycloalkenone systems was also exemplified in the synthesis of the natural products by Liu's group, including (-)-xenetorin C (**84**) [41], (±)-isoacanthodoral (**85**) [42], (-)-morphine (**86**) [43], (-)-qinghaosu IV (**87**) [44], (±)-solidago alcohol (**88**) [45], and (±)-montanin A (**89**) [46] as illustrated in Fig. **6**.

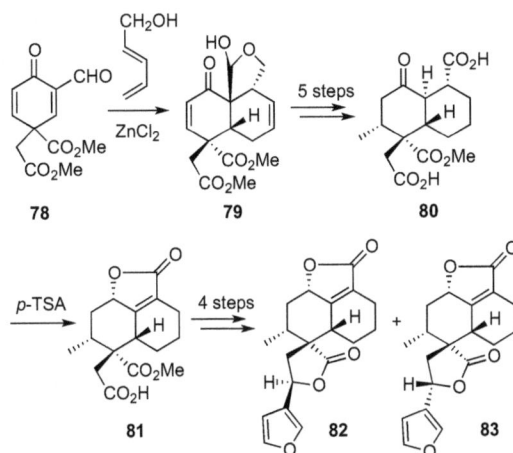

Scheme 32: Total Synthesis of Teucvin (**82**) and 12-*epi*-Teucvin (**83**) *via* a Diels-Alder approach.

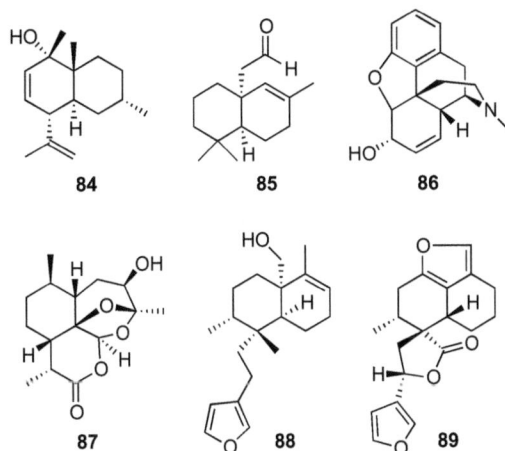

Figure 6: Structures of natural products (−)-xenetorin C (**84**), (±)-isoacanthodoral (**85**), (−)-morphine (**86**), (−)-qinghaosu IV (**87**), (±)-solidago Alcohol (**88**), and (±)-montanin A (**89**).

3. POLYENE CYCLIZATION

Polyene cyclization, also known as cationic cyclization, is a powerful synthetic tool for the preparation of polycyclic compounds. To facilitate this multi-step cyclization process, setting an appropriate functional group as the initiator is always one of the most important decisions. Functionalities reported to be effective initiators to varying degrees include epoxide, acetal, allylic alcohol, and α,β-unsaturated carbonyl group (aldehyde or ketone) [47]. Not unlike previously reported α,β-unsaturated carbonyl systems, α-activated cross conjugated cycloalkenones have also been found to be effective promoters of cationic cyclization with high regio- and stereoselectivity. An overview of this polyene cyclization chemistry will be presented as follows.

3.1. Approaches to Polycyclic Derivatives

Enone ester **90**, initially intended as an alternate dienophile toward the total synthesis of natural product **77**, was found to not provide the desired cycloadduct **93** regardless of reaction conditions. Interestingly, a pair of unexpected cyclic products **91** and **92** were constantly obtained in a variable ratio subject to Lewis acids employed (Scheme **33**) [48, 49]. In addition to the high regio- and

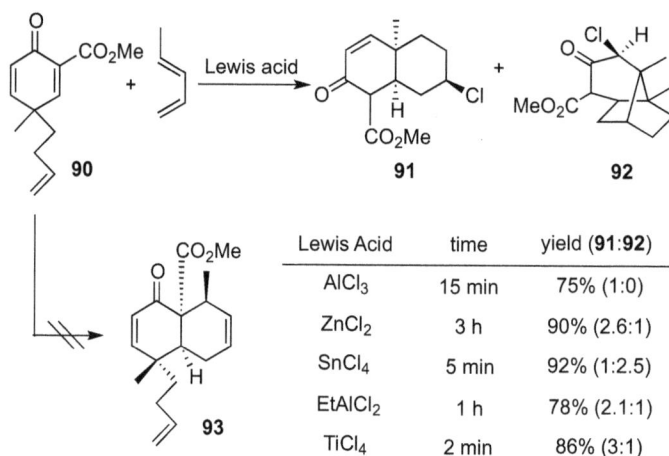

Lewis Acid	time	yield (**91:92**)
AlCl₃	15 min	75% (1:0)
ZnCl₂	3 h	90% (2.6:1)
SnCl₄	5 min	92% (1:2.5)
EtAlCl₂	1 h	78% (2.1:1)
TiCl₄	2 min	86% (3:1)

Scheme 33: Unexpected polyene cyclization of dienophile **90** under catalysis with various Lewis acids.

stereoselectivity, taking an unusual chloride-inserting mode as a termination step for this cyclization process is also of considerable interest. A mechanistic rationale was proposed and depicted in Scheme **34** [48]. It was assumed that the stereochemical control observed with both **91** and **92** was a result of the intramolecular transfer of the chloride ion from the complex metal to the incipient carbocation (**94**→**97** and **95**→**96**) with excellent selectivity from the *syn* face. Also noteworthy is that the formation of tricyclic compound **92** requires the participation of the β-carbon of the conjugated enone system. This is rather unusual, but could be rationalized by invoking the intermediacy of allylic carbocation **96** as part of the stabilizing force.

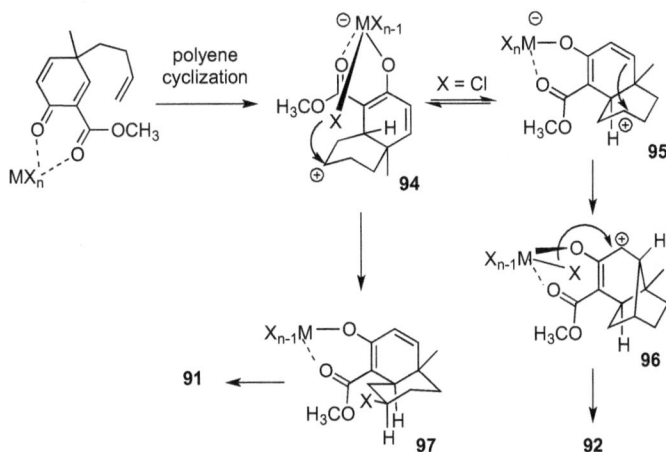

Scheme 34: Proposed mechanism for the formation of unexpected polyene cyclization products **91** and **92**.

The aforementioned finding was immediately extended to the formal synthesis of the natural product dehydrochamaecynenol (**99**) in optically active form (Scheme **35**) [50].

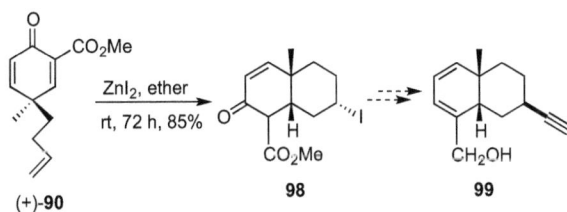

Scheme 35: Formal synthesis of dehydrochamaecynenol (**99**) *via* a polyene cyclization process.

The cyclization of enone ester **100**, lacking an extra conjugated double bond in the ring system compared to dienone ester **90**, was also carried out to determine the potential influence of the spectator double bond [49]. As a result, the ring closure of **100**, leading to product **101**, was found to be equally efficient to **90** (Scheme **33**) upon exposure to stannic chloride as catalyst is illustrated in Scheme **36**.

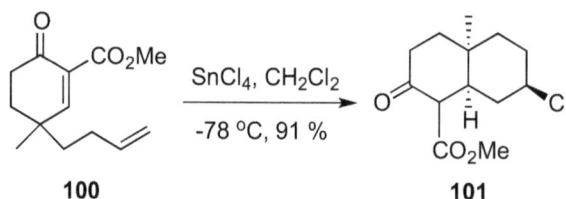

Scheme 36: Polyene cyclization of α-ester cyclohexenone **100**.

The use of the cross conjugated β-keto ester unit as a promoter for the polyene cyclization would also facilitate the rapid formation of highly functionalized hydrophenanthrene systems present as the core structure of a large number of naturally occurring compounds. Accordingly, the total synthesis of pygmaeocin C (**104**) was accomplished using **102** as an effective initiator (Scheme **37**) [51]. Upon treatment with trifluoacetic acid, **102** was transformed to **103** exclusively by a tandem ring closure, involving a double-bond isomerization and rapid Friedel-Crafts alkylation. This was followed by further structural modifications into the target **104** in three steps.

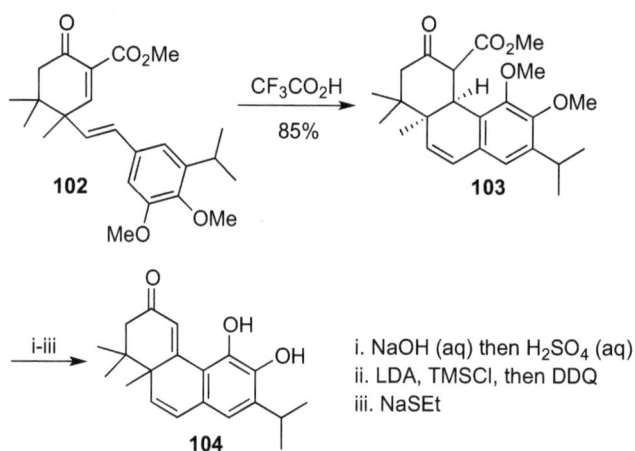

Scheme 37: Total synthesis of pygmaeocin C (**104**) *via* a polyene cyclization process.

By introducing a suitable aromatic appendage to the titled systems, theoretically, a polyene cyclization process might be effected to prepare compounds containing various numbers of aromatic rings with pre-existing functionalities. Along this axis, a complex hydrochrysene system, in practice, could be easily constructed by a designed polyene initiator **105** as shown in Scheme **38** [52].

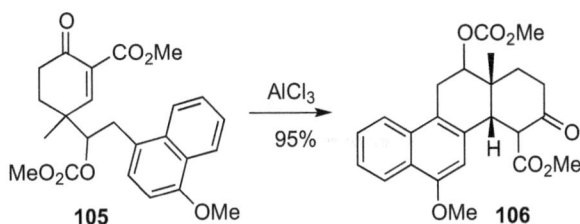

Scheme 38: Polyene cyclization of α-ester cyclohexenone **105**.

Enone ester **107** was also found to be a good chemical entity for polyene cyclization (Scheme **39**) [53]. Among several Lewis acids examined, zinc iodide was recognized as the most effective one, giving rise to a mixture of epimers **108** in high yields. Upon treatment of DDQ, **108** was further oxidized to dienone ester **109**, which proved to be an effective dienophile to undergo facile Diels-Alder reaction to give adduct **110** with complete face selectivity from the opposite side of the angular methyl group. Besides, an exclusive participation of the less substituted γ,δ-double bond was observed for the above cycloaddition process. This protocol might find broad synthetic utility, particularly in the preparation of steroid mimetics serving as potential therapeutic agents.

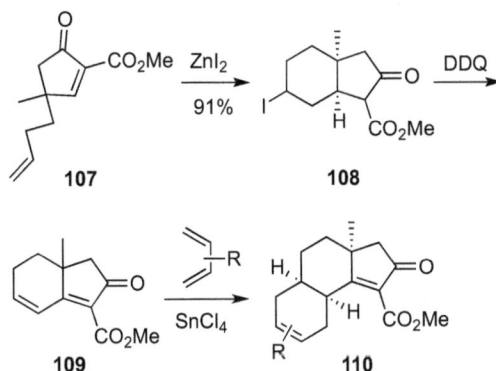

Scheme 39: Three-step synthesis of steroid-like compound **110**.

Thiophene containing polycycles often possess desirable medicinal properities [54]. Under catalysis of stannic chloride, enone ester **111** was found to cyclize effectively to give tricyclic **112** (Scheme **40**) [55] in virtually quantitative yield. Similarly, treatment of bromo dienone **113** with aluminum chloride resulted in the formation of the tricyclic product **114** as a sole product in excellent yield. The above experiments revealed that even poorly nucleophilic thiophenes could undergo cyclization with high efficiency and significant selectivity, again indicating that the α-activated cross conjugated cycloalkenone unit is an extremely active promoter to effect intramolecular polyene cyclization.

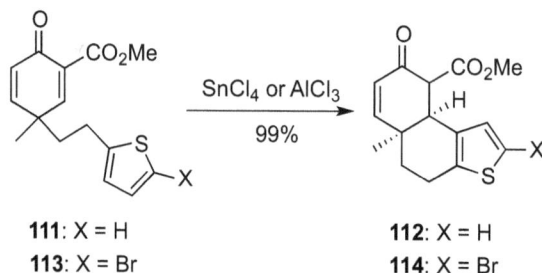

111: X = H
113: X = Br

112: X = H
114: X = Br

Scheme 40: Polyene cyclization of compounds **111** and **113**.

3.2. An Unusual Through-Space σ-Bond Tandem Rearrangement

Recently, Liu and co-workers reported a novel and unanticipated polyene cyclization/rearrangement process leading to polycarbocycles possessing a high degree of structural complexity. As listed in Table **1**, under catalysis with Lewis acid (AlCl$_3$ or SnCl$_4$), polyene cyclization of **115** and **116**, and their corresponding regioisomers **117** and **118** gave rise to complex products in moderate to high yields, respectively [56, 57]. Surprisingly, the anticipated cationic cyclization product of the individual reactions was not formed predominantly; instead, the major product was determined to possess a highly unexpected structure which was unambiguously confirmed by X-ray crystallographic analyses. Mechanistically, a generally plausible rationale was proposed to explain these unusual observations (Scheme **41**). As demonstrated by a specific example with **115** as a substrate (Scheme **42**), following an incipient aluminoxy complex in the presence of aluminum trichloride, a cascade of unusual σ-bond migrations are highly envisioned to furnish the unexpected product **119**. As well, it is noteworthy that the product formation was found to be catalyst dependent; as **115** was treated with SnCl$_4$ instead of AlCl$_3$, cyclic products **120** and **121**, evidently derived from different mechanistic pathways, were obtained. More encouragingly, when compound **117** was catalyzed with AlCl$_3$ in the presence of TMSCl, an effective enolate-trapping agent, only a pair of epimeric chlorides **124** were formed, implying that the ensuing enolate might play a critical role in promoting the 1,2-hydride shift leading to the formation of **123**. However, exactly what driving force triggers the 1,2-hydride shift, a key mechanistic step proposed for product formation in Table **1** is still not fully understood, and more evidence is required before we can come to any conclusions.

Schemes 41: General scheme of tandem polyene cyclization/rearrangement sequence.

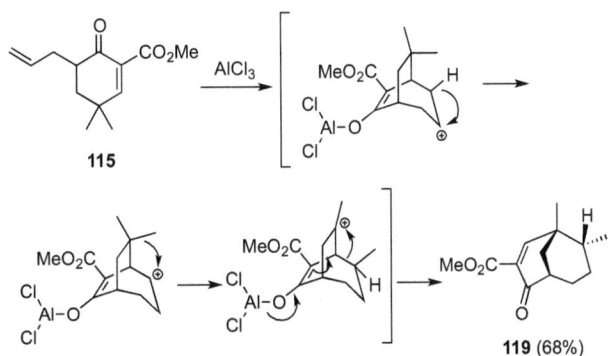

Schemes 42: Proposed mechanism of the bridged bicyclic compound **119**.

The mechanistic rationale invoked for the generation of the above products may be atypical for the conditions applied but not entirely unprecedented as similar mechanistic pathways have been reported for some electron-deficient olefinic and structurally highly constrained systems [58-62]; however, the hydride/alkyl shift observed by Liu's laboratory is quite exotic in that those described in above literatures are limited to substrates containing a $C(sp^3)-H$ bond adjacent to a heteroatom (N, O) or an activated tertiary benzylic $C(sp^3)-H$ bond as a hydride donor so that the ensuing carbocation can be stabilized by electron delocalization; this carbocation-stabilizing capacity might serve, in part, as the essential driving force to facilitate the through-space 1,5-hydride shift in these systems [63]. Nevertheless, the mechanistic pathways proposed by Liu *et al.*, though occurring without intermediacy of the extra stabilizing force as with many historical cases, are considered quite normal in the living system and commonly adopted in elucidating the formation of many secondary metabolites.

Table 1: Polyene cyclization promoted by the cross conjugated enone esters

Enone ester	Lewis acid	Products (yield%)
115	AlCl$_3$	**119** (68%)
115	SnCl$_4$	**120** (40%) **121** (22%)
116	AlCl$_3$ or SnCl$_4$	**122** (75%)
117	AlCl$_3$	**123** (39%) **124** (39%)
117	AlCl$_3$/TMSCl	**124** (72%)
118	AlCl$_3$	**125** (68%)

4. SEQUENTIAL MICHAEL REACTION/ANNULATION REACTION PROCESSES

Owing to the additional electron-withdrawing conjugation, the β-carbon of the titled system is intrinsically more electrophilic than the parent enone, allowing facile Michael-type reactions. If a side chain introduced at the β-position *via* 1,4-addition was installed with proper functional groups, that pendant reaction center could potentially react with activated α-carbon under treatment with Lewis acids or bases, or transition metals, a process that can be deliberated as a bimolecular annulation sequence (Scheme **43**). Recent developments of that strategy exploiting α-activated cycloalkenones will be discussed as follows.

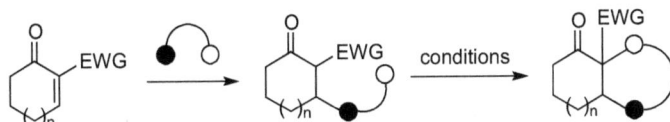

Scheme 43: General scheme for sequential Michael addition/annulation reaction process.

4.1. Sequential Michael Addition/Pd(II)-Mediated Oxidative Cyclization Process

2-Cyano-2-cycloalkenones have been recognized as highly reactive Michael acceptors, thus facilitating an uncatalyzed conjugate addition with a variety of Grignard reagents [64]. Knowing that salient feature of cross conjugated cyano ketones, Liu and co-workers devised a facile methylenecyclopentane annulation sequence involving nucleophilic addition of 3-butenylmagnesium bromide to **126** and palladium(II) acetate mediated oxidative cyclization of the resulting 1,4-adduct **127** (Scheme **44**) [65]. Remarkably, the cyclization, without prior activation of cyano ketones in a form of enol or enolate [66], proceeded smoothly with excellent regio- and stereocontrol, giving predominant *cis*-fused bicyclic products under relatively mild condition [67].

In addition, Liu's laboratories extended the aforementioned two-step sequence to the construction of homologous variants (Scheme **45**) [68]. While the transition metal catalyzed hexannulation of related systems was impeded by the β-quarternary center [69], it was notable that the Pd(II)-mediated cyclization of **129** (R^1 = Me) offered annulated products with exclusive *cis*-ring junction in good yields.

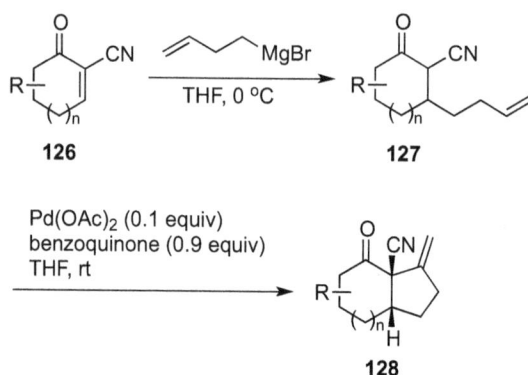

Scheme 44: Sequential Michael addition/Pd(II)-mediated methylenecyclopentane annulation process.

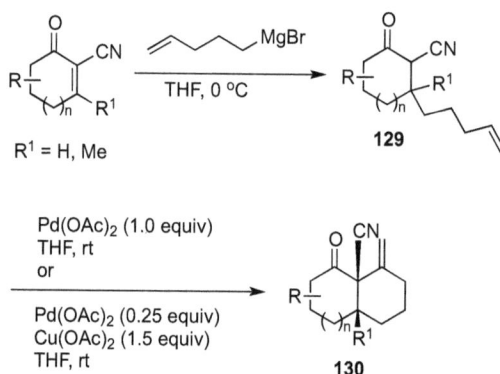

Scheme 45: Facile construction of methylenecyclohexane ring system *via* oxidative Conia-ene annulations.

Driven by the interests toward total synthesis of pinguisane sesquiterpenoids, Liu and co-workers realized that the previously developed method [65], with a simple structural modification to **126**, could provide a concise route toward the bicyclo[4.3.0] skeleton found in those natural products. As outlined in Scheme **46**, pentannulation of β-keto ester **131** in the presence of palladium(II) acetate lead to the formation of regioisomeric **132** and **133**, both of which could be converted to racemic acutifolone A [70]. Around the same time, Hibi and Toyota reported a related example comprising sequential conjugate addition and palladium-catalyzed cycloalkenylation under oxygen atmosphere [71].

Martin and co-workers recently described the total synthesis of lycopladine A highlighting sequential 1,4-addition and Pd(II)-catalyzed enolate arylation to establish the tricyclic core (Scheme **47**) [72]. Compared to the results employing parent 5-methylcyclohex-2-en-1-one, an additional α-ester group within **134/135** not only significantly enhanced the diastereoselectivity of the conjugate addition (*trans*:*cis* >20:1 *vs.* 2.5:1), but also gave a cleaner reaction profile in the succeeding arylation step (**135**→**136**).

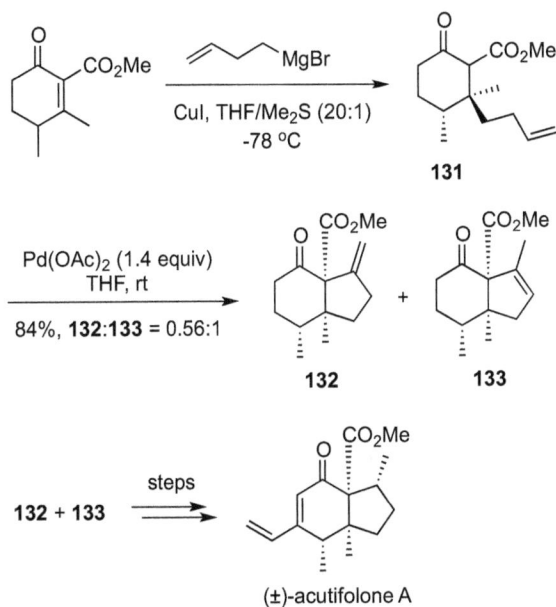

Scheme 46: Total synthesis of (±)-acutifolone A.

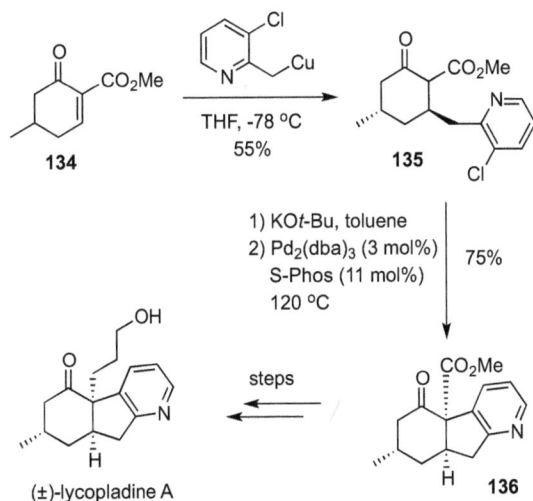

Scheme 47: Total synthesis of lycopladine A.

4.2. Sequential Michael Addition/Conia-Ene Cyclization Process

The Liu's laboratories explored the Conia-ene reaction of a diversed array of ω-alkynyl α-cyano ketones **137** that were readily prepared *via* Michael addition of 1-(trimethylsilyl)-1-butyn-4-yl magnesium chloride onto the corresponding 2-cyano-2-cycloalkenones **126** [73]. In the event, zinc iodide that has been used in related processes [74, 75] was found to be an effective promoter to achieve the desired annulation (Scheme **48**). Not unexpectedly, the ring fusion of resulting bicyclic systems **138** and **139** was dependent on the parent ring size. Although the α-ester counterparts displayed parallel reactivity for both 1,4-addition and Conia-ene reaction [73, 76], the demand for installing the α-cyano functionality was highly perceivable owing to its versatility for advanced synthetic elaborations [77-81].

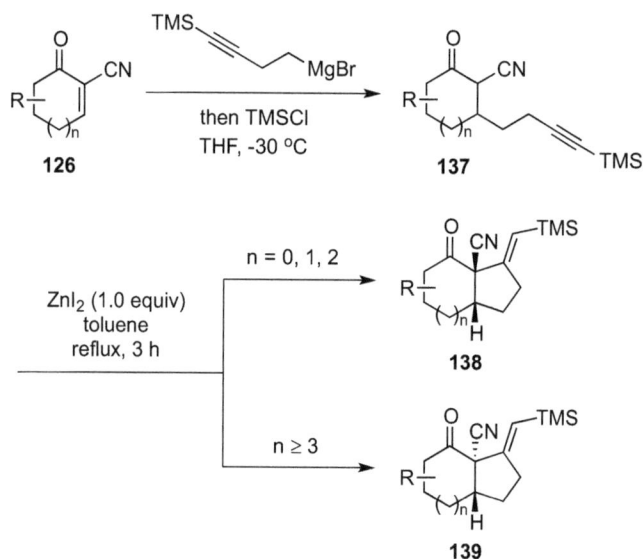

Scheme 48: Sequential Michael addition/Conia-ene cyclization process.

4.3. Sequential Michael Addition/Autoxidative Annulative Cascade Process

Shia and co-workers serendipitously discovered that ω-silylacetylenic α-cyano cycloalkenone **137** (R = H, n = 1) partially converted into a peculiar product type **140** upon exposure to air and light (Scheme **49**) [82]. After considerable optimization of rection conditions, a series of α-cyano cycloalkenone with appendant TMS-capped alkynyl moiety swiftly underwent a domino autoxidative annulation sequence in the presence of pyridine (10 mol%) under one oxygen atmosphere. To probe the subtle substitution effect, synthetic analogues **141** and **142** (Fig. **7**) were prepared and subjeted to the same reaction condition, in which starting materials were recovered intact even after prolonged reaction time. As such, authors presumed that α-cyano moiety functioned beyond a simple electron-withdrawing group; moreover, the silicon hyperconjugation likely exerted an extra stabalization to putative radical intermediates during the ring-closure event.

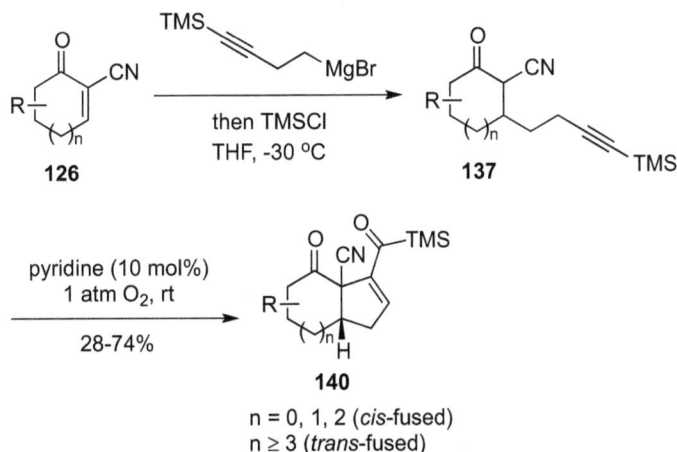

Scheme 49: Unusual aerobic annulative cascade of α-cyano TMS-capped alkynyl ketone **137**.

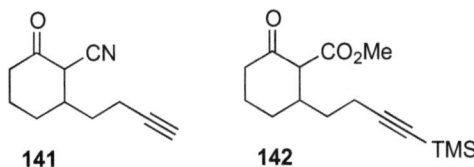

Figure 7: Compounds **141** and **142** inert to autoxidation.

4.4. Sequential Michael Addition/Alkylative Annulation Process

Fleming *et al.* reported that the uncatalyzed 1,4-addition of chlorobutyl Grignard reagent **143** to α-cyano cycloalkenone followed by capture of the enolate with silyl chloride afforded β-siloxy unsaturaed nitiles **144** with no detection of 1,2-adduct [83]. Upon treatment with *n*-Bu₄NF, intramolecular alkylation of **144** occured smoothly to provide the *cis*-decalin product **145** in good yield (Scheme **50**). Intriguingly, the *trans*-decalin **146** could be prepared from a common intermediate **144** in a three-step sequence. The complementary stereoselectivity was attributed to the difference in conformation of cyclic enolate and nitrile anion.

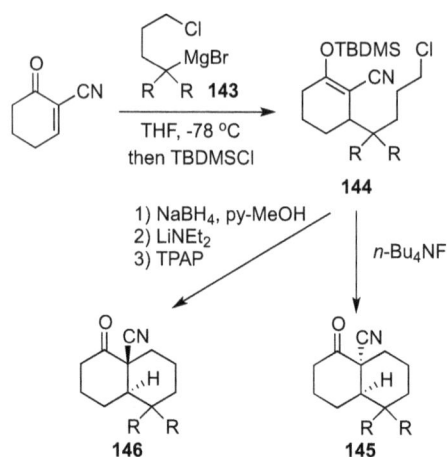

Scheme 50: Divergent synthesis of *cis*- and *trans*-decalin systems.

Johnson and co-workers disclosed that vinylidene cyclopropanes (VCPs) **148** could be derived from α-ester cycloalkenone through sequential conjugate acetylide addition and 3-*exo-dig* S_N2' cyclization (Scheme **51**) [84]. Since conventional carbene-type chemistry for making VCPs was not applicable to the electron-deficient enone systems, this method for accessing geminal dicarbonyl VCPs is synthetically substantial.

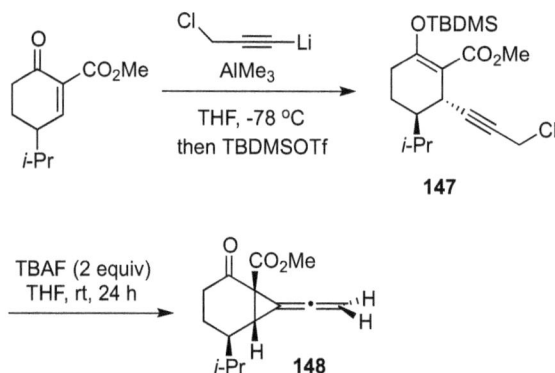

Scheme 51: Two-step synthesis of vinylidene cyclopropane **148**.

More recently, Liu's group reported a highly atom-economical sequence to construct bicyclo[n.4.0]derivatives involing $ZnCl_2$-catalyzed conjugate addition of lithium enolate onto α-cyano cycloalkenones followed by K_2CO_3/NaI-induced alkylative annulation (Scheme **52**) [85]. Despite the diastereoselectivity for the

Michael reaction required further improvement, the resulting bicyclic products **149** possessing differential carbonyl reactivities were valuable building blocks for subsequent synthetic elaboration.

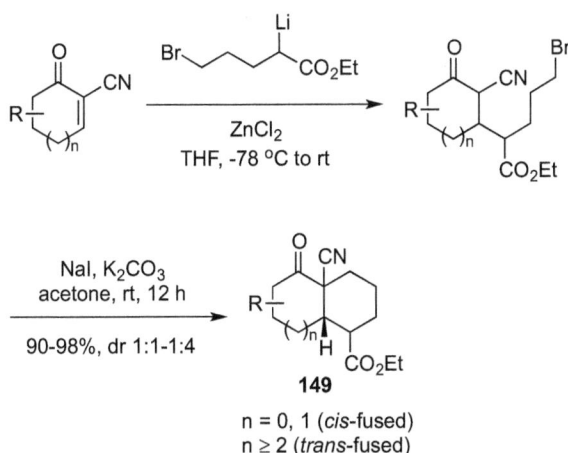

Scheme 52: Construction of highly functionalized bicyclic ring systems.

CONCLUSIONS

α-Activated cross conjugated cycloalkenone systems have proven to be versatile in synthetic applications, particularly serving as effective dienophiles in intermolecular Diels-Alder cycloadditions, active promoters in intramolecular polyene cyclizations, or reactive Michael acceptors in intermolecular copper-free conjugate reactions. These processes proceeded consistently with a high degree of stereochemical control, leading to a variety of complex structural motifs which are difficult to access by other synthetic methods. Mechanistically, an additional conjugated double bond incorporated in the cycloalkenone core of the titled systems might contribute to the extra secondary orbital effects, resulting in significant enhancement of the *endo*-to-ketone addition. However, a tandem multiple σ-bond migration process, an enzymatic pathway occurring in nature rather than in the pure chemical environment, was proposed to rationalize the formation of structurally unprecedented polyene-cyclization products, implying that enzyme-mimicking pathways might potentially occur under typical chemical reaction conditions and thus should not be entirely excluded nor considered aberrant in the elucidation of product formation. On the other hand, the titled system was recently exploited in various sequential Michael reaction/annulation reaction processes furnishing a diverse array of densely functionalized bicyclic systems.

ACKNOWLEDGEMENTS

We are grateful to the National Science Council (NSC 100-2113-M-400-002) and National Health Research Institutes, Taiwan, R.O.C. for financial support.

CONFLICT OF INTEREST

The author(s) confirm that this chapter content has no conflict of interest.

DISCLOSURE

The chapter submitted for series eBook titled **"Advances in Organic Synthesis, Volume 5"**is an update of our article published in **CURRENT ORGANIC SYNTHESIS, Volume 7, Number 1, February Issue 2010,** with additional text and references.

ABBREVIATIONS

9-BBN   = 9-Borobicyclo[3.3.1]nonane

Ac   = Acetyl

aq   = Aqueous

Bn   = Benzyl

DBU   = 1,5-Diazabicyclo[5.4.0]undec-7-ene

DDQ   = 2,3-Dichloro-5,6-dicyano-1,4-benzoquinone

ee   = Enantiomeric excess

eq   = Equivalent

Et   = Ethyl

EWG   = Electron-withdrawing group

hv   = Light

i-Pr   = *iso*-Propyl

LA   = Lewis acid

LAH   = Lithium aluminium hydride

LDA   = Lithium diisopropylamide

LN   = Lithium naphthalenide

m-CPBA   = *meta*-Chloroperbenzoic acid

Me   = Methyl

Ms   = Methanesulfonyl

Ph   = Phenyl

PhH   = Benzene

PNNP   = (1S,2S)-*N*,*N*′-bis[*o*-
                        (diphenylphosphino)benzylidene]cyclohexane-1,2-diamine

p-TSA   = *para*-Toluenesulfonic acid

pybox   = Pyridyl-*bis*(oxazoline)

rt   = Ambient temperature

TBDMS   = *tert*-Butyldimethylsilyl

TMS   = Trimethylsilyl

Tol   = Toluene

TPAP   = Tetrapropylammonium perruthenate

TPP   = 5,10,15,20-Tetraphenyl-21H,23H-porphine

REFERENCES

[1]   Schotes, C.; Mezzetti, A. Alkylidene β-ketoesters in asymmetric catalysis: Recent developments. *ACS Catal.*, **2012**, *2*, 528-538.
[2]   Fringuelli, F.; Taticchi, A.; Wenkert, E. Diels-Alder reactions of cycloalkenones in organic synthesis. *Org. Prep. Proc. Int.*, **1990**, *22*, 133-165.
[3]   Bartlett, P.D.; Woods, G.F. Some Reactions of Δ^2-Cyclohexenone, Including the Synthesis of Bicyclo(2,2,2)-octanedione-2,6. *J. Am. Chem. Soc.*, **1940**, *62*, 2933-2938.
[4]   Liu, H.J.; Browne, E.N.C. 2-Carbomethoxy-4,4-dimethyl-2,5-cyclohexadien-1-one as a dienophile. A convenient approach to the 4,4-dimethyl-1-decalone system. *Tetrahedron Lett.*, **1977**, *18*, 2919-2922.
[5]   Liu, H.J.; Browne, E.N.C. Diels–Alder reactions of 4,4-dimethyl-2-cyclohexenones. A direct route to the 4,4-dimethyl-1-decalones. *Can. J. Chem.*, **1987**, *65*, 1262-1278 and references therein.
[6]   Fringuelli, F.; Pizzo, F.; Taticchi, A.; Halls, T.D.J.; Wenkert, E. Diels-Alder reactions of cycloalkenones. 1. Preparation and structure of the adducts. *J. Org. Chem.*, **1982**, *47*, 5056-5065.
[7]   Fringuelli, F.; Minuti, L.; Pizzo, F.; Taticchi, A.; Halls, T.D.J.; Wenkert, E. Diels-Alder reactions of cycloalkenones. 2. Preparation and structure of cyclohexadienone adducts. *J. Org. Chem.*, **1983**, *48*, 1810-1813.
[8]   Liu, H.J.; Ngooi, T.K.; Browne, E.N.C. Diels–Alder reactions of 2-carbomethoxy-2-cyclohexen-1-one. *Can. J. Chem.*, **1988**, *66*, 3143-3152.
[9]   Liotta, D.; Saindane, M.; Barnum, C. Diels-Alder reactions involving cross-conjugated dienones. Effects of substitution on reactivity. *J. Am. Chem. Soc.*, **1981**, *103*, 3224-3226.
[10]   Liu, H.J.; Han, Y. Facial selectivity in Diels-Alder reaction of 4,4-disubstituted 2,5-cyclohexadienones. *Tetrahedron Lett.*, **1993**, *34*, 423-426.
[11]   Forster, H.; Vogtle, F. Steric Interactions in Organic Chemistry: Spatial Requirements of Substituents. *Angew. Chem. Int. Ed.*, **1977**, *16*, 429-441.
[12]   Liu, H.J.; Ulibarri, G.; Browne, E.N.C. Diels–Alder reactions of 2-carbomethoxy-4,4-dimethyl-2-cyclopenten-1-one. *Can. J. Chem.*, **1992**, *70*, 1545-1554.
[13]   Schotes, C.; Mezzetti, A. Asymmetric Diels-Alder reactions of unsaturated β-ketoesters catalyzed by chiral ruthenium PNNP complexes. *J. Am. Chem. Soc.*, **2010**, *132*, 3652-3653.
[14]   Schotes, C.; Althaus, M.; Aardoom, R.; Mezzetti, A. Asymmetric Diels–Alder and Ficini reactions with alkylidene β-ketoesters catalyzed by chiral ruthenium PNNP complexes: Mechanistic insight. *J. Am. Chem. Soc.*, **2012**, *134*, 1331.

[15]   Huang, Y.; Iwama, T.; Rawal, V.H. Highly enantioselective Diels-Alder reactions of 1-amino-3-siloxy-dienes catalyzed by Cr(III)-Salen complexes. *J. Am. Chem. Soc.*, **2000**, *122*, 7843.

[16]   Ohfusa, T.; Nishida, A. Reactivity and stereoselectivity of the Diels-Alder reaction using cyclic dienophiles and siloxyaminobutadienes. *Tetrahedron*, **2011**, *67*, 1893.

[17]   Liotta, D.; Barnum, C.; Puleo, R.; Zima, G.; Bayer, C.; Kezar, H.S.III A simple method for the efficient synthesis of unsaturated β-dicarbonyl compounds. *J. Org. Chem.*, **1981**, *46*, 2920-2923.

[18]   Snider, B.B. Diels-Alder reactions of 2-acetyl-2-cyclohexenone with enol ethers and emamines. *Tetrahedron Lett.*, **1980**, *21*, 1133-1136.

[19]   Liu, H.J.; Yeh, W.L.; Browne, E.N.C. Activated cycloheptenone dienophiles. A versatile approach to 6,7-fused ring targets. *Can. J. Chem.*, **1995**, *73*, 1135-1147.

[20]   Liu, H.J.; Wang, D.X.; Kim, J.B.; Browne, E.N.C.; Wang, Y. Activated cyclooctenones are effective dienophiles. *Can. J. Chem.*, **1997**, *75*, 899-912.

[21]   Liu, H.J.; Chew, S.Y.; Browne, E.N.C.; Kim, J.B. Facial selective Diels–Alder reactions of (1*R*,5*R*)-(+)-3-carbomethoxy-6,6-dimethylbicyclo[3.1.1]hept-3-en2-one.   Unusual ketalization–fragmentation reaction of adducts. *Can. J. Chem.*, **1994**, *72*, 1193-1210.

[22]   Liu, H.J.; Li, Y.; Browne, E.N.C. Face-selective Diels–Alder reactions of (1*R*,5*R*)-3-formyl-6,6-dimethylbicyclo[3.1.1]hept-3-en-2-one. *Can. J. Chem.*, **1994**, *72*, 1883-1893.

[23]   Evans, D.A.; Wu, J. Enantioselective rare-earth catalyzed quinone Diels–Alder reactions. *J. Am. Chem. Soc.*, **2003**, *125*, 10162-10163.

[24]   Liu, H.J.; Yip, J. Diels-Alder reactions of 4,4-disubstituted 2-cyano-2,5-cyclohexadienones. A facile approach to the angularly substituted cis-decalin system. *Synlett*, **2000**, 1119-1122.

[25]   Zhu, J.L.; Shia, K.S.; Liu, H.J. Diels–Alder chemistry of 2-cyanoalk-2-enones. A convenient general approach to angularly substituted polycyclic systems. *Chem. Commun.*, **2000**, 1599-1600.

[26]   Knapp, S.; Lis, R.; Michna, P. Diels-Alder reactions of 2-(phenylthio)cyclopentenone. Synthesis of dihydro-1-indanones. *J. Org. Chem.*, **1981**, *46*, 624-626.

[27]   Alonso, I.; Carretero, J.C.; Ruano, J.L.G.; Cabrejas, L.M.M.; Lopez-Solera, I.; Raithby, P.R. Diels-Alder reaction of (*S*)-2-*p*-tolylsulfinyl-2-cyclopentenone with Dane's diene: an efficient   approach   to   the   enantioselective   preparation   of   perhydro-cyclopenta[a]phenanthrenes. *Tetrahedron Lett.*, **1994**, *35*, 9461-9464 and references therein.

[28]   Liotta, D.; Saindane, M.; Barnum, C.; Zima, G. Synthetic applications of 2-phenylselenenylenones—III: An overview. *Tetrahedron*, **1985**, *41*, 4881-4889.

[29]   Zhu, J.L.; Ko, Y.C.; Kuo, C.W.; Shia, K.S. Lithium naphthalenide induced reductive selenenylation of α-cyano ketones: A regiocontrolled process for α-phenylseleno ketones and one-pot conversion into enone system. *Synlett*, **2007**, 1274-1278.

[30]   Liu, H.J.; Shia, K.S. Diels-Alder reactions of 2-bromo-2-cycloalkenones. A convenient approach to the doubly cisoid fully conjugated dienone system. *Tetrahedron Lett.*, **1995**, *36*, 1817-1820.

[31]   Lee, J.H.; Kim, W.H.; Danishefsky, S.J. Diels–Alder routes to angularly halogenated cis-fused bicyclic ketones: readily accessible cyclynone intermediates. *Tetrahedron Lett.*, **2010**, *51*, 4653-4654.

[32]   Lee, J.H.; Zhang, Y.; Danishefsky, S.J. A straightforward route to functionalized trans-Diels-Alder motifs. *J. Am. Chem. Soc.*, **2010**, *132*, 14330.

[33]   Chien, C.F.; Wu, J.D.; Ly, T.W.; Shia, K.S.; Liu, H.J. Diels–Alder chemistry of 2-diethoxyphosphinylcyclohex-2-enones. A new approach to complex phosphonates and synthetic applications of the β-keto phosphonate system. *Chem. Commun.*, **2002**, 248-249.

[34]   Liu, H.J.; Shang, X. Synthetic studies of Forskolin. A Diels-Alder approach to Corey's endoperoxide. *Heterocycles*, **1997**, *44*, 143-147.

[35]   Liu, H.J.; Shang, X. Synthetic studies of Forskolin. A formal synthesis *via* a Nicolaou's advanced intermediate. *Heterocycles*, **1999**, *50*, 1105-1113.

[36]   Liu, H.J.; Yeh, W.L.; Chew, S.Y. A total synthesis of the antimarial natural product (+)-qinghaosu. *Tetrahedron Lett.*, **1993**, *34*, 4435-4438.

[37]   Liu, H.J.; Shia, K.S.; Han, Y.; Sun, D.; Wang, Y. Synthetic studies on clerodane diterpenoids. The total synthesis of (±)-2-oxo-5α,8α-13,14,15,16-tetranorclerod-3-en-12-oic acid. *Can. J. Chem.*, **1997**, *75*, 646-655.

[38]   Liu, H.J.; Ho, Y.L.; Wu, J.D.; Shia, K.S. An efficacious synthetic strategy for cis-clerodane diterpenoids. application to the total synthesis of (±)-6β-acetoxy-2-oxokolavenool. *Synlett*, **2001**, 1805-1807.

[39]   Liu, H.J.; Shia, K.S. Synthetic studies on clerodane diterpenoids. 4. The total synthesis of (±)-6β-acetoxy-2-oxokolavenool. *Tetrahedron*, **1998**, *54*, 13449-13458.

[40]   Liu, H.J.; Zhu, J.L.; Chen, I.C.; Jankowska, R.; Han, Y.; Shia, K.S. The total synthesis of racemic teucvin and 12-*epi*-teucvin. *Angew. Chem. Int. Ed.*, **2003**, *42*, 1851-1853.

[41]   Chang, W.S.; Shia, K.S.; Liu, H.J.; Ly, T.W. The first total synthesis of xenitorins B and C: assignment of absolute configuration. *Org. Biomol. Chem.*, **2006**, *4*, 3751-3753.

[42]   Liu, H.J.; Ulibarri, G.; Nelson, L.A.K. The total synthesis of racemic isoacanthodoral. *Chem. Commun.*, **1990**, 1419-1421.

[43]   Boger, D.L.; Mullican, M.D.; Hellberg, M.R.; Patel, M. Preparation of optically active, functionalized *cis*-Δ⁶-1-octalones. *J. Org. Chem.*, **1985**, *50*, 1904-1911.

[44]   Liu, H.J.; Yeh, W.L. Total synthesis of (-)-qinghaosu IV (artemisinin D, arteannuin D). *Hetereocycles*, **1996**, *42*, 493-497.

[45]   Ly, T.W.; Liao, J.H.; Shia, K.S.; Liu, H.J. A highly effective Diels-Alder approach to cis-clerodane natural products: First total synthesis of solidago alcohol. *Synthesis*, **2004**, 271-275.

[46]   Chen, I.C.; Wu, Y.K.; Liu, H.J.; Zhu, J.L. Total syntheses of (±)-montanin A and (±)-teuscorolide. *Chem. Commun.*, **2008**, 4720-4722.

[47]   Sutherland, J. K. *Polyene Cyclizations.* In *Comprehensive Organic Synthesis: Selectivity, Strategy and Efficiency*; Trost, B. M.; Flemming, I., Eds.; Pergamon Press: Oxford, **1991**; Vol. 3, pp. 341-377.

[48]   Liu, H.J.; Sun, D.; Shia, K.S. Polyene cyclization promoted by the cross conjugated α-carbalkoxy enone system. *Tetrahedron Lett.*, **1996**, *37*, 8073-8076.

[49]   Liu, H.J.; Sun, D.; Shia, K.S. Polyene cyclization promoted by the cross conjugated α-carbalkoxy enone system. An efficient approach to highly functionalized decalins. *J. Chin. Chem. Soc.*, **1999**, *46*, 453-462.

[50]   Liu, H.J.; Sun, D. A formal synthesis of (-)-dehydrochamaecynenol. asymmetric synthesis of an advanced key intermediate. *Heterocycles*, **2000**, *52*, 1251-1260.

[51]   Chin, C.L.; Tran, D.D.P.; Shia, K.S.; Liu, H.J. The total synthesis of pygmaeocin C. *Synlett*, **2005**, 417-420.

[52]   Liu, H.J.; Tran, D.D.P. Intramolecular Friedel-Crafts alkylation promoted by the cross conjugated β-keto ester system. An efficient approach to highly functionalized hydrophenanthrenes and hydrochrysenes. *Tetrahedron Lett.*, **1999**, *40*, 3827-3830.

[53]   Liu, H.J.; Sun, D.; Roa-Gutierrez, F.; Shia, K.S. Formation of highly functionalized hydrindanes and hydroazulenes *via* polyene cyclization promoted by the cross conjugated α-carbomethoxy enone system. *Lett. Org. Chem.*, **2005**, *2*, 364-366.

[54]   Weissberger, A.; Taylor, E. C. In *Thiophene and its derivatives*; Gronowitz, S., Ed.; John Wiley & Sons Inc.: Toronto, **1985**; part IV, pp. 397-478.

[55]   Liu, H.J.; Tran, D.D.P. An efficient procedure for the synthesis of thiophene-containing polycyclic compounds *via* polyene cyclization promoted by the cross conjugated β-keto ester system. *Tetrahedron Lett.*, **1997**, *38*, 6501-6504.

[56]   Chou, H.H.; Wu, H.M.; Wu, J.D.; Ly, T.W.; Jan, N.W.; Shia, K.S.; Liu, H.J. Polyene cyclization promoted by the cross-conjugated α-carbalkoxy enone system. Observation on a putative 1,5-hydride/1,3-alkyl shift under Lewis acid catalysis. *Org. Lett.*, **2008**, *10*, 121-123.

[57]   Hsieh, M.T.; Chou, H.H.; Liu, H.J.; Wu, H.M.; Ly, T.W.; Wu, Y.K.; Shia, K.S. Polyene cyclization promoted by the cross-conjugated α-carbalkoxy cyclohexenone system. An unusual 1,2-hydride shift under Lewis acid catalysis. *Org. Lett.*, **2009**, *11*, 1673-1675.

[58]   Pastine, S.J.; McQuaid, K.M.; Sames, D. Room temperature hydroalkylation of electron-deficient olefins: sp³ C−H functionalization *via* a Lewis acid-catalyzed intramolecular redox event. *J. Am. Chem. Soc.*, **2005**, *127*, 12180-12181.

[59]   Pastine, S.J.; Sames, D. Room temperature intramolecular hydro-O-alkylation of aldehydes: sp³ C−H functionalization *via* a Lewis acid catalyzed tandem 1,5-hydride transfer/cyclization. *Org. Lett.*, **2005**, *7*, 5429-5431.

[60]   Rademacher, P.; Mohr, P.C. Transannular 1,5-hydride shift in 5-hydroxycyclooctanone: An experimental and theoretical investigation. *Org. Biomol. Chem.*, **2007**, *5*, 2698-2703.

[61]   McQuaid, K. M.; Sames, D. C−H bond functionalization *via* hydride transfer: Lewis acid catalyzed alkylation reactions by direct intramolecular coupling of sp3 C−H bonds and reactive alkenyl oxocarbenium intermediates. *J. Am. Chem. Soc.*, **2009**, *131*, 402-403.

[62]   Jin, T.; Himuro, M.; Yamamoto, Y. Brønsted acid-catalyzed cascade cycloisomerization of enynes *via* acetylene cations and sp3-hybridized C−H bond activation. *J. Am. Chem. Soc.*, **2010**, *132*, 5590-5591.

[63]   For a recent review, see: Tobisu, M.; Chatani, N. A catalytic approach for the functionalization of C(sp³)−H bonds. *Angew. Chem. Int. Ed.*, **2006**, *45*, 1683-1684.

[64]   Fleming, F.F.; Pu, Y.; Tercek, F. Unsaturated nitriles: conjugate addition−silylation with Grignard reagents. *J. Org. Chem.*, **1997**, *62*, 4883-4885.

[65]   Kung, L.R.; Tu, C.H.; Shia, K.S.; Liu, H.J. Palladium(II) acetate mediated oxidative cyclization of ω-unsaturated α-cyano ketones. A facile methylenecyclopentane annulation process. *Chem. Commum.*, **2003**, 2490-2491.

[66]   Dénès, F.; Pérez-Luna, A.; Chemla, F. Addition of metal enolate derivatives to unactivated carbon−carbon multiple bonds. *Chem. Rev.*, **2010**, *110*, 2366-2447.

[67]   Widenhoefer, R.A. Palladium-catalyzed alkylation of unactivated olefins. *Pure Appl. Chem.*, **2004**, *76*, 671-678.

[68]   Hsieh, M.T.; Shia, K.S.; Liu, H.J.; Kuo, S.C. Palladium(II) acetate mediated oxidative cyclization of ω-unsaturated α-cyano ketones for facile construction of methylenecyclohexane ring system. *Org. Biomol. Chem.*, **2012**, *10*, 4609-4617.

[69]   Chen, K.; Ishihara, Y.; Galan, M.M.; Baran, P.S. Total synthesis of eudesmane terpenes: Cyclase phase. *Tetrahedron*, **2010**, *66*, 4738-4744.

[70]   Hsieh, M.T.; Liu, H.J.; Ly, T.W.; Shia, K.S. A concise total synthesis of (±)-acutifolone A. *Org. Biomol. Chem.*, **2009**, *7*, 3285-3290.

[71]   Hibi, A.; Toyota, M. Development of palladium(II)-catalyzed oxidative cyclization of olefinic keto and/or lactone esters. *Tetrahedron Lett.*, **2009**, *50*, 4888-4891.

[72]   DeLorbe, J.E.; Lotz, M.D.; Martin, S.F. Concise total synthesis of (±)-lycopladine A. *Org. Lett.*, **2010**, *12*, 1576-1579.

[73]   Chin, C.L.; Liao, C.F.; Liu, H.J.; Wong, Y.C.; Hsieh, M.T.; Amancha, P.K.; Chang, C.P.; Shia, K.S. Conia-ene annulation of the α-cyano β-TMS-capped alkynyl cycloalkanone system and its synthetic application. *Org. Biomol. Chem.*, **2011**, *9*, 4778-4781.

[74]   Jackson, W.P.; Ley, S.V. Synthesis of substituted *cis*-decalins as potential insect antifeedants. *J. Chem. Soc., Perkin Trans. 1*, **1981**, 1516-1519.

[75]   Li, W.; Liu, X.Z.; Zhou, X.F.; Lee, C.S. Amine-induced Michael/Conia-ene cascade reaction: Application to a formal synthesis of (±)-clavukerin A. *Org. Lett.*, **2010**, *12*, 548-551.

[76]   Renaud, J.L.; Petit, M.; Aubert, C.; Malacria, M. Synthetic usefulness of the cobalt(I)-mediated ene type reaction for the diastereoselective construction of bicyclo[n.3.0]derivatives. *Synlett*, **1997**, 931-932.

[77]   Peng, F.; Grote, R.E.; Danishefsky, S.J. Further expansion of the *trans*-Diels-Alder paradigm: Reductive alkylation of α-cyanoketones. *Tetrahedron Lett.*, **2011**, *52*, 3957-3959.

[78]   Chu, K.C.; Liu, H.J.; Zhu, J.L. A new and efficient total synthesis of (±)-laurencenone C. *Synlett*, **2010**, 3061-3064.

[79]   Zhu, J.L.; Huang, P.W.; You, R.Y.; Lee, F.Y.; Tsao, S.W.; Chen, I.C. Total syntheses of (±)-(Z)- and (±)- (E)-9-(bromomethylene)-1,5,5-trimethylspiro[5.5]-undeca-1,7-dien-3-one and (±)-majusculone. *Synthesis*, **2011**, 715-722.

[80]   Amancha, P.K.; Liu, H.J.; Ly, T.W.; Shia, K.S. General approach to 2,3-dibenzyl-γ-butyrolactone lignans: Application to the total synthesis of (±)-5'-methoxyyatein, (±)-5'-methoxyclusin, and (±)-4'-hydroxycubebinone. *Eur. J. Org. Chem.*, **2010**, 3473-3480.

[81]   Amancha, P.K.; Lai, Y.C.; Chen, I.C.; Liu, H.J.; Zhu, J.L. Diels-Alder reactions of acyclic α-cyano α,β-alkenones: A new approach to highly substituted cyclohexene system. *Tetrahedron*, **2010**, *66*, 871-877.

[82]   Wong, Y.C.; Hsieh, M.T.; Amancha, P.K.; Chin, C.L.; Liao, C.F.; Kuo, C.W.; Shia, K.S. Autoxidative annulation cascade of the α-cyano β-TMS-capped alkynyl cycloalkanone system. *Org. Lett.*, **2011**, *13*, 896-899.

[83]   Fleming, F.F.; Shook, B.C.; Jiang, T.; Steward, O.W. β-Siloxy unsaturated nitriles: Stereoselective cyclizations to *cis*- and *trans*-decalins. *Org. Lett.*, **1999**, *11*, 1547-1550.

[84]   Campbell, M.J.; Pohlhaus, P.D.; Min, G.; Ohmatsu, K.; Johnson, J.S. An "anti-Baldwin" 3-*exo-dig* cyclization: Preparation of vinylidene cyclopropanes from electron-poor alkenes. *J. Am. Chem. Soc.*, **2008**, *130*, 9180-9181.

[85]   Tu, C.H.; Shia, K.S.; Kuo, S.C.; Liu, H.J.; Hsieh, M.T. K_2CO_3/NaI-induced cyclization of ω-bromo α-cyano ketones: A new annulation approach for the formation of carbalkoxycyclohexane ring system. *Synlett*, **2012**, *23*, 1653-1656.

Send Orders of Reprints at bspsaif@emirates.net.ae

Advances in Organic Synthesis, Vol. 5, 2013, 279-308   279

CHAPTER 6

An Update on Trifluoromethylation of Carbonyl Compounds

Nubia Boechat[*] and Monica Macedo Bastos

Fundação Oswaldo Cruz, Farmanguinhos, Rua Sizenando Nabuco, 100 CEP 21041-250 Manguinhos - Rio de Janeiro, RJ, Brazil

Abstract: This chapter reviews the recent findings on the synthetic methods developed for the direct introduction of a trifluoromethyl group in carbonyl compounds. It was organized by type of reagent highly considering the description of the trifluoromethylating agents, their activation mode, their asymmetric approaches, as well as applications in organic synthesis.

Keywords: DAST, fluorine chemistry, trifluoromethyl halides, sulfur tetrafluoride, deoxo-fluor, bromine trifluoride, organoboron reagent, diethyl trifluoromethylphosphonate, trifluoromethyl bromide, fluoroform, hemiaminals of fluoral, sodium trifluoroacetate, sodium trifluoromethylacetophenone-N,N-dimethyltrimethylsilylamine adduct.

1. INTRODUCTION

Fluorine is the 13th most abundant element found on the earth's crust where it exists predominantly in the form of cryolite, calcium fluorspar, and fluorapatite. Despite this abundance in nature, only a few naturally occurring organic compounds have been identified. Fluoroorganic chemistry is considered as one of the most interesting fields in chemistry. This fact reflects the exceptionally high number of publications and percentage of new fluorinated molecules in recent decades [1]. Presently, the significant increase in the use of fluorinated chemicals has been applauded by the organic, agricultural, medicinal, and material chemists [2-5]. Nowadays it is estimated that up to 20% of prescribed drugs contain fluorine atom, whereas 30% of leading blockbuster sales are fluorinated ones [6].

Among fluoroorganic compounds, trifluoromethyl-substituted molecules have highly been recognized in the past decades [3, 7-14]. The introduction of a trifluoromethyl group can boost significant changes in the physical, chemical, and

---

*Address correspondence to Nubia Boechat: Fundação Oswaldo Cruz, Farmanguinhos, Rua Sizenando Nabuco, 100 CEP 21041-250 Manguinhos - Rio de Janeiro, RJ, Brazil; Tel: +552139772465; Fax: +55 21-25602518; E-mail: boechat@far.fiocruz.br

Atta-ur-Rahman (Ed)

biological properties of the molecules. The use of the trifluoromethyl group in biologically active molecules is often linked with the increased lipophilicity this substituent provides, its electronegativity and that the bulkiness of a CF_3 group is similar to that of a $(CH_3)_2CH$ group. These factors idealize the incorporating of the trifluromethyl group as a powerful tool in drug development. Consequently, there are many trifluoromethylated compounds with high biological activities [15-21]. Some examples of active trifluoromethylated compounds have been shown in Fig. **1**.

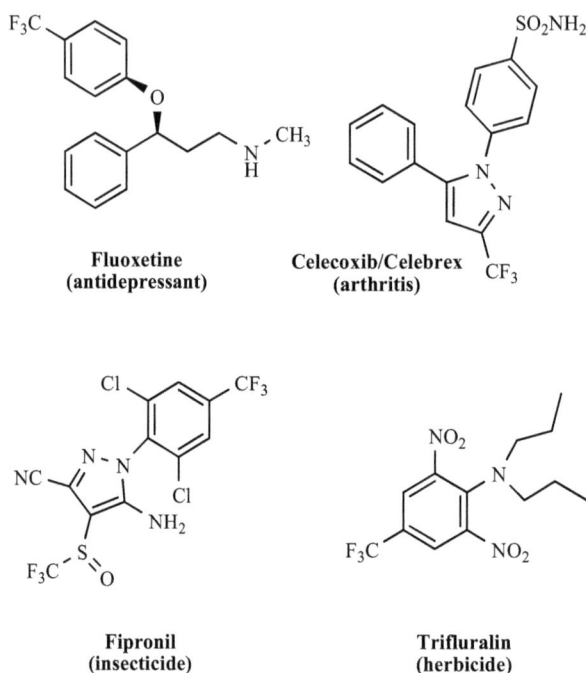

Figure 1: Examples of active trifluoromethylated compounds.

The decision for obtaining a trifluoromethylated compound is generally considered as an important synthetic challenge [22] since it can exploit readily available fluorinated building blocks in a multi-step process, or use the direct introduction of a trifluoromethyl moiety. This second strategy appears to be more powerful as it can be carried out on an elaborated chemical structure close to the final of the synthetic process. In a trifluoromethylation reaction the introduction of a CF_3 group is carried out through the formation of a carbon-carbon bond executed by electrophilic, nucleophilic, or free radical approaches [7, 22, 23].

The most promising strategy for the direct introduction of a trifluoromethyl group into organic molecules is the nucleophilic trifluoromethylation. In their earlier stages, trifluoromethyl metal species (Hg, Cu, Zn, and Cd) were extensively investigated for synthetic purposes [7]. Such trifluoromethyl reagents were usually employed for the substitution of aromatic iodides or bromides under thermal activation and shown to be not practical for trifluoromethylation of carbonyl compounds. Many of these reactions suffered from low yields giving undesirable fluorinated side products.

The trifluoromethy cation was observed in gas phase when electron ionization of CF_4 generated CF_3^+ ions in high yield that could attack saturated and acyclic ketones to make vibrationally excited adduct ions. Most of the C5-C7 ketones yielded $CH_3CH=OCF_3^+$ as the principal ion-molecule reaction product [24]. Recently, Umemoto reported *in situ* synthesis of the first CF_3 oxonium salts applied as a real CF_3^+ species source to the direct *O*- and *N*-trifluoromethylations of alcohols, phenols, amines, anilines, and pyridines under very mild conditions. These results played a vital role in eliminating the common concept about the CF_3^+ species as being extremely difficult to obtain in a solution [25, 26].

The trifluoromethyl radical can be generated under oxidative, reductive, photochemical, thermal, and electrochemical conditions. A wide variety of molecules have been applied as precursors to trifluoromethyl radicals. Mainly, electrophilic trifluoromethyl radicals were reacted with electron rich aromatics and heteroaromatics, whereas trifluoromethylation in non-aromatic sites has been less studied [7]. As example, the first enantioselective organocatalytic α-trifluoromethylation and α-perfluoroalkylation of aldehydes have been obtained by MacMillan and co-workers using a readily available iridium photocatalyst and a commercial imidazolidinone catalyst [27]. Nagib and MacMillan in analogy to the single-electron aryl modification processes employed by enzymes have also developed a trifluoromethylation in site-specific incorporation of electrophilic radicals at metabolically susceptible positions of arenes and heteroarenes [27b].

In this paper, we reviewed the synthetic methods developed for the direct introduction of a trifluoromethyl group in carbonyl compounds.

2. TRIFLUOROMETHYLATION OF CARBONYL COMPOUNDS

The direct introduction of the trifluoromethyl group in carbonyl compounds has been described in the literature for years. This review was organized by different types of reagent, their applications, as well as asymmetric approaches.

2.1. Methods Using the Trifluoromethyl Halides in the Presence of Metals

Amongst numerous methods for trifluoromethyl group incorporation into organic compounds, one of the most useful involves the use of reagents that effectively generates unstable CF_3^- anion as an *in situ* species for nucleophilic substrates, such as aldehydes and ketones.

Both (trifluoromethyl) lithium and Grignard reagents are not synthetically useful as they readily decompose, apparently to difluorocarbene, even when formed at low temperature in the presence of a suitable electrophilic agent. Gassman and co-workers [28] used (pentafluoroethyl) lithium for introducing perfluoroalkyl groups in carbonyl compounds, while Denson and co-workers reacted various perfluoroaliphatic Grignard with ketones [29]. In both cases the (perfluoroalkyl) lithium or Grignard reagent were yielded at low temperature when alkyl or aryl group were bigger than trifluoromethyl.

Similarly, perfluoroalkylcalcium reagents were used as perfluoroalkylating agents in reactions with carbonyl compounds [30]. Riess and co-workers reacted many perfluoroalkyl iodides as $R_F=C_2F_5$; C_6F_{13} at room temperature with finely divided calcium and carbonyl compounds, according to Scheme 1. When pure calcium metal was used the reaction started vigorously after an induction period of *ca.* 30 min, and large proportions of undesired R_FH were produced, the proportion increased with temperature. This shortcoming was partially tackled by using amalgamated calcium. Ca/Hg amalgam was prepared by heating the two metals together under an argon atmosphere. The consumption of R_FI was completed after *ca.* 2 h at room temperature or *ca.* 10 h at -20°C. No reaction was observed in more basic solvents such as triethylamine, ethylenediamine, or tetramethylethylenediamine likely to form complexes with R_FI [31].

Ishikawa and co-workers studied the introduction of perfluoroalkyl group to an organic molecule with perfluoroorganometallic compounds [32], and reported a trifluoromethylation of carbonyl compounds with trifluoromethyl iodide by ultrasonically dispersed zinc in N,N-dimethyl formamide (DMF) [33]. This ultrasonic technique was also employed by Luche and his co-workers, who reported the formation of the organometallic compounds [34]. Aliphatic and aromatic aldehydes and ketones were transformed into trifluoromethylated carbinols in less than one hour at room temperature (Scheme **2**).

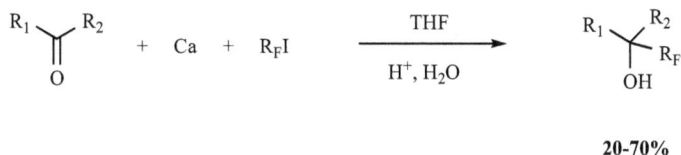

20-70%

Scheme 1: Perfluoroalkylation *via* perfluoroalkylcalcium derivatives.

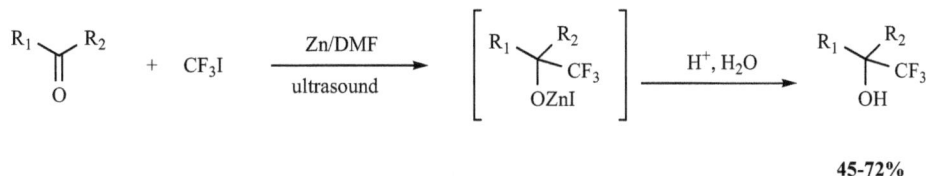

45-72%

Scheme 2: Trifluoromethylation using zinc and ultrasonic activation.

Searching for more versatile perfluoroalkyl organometallic species Ishikawa and co-workers developed another methodology for trifluoromethylation of aldehydes once the ultrasonic irradiation of larger scale reactions was impractical. With a view to solve these problems the authors introduced a method for the perfluoroalkylation of aldehydes with perfluoroalkyl halides and zinc in a catalytic cycle using palladium or nickel catalysts, ignoring the use of ultrasound (Scheme **3**) [35]. Divalent palladium and nickel complexes were used for their stability and ease of handling, and were readily reduced to the active zero valent species by zinc in the reaction media. In the absence of any catalyst either no

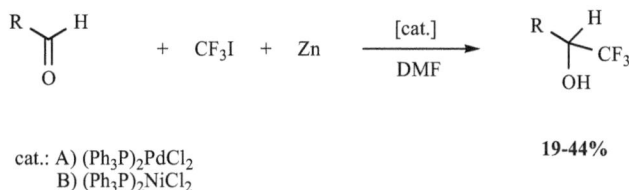

cat.: A) $(Ph_3P)_2PdCl_2$
B) $(Ph_3P)_2NiCl_2$

19-44%

Scheme 3: Trifluoromethylation using zinc in a catalytic cycle.

reaction occurred at all or a low yield of the carbinol was observed. When CF$_3$Br was used due to its very low solubility in DMF low yields were observed and the reactions were therefore carried out in a sealed tube [35].

A simple Barbier procedure, using trifluoromethyl bromide (CF$_3$Br) under slight pressure, allowed the preparation of trifluoromethyl-substituted methanols from aldehydes and preparation of trifluoromethyl ketones from some activated esters [36]. In previous studies Wakselman and co-workers had shown that the reaction between carbonyl compounds and trifluoromethyl bromide in the presence of Zn, Mn, Al, or Cd can be observed simply under a slight pressure of CF$_3$Br [37]. The first experiments were made in DMF. No reaction was observed between trifluoromethyl bromide and zinc in this solvent. Consequently, it became impossible to prepare trifluoromethylzinc derivatives from CF$_3$Br by a simple Grignard procedure. However, a partial consumption of the reagents when benzaldehyde was present at the beginning of the condensation was observed. The yield was very poor when the halide was simply bubbled into the solvent at atmospheric pressure. Nevertheless, the yield was found to be higher under slight pressure (2-4 bars) in a glass apparatus. This condensation proceeded slowly in DMF or in dimethyl sulphoxide (DMSO), with an induction period (15 min to 3 h). It was noticed that, in pyridine, the reaction started almost immediately increasing the yield of purified alcohol up to 52%. Under these conditions various aldehydes were transformed to fluorinated alcohols (Scheme **4**). The reaction proved to be more difficult with ketones and the yields of addition products were limited to 20%. In the case of acetone itself no product was obtained. The Barbier procedure was found to be effective even at atmospheric pressure when the ester was activated by an electron-withdrawing group [36].

Scheme 4: Barbie procedure for trifluoromethylation of carbonyl compounds.

Perfluoroalkyl organometallic reagents in organic synthesis were reviewed by Burton and Yang [38]; they were not easy to handle and their low reactivity reduced their synthetic utilities. Ishikawa and Kitazume found it possible to introduce a perfluoroalkyl group into carbonyl compounds by the reaction of a perfluoroalkyltin (IV) dihaloiodide (Scheme **5**) [39].

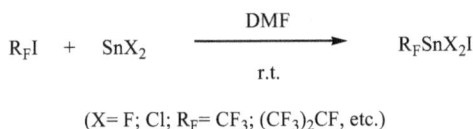

$$R_FI \ + \ SnX_2 \xrightarrow[\text{r.t.}]{\text{DMF}} R_FSnX_2I$$

(X= F; Cl; R_F= CF_3; $(CF_3)_2CF$, etc.)

Scheme 5: Preparation of perfluorinatedtin (IV) dihaloiodine.

The compound R_FSnX_2I could not be isolated in a pure form. The presence of these compounds in DMF solution, however, was evident from their ^{19}F NMR spectrum. Furthermore, it was found that the presence of pyridine boosted the nucleophilic reaction of R_FSnX_2I (X=Cl) with aldehydes and ketones, extracting perfluoroalkylated carbinols in good yields. Pyridine probably accelerated the releasing of a perfluoroalkyl anion by coordinating on the tin atom in R_FSnX_2I. Reactions were proceeded as shown in Scheme **6** [39].

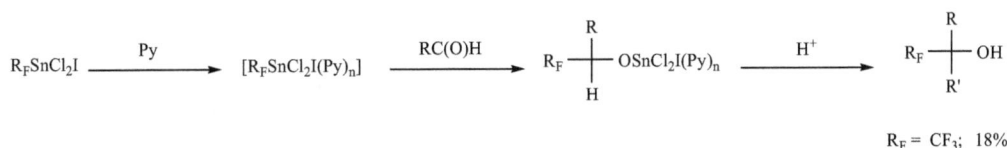

R_F = CF_3; 18%

Scheme 6: Trifluoromethylation using R_FSnX_2I.

2.2. Methods Using the Tetrakis(Dimethylamino)Ethylene (TDAE)

An interesting reagent for the introduction of trifluoromethyl group in carbonyl compounds *via* perfluoroalkyl iodides is tetrakis(dimethylamino)ethylene (TDAE). It was incidentally discovered by Pruett and co-workers at the DuPont Company in 1950 [40] from the condensation of chlorotrifluoroethylene (CTFE) and dimethylamine (both gas), under pressure, using an autoclave (Scheme **7**).

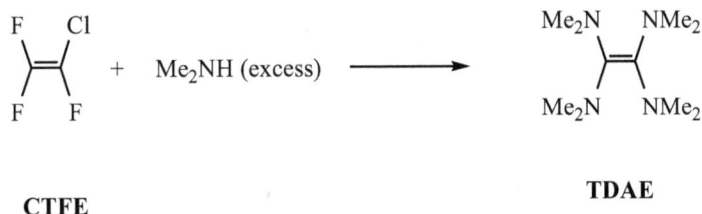

CTFE   **TDAE**

Scheme 7: Preparation of TDAE.

TDAE is a highly electron-rich tetra-aminoethylene molecule and a powerful organic reducing agent. It is a compound that readily reacts with oxygen, Lewis acids, strong bases, and acids. It reacts easily with some oxidants such as Br_2, I_2 to make the corresponding dication salts, and can also act as a carbon nucleophile [41].

In 1988, Pawelke described that the use of TDAE with CF_3I forms a charge transfer complex at low temperatures acting as a nucleophilic trifluoromethylating reagent in polar solvents. They prepared trifluoromethylsilicon and boron derivatives such as $TMSCF_3$ [42].

Initial results of adding TDAE to a solution of benzaldehyde and trifluoromethyl iodide in dry DMF were quite discouraging due to the poor yield obtained (Scheme **8**).

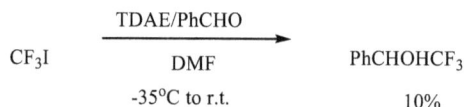

$$CF_3I \xrightarrow[\substack{DMF \\ -35°C \text{ to r.t.}}]{TDAE/PhCHO} PhCHOHCF_3$$
$$\qquad\qquad\qquad\qquad\qquad 10\%$$

Scheme 8: The first trifluoromethylation reaction using TDAE/CF_3I.

The yield was remarkably raised up to 80% on applying photochemistry. Subsequent reactions carried out in this manner with a large number of aldehydes and ketones (Scheme **9**) showed similar success. Although DMF was the preferred solvent, the reaction also gave good results even when other solvents were used [43]. The use of this procedure was extended to obtain other perfluoroalkylated compounds [44].

R_1=Ph, R_2=H; 78%

R_1=Ph, R_2=CH_3; 18%

R_1=*o*-Br-Ph, R_2=H; 86%

R_1=4-Me_2N-1-naphthyl, R_2=H; 86%

Scheme 9: Trifluoromethylation reactions using TDAE/CF_3I.

The reaction of trifluoromethyl anion with acyl chlorides accounted little synthetic use, since half of the acyl chloride was consumed in acylating the alcoholate forming the ester product. However, the chemoselective reaction with acyl chlorides (**1**) using CF_3I/TDAE [45] was investigated. The preferred solvent was 1,2-dimethoxyethane (DME) that produced a quantitative yield of benzoate ester (**2a,b**). These esters were readily converted quantitatively to the respective bis-trifluoromethyl-substituted alcohols (**3**) by transesterification using methanolic KOH (Scheme **10**).

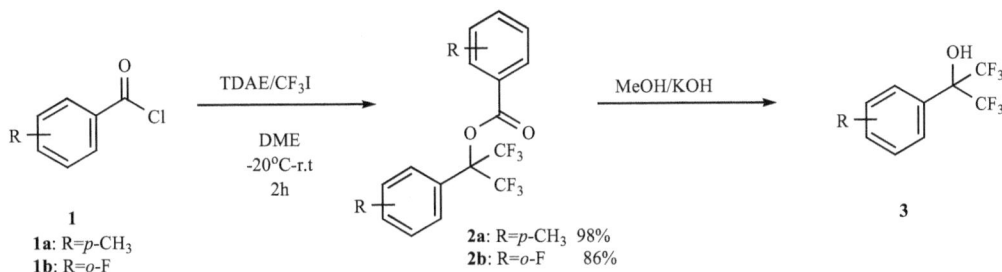

1
1a: R=*p*-CH_3
1b: R=*o*-F

2a: R=*p*-CH_3 98%
2b: R=*o*-F   86%

3

Scheme 10: Trifluoromethylation of acyl chlorides using TDAE/CF_3I.

2.3. Methods Using the Organosilicon Reagent: TMSCF_3

In 1989, Prakash and co-workers classified trifluoromethyltrimethylsilane (TMSCF_3) as a very efficient reagent for nucleophilic trifluoromethylation reaction of carbonyl compounds [46]. Several nucleophilic trifluoromethylation reagents containing silicon element were reported earlier by Ruppert group, such as TMSCF_3 being prepared from the trifluoromethyl bromide, according to Scheme **11** [47, 48].

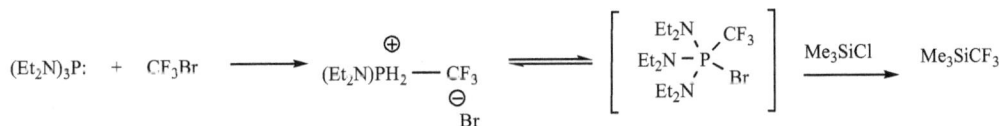

Scheme 11: Process of preparation of TMSCF_3 from CF_3Br.

Although after the initial discovery of TMSCF_3 as trifluoromethylating agent by Prakash, the main hindrance was its synthesis which typically used the ozone-depleting substance CF_3Br. In their efforts to replace it Prakash, Hu, and Olah, reported in 2003 a procedure for the preparation of TMSCF_3 from nonozone depleting trifluoromethane *via* phenyl trifluoromethyl sulfide (**4**) (Scheme **12**) [49].

Since then TMSCF$_3$ became the most important and popular reagent for trifluoromethylation of carbonyl compounds. Some comprehensive reviews have covered almost all the trifluoromethylation reactions of organic molecules with TMSCF$_3$ [50-54].

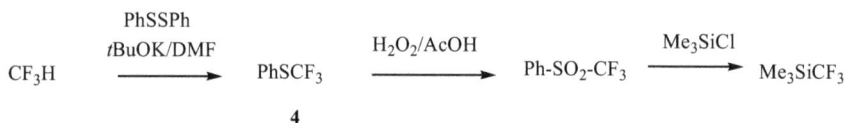

$$CF_3H \xrightarrow[\substack{tBuOK/DMF}]{PhSSPh} PhSCF_3 \xrightarrow{H_2O_2/AcOH} Ph\text{-}SO_2\text{-}CF_3 \xrightarrow{Me_3SiCl} Me_3SiCF_3$$

4

Scheme 12: Process of preparation of TMSCF$_3$ from CF$_3$H.

Consequently, a rather impressive number of ketones, aldehydes, esters, and activated imines can undergo reactions with TMSCF$_3$ under smooth conditions in different activations. In general, aldehydes and ketones can readily react to give trifluoromethylcabinols while esters can provide trifluoromethylketones and the corresponding hydrates. The trifluoromethylation of methyl (S)-N-Boc-pyrrolidine-2-carboxylate (**5**) with TMSCF$_3$ in the presence of a catalytic amount (3 mol %) of tetrabutylammonium fluoride (TBAF) yielded the trifluoromethylketone (**6**) and its hydrate (**7**) up to 28% and 40% respectively (Scheme **13**) [55].

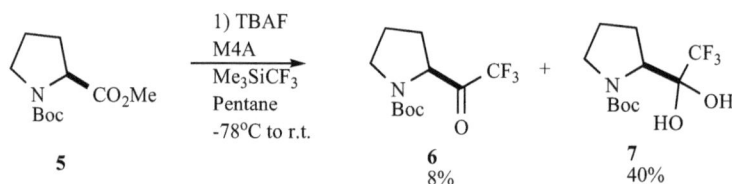

Scheme 13: Trifluoromethylation of methyl (S)-N-Boc-pyrrolidine-2-carboxylate (**5**) with TMSCF$_3$.

Considerable attention has been devoted to the development of different catalytic systems for the activation of the so-called Ruppert-Prakash reagent. TMSCF$_3$ itself does not react with carbonyl compounds. The trifluoromethide anion must be liberated by activation with a nucleophilic initiator such as a fluoride source from: CsF, n-Bu$_4$NF, KF, KF·t-BuOK, Ph$_3$SnF·KF, and other salts (acetates, phosphates, carbonates), and phenoxide in DMF solution, amines (Et$_3$N, pyridine), and phosphines (Ph$_3$P, t-Bu$_3$P, tris-(2,4,6-trimethoxyphenyl)phosphine (TTMPP)), phosphine oxides (Ph$_3$PdO), amine oxides (Me$_3$NO), and molecular sieves [56-67]. For example, upon addition of a catalytic amount of TBAF to the reaction mixture of a carbonyl compound and TMSCF$_3$ in a suitable solvent, the process starts with the initial formation of Me$_3$SiF (Scheme **14**). The reaction between TMSCF$_3$ and Me$_3$SiF initiates the formation of the pentavalent complex followed by the transfer of the trifluoromethyl group to the electrophilic carbon of the carbonyl function until all of the starting material has reacted [53].

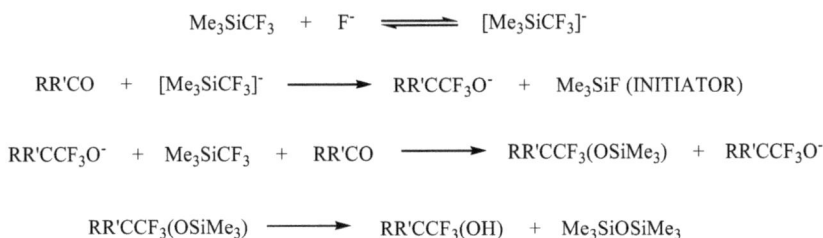

$$Me_3SiCF_3 \ + \ F^- \ \rightleftharpoons \ [Me_3SiCF_3]^-$$

$$RR'CO \ + \ [Me_3SiCF_3]^- \ \longrightarrow \ RR'CCF_3O^- \ + \ Me_3SiF \ (INITIATOR)$$

$$RR'CCF_3O^- \ + \ Me_3SiCF_3 \ + \ RR'CO \ \longrightarrow \ RR'CCF_3(OSiMe_3) \ + \ RR'CCF_3O^-$$

$$RR'CCF_3(OSiMe_3) \ \longrightarrow \ RR'CCF_3(OH) \ + \ Me_3SiOSiMe_3$$

Scheme 14: Mechanism of activation of reaction with TMSCF$_3$.

We described the synthesis of new 3-trifluoromethylindoles (**8a-e**) using isatins as a starting material. Isatins (**8**) were trifluoromethylated using TMSCF$_3$ as a nucleophilic agent giving new 3-hydroxy-3-(trifluoromethyl) indolin-2-one (**9a-e**). Different one-step procedures attempted to transform the latter compounds into the reduced indoles were failed. For the synthesis of the new trifluoromethylindoles (**10a-e**) the corresponding **9a-e** were reduced using borane/THF complex to furnish 3-(trifluoromethyl)indolin-3-ol (**10a-e**) that additionally were dehydrated using thionyl chloride in pyridine for obtaining excellent yields of the desired products [57] (Scheme **15**).

Fuchikami and co-workers reported trifluoromethylation of aminoketones using TMSCF$_3$ without using any catalyst. They found that the amino group can activate the silane reagent intramolecularly making the trifluoromethyl group easier to attack carbonyl moieties (Scheme **16**). The reactivity of the substrates was found to be strongly dependent on carbon chain length between the amino group and carbonyl group suggesting a cyclic transition state that may play an important role. Moderates diasteroselectivities of the trifluoromethyl- alcohols were probed [68].

The chiral auxiliary-controlled asymmetric nucleophilic trifluoromethylations were achieved using TMSCF$_3$ in the presence of a chiral ammonium derivative [69]. Shibata and co-workers reported the enantioselective trifluoromethylation of aryl ketones using TMSCF$_3$ and TMAF in the presence of *Cinchone* alkaloid derivative as the nucleophilic trifluoromethylating system, as shown in Scheme **17** [70].

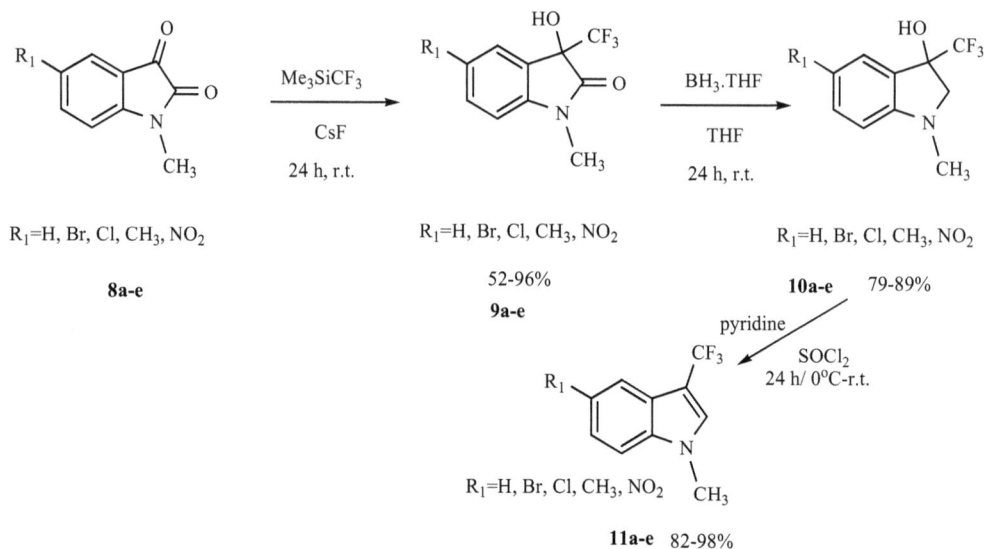

Scheme 15: Synthesis of trifluoromethylindoles from isatins using TMSCF$_3$.

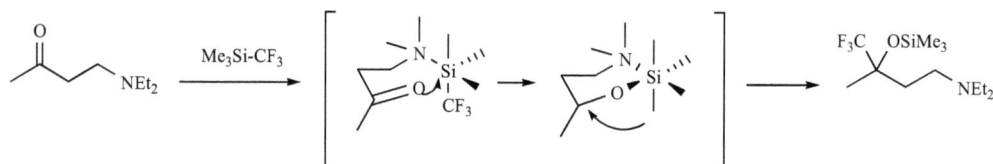

Scheme 16: Trifluoromethylation of aminoketones.

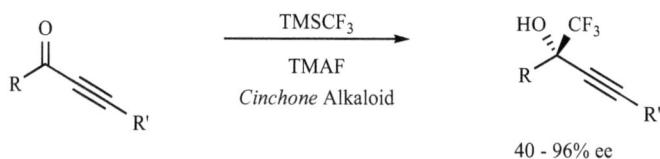

Scheme 17: Asymmetric trifluoromethylation of arylketone using TMSCF$_3$.

Significant progress has been marked in the development of asymmetric trifluoromethylation reactions [71-77]. However, the direct enantioselective trifluoromethylation reaction still remains a challenge as recently observed when chiral crown ethers were used to the enantioselective trifluoromethylation of aldehydes and ketones [77]. Moderate enantioselectivities were observed for the trifluoromethylation of both aryl or alkyl aldehydes, and alkyl or aryl ketones in 21-44% ee (Scheme **18**).

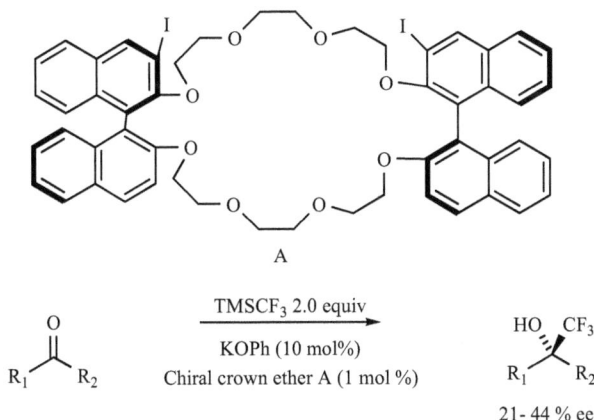

Scheme 18: Enantioselective trifluoromethylation using crown ether.

A novel synthetic approach to obtain trifluoromethylalkene dipeptide isosteres (CF$_3$-ADI) (**11**) has recently been introduced by Fujii and co-workers [76]. These isosteres can be considered as more ideal peptide bond mimetics when the carbonyl oxygen of the peptide bond is replaced by a highly electronegative trifluoromethyl group. Starting from L-phenylalanine and L-alanine, several CF$_3$-ADIs were obtained through nucleophilic trifluoromethylation of γ-keto esters (**9**), giving **10** and subsequent SN$_2'$ alkylation of trifluoromethylated mesylate derivatives (Scheme **19**). They investigated the nucleophilic trifluoromethylation of the ketoester (**9**) using TMSCF$_3$ in the presence of K$_2$CO$_3$.

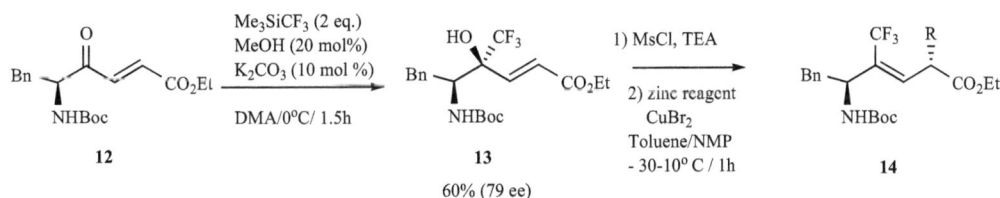

Scheme 19: Trifluoromethylation of α,β-unsaturated-γ-keto using TMSCF$_3$ in the presence of K$_2$CO$_3$.

The use of Ruppert-Prakash reagent for obtaining trifluoromethylketones from esters [76] pointed out important application for the enantioselective construction of stereogenic tetrasubstituted carbon centers. These are very challenging goals and the subject of recent curiosity in the synthesis of drugs such as a neurokinin 1 receptor antagonist [78, 79] or efavirenz [80] two chiral R-trifluoromethyl tertiary alcohols in which the CF_3 moiety is located in a stereogenic tetrasubstituted carbon atom (Fig. **2**).

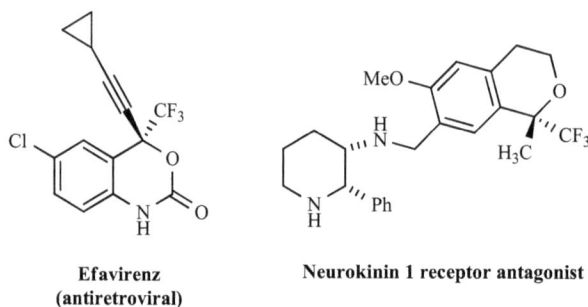

Efavirenz
(antiretroviral)

Neurokinin 1 receptor antagonist

Figure 2: Drugs containing chiral trifluoromethyl tertiary alcohols in which the CF_3 moiety is located in a stereogenic tetrasubstituted carbon atom.

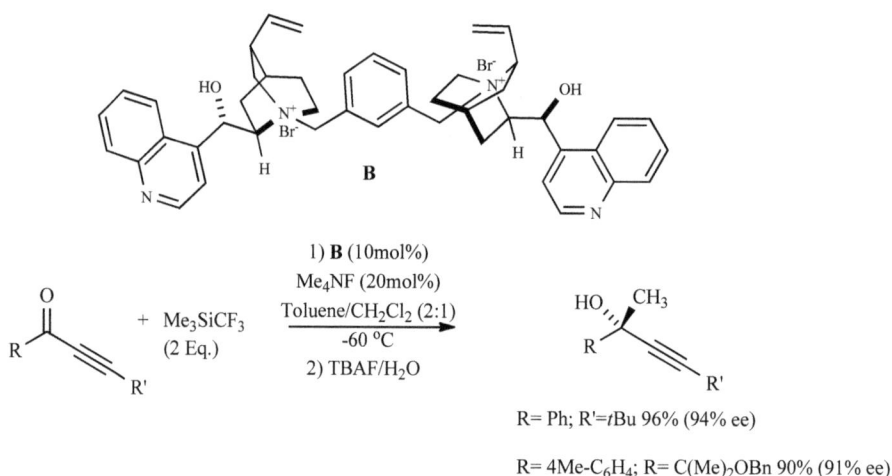

1) **B** (10mol%)
Me_4NF (20mol%)
Toluene/CH_2Cl_2 (2:1)
-60 °C
2) TBAF/H_2O

R= Ph; R'=tBu 96% (94% ee)

R= 4Me-C_6H_4; R= C(Me)$_2$OBn 90% (91% ee)

Scheme 20: Catalytic enantioselective trifluoromethylation of alkynyl ketones with TMSCF$_3$.

With the same proposal, in 2010 Shibata and co-workers [81] developed the first catalytic enantioselective trifluoromethylation of alkynyl ketones with TMSCF$_3$ by combining ammonium bromide and bis-*cinchona* alkaloids (**B**) with Me$_4$NF. This yielded trifluoromethyl-substituted tertiary propargyl alcohols (up to 96% ee) being important chiral building blocks for pharmaceuticals (Scheme **20**). This methodology was extended to synthesize some biologically attractive heteroaryl trifluoromethyl carbinols.

The authors utilized this same approach to the first example of the enantioselective synthesis of efavirenz in five steps from a commercially available and operationally simple substrate [82].

2.4. Methods Using Sulfur Reagents

2.4.1. Sulfur Tetrafluoride SF₄

A classic and efficient method for the transformation of the carboxylic group into a trifluoromethyl group is the treatment with SF_4. Since it is a toxic gas therefore it must be handled with caution [83]. As this chemical is highly reactive it can react with carbonyl compounds, halides, and alcohols. The carboxylic acid derivatives are important synthons in organic synthesis, as the carboxylic group can be easily introduced into organic molecules and can be transformed into other functional groups. Thus, carboxylic acids (aliphatic, aromatic, and heterocyclic) react with SF_4 in two steps and remains as the main industrial method for obtaining trifluoromethyl derivatives [13] (Scheme **21**).

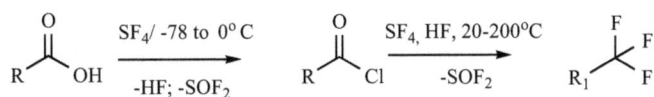

Scheme 21: General reaction of SF_4 with carboxylic acid.

2.4.2. (Diethylamino) Sulfur Trifluoride (DAST) and Bis(2-Methoxyethyl)Aminosulfur Trifluoride

DAST was first reported by Middleton [84] and emerged as one of the most important fluorinating agents. It is a commercially available liquid that mimics the chemistry of SF_4 while avoiding the high pressure and toxicity problems associated with the use of SF_4. DAST is mainly used to fluorinate alcohols giving monofluorinated compounds or carbonyl groups those help in yielding *gem*-difluoroderivatives. An important congener of DAST is bis (2-methoxyethyl) aminosulfur trifluoride, known as Deoxo-Fluor reagent that was first reported by Lal and co-workers and is considered to be much more thermally stable than DAST [85].

Although both reagents have been largely used in difluorination of aldehydes and ketones, some conversions of carboxylic acids to trifluoromethyl group derivatives have been reported in Scheme **22** [86]. These are some of the examples of the extensive applications of Deoxo-Fluor described in subsequent reports from many research groups (Scheme **22**) [87, 88].

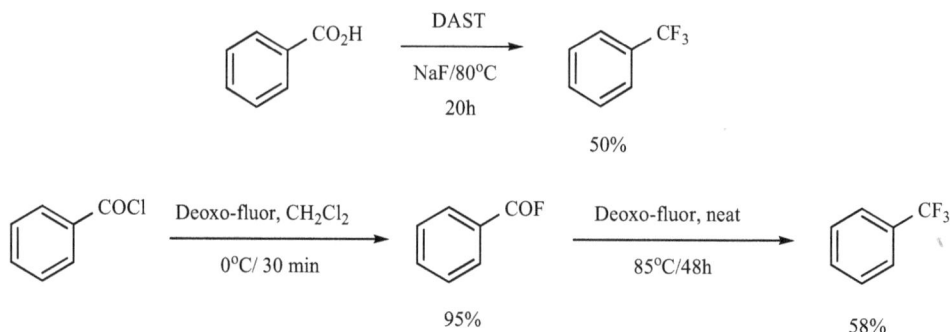

Scheme 22: Use of DAST and Deoxo-Fluor for trifluoromethylation of carbonyl compounds.

2.4.3. Methods Using Sulfones, Sulfoxides and Sulfonamides

During the process of obtaining TMSCF$_3$ from CF$_3$H *via* phenyl trifluoromethyl sulfide [49] Prakash, Hu and Olah reported a reductive trifluoromethylation using trifluoromethyl sulfides, sulfoxides, and sulfones as trifluoromethyl group precursors. The method was extended showing the first alkoxide- and hydroxide-induced nucleophilic trifluoromethylation of carbonyl compounds, disulfides, and other electrophiles, using phenyl trifluoromethyl sulfone and sulfoxide (Scheme **23**). The trifluoromethyl sulfone or sulfoxide acted as a "CF$_3$" synthon [89].

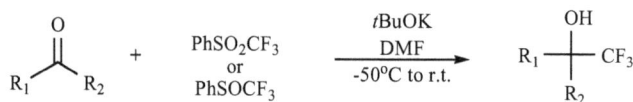

Scheme 23: Trifluoromethylation using trifluoromethyl sulfone or sulfoxide.

2.4.4. Methods Using Dithionic Esters with BrF$_3$

The bromine trifluoride (BrF$_3$) undergoes strong and frequently uncontrolled reactions with water and hydroxyllic solvents but can be conveniently and safely handled in halogenated solvents. Thus, Rozen studied the chemistry of BrF$_3$ and developed a process of the trifluoromethylation of aryl carboxylic acids (**15**) by their dithionic esters (**17**) with BrF$_3$ under very mild conditions [90, 91]. Trifluoromethyl derivatives (**18**) were prepared by reacting the appropriate acyl halide with ethanethiol followed by reaction with Lawenson's reagent (**16**) [92]. The resulting product (**14**) was then dissolved in a solution of dried BrF_3 in $CHCl_3$ or CCl_4 at low temperature (Scheme **24**).

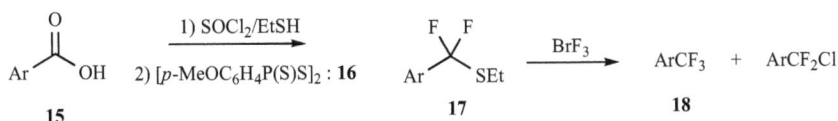

Scheme 24: Trifluoromethylation of aryl carboxylic acids by their dithionic esters with BrF_3 compounds.

The best results in the aromatic field were generally obtained with compounds having a deactivated ring. The reaction was not limited to aromatic carboxylic acids since many aliphatic ones served as substrates as well [93].

2.5. Methods Using Organoboron Reagent

After some foiled attempts Molander and Hoag succeeded in developing a high-yielding route to potassium (trifluoromethyl) trifluoroborate. They reacted Ruppert's reagent with trimethoxyborane in the presence of potassium fluoride. Aqueous hydrogen fluoride was added to the resulting intermediate, and K[CF_3-BF_3] was isolated in overall yield that was 85%. They believed to have found a trifluoromethylating agent for selective organic synthesis [94].

In 2011, Dilman and co-workers synthesized some shelf-stable and easily accessible organoboron reagents (**19**, **20**) and K[CF_3-BF_3] (**21**) for nucleophilic trifluoromethylation (Scheme **25**) [95]. However, contrary to what Molander and Hoag thought **21** did not yield desired trifluoromethylated product.

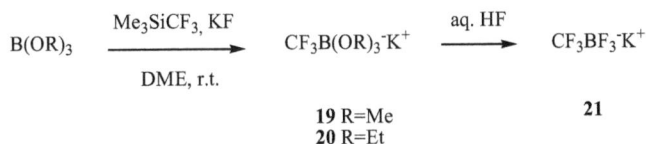

Scheme 25: Preparation of potassium (trifluoromethyl) trifluoroborate.

Trifluoromethylations performed with the reactants **19** and **20** were found to be effective (Scheme **26**). Reaction using **20** was slightly faster, however, this reagent was more hygroscopic and difficult to handle. The optimization of the reaction conditions was thus carried out using $CF_3B(OCH_3)_3^-K^+$. Trifluoromethylated alcohol was isolated in 97% yield from the complete conversion of benzaldehyde using **20** after one hour at 50°C in DMF [95].

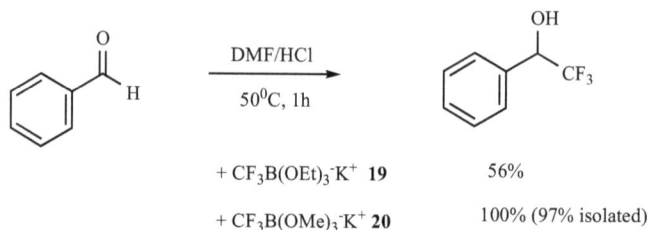

$$+ CF_3B(OEt)_3^-K^+ \; \mathbf{19} \qquad\qquad 56\%$$

$$+ CF_3B(OMe)_3^-K^+ \; \mathbf{20} \qquad 100\% \; (97\% \; \text{isolated})$$

Scheme 26: Trifluoromethylation of benzaldehyde with organoboron reagents.

Other substrates were subjected to the trifluoromethylation reaction with this reagent. The results showed the enolizable compounds undergoing deprotonation in the reaction conditions applied [95].

2.6. Methods Using Diethyl Trifluoromethylphosphonate

In 1977, Paulin and Tomlinson prepared trifluoromethyltris(dialkylamino) phosphonium halides, $[CF_3P(NR_2)_3]^+X^-$ (R= Me, X= Cl), (R = Et, X = Br) [96] and in 1983, they evidenced the generation of CF_3^- from $[CF_3P(NMe_2)_3]^+Br^-$ by trapping experiments with benzaldehyde and copper(I) iodide [97].

Recently, Cherkupally and Beier [98] used diethyl trifluoromethylphosphonate (**22**) as CF_3^- transfer reagent. **22** was prepared by reaction of CF_3I with $P(OEt)_3$ under photolytic conditions and then used for the trifluoromethylation of non-enolizable ketones and aldehydes. The reaction with benzophenone in the presence of CsF (2 equiv) in DMF at room temperature or with *t*-BuOK only recovered the starting material. Trifluoromethylated alcohol product was amply yielded using 2 equivalents of *t*-BuOK in DMF at -40°C, and slowly warmed to room temperature over 1 h (Scheme **27**) [98]. However the use of this reagent did

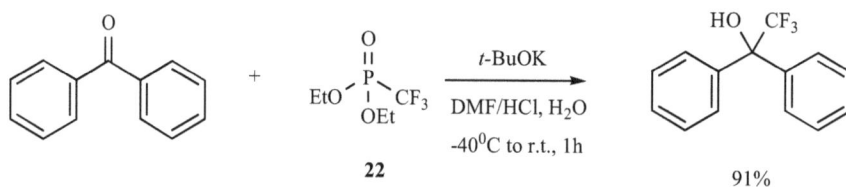

Scheme 27: Trifluoromethylation with diethyl trifluoromethylphosphonate.

not present satisfactory results as compared to those by TMSCF$_3$.

2.7. Other Methods of Trifluoromethylation of Carbonyl Compounds

2.7.1. Methods Using Electroreduction of Trifluoromethyl Bromide (CF$_3$Br) in DMF

In 1989, Sibille and co-workers [99] investigated the synthetic utility of an electrochemical procedure applied to trifluoromethylation compounds. They described the electroreduction of trifluoromethyl bromide (CF$_3$Br) in DMF and its reductive coupling with other aldehydes as well as ketones. With aldehydes, the anticipated alcohols were obtained in almost quantitative yields however poor yields were observed with ketones.

When CF$_3$Br was allowed to reduce at a nickel cathode in a 0.2-1M aldehyde solution in DMF, using a zinc anode, fluorinated alcohols were formed with excellent chemical and faradic yields (Scheme **28**). The use of zinc as anodic material was considered essential for the coupling process; the replacement of zinc by magnesium or aluminium gave very poor results. In the same conditions, trifluoromethylated alcohols could not be obtained with good yields from ketones, except for the easily reducible ones, such as benzophenone [99].

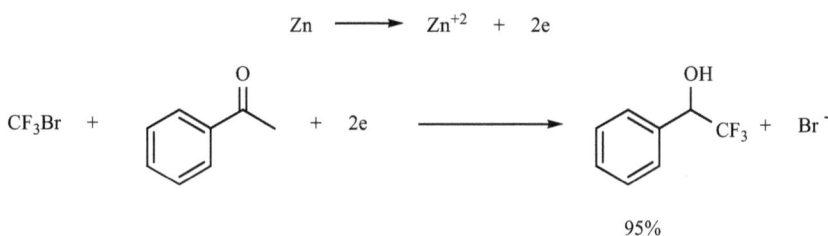

Scheme 28: Use of CF$_3$Br for electrochemical trifluoromethylation of carbonyl compounds.

The differences in reactivity between aldehydes and ketones were explained by ^{19}F NMR in DMF. When CF$_3$Br was allowed to reduce at a nickel cathode, using a zinc anode, the first specie to arise was CHF$_3$ (δ=78.4 ppm, J$_{H-F}$=85 Hz), from residual water, then the organometallic species (CF$_3$)$_2$Zn (δ=38.7 ppm) and CF$_3$ZnBr (δ=39.4 ppm), together with fluorinated products arose from the attack on DMF during electrolysis. Upon addition of ZnBr$_2$ to the DMF solution, slow equilibration was observed between species. The main difference between aldehydes and ketones was that the organozinc species were formed after the alcohol, when the substrate being an aldehyde ($RCH(O^-)CF_3$), and before the alcohol when the substrate being a ketone ($R_1R_2C(O^-)CF_3$) [99].

2.7.2. Methods Using Trifluoromethane (Fluoroform) as a Source of CF_3^-

The use of trifluoromethane (fluoroform) as a source of CF_3^- has been reported since 2000. The easy availability of this cheap and environmentally benign reagent has advocated its use for synthetic purposes.

The first result concerning trifluoromethylation with fluoroform was reported by Shono and co-workers. They found that a trifluoromethyl anion equivalent could be efficiently formed by deprotonation of trifluoromethane with the base from 2-pyrrolidone by electrochemical reduction (Scheme **29**). The reaction of this specie with aldehydes and ketones gave (trifluoromethyl) carbinols in a fair to good yield [100]. The chemical reactivity of electrogenerated bases from pyrrolidone was observed in their earlier studies [101, 102].

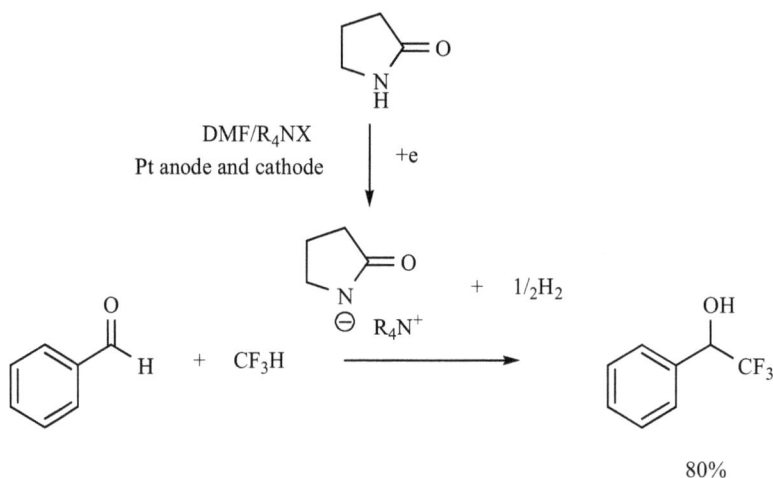

Scheme 29: Trifluoromethylation of benzaldehyde by deprotonated trifluoromethane through electroreduction of 2-pyrrolidone.

Another trifluoromethylating system using trifluoromethane was introduced by Langlois and co-workers. They showed the system formed by four-components HCF_3/N (TMS)$_3$/catalytic F-/catalytic DMF behaved similarly to Ruppert's reagent.

The process involved an adduct formation between DMF and CF_3^- as the true trifluoromethylating agent of nonenolizable carbonyl compounds [103] (Scheme **30**).

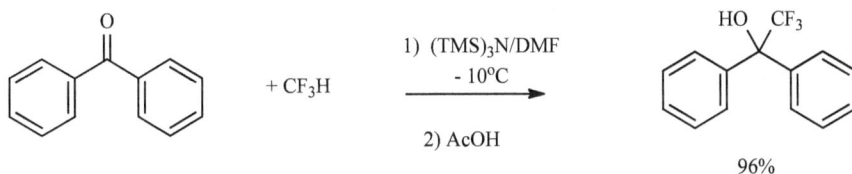

Scheme 30: Trifluoromethylation using HCF_3/N $(TMS)_3$/catalytic F-/catalytic DMF system.

They also described the preparation of a silylated stable reagent formed by the reaction of N-formylmorpholine with trifluoromethane in the presence of $(TMS)_3N$/TBAF being able to react with aldehydes and ketones in DMF [104].

2.7.3. Methods Using Hemiaminals of Fluoral

During the investigations to design new nucleophilic trifluoromethylating reagents, Langlois and co-workers reported a family of suitable reagents, namely the hemiaminals of fluoral [105, 106]. Trifluoroacetaldehyde (fluoral), in its hemiketal form (**23**), reacted with N-benzyl piperazine (**24**) giving piperazino hemiaminal (**25**) [106]. This was a stable compound which, after silylation, gave the corresponding *O*-silylated hemiaminal (**26**) that could trifluoromethylate nonenolizable carbonyl compounds (**27**) into trifluoromethyl carbinols (**28**) [106] (Scheme **31**).

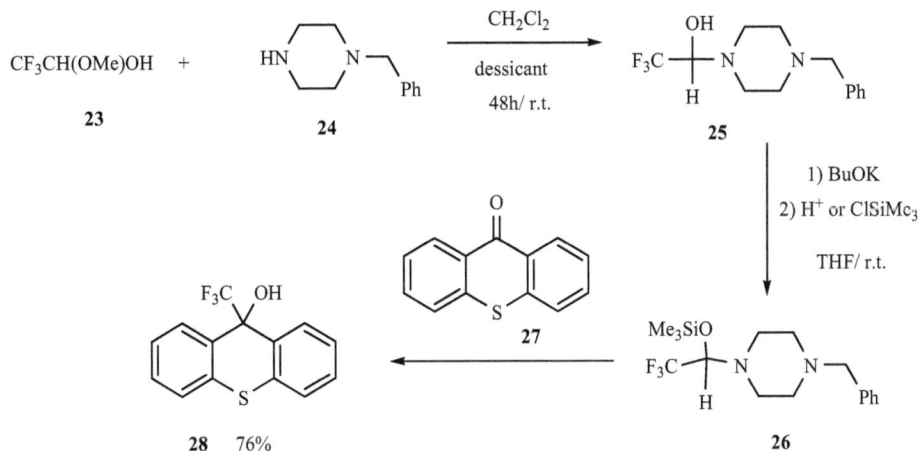

Scheme 31: Trifluoromethylation of carbonyl compounds using stable piperazino hemiaminal of trifluoroacetaldehyde.

2.7.4. Methods Using Sodium Trifluoracetate

Matsui and co-workers described the use of sodium trifluoroacetate as a powerful tool for the introduction of a trifluoromethyl moiety into organic substrates through halogen substitution [107]. Years later, Chang and Cai used this trifluoromethylating reagent catalyzed by copper I with carbonyl compounds. Aryl aldehydes, aliphatic aldehydes and ketones could smoothly be reacted with sodium trifluoroacetate to give the corresponding trifluoromethylated alcohols in moderate to good yields (Scheme **32**). When aliphatic ketones were applied as the substrates, the yields declined with the increasing of the carbon chain. This result was attributed to be partly caused by the steric hindrance of the alkyl group [108].

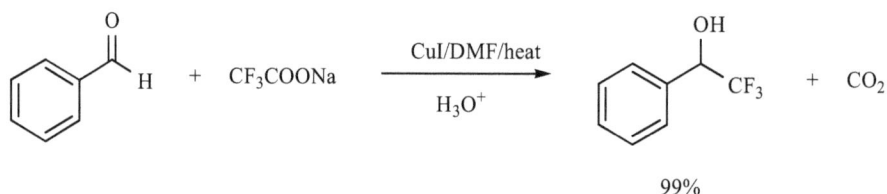

Scheme 32: Trifluoromethylation of benzaldehyde with sodium trifluoroacetate.

They maximized the use of this readily available, cheap, and environmental "friendly" reagent with other carbonyl compounds, such as aldehydes, ketones, acetyl chloride, and acid anhydrides [109]. Different types of copper compounds, such as CuI, CuBr, CuBr$_2$ and copper and zinc powder could effectively catalyze the reaction. Benzoyl chloride (**29**) gave trifluoromethyl-phenyl ketone (**30**). This product was found to depend on the treatment of the reaction that could suffer hydrolysis, giving a mixture of **31** and **32** (Scheme **33**). Acid anhydrous also showed good reactivity *via* the hydrolysis of the reaction mixture, and surprisingly, only the mono-adducts were obtained.

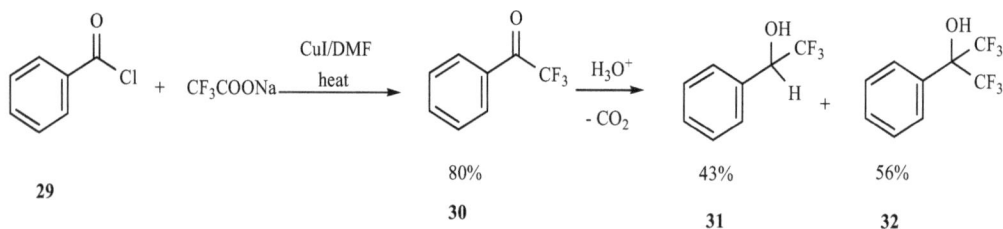

Scheme 33: Trifluoromethylation of benzoyl chloride with trifluoromethyl acetate.

2.7.5. Methods Using Sodium Trifluoromethylacetophenone-N,N-Dimethyl-trimethylsilylamine Adduct

The reaction of trifluoroacetophenone (**33**) with N,N-dimethytrimethylsilylamine (**34**) under simple thermal conditions furnished the stable reagent N,N-Dimethyl-(1-phenyl-2,2,2-trifluoroethoxytrimethylsilyl)-amine (**35**) in excellent yield (Scheme **34**) [110].

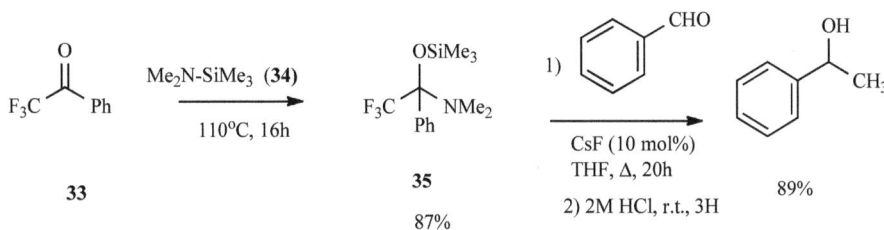

Scheme 34: Preparation of N,N-Dimethyl-(1-phenyl-2,2,2-trifluoroethoxytrimethylsilyl)-amine (**35**).

Non-enolisable aldehydes and ketones reacted with two equivalents of nucleophilic shelf-stable reagent **35** in the presence of CsF (10% mol) as the preferred initiator in THF giving a number of trifluoromethyl carbinols (Scheme **34**). Moreover, the same reagent, in combination with 2-trimethylsiloxy pyridine, as a potentially versatile additive for silicon demonstrated the addition to simple aromatic imine derivatives possible [110].

CONCLUSIONS

The direct introduction of a trifluoromethyl group on carbonyl compounds has been described in the literature for years. Although there are many applicable methods for the introduction of this group, some of them not suitable for practical purposes due to the reagent characteristic or low yields.

The trifluoromethylation of carbonyl compounds was described on the basis of the kind of reagent used for it. Many methods have been described, such as: addition of a trifluoromethyl group to the carbonyl carbon through Ruppert-Prakash reagents, use of TDAE, and CF$_3$H, Barbier procedure, electrochemical methods, and the decarboxylation of trifluoromethyl acetate. These methods, however, generally suffer from the use of toxic and expensive reagents or severe experimental conditions requirements.

The perfluoroalkyl metallic reagents are not easy to handle and their low reactivity reduces their synthetic utilities. The use of CF_3Br suffers with the serious environmental problems and has been substituted by fluoroform. SF_4 is the most reactive of them but is a gas, toxic, and difficult to handle, although it is still used in industrial applications. It is the main method to obtain trifluoromethyl derivatives from carboxylic acids.

A recent development of shelf-stable CF_3-borates appears as a new option for trifluoromethylation. They react with non-enolizable carbonyl compounds and N-tosylimines under mild conditions to give products in good isolated yields.

Amongst all available methods, the trifluoromethylation using the Ruppert-Prakash reagent is considered as the most popular one. However, direct enantioselective trifluoromethylation reaction using this reagent remains a challenge.

ACKNOWLEDGEMENTS

The authors thank the National Council of R&D of Brazil (CNPq) and Foundation for Research of the State of Rio de Janeiro (FAPERJ) for the fellowships granted.

CONFLICT OF INTEREST

The author(s) confirm that this chapter content has no conflict of interest.

DISCLOSURE

The chapter submitted for series eBook titled **"Advances in Organic Synthesis, Volume 5"**is an update of our article published in **CURRENT ORGANIC SYNTHESIS, Volume 7, Number 5, October Issue 2010,** with additional text and references.

REFERENCES

[1]   Schofield, H. Fluorine chemistry statistics: numbers of organofluorine compounds and publications associated with fluorine chemistry. *J. Fluorine Chem.*, **1999**, *100*(1-2), 7-11.
[2]   Filler, R.; Kobayashi, Y.; Yagupolskii, Y.L. *Organofluorine Compounds in Medicinal Chemistry and Biological Applications*, Elsevier: Amsterdam, **1993**.
[3]   Ojima, I. *Fluorine in Medicinal Chemistry and Chemical Biology*, John Wiley & Sons: Chichester, **2009**.

[4]   Banks, R.E.; Smart, B.E.; Tatlow, J.C. *Organofluorine Chemistry: Principles and Commercial Applications*, Plenum Press: New York, **1994**.

[5]   Petrov, V.A. *Fluorinated Heterocyclic Compounds: Synthesis, Chemistry and Applications*, John Wiley & Sons Inc: New Jersey, **2009**.

[6]   O' Hagan, D. Fluorine in health care: Organofluorine containing blockbuster drugs. *J. Fluorine Chem.*, **2010**, *131*(11), 1071-1081.

[7]   McClinton, M.A.; McClinton, D.A. Trifluoromethylations and related reactions in organic chemistry. *Tetrahedron*, **1992**, *48*(32), 6555-6666.

[8]   Lin, P.; Jiang, J. Synthesis of mono(trifluoromethyl)-substituted saturated cycles. *Tetrahedron*, **2000**, *56*(23), 3635-3671.

[9]   Tomashenko, O.A.; Grushin, V.V. Aromatic trifluoromethylation with metal complexes. *Chem. Rev.*, **2011**, *111*(8), 4475-4521.

[10]   Nie, J.; Guo, H.C.; Cahard, D.; Ma, J.A. Asymmetric construction of stereogenic carbon centers featuring a trifluoromethyl group from prochiral trifluoromethylated substrates. *Chem. Rev.*, **2011**, *111*(2), 455-529.

[11]   Valero, G.; Company, X; Rios, R. Enantioselective organocatalytic synthesis of fluorinated molecules. *Chem. Eur. J.*, **2011**, *17*(7), 2018-2037.

[12]   a) Tur, F.; Mansilla, J.; Lillo, V.J.; Saa, J.M. Constructing quaternary centers of chirality: the lanthanide way to trifluoromethyl-substituted tertiary alcohols. *Synthesis*, **2010**, *11*, 1909-1923. b) Dhara, M.G.; Banerjee, S. Fluorinated high-performance polymers: poly(arylene ethers) and aromatic polyimides containing trifluoromethyl groups. *Prog. Polym. Sci.*, **2010**, *35*(8), 1022-1077.

[13]   Quirmbach, M.; Steiner, H. The trifluoromethyl group: an overview of available synthetic methods at Solvias AG. *Chimica Oggi / CHEMISTRY TODAY*, **2009**, *27*(3), 23-26.

[14]   Prakash, G.K.S.; Chacko, S. Novel nucleophilic and eletrophilic fluoroalkylation methods. *Curr. Opin. Drug Disc.*, **2008**, *11*, 793-802.

[15]   Khomenko, T.M.; Tolstikova, T.G.; Bolkunov, A.V.; Dolgikh, M.P.; Pavlova, A.V.; Korchagina, D.V.; Volcho, K.P.; Salakhutdinov, N.F. 8-(Trifluoromethyl)-1,2,3,4,5-benzopentathiepin-6-amine: novel aminobenzopentathiepine having *in vivo* anticonvulsant and anxiolytic activities. *Lett. Drug Des. Disc.*, **2009**, *6*(6), 464-467.

[16]   Agbaje, O.; Fadeyi, O.O.; Fadeyi, S.A.; Myles, L.E.; Okoro, C.O. Synthesis and *in vitro* cytotoxicity evaluation of some fluorinated hexahydropyrimidine derivatives. *Bioorg. Med. Chem. Lett.*, **2011**, *21*(3), 989-992.

[17]   Zhu, L.; Miao, Z.; Sheng, C.; Guo, W.; Yao, J.; Liu, W.; Che, X.; Wang, W.; Cheng, P.; Zhang, W. Trifluoromethyl-promoted homocamptothecins: synthesis and biological activity. *Eur. J. Med. Chem.*, **2010**, *45*(7), 2726-2732.

[18]   Huang, X.; Dong, M.; Liu, J.; Zhang, K.; Yang, Z.; Zhang, L. Concise syntheses of trifluoromethylated cyclic and acyclic analogues of cADPR. *Molecules*, **2010**, *15*(12), 8689-8701.

[19]   Boechat, N.; Kover, W.B.; Bastos, M.M.; Romeiro, N.C.; Valverde, A.L.; Silva, A.S.; Santos, F.C.; Guida, M.L.; Wollinger, W.; Frugulhetti, I.C.P. Design, synthesis and biological evaluation of new 3-hidroxy-2-oxo-3-trifluoromethylindole as potential HIV-1 reverse transcriptase inhibitors. *Med. Chem. Res.*, **2007**, *15*(9), 492-507.

[20]   Boechat, N.; Pinheiro, L.C.S.; Silva, T.S.; Aguiar, A.C.; Carvalho, A.S.; Bastos, M.M.; Costa, C.C.P.; Pinheiro, S.; Pinto, A.C.; Mendonça, J.S.; Dutra, K.D.B.; Valverde, A.L.; Santos-Filho, O.A.; Krettli, A.U. New trifluoromethyl triazolopyrimidines as anti-*Plasmodium falciparum* agents. *Molecules*, **2012**, *17*(7), 8285-8302.

[21] Wan, Z.K.; Chenail, E.; Li, H.Q.; Kendall, C.; Wang, Y.; Gingras, S.; Xiang, J.; Massefski, W.W.; Mansour, T.S.; Saiah, E. Synthesis of potent and orally efficacious 11β-hydroxysteroid dehydrogenase type 1 inhibitor HSD-016. *J. Org. Chem.*, **2011**, *76*(17), 7048-7055.

[22] Ma, J.A.; Cahard, D. Strategies for nucleophilic, electrophilic and radical trifluoromethylations. *J. Fluorine Chem.*, **2007**, *128*(9), 975-996.

[23] Langlois, B.R.; Billard, T.; Roussel, S. Nucleophilic trifluoromethylation of some reagents and their stereoselective aspect. *J. Fluorine Chem.*, **2005**, *126*(2), 173-179.

[24] Mayer, P.S.; Leblanc, D.; Morton, T.H. Gas-phase halonium metathesis and its competitors. Skeletal rearrangements of cationic adducts of saturated ketones. *J. Am. Chem. Soc.*, **2002**, *124*(47), 14185-14194.

[25] Umemoto, T.; Adachi, K.; Ishihara, S. CF$_3$ Oxonium salts, *o*-(trifluoromethyl)dibenzofuranium salts: *in situ* synthesis, properties and application as a real CF$_3^+$ species reagent. *J. Org. Chem.*, **2007**, *72*(18), 6905-6917.

[26] Umemoto, T. In: *Recent Advances in Perfluoroalkylation Methodology.* ACS Symposium Series: Washington, DC, **2005**, Vol. 911, pp. 2-15.

[27] (a) Nagib, D.A.; Scott, M.E.; MacMillan, D.W.C. Enantioselective α-trifluoromethylation of aldehydes *via* photoredox organocatalysis. *J. Am. Chem. Soc.*, **2009**, *131*(31), 10875-10877. (b) Nagib, D.A.; MacMillan, D.W.C. Trifluoromethylation of arenes and heteroarenes by means of photoredox catalysis. *Nature*, **2011**, *480*(7376), 224-228.

[28] Gassman, P.G.; O'Reilly, N.J. Nucleophilic addition of the pentafluoroethyl group to aldehydes, ketones and esters. *J. Org. Chem.*, **1987**, *52*(12), 2481-2490.

[29] Denson, D.D.; Smith, C.F.; Tamborski, C. Synthesis of perfluoroaliphatic Grignard reagents. *J. Fluorine Chem.*, **1973**, *3*(3-4), 247-258.

[30] Santini, G.; Le Blanc, M.A.; Riess, J.G. Reactions of perfluoroalkylcalcium derivatives with ketones and aldehydes. *J. Organomet. Chem.*, **1977**, *140*(1), 1-9.

[31] Santini, G.; Le Blanc, M.A.; Riess, J.G. Perfluoroalkyl calcium derivatives: reactions with carbonyl compounds. *J. Chem. Soc. Chem. Commun.*, **1975**, *16*, 678-679.

[32] Sekiya, A.; Ishikawa, N. Reaction of heptafluoro-1-methylethylzinc iodide with halides and anhydrides of carboxylic acids. *Chem. Lett.*, **1977**, *6*(1), 81-84.

[33] a) Kitazume, T.; Ishikawa, N. Trifluoromethylation of carbonyl compounds with trifluoromethylzinc iodide under ultrasonic irradiation. *Chem. Lett.*, **1981**, *10*(12), 1679-1680. b) Kitazume, T.; Ishikawa, N. Ultrasound-promoted selective perfluoroalkylation on the desired position of organic molecules. *J. Am. Chem. Soc.*, **1985**, *107*(18), 5186-5191.

[34] Luche, J.L.; Damiano, J.C. Ultrasounds in organic syntheses. 1. Effect on the formation of lithium organometallic reagents. *J. Am. Chem. Soc.*, **1980**, *102*(27), 7926-7927.

[35] Ishikawa, N.; Maruta, M.; O'Reilly, N.J. Palladium and nickel-catalyzed perfluoroalkylation of aldehydes using zinc and perfluoroalkyl halides. *Chem. Lett.*, **1984**, *13*(4), 517-520.

[36] Tordeux, M.; Francese, C.; Wakselman, C. Reactions of trifluoromethyl bromide and related halides: part 9. Comparison between additions to carbonyl compounds, enamines and sulphur dioxide in the presence of zinc. *J. Chem. Soc. Perkin Trans.*, **1990**, *7*, 1951-1957.

[37] Wakselman, C.; Francese, C.; Tordeux, M. Synthesis of trifluoromethyl-substituted methanols: a Barbier procedure under pressure. *J. Chem. Soc. Chem. Commun.*, **1987**, *9*, 642-643.

[38] Burton, D.J.; Yang, Z.Y. Fluorinated organometallics: perfluoroalkyl and functionalized perfluoroalkyl organometallic reagents in organic synthesis. *Tetrahedron*, **1992**, *48*, 189-275.

[39] Ishikawa, N.; Kitazume, T. Perfluoroakyltin(IV) halides: a novel perfluoroalkylating agent for carbonyl compounds. *Chem. Lett.*, **1981**, *10*(10), 1337-1338.

[40] Pruett, R.L.; Barr, J.T.; Rapp, K.E.; Bahner, C.T.; Gibson, J.D.; Lafferty Jr., R.H. Reactions of polyfluoro olefins. II. Reactions with primary and secondary amines. *J. Am. Chem. Soc.*, **1950**, *72*, 3646-3650.

[41]   Médebielle, M.; Dolbier Jr., W.R. Nucleophilic difluoromethylation and trifluoromethylation using tetrakis(dimethylamino)ethylene (TDAE) reagent. *J. Fluorine Chem.*, **2008**, *129*(10), 930-942.

[42]   Pawelke, G. Tetrakis(dimethylamino)ethylene-trifluoroiodomethane, a specific novel trifluoromethylating agent. *J. Fluorine Chem.*, **1989**, *42*(3), 429-433.

[43]   Aıt-Mohand, S.; Takechi, N.; Medebielle, M.; Dolbier Jr., W.R. Nucleophilic trifluoromethylation using trifluoromethyl iodide. A new and simple alternative for the trifluoromethylation of aldehydes and ketones. *Org. Lett.*, **2001**, *3*(26), 4271-4273.

[44]   Pooput, C.; Dolbier Jr., W.R.; Medebielle, M. Nucleophilic perfluoroalkylation of aldehydes, ketones, imines, disulfides and diselenides. *J. Org. Chem.*, **2006**, *71*(9), 3564-3568.

[45]   Takechi, N.; Aıt-Mohand, S.; Medebielle, M.; Dolbier Jr., W.R. Nucleophilic trifluoromethylation of acyl chlorides using the trifluoromethyl iodide/TDAE reagent. *Tetrahedron Lett.*, **2002**, *43*(24), 4317-4319.

[46]   Prakash, G.K.S.; Krishnamurti, R.; Olah, G.A. Synthetic methods and reactions. Fluoride-induced trifluoromethylation of carbonyl compounds with trifluoromethyltrimethylsilane (TMS-CF$_3$). A trifluoromethide equivalent. *J. Am. Chem. Soc.*, **1989**, *111*(1), 393-395.

[47]   Ruppert, I.; Schlich, K.; Volbach, W. Fluorinated organometallic compounds. 18. First trifluoromethyl-substituted organyl(chloro)silanes. *Tetrahedron Lett.*, **1984**, *25*(21), 2195-2198.

[48]   Beckers, H.; Bürger, H.; Bursch, P.; Ruppert, I. Synthesis and properties of (trifluoromethyl)trichlorosilane, a versatile precursor for (trifluoromethyl)silyl compounds. *J. Organomet. Chem.*, **1986**, *316*(1-2), 41-50.

[49]   Prakash, G.K.S.; Hu, J.; Olah, G.A. Preparation of tri- and difluoromethylsilanes *via* an unusual magnesium metal-mediated reductive tri- and difluoromethylation of chlorosilanes using tri- and difluoromethyl sulfides, sulfoxides and sulfones. *J. Org. Chem.*, **2003**, *68*(11), 4457-4463.

[50]   Prakash, G.K.S.; Yudin, A.K. Perfluoroalkylation with organosilicon reagents. *Chem. Rev.*, **1997**, *97*(3), 757-786.

[51]   Bastos, R.S. (Trifluoromethyl)trimethylsilane (TMSCF$_3$) - Ruppert's reagent: an excellent trifluoromethylation agent. *Synlett*, **2008**, *9*, 1425-1426.

[52]   Singh, R.P.; Shreeve, J.M. Nucleophilic trifluoromethylation reactions of organic compounds with (trifluoromethyl)trimethylsilane. *Tetrahedron*, **2000**, *56*(39), 7613-7632.

[53]   Prakash, G.K.S.; Mandal, M. Nucleophilic trifluoromethylation tamed. *J. Fluorine Chem.*, **2001**, *112*(1), 123-131.

[54]   Vuong, T.M.H. (Trifluoromethyl)trimethylsilane. *Synlett*, **2012**, *23*, 1409-1410.

[55]   Funabiki, K.; Shibata, A.; Iwata, H.; Hatano, K.; Kubota, Y.; Komura, K.; Ebihara, M.; Matsui, M. Asymmetric synthesis of (αR)-polyfluoroalkylated prolinols based on the perfluoroalkyl-induced highly stereoselective reduction of perfluoroalkyl *N*-Boc-pyrrolidyl ketones *J. Org. Chem.*, **2008**, *73*(12), 4694-4697.

[56]   Singh, R.P.; Leitch, J.M.; Twamley, B.; Shreeve, J.M. Diketo compounds with (trifluoromethyl)trimethylsilane: double nucleophilic trifluoromethylation reactions. *J. Org. Chem.*, **2001**, *66*(4), 1436-1440.

[57]   Boechat, N.; Bastos, M.M.; Mayer, L.M.U.; Figueira, E.C.S.; Soares, M.; Kover, W.B. Synthesis of new 3-(trifluoromethyl)-1H-indoles by reduction of trifluoromethyloxoindoles. *J. Heterocycl. Chem.*, **2008**, *45*(4), 969-973.

[58]   Singh, P.R.; Cao, G.; Kirchmeier, R.L.; Shreeve, J.M. Cesium fluoride catalyzed trifluoromethylation of esters, aldehydes and ketones with (trifluoromethyl)trimethylsilane. *J. Org. Chem.*, **1999**, *64*(8), 2873-2876.

[59] Shibata, M.S.; Mizuta, S.; Hibino, M.; Nagano, S.; Nakamura, S.; Toru, T. Ammonium bromides/KF catalyzed trifluoromethylation of carbonyl compounds with (trifluoromethyl) trimethylsilane and its application in the enantioselective trifluoromethylation reaction. *Tetrahedron*, **2007**, *63*(35), 8521-8528.

[60] Song, J.J.; Tan, Z.; Reeves, J.T.; Gallou, F.; Yee, N.K.; Senanayake, C.H. N-heterocyclic carbene catalyzed trifluoromethylation of carbonyl compounds. *Org. Lett.*, **2005**, *7*(11), 2193-2196.

[61] Prakash, G.K.S.; Mandal, M.; Panja, C.; Mathew, T.; Olah, J.A. Preparation of TMS protected trifluoromethylated alcohols using trimethylamine N-oxide and trifluoromethyltrimethylsilane (TMSCF₃). *J. Fluorine Chem.*, **2003**, *123*(1), 61-63.

[62] Matsukawa, S.; Saijo, M. TTMPP-catalyzed trifluoromethylation of carbonyl compounds and imines with trifluoromethylsilane. *Tetrahedron Lett.*, **2008**, *49*(31), 4655-4657.

[63] Mukaiyama, T.; Kawano, Y.; Fujisawa, H. Lithium acetate-catalyzed trifluoromethylation of carbonyl compounds with (trifluoromethyl)trimethylsilane. *Chem. Lett.*, **2005**, *34*(1), 88-89.

[64] Mizuta, S.; Shibata, N.; Sato, T.; Fujimoto, H.; Nakamura, S.; Toru, T. Tri-*tert*-butylphosphine is an efficient promoter for the trifluoromethylation reaction of aldehydes, ketones, imides and imines. *Synlett*, **2006**, *2*, 267-270.

[65] Prakash, G.K.S.; Panja, C.; Vaghoo, H.; Surampudi, V.; Kultyshev, R.; Mandal, M.; Rasul, G.; Mathew, T.; Olah, G.A. Facile synthesis of TMS-protected trifluoromethylated alcohols using trifluoromethylsilane (TMSCF₃) and various nucleophilic catalysts in DMF. *J. Org. Chem.*, **2006**, *71*(18), 6806-6813.

[66] Kawano, Y.; Kaneko, N.; Mukaiyama, T. Lewis base-catalyzed perfluoroalkylation of carbonyl compounds and imines with (perfluoroalkyl)trimethylsilane. *Bull. Chem. Soc. Jpn.*, **2006**, *79*(7), 1133-1145.

[67] Iwanami, K.; Oriyama, T. A new and efficient method for the trifluoromethylation of carbonyl compounds with trifluoromethylsilane in DMSO. *Synlett*, **2006**, *1*, 112-114.

[68] Hagiwara, T.; Mochizuki, H.; Fuchikami, T. Unique reactivity of aminoketones in the trifluoromethylation with trialkyl(trifluoromethyl)silanes. *Synlett*, **1997**, *5*, 587-588.

[69] Gawronski J.; Wascinska N.; Gajewy J. Recent progress in Lewis base activation and control of stereoselectivity in the addition of trifluorosilyl nucleophiles. *Chem. Rev.*, **2008**, *108*(12), 5227-5252.

[70] Mizuta, S., Shibata, S.; Akiti, S.; Fugimoto, H.; Nakamura, S.; Toru, T. Cinchona alkaloid/TMAF combination-catalysed nucleophilic enantioselective. *Org. Lett.*, **2007**, *9*(18), 3707-3710.

[71] Shibata, N.; Mizuta, S.; Kawai, H. Recent advances in enantioselective trifluoromethylation reactions. *Tetrahedron: Asymmetry*, **2008**, *19*(23), 2633-2644.

[72] Nagao, H.; Kawano Y.; Mukaiyama, T. Enantiosselective trifluoromethylation of ketones with (trifluoromethyl) trimethylsilane catalazed by chiral quaternary ammonium phenoxides. *Bull. Chem. Soc. Jpn.*, **2007**, *80*(12), 2406-2412.

[73] Massicot, F.; Monnier-Benoit, N.; Deka, N.; Plantier-Royon, R.; Portella, C. Synthesis of enantiopure trifluoromethyl building blocks *via* a highly chemo and diasteroselective nucleophilic trifluoromethylation of tartaric acid-derived diketones. *J. Org. Chem.*, **2007**, *72*(4), 1174-1180.

[74] Zhao, H.; Qin, B.; Liu, X.; Feng, X. Enantioselective trifluoromethylation of aromatic aldehydes catalyzed by combinatorial catalysts. *Tetrahedron*, **2007**, *63*(29), 6822-6826.

[75] Nonnenmacher, J.; Massicot, F.; Grellepois, F.; Portella, C. Enantiopure quaternary α-trifluoromethyl-α-alkoxyaldehydes from tartaric acid derived ketoamides. *J. Org. Chem.*, **2008**, *73*(20), 7990-7995.

[76] Kobayashi, K.; Narumi, T.; Oishi, S.; Ohno, H.; Fujii, N. Amino acid-based synthesis of trifluoromethylalkene dipeptide isosteres by alcohol-assisted nucleophilic trifluoromethylation and organozinc-copper-mediated SN$_2$' alkylation. *J. Org. Chem.*, **2009**, *74*(12), 4626-4629.

[77] Kawai, H.; Kusuda, A.; Mizuta, S.; Nakamura, S.; Funahashi, Y.; Masuda, H.; Shibata, N. Synthesis of novel C2-symmetric chiral crown ethers and their application to enantioselective trifluoromethylation of aldehydes and ketones. *J. Fluorine Chem.*, **2009**, *130*(8), 762-765.

[78] Blay, G.; Fernández, I.; Monleón, A.; Pedro, J.R.; Vila, C. Enantioselective zirconium-catalyzed Friedel-Crafts alkylation of pyrrole with trifluoromethyl ketones. *Org. Lett.,* **2009**, *11*(2), 441-444.

[79] Caron, S.; Do, N.M.; Sieser, J.E.; Arpin, P.; Vazquez, E. Process research and development of an NK-1 receptor antagonist. Enantioselective trifluoromethyl addition to a ketone in the preparation of a chiral isochroman. *Org. Process Res. Dev.*, **2007**, *11*(6), 1015-1024.

[80] Pierce, M.E.; Parsons Jr., R.L.; Radesca, L.A.; Lo, Y.S.; Silverman, S.; Moore, J.R.; Islam, Q.; Choudhury, A.; Fortunak, J.M.D.; Nguyen, D.; Luo, C.; Morgan, S.J.; Davis, W.P.; Confalone, P.N. Practical asymmetric synthesis of efavirenz (DMP 266), an HIV-1 reverse transcriptase inhibitor. *J. Org. Chem.*, **1998**, *63*(23), 8536-8543.

[81] Kawai, H.; Tachi, K.; Tokunaga, E.; Shiro, M.; Shibata, N. Cinchona alkaloid-catalyzed asymmetric trifluoromethylation of alkynyl ketones with trimethylsilyl trifluoromethane. *Org. Lett.*, **2010**, *12*(22), 5104-5107.

[82] Shibata, N.; Kawai, H.; Kitayama, T.; Tokunaga, E. A new synthetic approach to efavirenz through enantioselective trifluoromethylation by using the Ruppert-Prakash reagent. *Eur. J. Org. Chem.*, **2011**, *2011*(30), 5959-5961.

[83] Smith, W.C. Chemie des Schwefeltetrafluorids. *Angew. Chem.*, **1962**, *74*(19), 742-751.

[84] Middleton, W.J. New fluorinating reagents. Dialkylaminosulfur fluorides. *J. Org. Chem.*, **1975**, *40*(5), 574-578.

[85] a) Lal, G.S.; Pez, G.P.; Pesaresi, R. J.; Prozonic, F.M. Bis(2-methoxyethyl)aminosulfur trifluoride: a new broad-spectrum deoxofluorinating agent with enhanced thermal stability. *J. Chem. Soc. Chem. Commun.*, **1999**, *2*, 215-216. b) Lal, G.S.; Pez, G.P.; Pesaresi, R.J.; Prozonic, F.M.; Cheng, H. Bis(2-methoxyethyl)aminosulfur trifluoride: a new broad-spectrum deoxofluorinating agent with enhanced thermal stability. *J. Org. Chem.*, **1999**, *64*(19) 7048-7054.

[86] Hudlicky, M. Fluorination with diethylaminosulfur trifluoride and related aminofluorosulfuranes. *Org. React.*, **1988**, *35*, 513-641.

[87] Singh, R.P.; Shreeve, J.M. Recent advances in nucleophilic fluorination reactions of organic compounds using deoxofluor and DAST. *Synthesis*, **2002**, *17*, 2561-2578.

[88] Singh, R.P.; Chakraborty, D.; Shreeve, J.M. Nucleophilic trifluoromethylation and difluorination of substituted aromatic aldehydes with Ruppert's and Deoxofluor reagents. *J. Fluorine Chem.*, **2001**, *111*(2), 153-160.

[89] Prakash, S.G.K.; Hu, J.; Olah, G.A. Alkoxide and hydroxide-induced nucleophilic trifluoromethylation using trifluoromethyl sulfone or sulfoxide. *Org. Lett.,* **2003**, *5*(18), 3253-3256.

[90] Rozen, S.; Mishani, E. A novel method for converting aromatic acids into trifluoromethyl derivatives using BrF$_3$. *J. Chem. Soc. Chem. Comm.,* **1994**, *18*, 2081.

[91] Rozen, S. Attaching the fluorine atom to organic molecules using BrF$_3$ and other reagents directly derived from F$_2$. *Acc. Chem. Res.*, **2005**, *389*(10), 803-812.

[92] Ozturk, T.B.; Ertas, E.; Mert, O. Use of Lawesson's reagent in organic synthesis. *Chem. Rev.*, **2007**, *107*(11), 5210-5278.

[93] Cohen, O.; Mishani, E.; Rozen, S. From carboxylic acids to the trifluoromethyl group using BrF$_3$. *Tetrahedron*, **2010**, *66*(20), 3579-3582.

[94]   Molander, G.A.; Hoag, B.P. Improved synthesis of potassium (trifluoromethyl)trifluoroborate [K(CF$_3$BF$_3$)]. *Organometallics*, **2003**, *22*(16), 3313-3315.

[95]   Levin, V.V.; Dilman, A.D.; Belyakov, P.A.; Struchkova, M.I.; Tartakovsky, V.A. Nucleophilic trifluoromethylation with organoboron reagents. *Tetrahedron Lett.*, **2011**, *52*(2), 281-284.

[96]   Paulin, D.S.; Tomlinson, A.J.; Cavell, R.G. Aminolysis of trifluoromethylchlorophosphoranes. Preparation and characterization of the trifluoromethyltris(dimethylamino)phosphonium ion CF$_3$P[N(CH$_3$)$_2$]$_3^+$. *Inorg. Chem.,* **1977**, *16*(1), 24-27.

[97]   Volbach, W.; Ruppert, I. CF$_3$-P(NEt$_2$)$_2$ als schlüsselsubstanz für CF$_3$-substituierte phosphane - direktsynthese und verwendung. *Tetrahedron Lett.,* **1983**, *24*(49), 5509-5512.

[98]   Cherkupally, P.; Beier, P. Alkoxide-induced nucleophilic trifluoromethylation using diethyl trifluoromethylphosphonate. *Tetrahedron Lett.*, **2010**, *51*(2), 252-255.

[99]   Sibille, S.; Mcharek, S.; Perichon, J. Electrochemical trifluoromethylation of carbonyl compounds. *Tetrahedron*, **1989**, *45*(2), 1423-1428.

[100]   Shono, T.; Kashimura, S.; Ishifune, M.; Okada, T. Electroorganic chemistry. A novel trifluoromethylation of aldehydes and ketones promoted by an electrogenerated base. *J. Org. Chem.*, **1991**, *56*(1), 2-4.

[101]   Shono, T.; Kashimura, S.; Nogusa, H. A novel electrogenerated base. Alkylation of methyl arylacetates at the alpha-methylene group. *J. Org. Chem.*, **1984**, *49*(11), 2043-2045.

[102]   Shono, T.; Kashimura, S.; Ishige, O. Electroorganic chemistry. A novel base useful for synthesis of esters and macrolides. *J. Org. Chem.*, **1986**, *51*(4), 546-549.

[103]   Large, S.; Roques, N.; Langlois, B.R. Nucleophilic trifluoromethylation of carbonyl compounds and disulfides with trifluoromethane and silicon-containing bases. *J. Org. Chem.*, **2000**, *65*(26), 8848-8856.

[104]   Billard, T.; Bruns, S.; Langlois, B.R. New stable reagents for the nucleophilic trifluoromethylation. Trifluoromethylation of carbonyl compounds with *N*-formylmorpholine derivatives. *Org. Lett.*, **2000**, *2*(14), 2101-2103.

[105]   Billard, T.; Langlois, B.R.; Blond, G. New stable reagents for nucleophilic trifluoromethylation. Part 2: Trifluoromethylation with silylated hemiaminals of trifluoroacetaldehyde. *Tetrahedron Lett.*, **2000**, *41*(45), 8777-8780.

[106]   Billard, T.; Langlois, B. R.; Blond, G. Trifluoromethylation of nonenolizable carbonyl compounds with a stable piperazino hemiaminal of trifluoroacetaldehyde. *Eur. J. Org. Chem.*, **2001**, *8*, 1467-1471.

[107]   Matsui, K.; Tobita, E.; Ando, M.; Kondo, K. A convenient trifluoromethylation of aromatic halides with sodium trifluoroacetate. *Chem. Lett.*, **1981**, *10*(12), 1719-1720.

[108]   Chang, Y.; Cai, C. Trifluoromethylation of carbonyl compounds with sodium trifluoroacetate. *Chin. Chem. Lett.*, **2005**, *16*(10), 1313-1316.

[109]   Chang, Y.; Cai, C. Trifluoromethylation of carbonyl compounds with sodium trifluoroacetate. *J. Fluorine Chem.*, **2005**, *126*(6), 937-940.

[110]   Motherwell, W.B.; Storey, L.J. Some studies on nucleophilic trifluoromethylation using the shelf-stable trifluoromethylacetophenone-N,N-dimethyltrimethylsilylamine adduct. *J. Fluorine Chem.*, **2005**, *126*(4), 491-498.

Send Orders of Reprints at bspsaif@emirates.net.ae

CHAPTER 7

Synthetic Chemistry with *N*-Acyliminium Ions derived from Piperazine-2,5-diones and Related Compounds

Carmen Avendaño and Elena de la Cuesta[*]

Departamento de Química Orgánica y Farmacéutica, Facultad de Farmacia, Universidad Complutense, 28040 Madrid, Spain

Abstract: The ubiquity of the piperazine-2,5-dione core (2,5-diketopiperazine, DKP) in natural products and the preponderance of this heterocyclic framework in many bioactive compounds have encouraged the development of methods for the selective functionalization of readily available piperazine-2,5-dione substrates. C-functionalization using DKPs as electrophilic glycine templates is generally mediated by *N*-acyliminium ions. These intermediates show highly versatile reactivities, which are reflected in an impressive number of synthetic applications. However, even in some highly comprehensive treatments of *N*-acyliminium ions, references to the construction of these species on piperazine-2,5-dione frameworks are scarce. The present review aims at filling this gap, placing emphasis upon the construction of endocyclic and exocyclic acyliminium ions derived from piperazine-2,5-diones and their synthetic applications. More complex structures that include this framework as a structural fragment, such as pyrazino[1,2-*b*]isoquinoline-1,4-diones or pyrazino[2,1-*b*]quinazoline-3,6-diones, are also overviewed.

Keywords: Piperazine-2,5-dione, 2,5-diketopiperazine, DKP, *N*-acyliminium ions, privileged structures, dipeptide dimer, electrophilic glycine templates, protein β-turn mimics, natural products, pyrazino[2,1-*b*]quinazoline-3,6-diones, 1,5-imino-3-benzazocins, *N*-acetylardeemin, bicyclomycin, dragmacidin B, phthalascidin analogs, antitumor tetrahydroisoquinoline alkaloids, trabectedin, saframycins A and B, quinocarcin, *meta*- and *para*-2,6- diazacyclophanes, fumiquinazoline C, alantrypinone.

1. INTRODUCTION

The use of heterocyclic frameworks as privileged structures in many drugs has widely been employed in lead discovery programs [1]. One of these privileged structures is the piperazine-2,5-dione core (2,5-diketopiperazine, DKP), the smallest head-to-tail dipeptide dimer [2-5]. 2,5-Diketopiperazines can be regarded as protein β-turn mimics that contain constrained amino acids embedded within

---

[*]**Address correspondence to Elena de la Cuesta:** Departamento de Química Orgánica y Farmacéutica, Facultad de Farmacia, Universidad Complutense, 28040 Madrid, Spain; Tel: 00-34-913941823; Fax: 00-34-913941822; E-mail: ecuestae@farm.ucm.es

their structures and lack most of the bioavailability issues associated to peptides, due to poor absorption and high metabolic rates [6]. Due to their chirality, conformational rigidity and densely functionalized structures, many natural or synthetic DKPs bind with high affinity to a large variety of receptors, showing a broad range of biological activities [6-9]. They have also been used as starting materials for the synthesis of a large number of natural products [10-13] and as templates for amino acid synthesis [14]. DKPs have also found broad acceptance in combinatorial chemistry because they are simple heterocyclic scaffolds that can be prepared from readily available α-amino acids using very robust chemistry and which allow the introduction of structural diversity at up to four positions with good stereocontrol [15, 16]. They are also involved in self-assembly processes because their bis-*cis*-amide functions contain hydrogen-bond acceptor and donor sites that allow the formation of a variety of supramolecular structures [17].

The ubiquity of this heterocyclic motif in several biologically active natural products has encouraged the development of many methods for its synthesis. The most common strategy takes advantage of the large pool of natural and unnatural enantiopure α-amino acids available as starting materials. Thus, the simplest route to DKPs involves the intramolecular cyclization, which may be microwave-assisted [18, 19], of a suitable linear dipeptide by formation of an amide bond between the terminal amino and carboxylic groups following N-deprotection. The ease with which this process occurs is responsible for the frequent generation of DKPs as unwanted by-products in the synthesis of oligopeptides [16]. Alternative intramolecular cyclizations involving the formation of other bonds, although less represented in the literature, are also possible, and some of them have been adapted to solid-phase synthesis [20, 21]. The selective functionalization of readily available piperazine-2,5-diones is also possible [22, 23]. In these approaches, these precursors are used as nucleophilic, radical, or electrophilic glycine templates, enabling the assembly of the required carbon and heteroatom framework without the need for developing individual strategies for the synthesis of uncommon amino acids.

C-Functionalization protocols using DKPs as nucleophilic glycine templates (see for instance references [24-27]) is limited in scope due to regioselectivity problems with unsymmetrical compounds, and also to the fact that the strong bases required to generate enolate ions may affect the integrity of the enantiomeric carbon centres. Other methods for *C*-functionalization use radical chemistry, which relies on the availability of suitable precursors and the sensitivity of radicals to polar and steric effects [28]. Pericyclic reactions on DKP substrates, including Diels-Alder [29], hetero Diels-Alder [30, 31] and 1,3-dipolar cycloadditions, are especially suitable for the synthesis of DKP-derived heterocyclic systems [32, 33].

2. *C*-FUNCTIONALIZATION USING DKPS AS ELECTROPHILIC GLYCINE TEMPLATES: *N*-ACYLIMINIUM ION CHEMISTRY

C-Functionalization reactions using DKPs as electrophilic glycine templates are generally mediated by *N*-acyliminium ions. These species have an enhanced cationic character, and are considerably more reactive than iminium species **1** due to the electron-attracting group on the nitrogen atom. *N*-Acyl and carbamate derivatives **2** and **3**, which will be considered along with this overview as *N*-acyliminium ions, have more widely been exploited than carbamoyl and tosyl derivatives **4** and **5**, while hydrazonium cations **6** have only summarily explored (Fig. **1**).

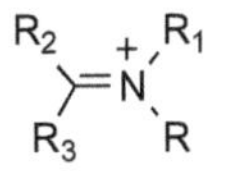

$$R_2 \quad + \, R_1$$
$$\searrow = N$$
$$R_3 \qquad R$$

1, R = H, alkyl   **4**, R = CONR$_2$
2, R = acyl   **5**, R = Ts
3, R = CO$_2$R   **6**, R = NR$_2$

Figure 1: Examples of different iminium ions.

N-Acyliminium ions show highly versatile reactivities reflected in an impressive number of synthetic applications, which have been comprehensively reviewed [34-42]. Both inter- and intramolecular carbon-carbon bond-forming reactions, including cyclizations onto olefins and arenes as well as reactions with heteroatom nucleophiles, are in continuous expansion [43-57]. However, even in some very comprehensive reviews dealing with *N*-acyliminium ions [38], references to the construction of these species on piperazine-2,5-dione frameworks are scarce. The present review aims at filling this gap, placing emphasis upon the construction of endocyclic and exocyclic acyliminium ions derived from piperazine-2,5-diones and their synthetic applications. More complex structures that include this framework as a structural fragment, such as pyrazino[1,2-*b*]isoquinoline-1,4-diones or pyrazino[2,1-*b*]quinazoline-3,6-diones, are also overviewed (Fig. **2**).

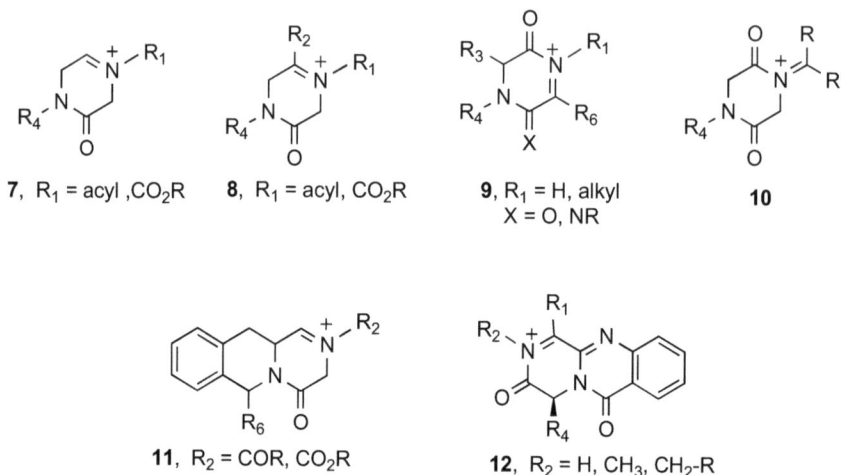

7, R$_1$ = acyl ,CO$_2$R   **8**, R$_1$ = acyl, CO$_2$R   **9**, R$_1$ = H, alkyl   **10**
                                                            X = O, NR

11, R$_2$ = COR, CO$_2$R   **12**, R$_2$ = H, CH$_3$, CH$_2$-R

Figure 2: Examples of acyliminium ions derived from piperazine-2,5-diones and other compounds that include this structural fragment.

3. METHODS FOR THE GENERATION OF PRECURSORS TO ENDOCYCLIC *N*-ACYLIMINIUM IONS OF TYPES 7-9 APPLICATIONS

3.1. Hydride Addition to One Carbonyl Group of a DKP

The selective reduction of one carbonyl group in a DKP system is simply achieved through the previous activation of the selected lactam function by *N*-acylation or *N*-alkoxycarbonylation to give the corresponding imides or carbamates **13**. Hydride addition to these derivatives affords α-hydroxylactams **14** [58]. Carbamates are more suitable than imides for this purpose, because in *N*-acyl derivatives (imides) the hydride transfer to the exocyclic carbonyl group competes with the desired addition to the more hindered endocyclic carbonyl, a problem that does not exist in the *N*-alkoxycarbonyl derivatives (carbamates) due to the lower reactivity of the exocyclic carbonyl. The most common reducing agent is NaBH$_4$ in methanol [59], and the stereochemistry of the α-oxylactams is usually not relevant, since a mixture of stereoisomers can be used in the subsequent generation of the corresponding *N*-acyliminium species **7** under acid catalysis

(Scheme **1**). Several applications of these strategies will be discussed in Sections 5 and 6.

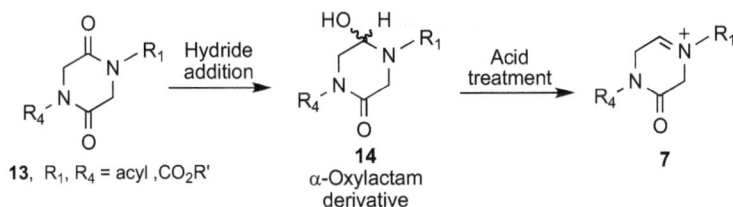

Scheme 1: Generation of endocyclic acyliminium ions by hydride addition to activated DKPs followed by acid treatment.

3.2. Addition of Organometallics to one Carbonyl Group of a DKP

Addition of organolithium or Grignard reagents to *N*-alcoxycarbonyl-DKPs leads to tertiary hydroxylactam derivatives **15**, which are less stable than the above mentioned secondary reduction products **14** [60]. These hemiaminals, that may be obtained by alternative procedures (see Section 5), often exist as mixtures of ring-closed hemiaminal and linear compounds **15** and **16** and usually give *N*-acyliminium species of type **8** under acid catalysis (Scheme **2**) [61-63]. Some synthetic applications of this method are shown in Section 8.

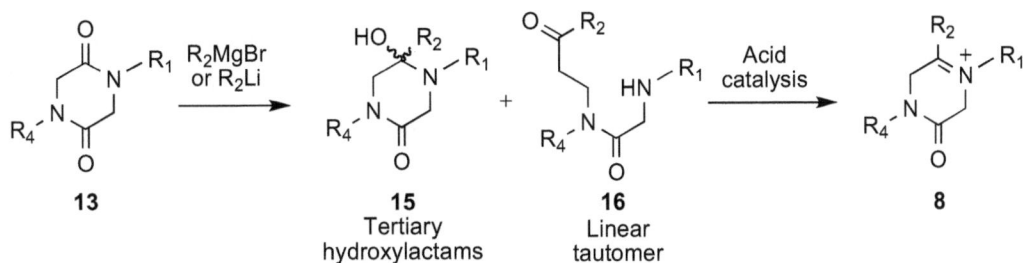

Scheme 2: Generation of endocyclic acyliminium ions by addition of organometallics to activated DKPs followed by acid catalysis.

3.3. Chemical or Electrochemical Oxidation at the α-CH of a DKP or an Imine Derivative

Elimination of hydride from a carbon atom α to a DKP nitrogen allows the generation of *N*-acyliminium species of type **9**. Thus, the selective oxidation of compounds **17** may be accomplished electrochemically, in a process that begins with the elimination of one electron from the nitrogen lone pair [64]. When the reaction is carried out in the presence of nucleophilic solvents, like alcohols or acids, these species are trapped to give alkoxy or acyloxy derivatives (**18**, R_6 = alkyl or acyl) [65-68] that behave as electrophilic glycine templates through the generation of species of type **9** under acid catalysis. Alternatively, chemical reagents such as hypervalent iodine compounds can be employed to oxidize these positions giving N-acyliminium precursors of type **18** (Scheme **3**) [69, 70]. Some synthetic applications of this method are shown in Section 8.

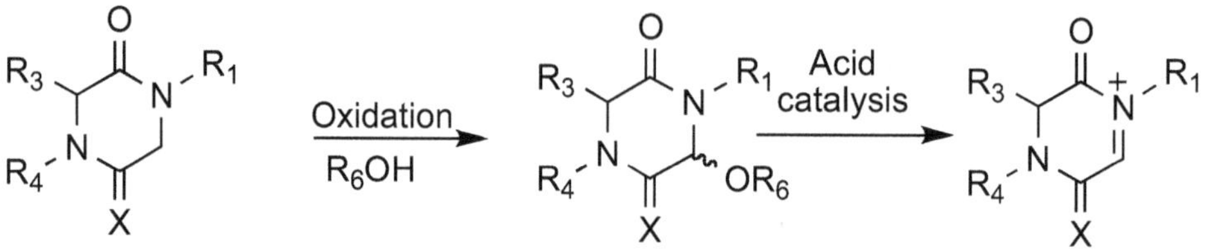

17, R_1 = H, alkyl, R_4 = acyl,
      X = O, NR

18

9

Scheme 3: Generation of endocyclic acyliminium ions by oxidation at a carbon atom α to a DKP nitrogen followed by acid catalysis.

3.4. Radical Oxidation/Decarboxylation and Oxidation/Deformylation of Carboxy- and Hydroxymethyl-DKPs

A mild and efficient method for the oxidative generation and trapping of N-acyliminium species uses as starting materials carboxylic derivatives of DKPs such as **19**, and represents an extension of chemistry previously developed on α-amino acid derivatives [71-73]. Treatment of these compounds with iodosylbenzene or (diacetoxyiodo)benzene (DIB) and iodine generates the corresponding carboxyl radicals **20**, which evolve to carbon radicals by loss of carbon dioxide and, through a further oxidation by loss of another electron, give N-acyliminium ions **21** that can be trapped by nucleophilic solvents affording derivatives **22** and **23** (Scheme **4**).

In general, the alkoxy or acyloxy derivatives thus obtained (see compounds **24** and **25** in Scheme **5** are precursors of acyliminium ions **26** that are subsequently trapped by oxygen, nitrogen and carbon nucleophiles to give compounds such as **27** and **28** [74] (Scheme **5**).

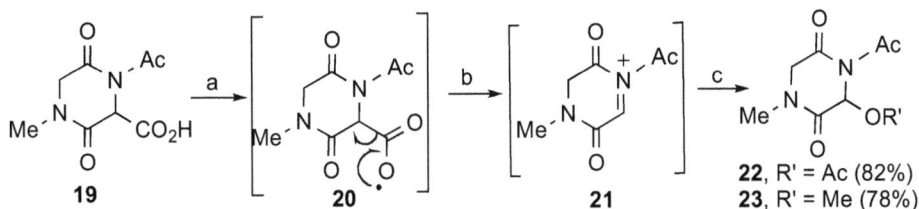

Reagents and Conditions: (a) DIB (0.5 eq), I$_2$ (0.5 eq), CH$_2$Cl$_2$, r.t.; (b) DIB (0.5 eq); (c) R'OH.

Scheme 4: Synthesis of acyliminium precursors by radical oxidation/decarboxylation of carboxy-DKPs.

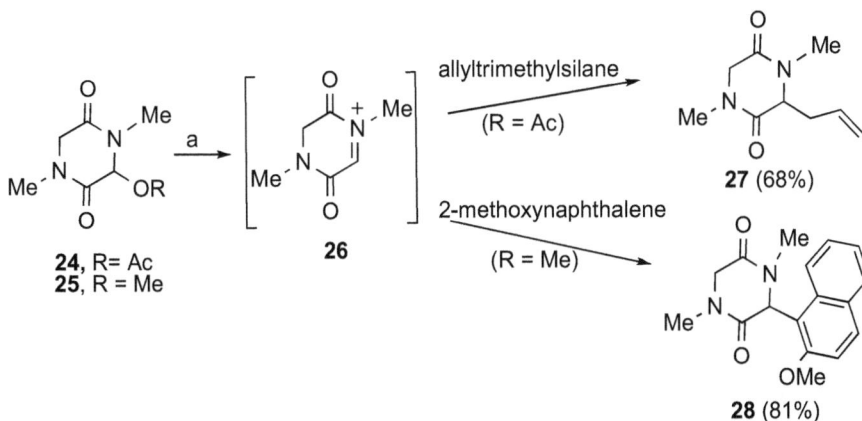

Reagents and Conditions: (a) BF$_3$.OEt$_2$, CH$_2$Cl$_2$, 0 °C.

Scheme 5: Trapping by carbon nucleophiles of acyliminium ions generated from alkoxy or acyloxy DKP derivatives.

Alkoxy derivatives can also be prepared from the corresponding enolates. Thus, the oxidation of the enolate prepared from the valine-derived DKP **29** with molecular oxygen provides a 2:1 *trans* to *cis* mixture of hydroxylated 2,5-diketopiperazines **30** and **31**, a result that is also obtained by reaction of the bis-*O*-trimethylsilyl lactim ether of **29** with (diacetoxy)iodobenzene Treatment of a 1:1 mixture of *trans*- and *cis*-acetates **32** and **33** with allyltrimethylsilane and BF$_3$.OEt$_2$ in CH$_2$Cl$_2$ afforded a clean mixture of *trans*-(3S,6R)-**35** and *cis*-(3S,6S)-**36** *via* a *N*-acylimiun ion intermediate **34** [14] (Scheme **6**).

Alternatively, hydroxymethylpiperazine-2,5-diones, such as **37**, can act as precursors to radical intermediates, such as **38**, that evolve to *N*-acyliminium ions (see **39**) through a sequence of radical deformylation and oxidation steps [75] (Scheme **7**).

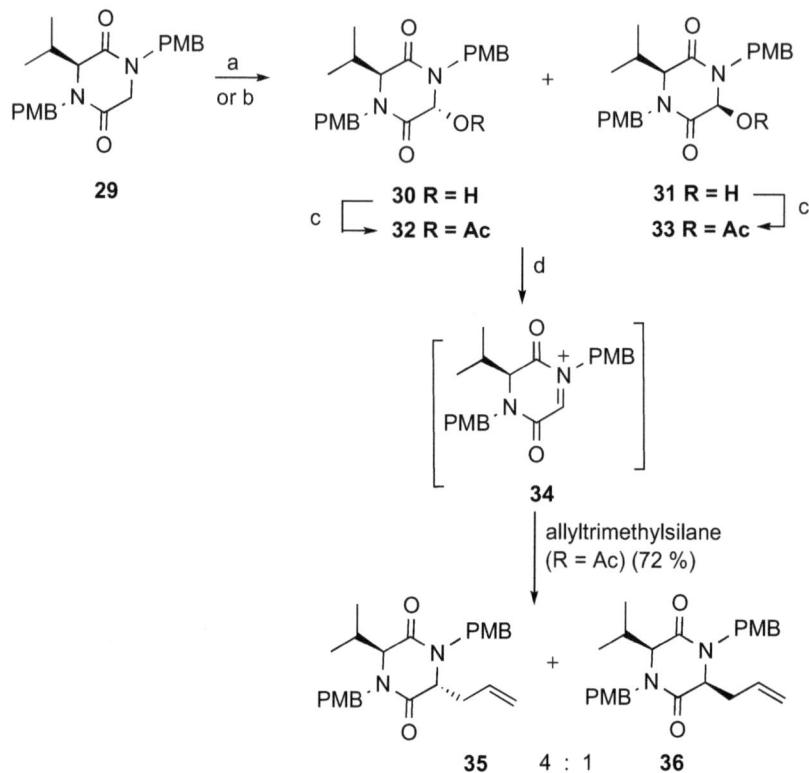

Reagents and Conditions: (a) LiHMDS, O$_2$, THF,-78 °C; (b) 1) TMSCl, THF, -78 °C, 2) PhI(OAC)$_2$, THF, -78 °C, 3) H$_2$O, rt; (c) Ac$_2$O, Py, DMAP; (d) BF$_3$.OEt$_2$, CH$_2$Cl$_2$, -78 °C.

Scheme 6: Generation of acyliminium ions by oxidation of DKP derived enolates and subsequent trapping with *C*-nucleophiles.

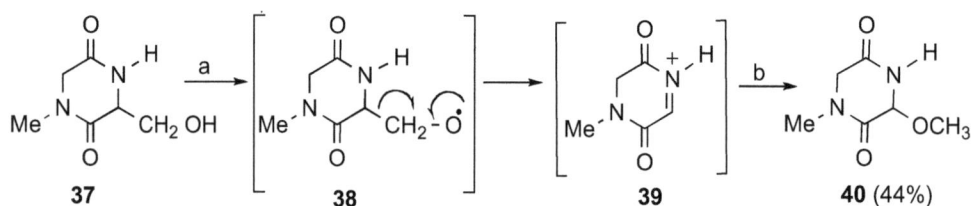

Reagents and Conditions: (a) DIB, I$_2$ (0.5 eq), CH$_2$Cl$_2$, r.t.; (b) CH$_3$OH, r.t..

Scheme 7: Oxidation/deformylation of hydroxymethyl derivatives of DKP to generate bis-acyliminium ions and subsequent trapping with *O*-nucleophiles.

3.5. Protonation of α-Alkylidene-DKPs

α-Alkylidene-DKPs **41**, which are often named as cyclic dehydrodipeptides, generate *N*-acyliminium species of type **9** by protonation of their enamine portion (Scheme **8**).

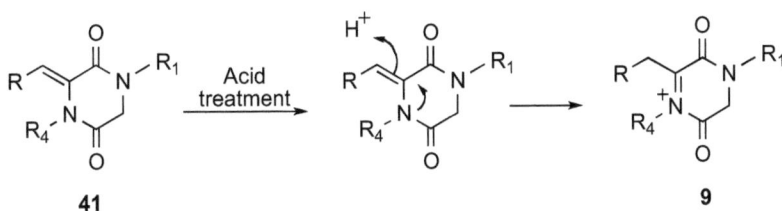

Scheme 8: Generation of acyliminium ions by protonation of α-alkylidene-DKPs.

For instance, compound **42** is an efficient precursor of acyliminium ion **43**, that forms an intramolecular C-C bond with the nucleophilic indole C(2)-position to give the fused tetracyclic derivative **44** [76] (Scheme **9**). Other applications of this approach can be found in Section 7 and also in references [51, 77].

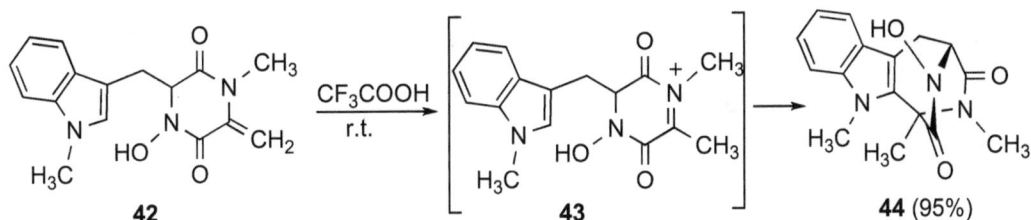

Scheme 9: Intramolecular cyclization of an acyliminium ion generated by protonation of an α-alkylidene-DKP.

3.6. Elimination of an Halide α to Nitrogen

Besides protonation, the exocyclic double bond of α-alkylidene-DKPs **45** undergoes addition of halogens (chlorine or bromine) to give compounds **46**. Subsequent treatment of these adducts with water, MeOH/NaAcO or thiols allows the easy substitution of the halogen α to nitrogen by OH, MeO, or RS groups to give compounds **48** through acyliminium intermediates **47** [78] (Scheme **10**). This chemistry was applied in several synthetic approaches to the 3,6-epidithio-2,5-piperazinedione skeleton present in gliotoxin, sporidesmin and related antibiotics [79].

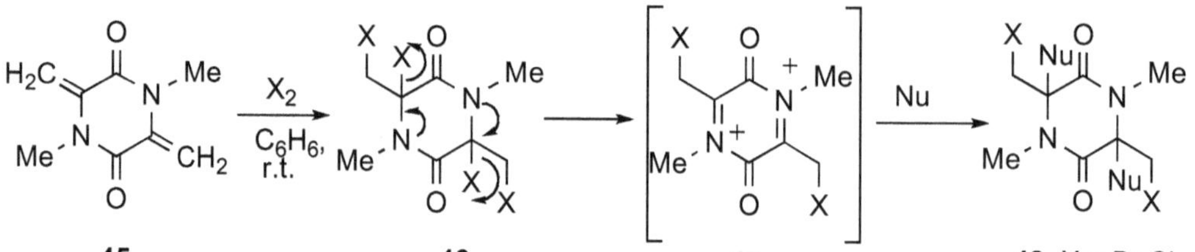

Scheme 10: Addition of halogens to a bis-α-alkylidene-DKP, elimination of the halide α to nitrogen with generation of bis-acyliminium ions, and subsequent trapping with nucleophiles.

Alternatively, bromination of glycine anhydride derivatives such as **49**, may give under radical conditions 3,6-dibromides [80] or 3-monobromides [81]. For instance, the *in situ* methanolysis of monobromide **50** afforded **52**, which was used in the synthesis of non-symmetric piperazinediones **53** [82]. These processes take place through the intermediacy of *N*-acyliminium ions **51** (Scheme 11).

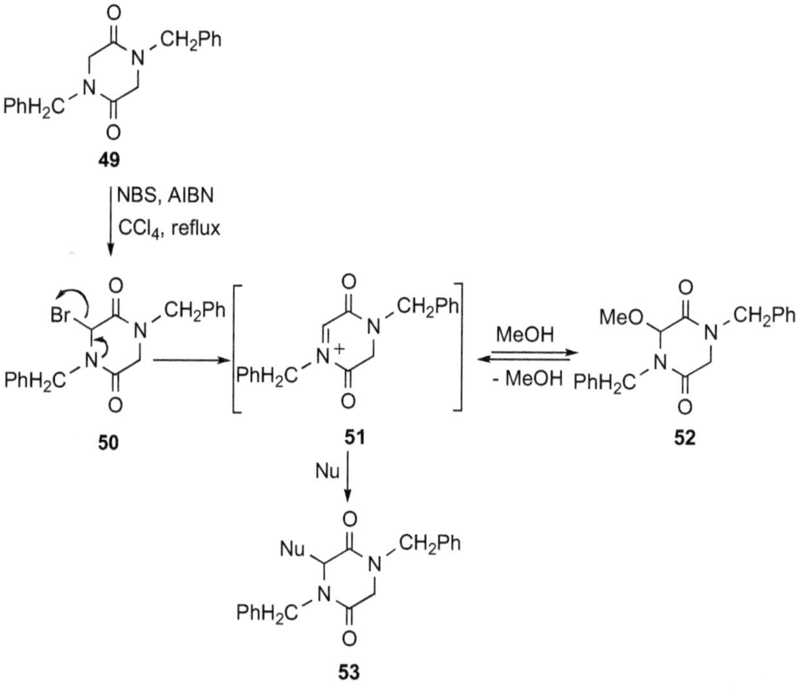

Scheme 11: Radical bromination of glycine anhydrides, halide elimination, and subsequent trapping with nucleophiles.

Besides methanol, a nucleophile frequently used to capture the *N*-acyliminium ions generated by the above mentioned bromide elimination is the sodium salt of

2-mercaptopyridine. This strategy was used by Williams in the total synthesis of (±)- and (+)-bicyclomycin, a clinically useful antibiotic produced on large scale from cultures of *Streptomyces sapporonensis*. Starting from **54**, the first adduct **55** was transformed into the corresponding *N*-acyliminium species **56**, which was trapped by butyrolactone trimethylsilyl enol ether to give **57**, in a remarkably chemoselective and stereoselective reaction that required precomplexation of **55** with silver triflate. Reduction of lactone **58** afforded diol **59**, which was cyclized to the bicyclic alcohol **60**, again in the presence of silver triflate, through the generation of the corresponding *N*-acyliminium ion **59**. Dehydration of the primary alcohol, bridgehead hydroxylation, aldol condensation of the bridgehead carbanion with (±) or (-)-2,2,4-trimethyl-1,3-dioxolane-4-carbaldehyde and oxidative removal of the *N*-*p*-methoxybenzyl protecting groups completed the synthesis of the target natural product **61** [83, 84] (Scheme **12**).

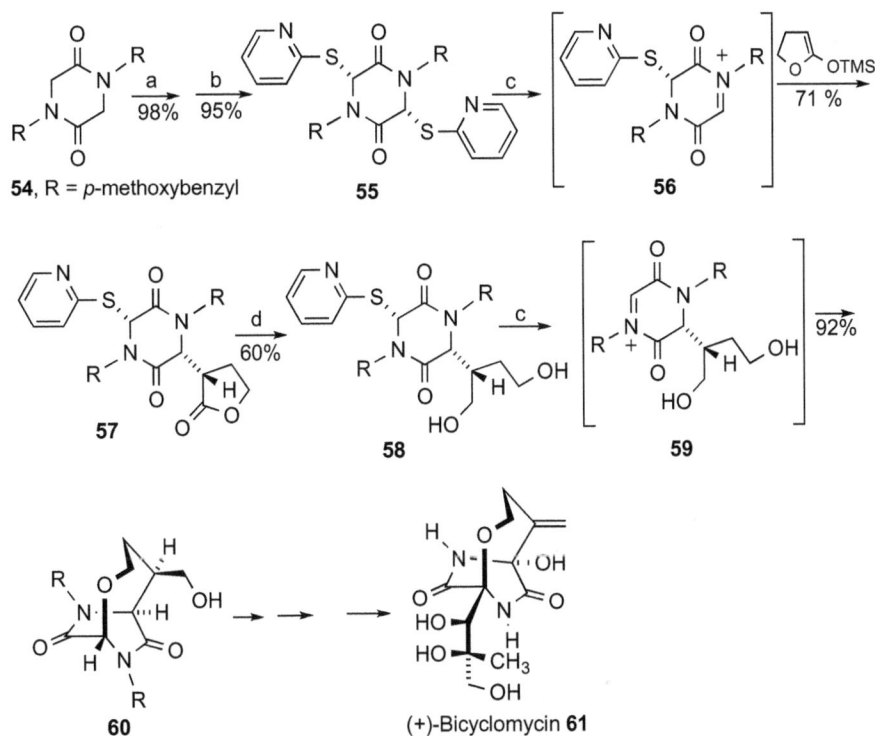

Reagents and Conditions: (a) NBS, (BzO)$_2$, CCl$_4$, reflux; (b) NaS (Py), THF, 25 °C; (c) AgOTf, THF, 25 °C; (d) LiAlH$_4$, THF, 25 °C, then Na$_2$SO$_4$. 10 H$_2$O.

Scheme 12: Synthesis of (+)-bicyclomycin through sequential nucleophilic trapping of acyliminium ions derived from a glycine anhydride precursor.

Cava reported the total synthesis of dragmacidin B **66**, a marine alkaloid belonging to a family of natural products that show anticancer, antifungal, and antiviral activities. This synthesis was accomplished through a very simple protocol starting by the 3,6-dibromination of the sarcosine anhydride **62**, followed by reaction of the crude product **63** with 6-bromoindole in DMF solution to give **65** *via* the generation of bis iminium ion **64** as an intermediate, and a final reduction with borane-THF complex to afford the piperazine ring of the natural product **66** [85] (Scheme **13**).

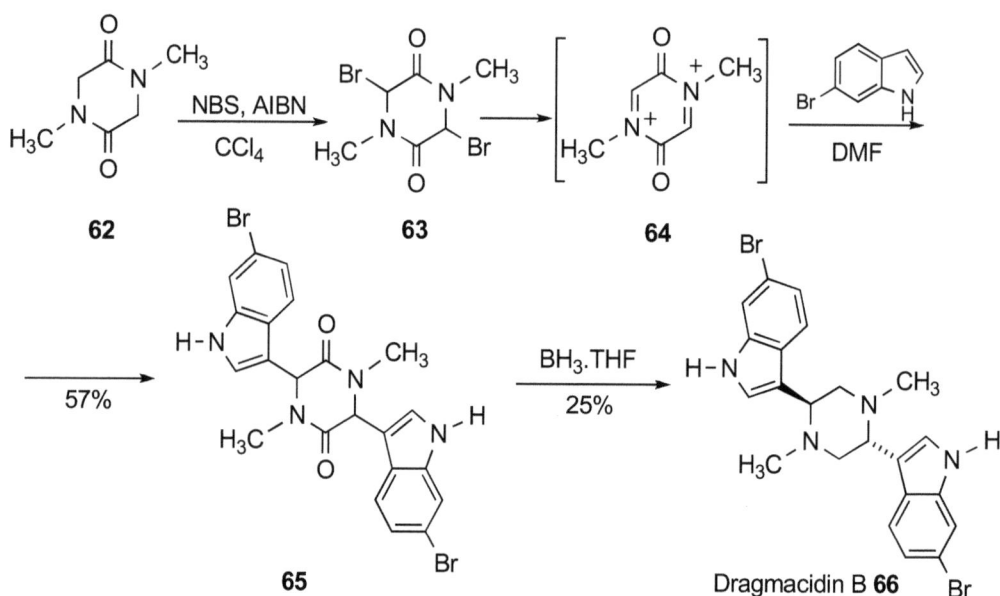

Scheme 13: Synthesis of dragmacidin B through radical bromination of sarcosine anhydride, halide elimination, and subsequent trapping with 6-bromo-indole as a *C*-nucleophile.

Bromination of unsymmetrical anhydrides formed by condensation of glycine and a second amino acid is diastereoselective, giving *trans*-3-alkyl-6-bromo isomers [86]. Furthermore, since radical bromination is favoured in carbon centers vicinal to one electron-realeasing and one electron-withdrawing group due to the captodative effect [87], *N*-acyl substituents render their C-α atoms much less reactive while *N*-alkyl groups favour the radical bromination. Accordingly, regioselectivity may be achieved in tandem bromination/nucleophilic substitution reactions with *N*-alkyl-*N'*-acetyl-2,5-piperazinediones [88]. Other synthetic applications of this method are shown in Section 8.

4. PRECURSORS OF EXOCYCLIC *N*-ACYLIMINIUM IONS OF TYPE 10 APPLICATION TO THE SYNTHESIS OF PYRAZINO[1,2-*b*]ISOQUINOLINE-1,4-DIONES

Exocyclic *N*-acyliminium ions of type **10** may be generated by condensation of one DKP-amide function with aldehydes in the presence of acids. This approach has been developed to obtain pyrazino[1,2-*b*]isoquinoline-1,4-diones, which are suitable precursors of natural antitumor alkaloids (see Section 5). Construction of the tetrahydroisoquinoline system in these compounds has been usually carried out from 3-arylmethylpiperazine-2,5-dione derivatives by reduction of their C(5)-carbonyl to a methylene to give a secondary amine, followed by a modified Pictet-Spengler cyclization (see, for instance, reference [56]). Alternatively, this amide function may react with aldehydes to give *N*-α-hydroxyalkyl derivatives that generate *N*-acyliminium species under acid conditions and cyclize through an intramolecular Friedel-Crafts reaction. This methodology is shown in Scheme **14** for the synthesis of compound **69** by condensation of **67** with acetaldehyde in the presence of trifluoroacetic and acetic acids. This cyclization is diastereoselective and proceeds mainly from the less hindered α-face of *N*-acyliminium species **68** but the yield is moderate and compound **70** was also obtained as a minor product [89].

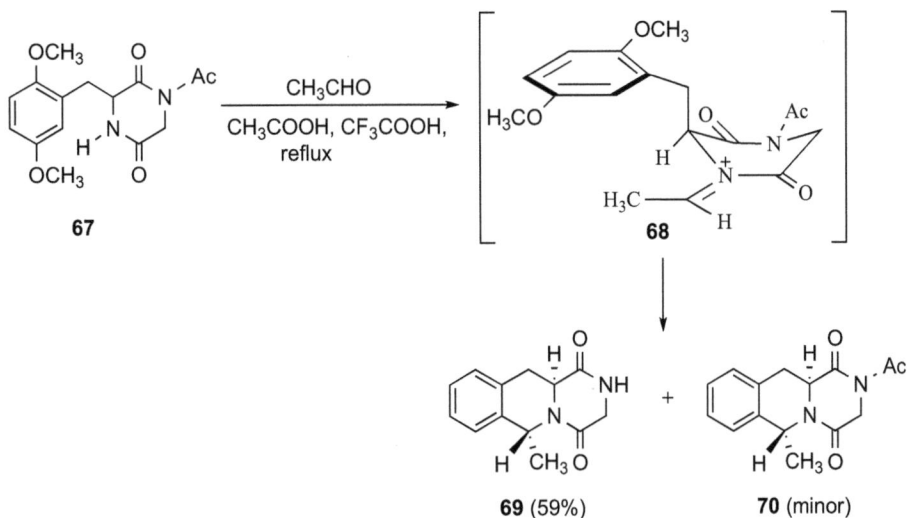

Scheme 14: Synthesis of a tetrahydroisoquinoline system through condensation of a 3-arylmethyl-piperazine-2,5-dione derivative with acetaldehyde, generation of an exocyclic acyliminium ion under acid conditions, and intramolecular Friedel-Crafts reaction.

Our group developed a significant improvement of this approach by achieving the previous activation of the N(4)-C(5)-amide as an *O*-trimethylsilyl derivative **71** [90, 91], and by using acetaldehyde dimethylacetal as electrophile. The *N*-α-alkoxyalkylated product **72** was isolated as a mixture of diastereomers that, under acid catalysis, undergo an *N*-acyliminium ion-mediated 6-*exo-trig* cyclization to give **70** in very good yields [92] (Scheme **15**).

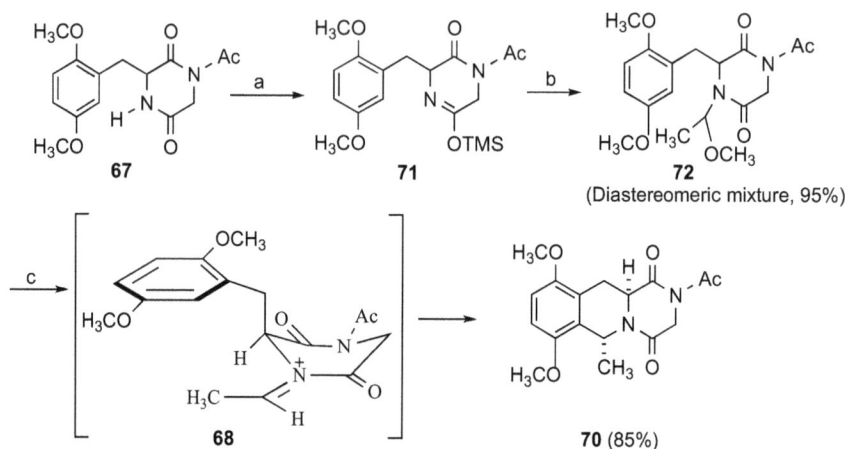

Reagents and Conditions: (a) TMSOTf, ^{i}Pr$_2$NEt, CH$_2$Cl$_2$, -78 °C; (b) CH$_3$CH(OCH$_3$)$_2$, TMSOTf (cat), -78 °C; (c) HCO$_2$H, reflux.

Scheme 15: Synthesis of a tetrahydroisoquinoline system through condensation of an *O*-trimethylsilyl DKP derivative and acetaldehyde dimethylacetal under acid catalysis.

The *trans*-relationship between the H-6 and H-11a protons in compound **70** is explained taking in consideration that **68** must adopt an *E*-configuration in order to minimize the steric interactions between the methyl side chain and the C(5)=O group of the piperazinedione ring. This stereochemistry may be generalized for most acetals in which *Z*-intermediates **73** are more unstable that their *E*-isomers **74** (Fig. **3**).

Although the *trans* configuration is opposite to the one found in natural products, it can be reverted to give the *cis*-isomers by radical bromination followed by catalytic hydrogenation [93]. This epimerization implies an initial chemoselective bromination at the captodative C(11a)-position, followed by spontaneous dehydrohalogenation (probably through an *N*-acyliminium species) and diastereoselective hydrogen addition to the face opposite to the C(6)-substituent.

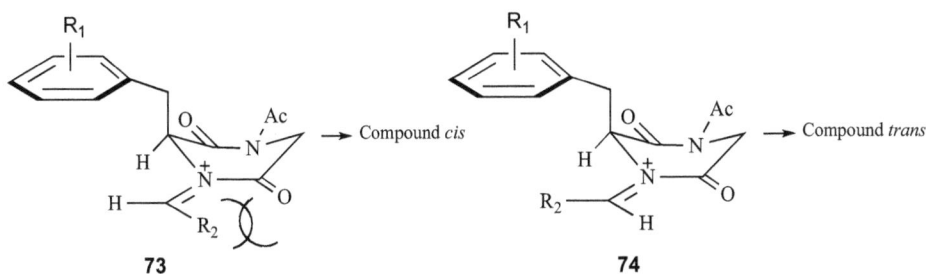

Figure 3: The more stable *E*-isomers of exocyclic iminium ion intermediates afford the tricyclic system with a *trans*-relationship between the H-6 and H-11a protons.

Rather surprisingly, cations of type **77** may be also intramolecularly trapped by an adjacent methoxy group, acting as an *O*-nucleophile. For instance, the 2,4,5-trimethoxyarylmethyl trimethylsilyl lactim **76** derived from **75**, gave with paraformaldehyde and $BF_3.Et_2O$ as catalyst the corresponding *N*-hydroxymethyl piperazinedione. When treated with *p*-toluenesulfonic acid, this compound cyclized by nucleophilic attack of the oxygen atom of the *ortho-* methoxy group onto the methylene of **77**, followed by demethylation, to give the oxazepine derivative **78** [92] (Scheme **16**).

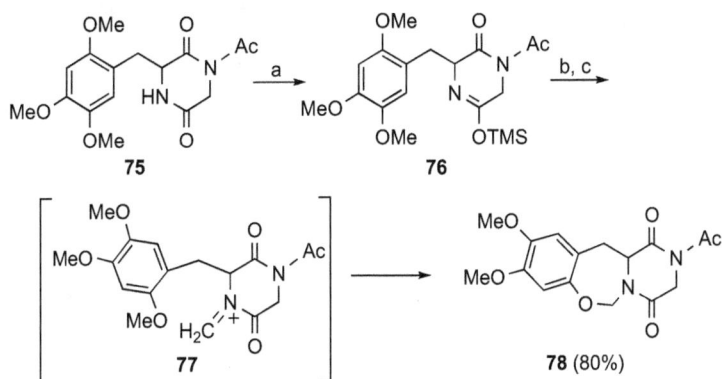

Reagents and Conditions: (a) TMSCI, Et$_3$N, CH$_2$Cl$_2$,r.t.; (b) (CH$_2$O)$_n$, BF$_3$.Et$_2$O; (c) *p*TsOH (10%), CH$_2$Cl$_2$, 80 °C.

Scheme 16: An alternative cyclization to give a tricyclic oxazepine derivative.

The previous protocol to generate pyrazino[1,2-*b*]isoquinoline-1,4-diones could not be extended to aldehydes less reactive than formaldehyde or acetaldehyde, but it worked in reactions catalyzed by trimethylsilyl triflate. It was reasoned that in these conditions acetals **79** form the triflate-acetal ion pairs **80**, which are cleaved to generate methyl alkylmethyleneoxonium triflates **81**. Subsequent trapping of these intermediates by the *O*-trimethylsilyl derivatives **82** gives **83** and, finally, **84** by loss of a molecule of trimethylsilyl triflate, which is thus recycled (Scheme **17**).

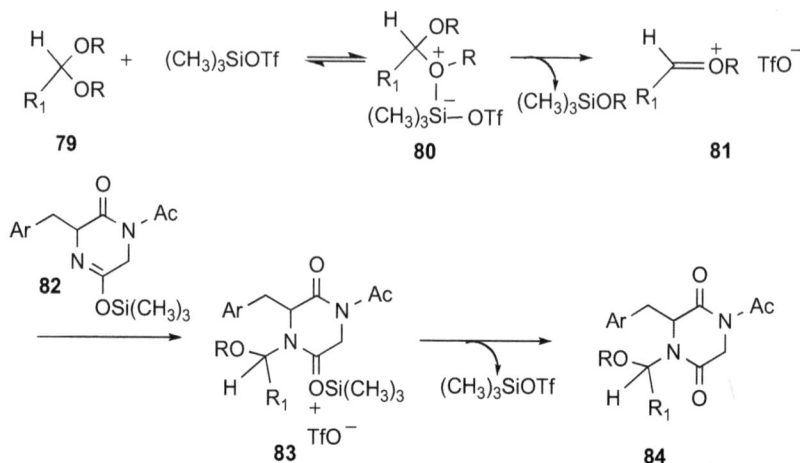

Scheme 17: Mechanism proposed for reactions catalyzed by trimethylsilyl triflate.

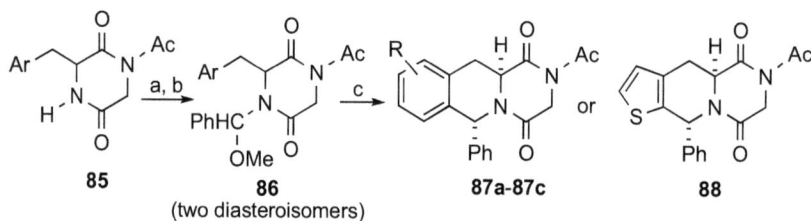

Ar	Comp.	Yield (%)	Comp.	Yield (%)
Phenyl	**86a**	80	**87a**	78
3-Methoxyphenyl	**86b**	91	**87b**	79
2,4,5-Trimethoxyphenyl	**86c**	87	**87c**	80
3-Methyl.2,4,5-trimethoxyphenyl	**86d**	95	**87d**	20[a]
3-Thienyl	**86e**	78	**88**	80

Reagents and Conditions: (a) TMSOTf, $^{i}Pr_2NEt$, CH_2Cl_2, -78 °C; (b) $PhCH(OCH_3)_2$, TMSOTf (cat), -78 °C; (c) *p*-TsOH, CH_2Cl_2, reflux; [a] Conc H_2SO_4, r. t. yielded 80% of *N*-deacetyl-**87d**.

Scheme 18: Reactions of different arylmethylpiperazinediones with benzaldehyde dimethyl acetal.

The scope of this methodology is shown in Scheme **18** for the reactions of benzaldehyde dimethyl acetal with different arylmethylpiperazinediones **85** to give *N*-methoxybenzyl intermediates **86** and finally the tricyclic products **87** and **88**, all of them with a 6,11a-*trans* relationship [92].

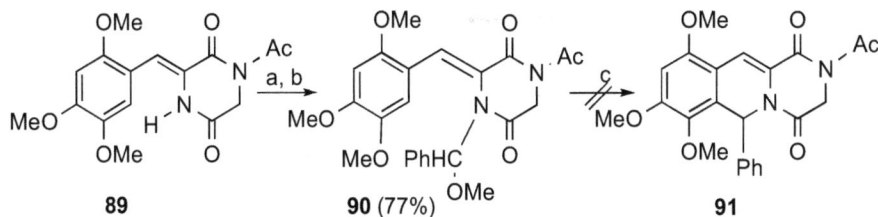

Reagents and Conditions: (a) TMSOTf, ^{i}Pr$_2$NEt, CH$_2$Cl$_2$, -78 °C; (b) PhCH(OCH$_3$)$_2$, TMSOTf (cat), -78 °C; (c) *p*-TsOH, CH$_2$Cl$_2$, reflux.

Scheme 19: Reactions with 3-arylmethylene unsaturated precursors.

3-Arylmethylene unsaturated precursors such as **89** may be also *N*-alkylated to give compound **90**, but the cyclization step does not take place due to the sensitivity of the enamine portion of **90** to acid media [92] (Scheme **19**).

For other acetals, one-pot procedures are the most convenient, as was shown in the reaction of **92** with *N*-phthalimidoacetaldehyde dimethyl acetal to give 6-phthalimidomethyl-pyrazinoisoquinolinedione **93**, which was used as a precursor of phthalascidin analogs [94] (Scheme **20**).

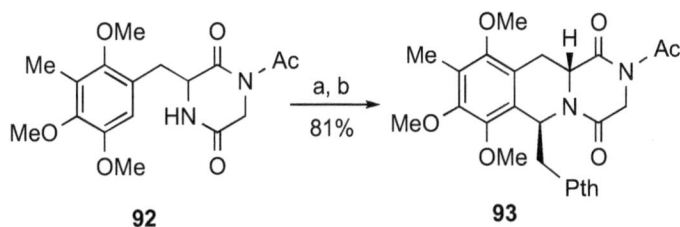

Reagents and Conditions: (a) TMSCl, Et$_3$N, CH$_2$Cl$_2$, r.t.; (b) PhtCH$_2$CH(OCH$_3$)$_2$, TMSOTf (5 eq), r.t.

Scheme 20: One-pot reaction of arylmethylpiperazinediones with *N*-phthalimidoacetaldehyde dimethyl acetal.

The greater conformational freedom of the side chain in the transition state of reactions with other acetals, as in the case of the 2-benzyloxyacetaldehyde dimethyl acetal, increases the stability of the Z-isomers of *N*-acyliminium species such as **94**, which cyclizes to compounds **95** with a *cis* configuration between the

C(6) and C(11a) stereocenters. Here, the cyclized product results from the attack of the aromatic ring from the *Re* face [95, 96] (Scheme **21**).

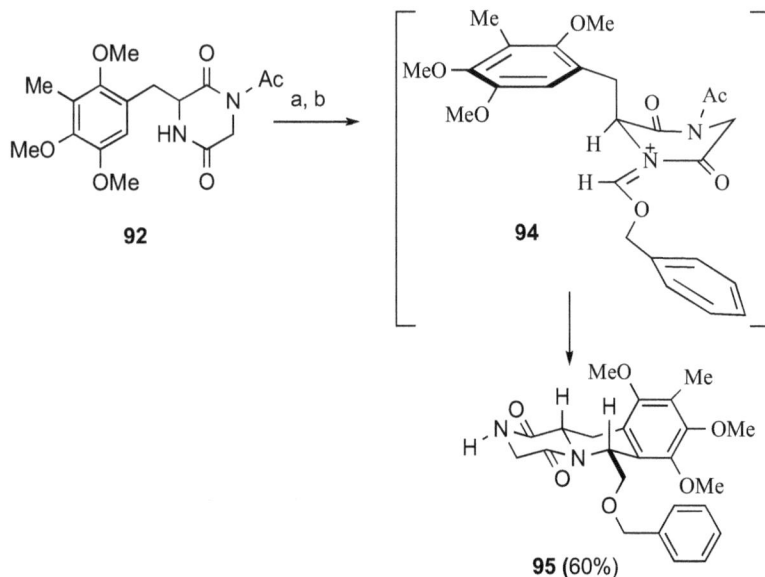

Reagents and Conditions: (a) TMSCl, Et$_3$N, CH$_2$Cl$_2$, r.t.; (b) BnOCH$_2$CH(OEt)$_2$ (2 eq), TMSOTf (3 eq), r.t.

Scheme 21: Rationalization of the found *cis* configuration between the C(6) and C(11a) stereocenters in certain cyclizations.

5. APPLICATION OF *N*-ACYLIMINIUM IONS TO THE SYNTHESIS OF 2,6-BRIDGED PIPERAZIN-3-ONES SYNTHESIS OF ANTITUMOR TETRAHYDROISOQUINOLINE ALKALOIDS

Substituted piperazines containing a 1-benzylpiperazine moiety may be considered as conformationally restricted 2-arylethylamines and, because of their analogy with neurotransmitters serotonin and dopamine, these compounds are common structural elements in many drugs with central nervous system (CNS) activity. Further introduction of a 2,6-bridge over the piperazine ring in these structures further diminishes their conformational freedom, and potentially provides valuable insights into their biologically active conformation. In spite of this pharmacological interest, most studies aiming at the construction of this bridge are related to the total synthesis of several natural products with antitumor

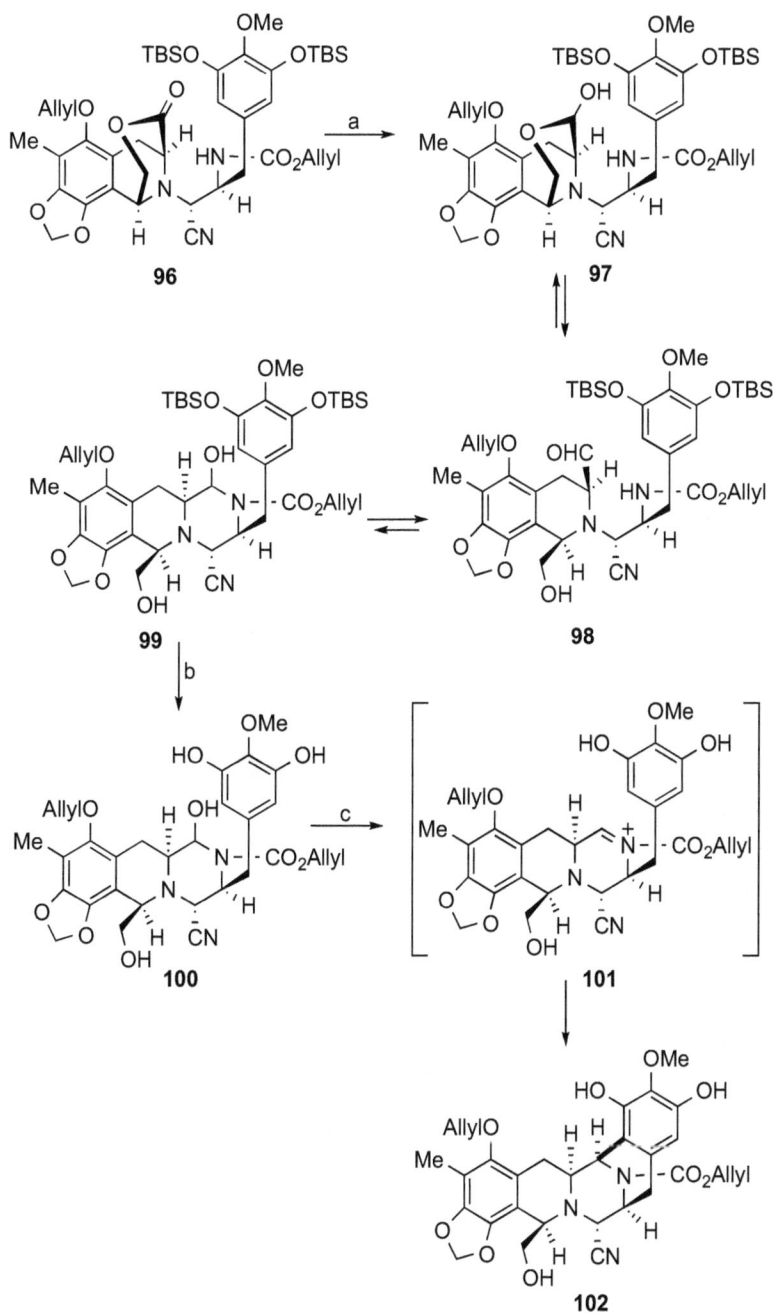

Reagents and Conditions: (a) DIBAL-H (1.2 eq), PhCH$_3$, -78 °C; (b) KF.2H$_2$O, CH$_3$OH, r.t.; (c) CH$_3$SO$_3$H (20 eq), CH$_2$Cl$_2$, 3 Å molecular sieves, r.t., 55% (3 steps).

Scheme 22: Formation of a 2,6-bridged piperazine moiety by intramolecular trapping of the acyliminium cation **101** in Corey's total synthesis of ecteinascidin 743.

or antimicrobial activity, and in these studies the 2,6-bridge in the piperazine moiety has been elaborated *via* N-acyliminium ion cyclizations. Quinocarcin A [97], saframycins [98], safracins [99], and cribrostatin IV [100] are representative examples of these polycyclic structures. In some cases, the ionic species have not been properly generated from a piperazine-2,5-dione but, since their precursors are in tautomeric equilibria with piperazine derivatives, we have considered their inclusion in this review appropriate. This is the case of the total synthesis developed by Corey [101, 102] of the most potent member of this family, namely ecteinascidin 743 (ET-743, trabectedin), a drug that has been approved under the trade name Yondelis® by the European Commission and the USA Food and Drug Administration for the treatment of advanced soft-tissues sarcoma and ovarian cancer. In Corey's process, the hydroxy compound **99** was formed from the ring-closed hemiaminal form of aldehyde **98**, obtained from lactone **96** through reduction to **97**. Conversion of hemiaminal **100** into acyliminium ion **101** by treatment with CH_3SO_3H/molecular sieves promotes ring closure onto the aromatic nucleophile, with formation of the 2,6-bridged piperazine moiety of compound **102** (Scheme **22**).

A similar cyclization was used by Fukuyama in the first reported total synthesis of (±)-saframycin B [103]. Here, the mixture of *cis* and *trans*-α,β-unsaturated aldehydes **104** and **105** formed by base-promoted elimination of acetic acid from **103**, was converted into **108** through the key intermediate **106** by heating in formic acid to generate acyliminium species **107** (Scheme **23**).

In contrast, hydroxylactam **111**, used as an intermediate in the alternative total synthesis of (±)-saframycin B **114** developed by Kubo, was obtained by hydride addition to the 3-arylmethylene-2,5-piperazinedione isopropyl carbamate precursor **110** [58, 98, 99]. This hydroxylactam cyclized by acid treatment [104] through attack of the aromatic nucleophile onto acyliminium **112** to give **113**, which was finally derived to **114** (Scheme **24**). Methyl, benzyl, and isobutyl carbamates were also suitable to activate the carbonyl function, but the cyclization of the reduced N-*tert*-butyloxycarbonyl derivative failed, possibly because of the steric bulk of this group. It is also interesting that the *Z*-exocyclic double bond in **111** isomerised to the *E*-configuration. The conversion into **113** possibly involves

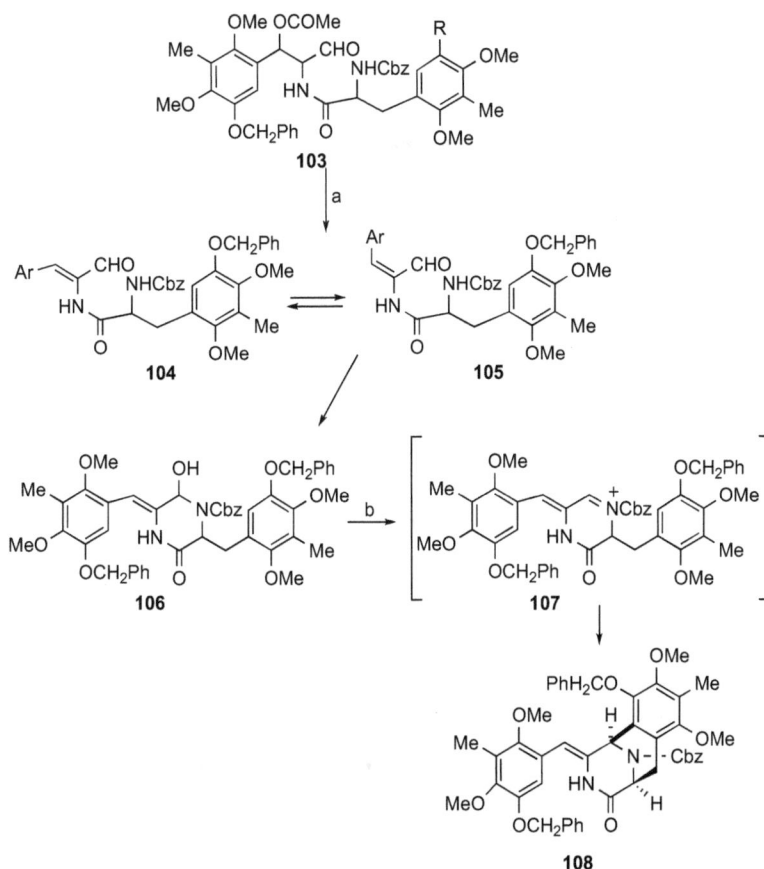

Reagents and Conditions: (a) DBU (1.5 eq), CH$_2$Cl$_2$, 0 °C; (b) HCO$_2$H, 60 °C, (74% overall yield).

Scheme 23: First Fukuyama's total synthesis of (±)-saframycin B starting from α,β-unsaturated aldehydes **104** and **105**.

the lone pair of the methoxy substituent in the *p*-position. This isomerization leads to competition between both aromatic rings at the cyclization stage but it was not observed in NH compounds lacking an alkyl substituent on the nitrogen atom [105-107].

Thus, in the first (See ref. [103] and Scheme **23**) total synthesis of (±)-saframycin A **120** published by Fukuyama [108], the 3-arylmethyl hydroxylactam **116** was obtained by hydride addition to the corresponding piperazinedione **115**, and this compound gave acyliminium **117**, which cyclized to **118**. After deprotection, compound **119** was transformed into **120** (Scheme **25**).

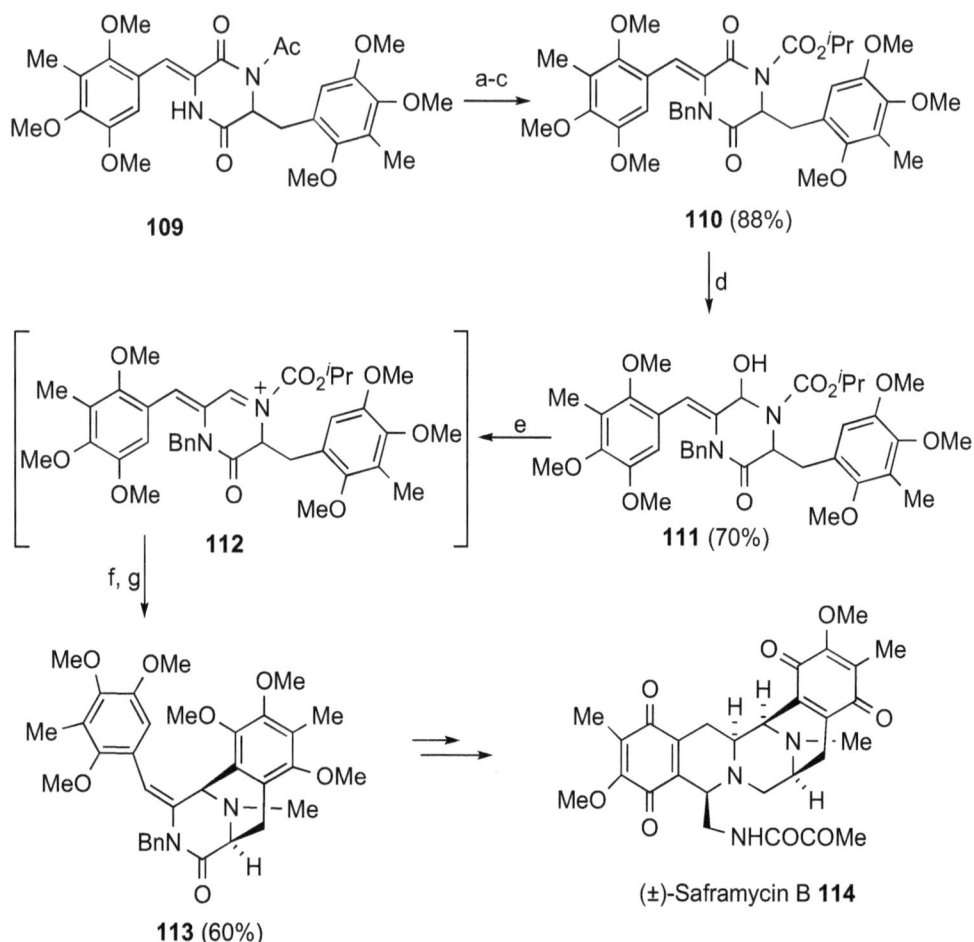

109 → **110 (88%)**

110 (88%) → (d) → **111 (70%)**

111 (70%) → (e) → **112**

112 → (f, g) → **113 (60%)**

113 (60%) → → (±)-Saframycin B **114**

Reagents and Conditions: (a) BnBr, NaH, DMF, r.t.; (b) NH$_2$-NH$_2$.H$_2$O, DMF, r.t.;(c) ^{i}PrOCOCl, DMAP, CH$_2$Cl$_2$, 0 °C; (d) LiAlH(O^{t}Bu)$_3$, THF, 0°C; (e) HCO$_2$H, 60 °C; (f) H$_2$SO$_4$, CF$_3$CO$_2$H, r.t.; (g) HCHO, HCO$_2$H, 70 °C.

Scheme 24: Kubo's total synthesis of (±)-saframycin B starting from the DKP **109**.

Liu has recently developed the asymmetric total synthesis of (-)-Saframycin A **127** by employing L-tyrosine as the starting chiral building block [109]. In this route, the 1,2,3,4-tetrahydroisoquinoline **121** and amino acid **122** were coupled to give the amide intermediate **123**. Selective deprotection and oxidation afforded the corresponding hemiaminal **124**, which was converted to the pentacyclic skeleton intermediate **126** through a Pictet-Spengler cyclization of the acyliminium intermediate **125**. The pentacyclic compound was transformed in (-)-Saframycin A **127** in five steps (Scheme **26**).

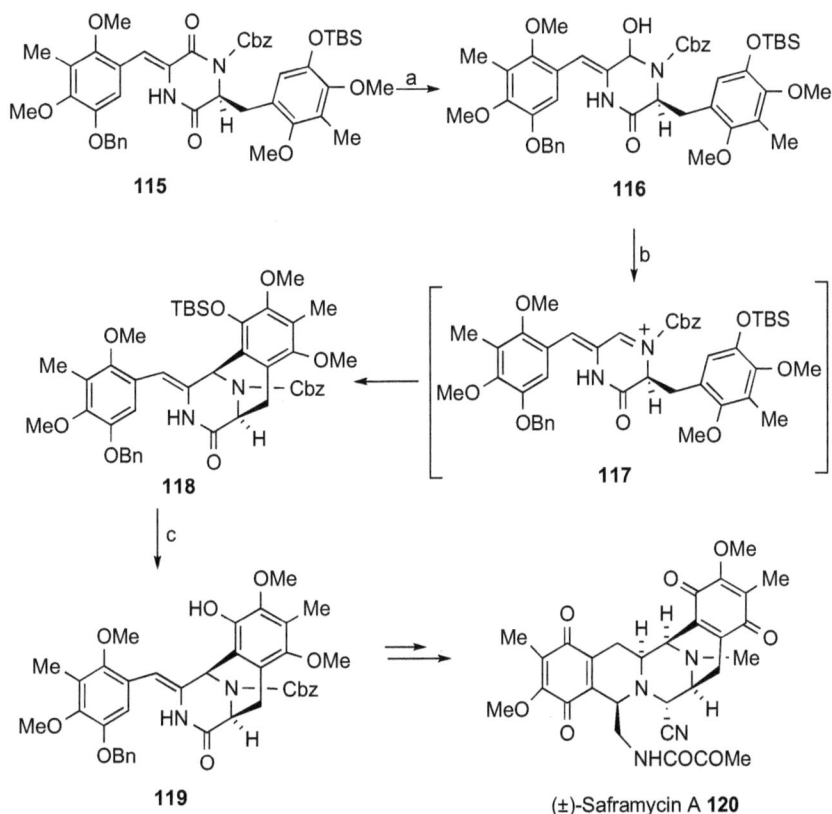

Reagents and Conditions: (a) NaBH₄, AcOH, EtOH, -25 °C; (b) HCO₂H, r.t.; (c) *n*-Bu₄NF, THF, r.t. 75% (3 steps).

Scheme 25: Fukuyama's total synthesis of (±)-saframycin A starting from the DKP **115**.

One of the first reports on this type of acyliminium-mediated cyclizations was directed to the synthesis of (±)-quinocarcin **133** [97]. The selective activation of one amide function gave the corresponding carbamate that, by partial amide carbonyl reduction to **129**, afforded the *N*-acyliminium ion **130**. Its cyclization to **131** and final reduction of the resultant aldehyde gave alcohol **132**, which was transformed into (±)-quinocarcin (Scheme **27**).

In the synthesis of octahydro-1,5-imino-3-benzazocin-4,7,10-trione derivatives developed by Kubo, the hydroxylactam **135** obtained in three steps from (-)-**134**, was exposed to formic acid at 70 °C for 30 minutes, but the acyliminium intermediate **136** thus formed was converted to the enamine **137** by loss of a proton. The cyclization of either **135** or **137** required reflux in TFA for 24 hours, giving in both cases a diastereomeric mixture of the 2,6-bridged piperazine-3-ones **138** and **139** [110] (Scheme **28**).

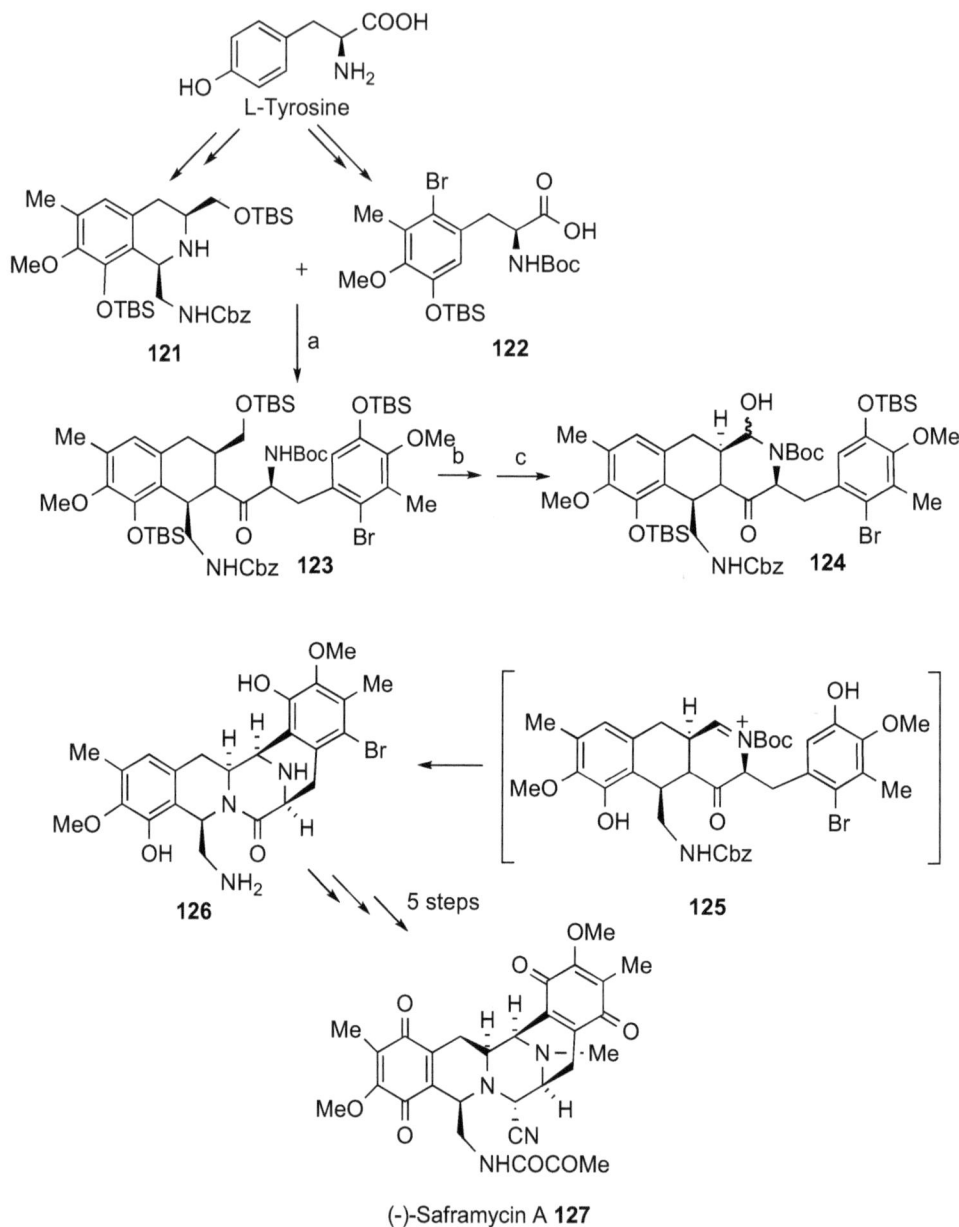

(-)-Saframycin A **127**

Reagents and Conditions: (a) BOPCl, Et$_3$N, CH$_2$Cl$_2$,r.t.; (b) HCO$_2$H, THF, H$_2$O, r.t.; (c) Dess-Martin periodinane,CH$_2$Cl$_2$,r.t; (d) TBAF, THF, 0 °C; (e) TfOH, r.t.

Scheme 26: Liu's asymmetric total synthesis of (-)-Saframycin A.

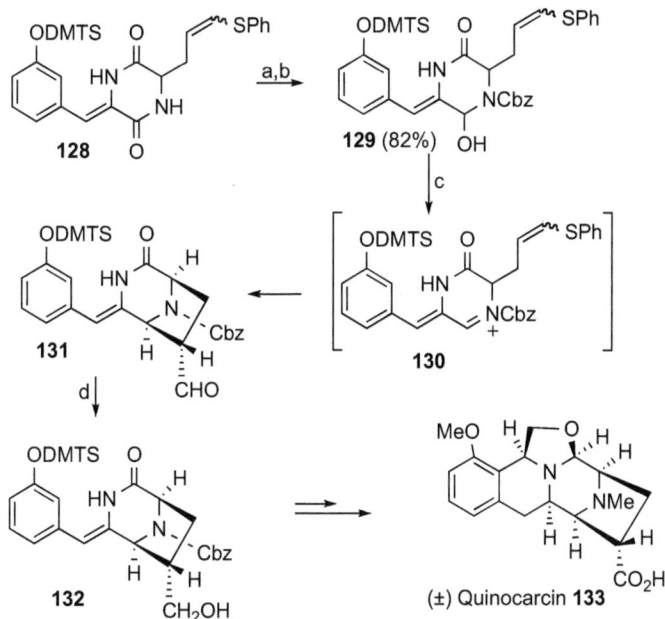

Reagents and Conditions: (a) CBzCl (1.5 eq), DMAP (0.5 eq), Et₃N (20 eq), CH₂Cl₂, -20 °C; (b) NaBH₄, MeOH/CH₂Cl₂ (1:1), -20 °C; (c) HgCl₂, CSA, CH₃CN/H₂O (9:1), 40 °C; (d) NaBH₄, MeOH/CH₂Cl₂ (1:1), -20 °C, (59% 3 steps).

Scheme 27: Synthesis of (±)-quinocarcin from the DKP **128**.

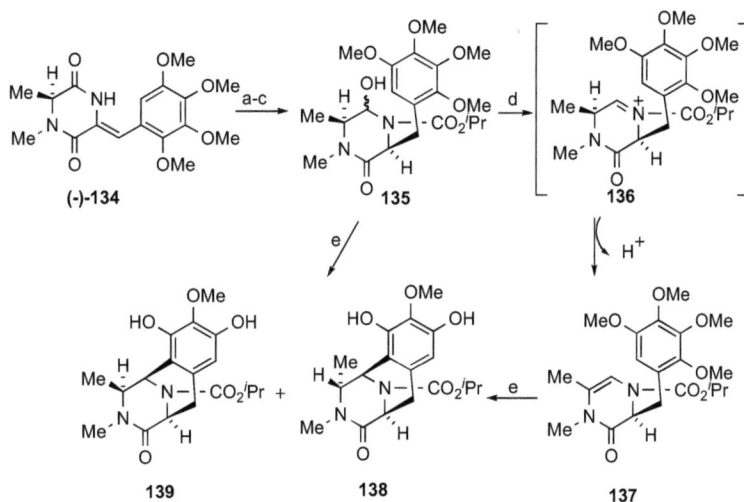

Reagents and Conditions: (a) H₂, Pd/C; (b) ClCO₂^{i}Pr, DMAP, Et₃N, CH₂Cl₂, -20 °C; (c) LiAlH(O^{t}Bu)₃, THF, 0°C; (d) HCO₂H, 70 °C; (e) TFA reflux.

Scheme 28: Synthesis of octahydro-1,5-imino-3-benzazocin-4,7,10-trione derivatives from the DKP **134**.

In a systematic study of these reactions, Hiemstra and coworkers determined the scope of π-nucleophiles in the cyclization of DKPs **140** to derivatives **143**, as well as the influence of a C-5 substituent on the stereochemical outcome of the reaction [111]. Activation of the C(6)-carbonyl group by *N*-methoxycarbonylation was followed by its chemoselective reduction with sodium borohydride in methanol, and the resulting α-hydroxylactams **141** were treated without further purification with TFA or formic acid to generate the reactive *N*-acyliminium ions **142**. These intermediates were trapped by the corresponding π-nucleophile at C-2 to yield the corresponding 2,6-bridged piperazin-3-ones **143** (Scheme **29**).

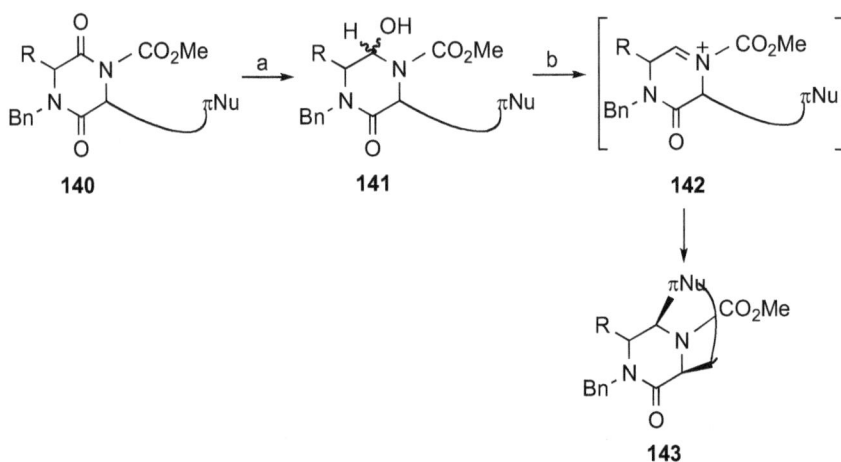

Reagents and Conditions: (a) NaBH$_4$, MeOH, 0 °C; (b) HCO$_2$H, r.t. or TFA, r.t.

Scheme 29: Chemoselective reduction of activated DKPs and acid-promoted cyclization to 2,6-bridged piperazin-3-ones.

The cyclization failed when some non activated arenes or heteroarenes and non aromatic π-nucleophiles were used. This was the case of the histidine-derived piperazinedione (*S*)-**144**, in which protonation of the imidazole ring results in a deactivated π-nucleophile at the side-chain of the acyliminium intermediate **145**. Consequently, no traces of the expected cyclic compound were observed, and enamide **146** was obtained instead by elimination of two protons. In the propargyl-substituted precursor *rac*-**147**, the acetylene moiety is too rigid to be in close proximity to the *N*-acyliminium ion in intermediate **148**, and the cyclic ketone **149** was formed in a very poor yield at 100 °C (Scheme **30**). Better results were obtained with more flexible allyl-substituted precursors.

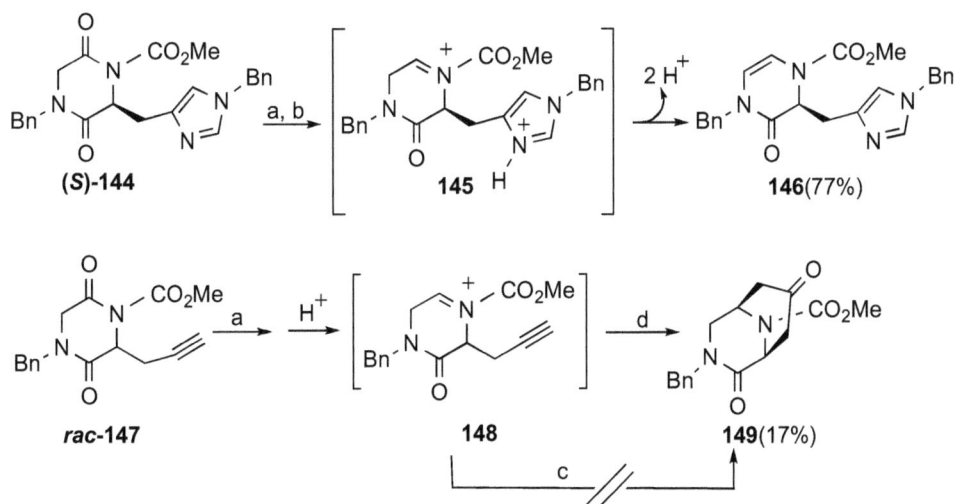

Reagents and Conditions: (a) NaBH$_4$, MeOH, 0 °C; (b) TFA, r.t.; (c) HCO$_2$H, r.t.; (d) HCO$_2$H, 100 °C.

Scheme 30: Importance of the nucleophilic group reactivity in the cyclization of acyliminium ions derived from DKPs.

Regarding the integrity of the stereocenter at the C(5)-position, the reduction/cyclization of the *trans*-2,5-disubstituted piperazinedione (*R,S*)-**150** proceeded smoothly through acyliminium intermediate **151** to give **152**, with the C(5)-methyl group positioned *trans* with respect to the 2,6-bridge. However, because of the steric hindrance of the methyl substituent in the *cis*-disubstituted diastereoisomer (*S,S*)-**153**, enamide **155** was obtained by deprotonation of the acyliminium intermediate **154** in the same reaction conditions. Higher temperatures (100 °C instead of 0 °C) furnished the *trans*-substituted piperazinone **157** in poor yield, probably because reprotonation of the enamide mainly takes place at the C(6)-position, leading to the less reactive tertiary *N*-acyliminium ion **156** (Scheme **31**).

Because the *N*-methylindolyl derivative **158** gave the all-*cis*-substituted piperazinone **160** in 71% yield through the acyliminium species **159**, without affecting the C(3)-stereocenter, it was postulated that cyclizations to give 2,6-bridged piperazine-3-one derivatives with an all-*cis*-stereochemistry required the presence of electron-rich arenes or heteroarenes (Scheme **32**). However, this assumption is not predictive for all *cis*-3,6-substituted piperazine-2,5-diones. See, for instance, the cyclization of **135** to **138** and **139** in Scheme **28**.

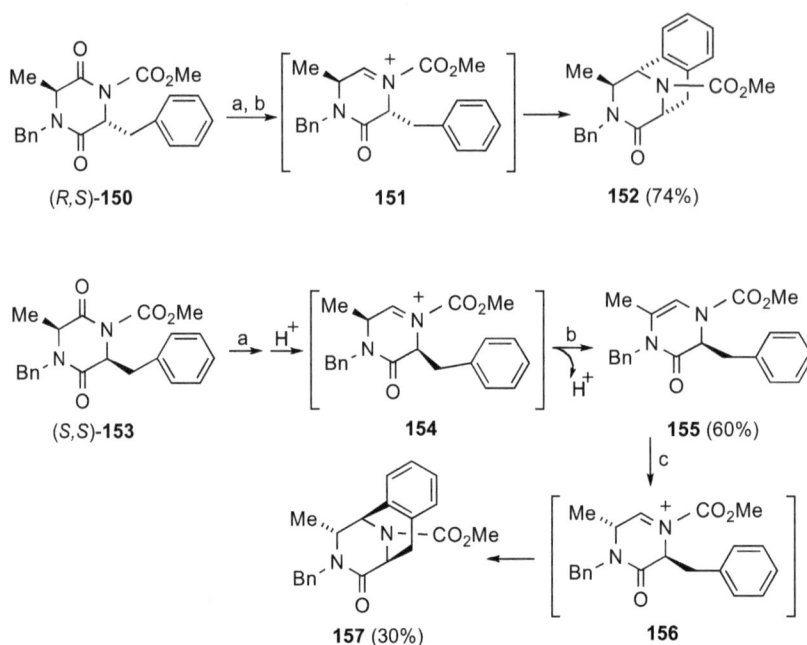

(R,S)-**150**   **151**   **152** (74%)

(S,S)-**153**   **154**   **155** (60%)

157 (30%)   **156**

Reagents and Conditions: (a) NaBH$_4$, MeOH, 0 °C; (b) TFA, r.t.; (c) TFA, 100 °C.

Scheme 31: Problems related to the integrity of the stereocenter at the C(5)-position.

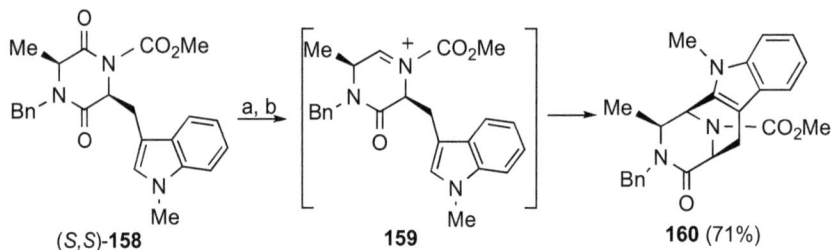

(S,S)-**158**   **159**   **160** (71%)

Reagents and Conditions: (a) NaBH$_4$, MeOH, 0 °C; (b) TFA, r.t.

Scheme 32: Synthesis of all-*cis*-2,6-bridged piperazine-3-ones requires electron-rich arenes or heteroarenes.

6. APPLICATION OF *N*-ACYLIMINIUM IONS TO THE SYNTHESIS OF THE 6,15-BRIDGE IN 6,15-IMINOISOQUINO[3,2-*b*]-3-BENZAZOCINE PENTACYCLIC COMPOUNDS

The 6,15-bridge of 9-alkyl-(6*S**,9*R**,15*S**)-5,6,9,15-tetrahydro-6,15-iminoisoquino[3,2-*b*]-3-benzazocine pentacyclic compound such us **167** (*N*-

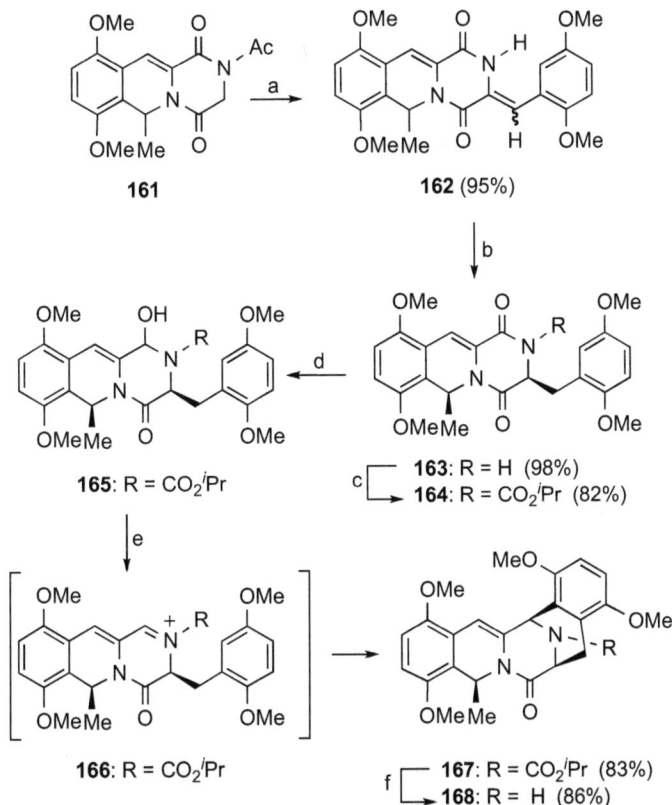

Reagents and Conditions: (a) ArCHO, dry DCM, K^{t}BuO/^{t}BuOH, r.t.; (b) H$_2$, Pd-C, MeOH, 1 bar; (c) Et$_3$N, DMAP, ClCO$_2^i$Pr, r.t.; (d) Li(^{t}BuO)$_3$AlH (3 eq), THF, r.t.; (e) HCO$_2$H, 80 °C; (f) TFA/H$_2$SO$_4$, r.t.

Scheme 33: Protocol applied to the synthesis of 6,15-iminoisoquino[3,2-*b*]-3-benzazocines from tricyclic compound **161**.

isopropiloxicarbonil), was built in our group by using 11,11a-dehydroderivatives such as **161** to circumvent the stereochemical problems related to the lability of the 11a-stereocenter in the saturated tricyclic compounds. These dehydroderivatives, through an aldol-type condensation with aromatic aldehydes, gave compounds such as **162** that afforded **163** by a regio- and diastereoselective hydrogenation of the hexocyclic double bond. Activation of the C(1)-N(2) amide *via* formation of carbamate **164**, partial reduction of the C(1)-carbonyl group to **165**, and generation of the *N*-acyliminium **166** ion with formic acid, allowed the cyclization that led to the desired pentacyclic compound **167**. Since the

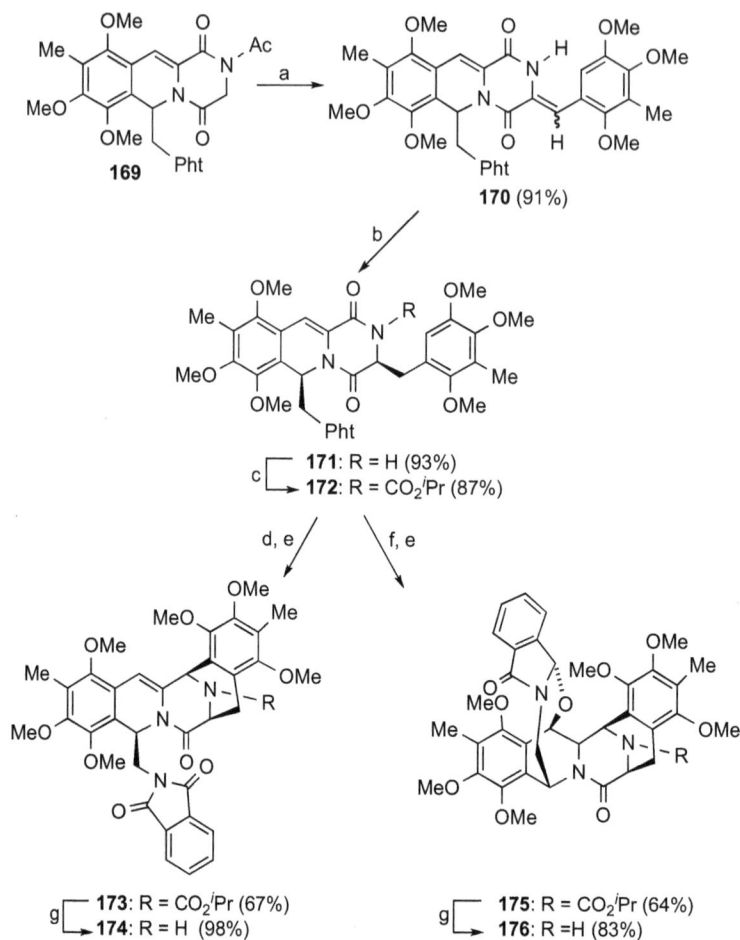

Reagents and Conditions: (a) ArCHO, dry DCM, K^{t}BuO/^{t}BuOH, r.t.; (b) H$_2$, Pd-C, MeOH, 3.5 bar; (c) Et$_3$N, DMAP, ClCO$_2$^{i}Pr, r.t.; (d) Li(^{t}BuO)$_3$AlH (3 eq), THF, r.t., 5 h; (e) HCO$_2$H, 80 °C; (f) Li(^{t}BuO)$_3$AlH (10 eq), THF, r.t., 20 h; (g) TFA/H$_2$SO$_4$, r.t.

Scheme 34: Found cyclizations in acyliminium ions derived from compound **172**.

stereochemistry of the C(3) stereocenter in the hydrogenation step that affords compound **163** is governed by the asymmetric induction of the C(6)-substituent, the cyclization products have an all-*cis* relative stereochemistry [94] (Scheme **33**). This reduction/cyclization sequence was previously described for 6-unsusbstituted pyrazinoisoquinoline-1,4-diones in which the 11,11a-double bond has been saturated [112] but, in our hands, the reduction/cyclization process with these compounds bearing a 6-substituent was not fully stereocontrolled.

When this protocol was applied to the 6-phthalimidomethyl unsaturated compound **169**, the expected pentacyclic product **173** was obtained from **172** in a near equimolecular mixture with the octacyclic compound **175**. Both products were obtained as single diastereoisomers. Full selectivity in favour of one of them could be achieved by using a smaller excess of reductive agent and shorter reaction times (**173**) or a greater excess of reductive agent and longer reaction times (**175**) [94] (Scheme **34**)

The octacyclic compound **175** is formed from **172** through the diastereoselective reduction of two carbonyl groups: the activated C(1)=O group and one carbonyl of the phthalimide moiety. The dihydroxy intermediate **177**, in the presence of acid, may give acyliminium ion **178**, which undergoes conjugate nucleophilic attack by the side-chain hemiaminal hydroxy group and is later captured by the arene ring of the arylmethyl side chain. Alternatively, **178** may give first the pentacycle and, subsequently, the phthalimide hydroxy group would attack the benzyl cation formed by protonation of the double bond (Scheme **35**).

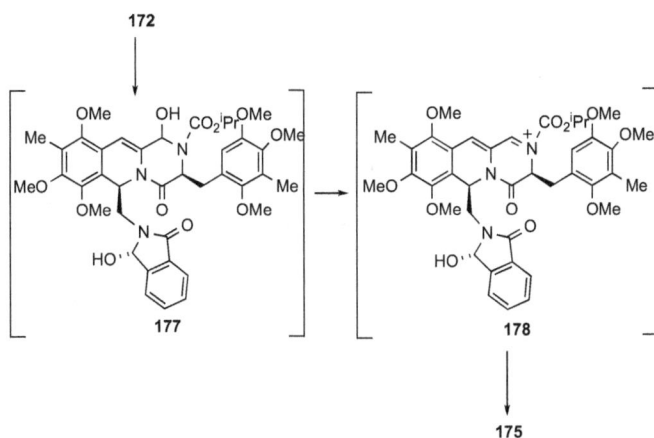

Scheme 35: Proposed mechanism to explain the formation of the octacyclic compound **175**.

7. APPLICATION OF *N*-ACYLIMINIUM IONS TO THE SYNTHESIS OF *N*-PYRUVYLAMINO ESTERS, *N*-INDOLE-2-CARBONYLAMINO ESTERS AND *META*- AND *PARA*-2,6-DIAZACYCLOPHANES

3-Arylmethylenepiperazinediones **179** are also starting materials for the synthesis of diverse compounds through generation of *N*-acyliminium cations. Thus, compounds **179** give *N*-pyruvoyl-amino esters **182** in a one-pot, three-step process that involves protonation of the enamine portion of **180**, ring-opening of *N*-acyliminium cation **181** thus formed by addition of the alcohol used as solvent and imine hydrolysis during the reaction work-up. When these compounds bear an *o*-nitro group, they afford the *N*-indole-2-carbonylamino esters **183** after catalytic hydrogenation, through a mechanism related to the Reissert indole synthesis [113] (Scheme **36**).

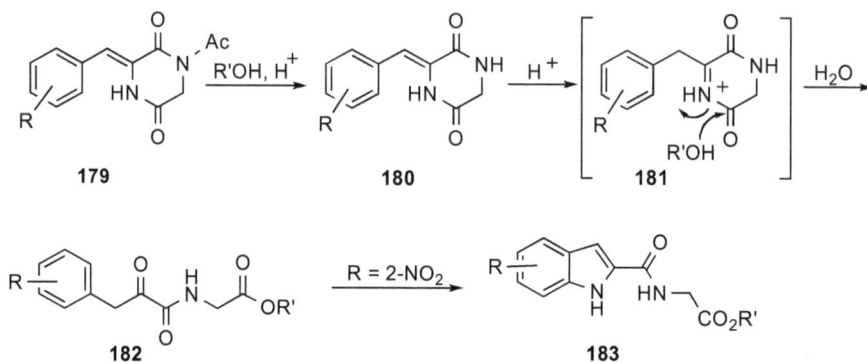

Scheme 36: *N*-acyliminium ions as intermediates in the synthesis of *N*-pyruvylamino esters and *N*-indole-2-carbonylamino esters.

Other nitro derivatives of compounds **182** gave by catalytic reduction the corresponding aminoaryl derivatives, as was the case of **182a** and **182b**, which afford compounds **184** and **185**. However, *m*- and *p*-nitro-compounds **182c** and **182d** gave, either the amino compounds **186** and **188**, or the *meta*- and *para*-2,6-diazacyclophanes **187** and **189**. These cyclophanes were formed through an intermolecular reductive amination at higher concentrations and have a relative *cis*-stereochemistry, according to a semiempirical study [114] (Scheme **37**).

8. APPLICATION OF *N*-ACYLIMINIUM IONS TO THE SYNTHESIS OF PYRAZINO[2,1-*b*]QUINAZOLINE-3,6-DIONE DERIVATIVES

The pirazino[2,1-*b*]quinazoline-3,6-dione system **190** is used by nature as an scaffold for constrained peptidomimetics, as demostrated by the structure of several fungal metabolites such as glyantrypine [115], the fumiquinazolines [116, 117], and the fiscalins [118, 119]. It contains an *N*-alkyl-*N*'-acyl-iminopiperazine-

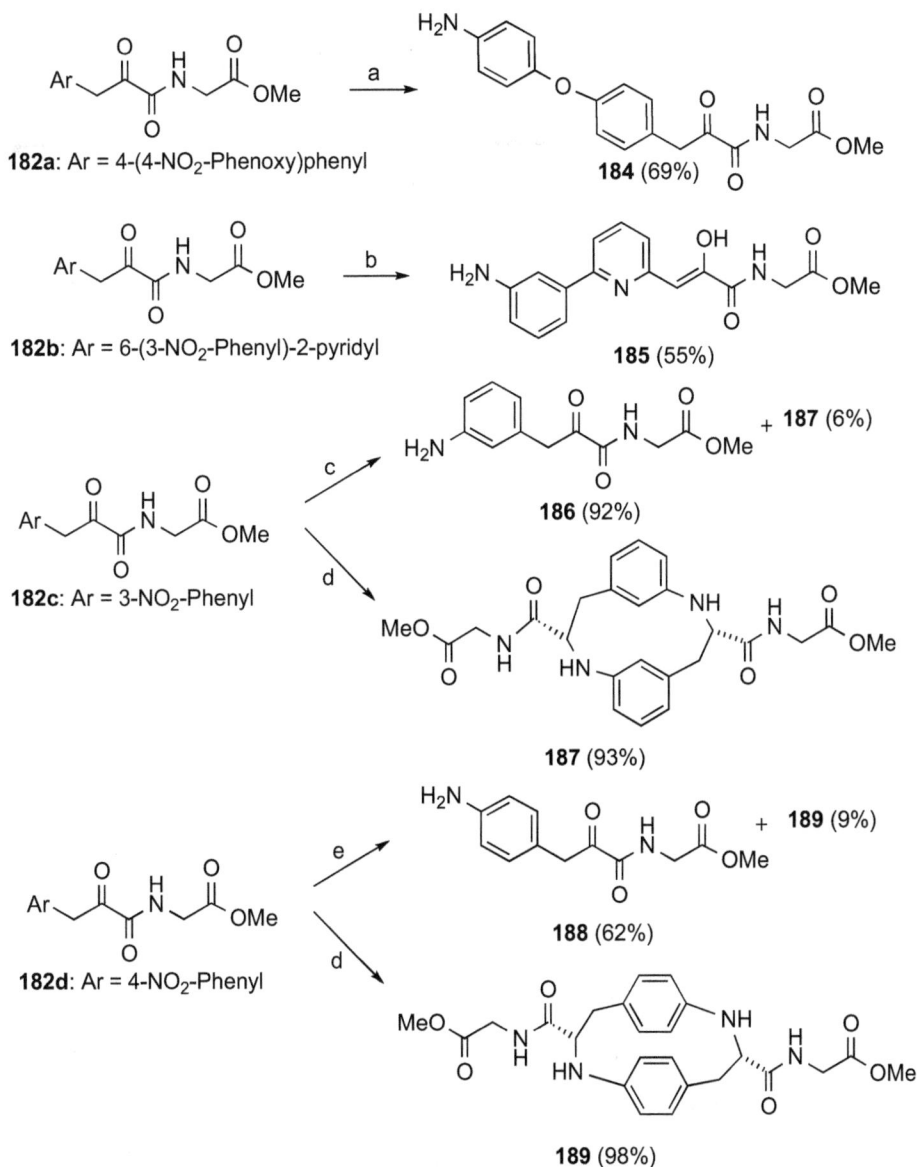

Reagents and Conditions: (a) AcOEt, 4×10^{-3} M, H_2, 10% Pd-C, r.t.; (b) AcOEt, 2×10^{-3} M, H_2, 26% Pd-C, r.t.; (c) AcOEt, 2×10^{-3} M, H_2, 10% Pd-C, r.t.; (d) AcOEt, 2×10^{-2} M, H_2, 10% Pd-C, r.t.; (e) AcOEt, 6×10^{-3} M, H_2, 10% Pd-C, r.t.;

Scheme 37: *N*-acyliminium ions as intermediates in the synthesis of *meta*- and *para*-2,6-diazacyclophanes.

2,5-dione moiety, and was shown to be a precursor of acyliminium cations **191** that may be trapped by nucleophiles to give compounds **192**. This chemistry was applied in the search for the pharmacophore of the fungal metabolite and multiple drug resistance (MDR) inhibitor *N*-acetylardeemin [120-122] (Scheme **38**).

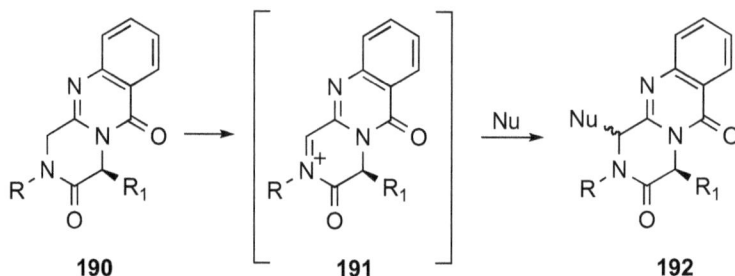

Scheme 38: Application of *N*-acyliminium ions to the synthesis of pyrazino[2,1-*b*]quinazoline-3,6-dione derivatives.

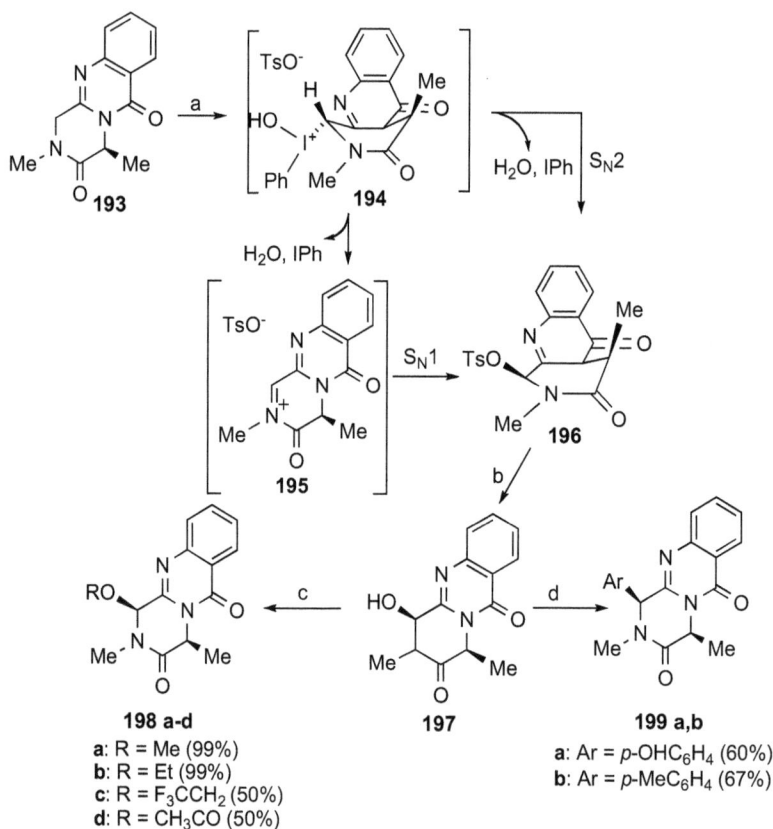

198 a-d

a: R = Me (99%)
b: R = Et (99%)
c: R = F$_3$CCH$_2$ (50%)
d: R = CH$_3$CO (50%)

197

199 a,b

a: Ar = *p*-OHC$_6$H$_4$ (60%)
b: Ar = *p*-MeC$_6$H$_4$ (67%)

Reagents and Conditions: (a) (PhIOH)OTs, AcOEt, reflux; (b) H$_2$O, NH$_4$Cl, CH$_2$Cl$_2$, r.t.; (c) TsOH, ROH, r.t.; (d) conc. H$_2$SO$_4$ (0.1 eq), ArH (2 eq), r.t..

Scheme 39: Generation of *N*-acyliminium ions by oxidation with HTIB.

For instance, treatment of **193** with the hypervalent iodine reagent [hydroxy(tosyloxy)iodo]benzene (HTIB) [123] gave the expected *cis*-tosylate at C(1) **196**, which was quantitatively transformed into the *cis*-1-hydroxy derivative **197** in aqueous ammonium chloride. Tosyloxylation takes place through a previous electrophilic attack leading to the unstable intermediate **194** that generates *N*-acyliminium ion **195**. This species is also generated in acid from the hydroxy derivative **197** and it can be trapped by O- or C nucleophiles to give adducts **198** and **199** [124] (Scheme **39**).

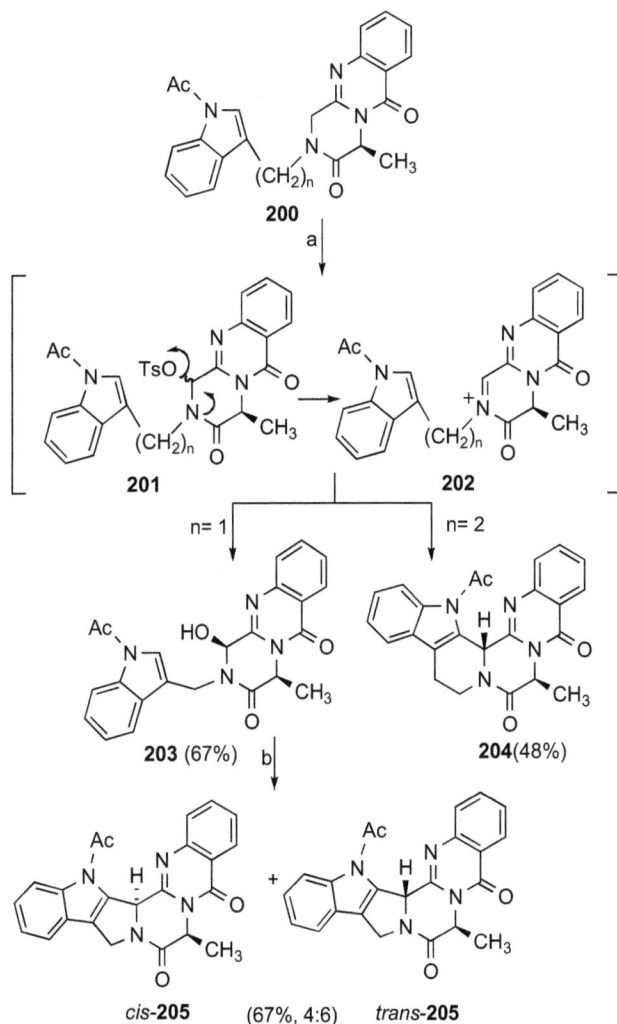

Reagents and Conditions: (a) (PhIOH)OTs, AcOEt, reflux; (b) H$_2$SO$_4$, 0 °C.

Scheme 40: Application of this methodology to achieve an intramolecular cyclizations.

The application of this methodology to suitable *N*-arylalkyl substituted compounds such as **200** permits intramolecular cyclizations through tosyloxy and acyliminium intermediates **201** and **202** [125, 126] (Scheme **40**).

N-Acyliminium ions **208** were also generated by radical bromination of compounds **206** through the corresponding 1-bromo derivatives **207** and, after treatment with strong Lewis acids and weak nucleophiles, gave compounds **209**-**211** [127] (Scheme **41**).

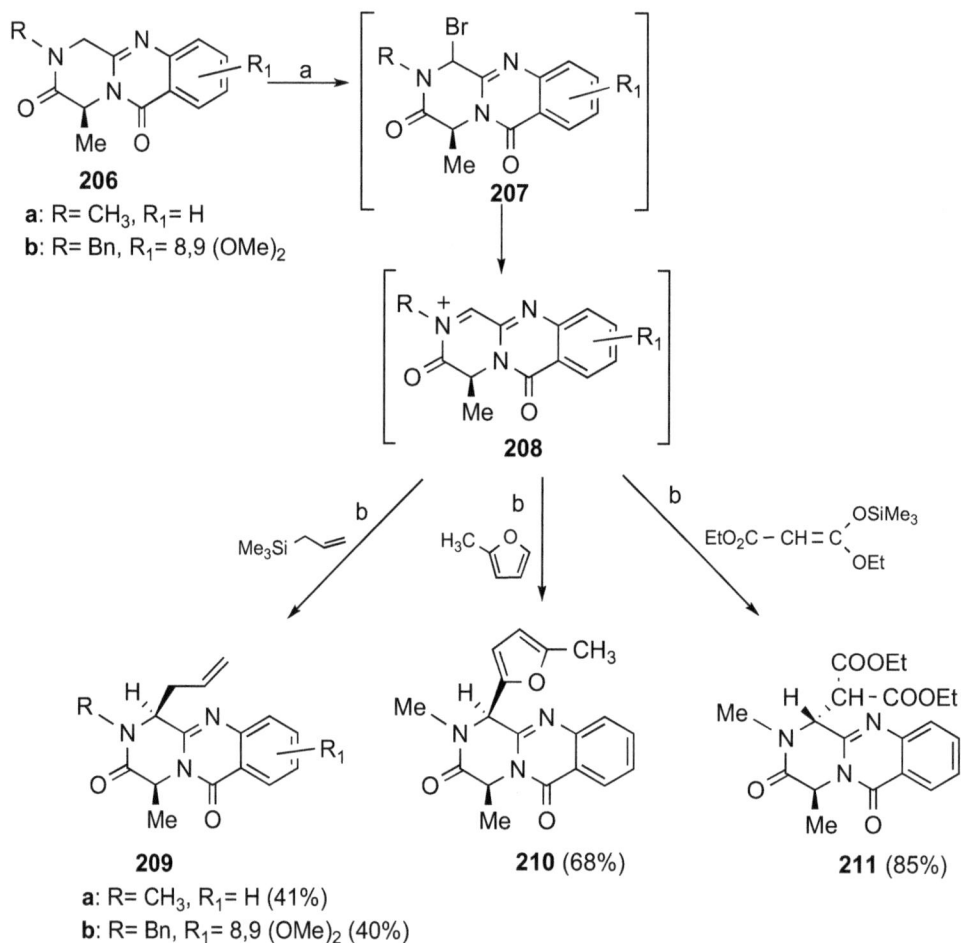

Reagents and Conditions: (a) NBS (1,1 eq), AIBN (1%), CCl$_4$, reflux; (b) NuH (4 eq), ZnCl$_2$ (2 eq), THF, argon atmosphere, r.t.

Scheme 41: Generation of *N*-acyliminium ions by radical bromination.

Tertiary *N*-acyliminium ions **214** were generated from 1-hydroxyderivatives **213** which were obtained by regioselective *syn*-addition of organometallics to 1-oxoderivatives **212**. These species are deprotonated to give 1-methylene derivatives (see compounds **215a** and **215b**) or may be trapped by the solvent (see compounds **216** and **217**) [128] (Scheme **42**).

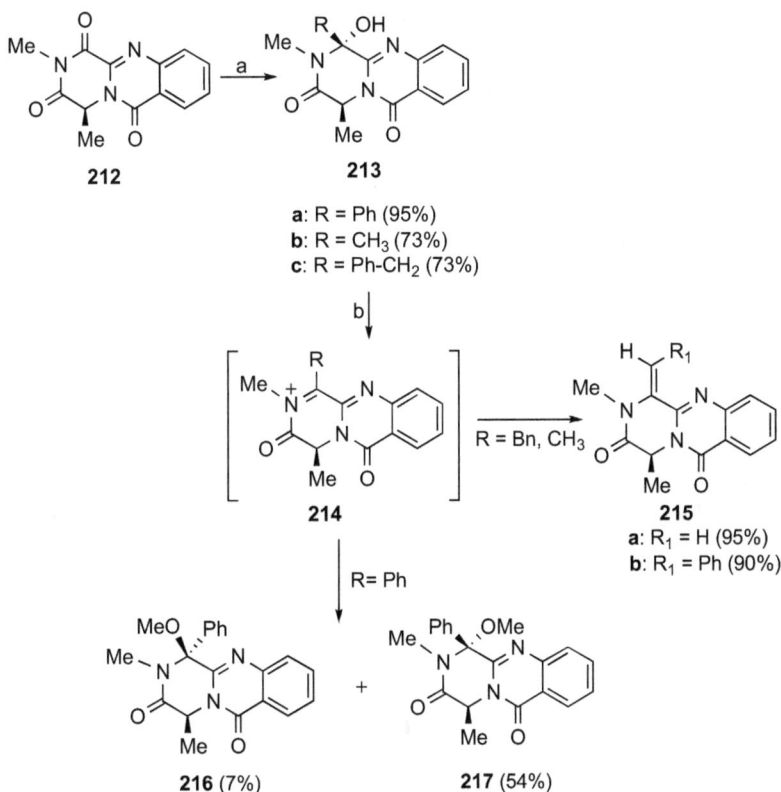

a: R = Ph (95%)
b: R = CH$_3$ (73%)
c: R = Ph-CH$_2$ (73%)

215
a: R$_1$ = H (95%)
b: R$_1$ = Ph (90%)

216 (7%)   **217** (54%)

Reagents and Conditions: (a) RLi or RMgBr, dry THF, -30 °C or -78 °C; (b) TsOH, MeOH, reflux.

Scheme 42: Generation of *N*-Acyliminium ions by addition of organometallics.

Also in this case, the use of suitable *N*-arylalkyl substituents, as in the case of **218**, permitted the access to cyclized products **221-224** through Pictet-Spengler type reactions [128] (Scheme **43**).

Alternatively, these iminium species may be trapped by a π-nucleophile at the C(4)-position, as exemplified by the transformation of **225** into **227** [129] (Scheme **44**).

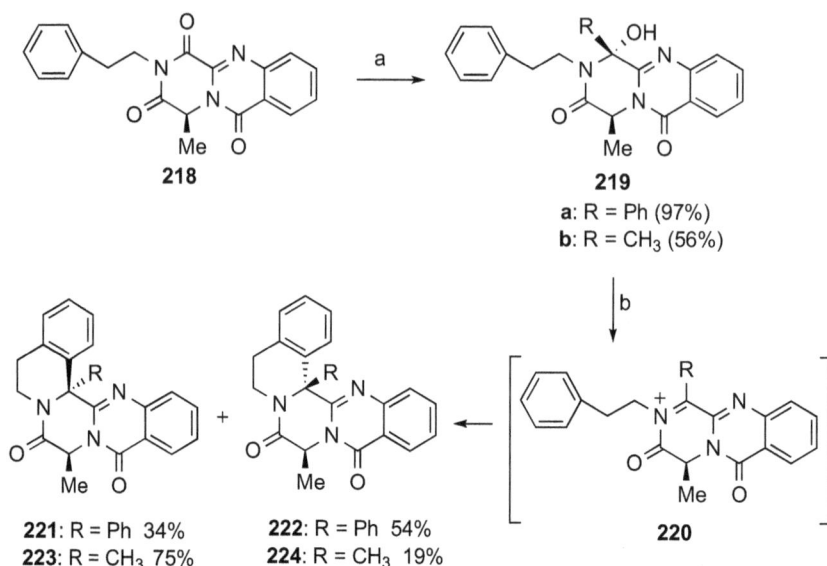

218   →(a)→   **219**

a: R = Ph (97%)
b: R = CH₃ (56%)

221: R = Ph 34%
223: R = CH₃ 75%

222: R = Ph 54%
224: R = CH₃ 19%

220

Reagents and Conditions: (a) RLi or RMgBr, dry THF, -30 °C or -78 °C; (b) H₂SO₄, r.t.

Scheme 43: Pictet-Spengler type reactions of *N*-acyliminium intermediates.

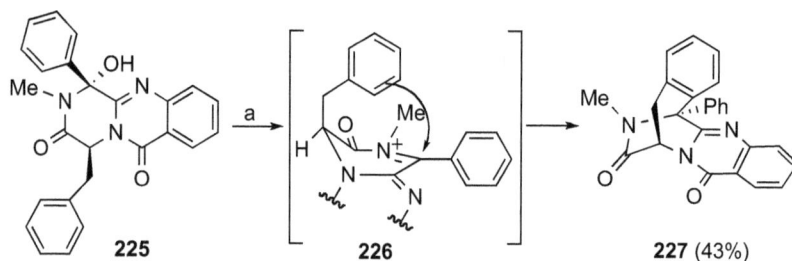

225   **226**   **227** (43%)

Reagents and Conditions: (a) F₃CCO₂H, reflux.

Scheme 44: Trapping of iminium species by a π-nucleophile at the C(4)-position.

The tertiary iminium ions generated by protonation of 1-methylene derivatives, such as **229**, could be trapped by the π-nucleophile at the C(4) position. The reaction only works with 2-unsubstituted compounds (see for instance the transformation to **232** through the intermediacy of **231** [129] (Scheme **45**).

A tertiary *N*-acyliminium cation **234** formed by protonation of the double bond of compound **233** was intramolecularly trapped by the hydroxy group to form the seven-membered ether ring of fumiquinazoline C **236** [130] (Scheme **46**).

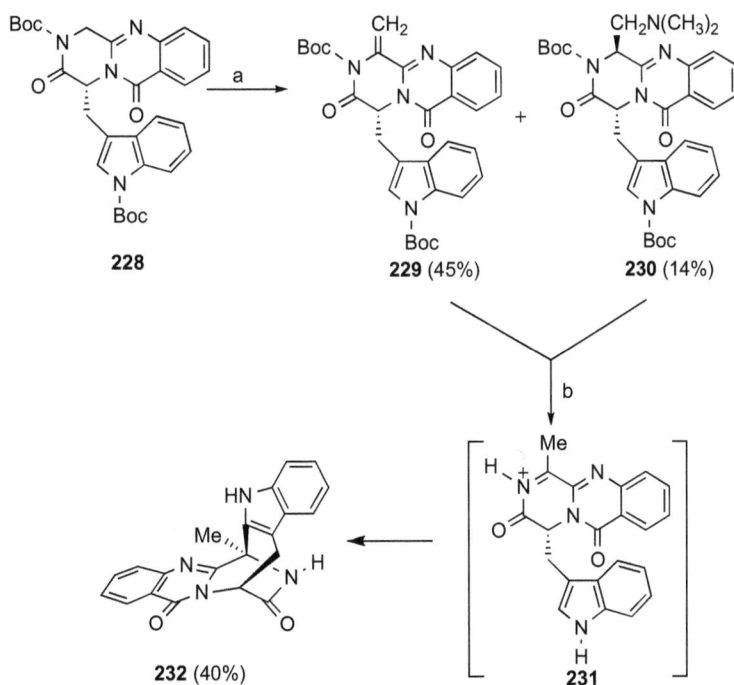

Reagents and Conditions: (a) [(CH₃)₂N]₂CH₂, F₃CCO₂H, CH₂Cl₂, r.t.; (b) F₃CCO₂H, reflux.

Scheme 45: Cyclization of tertiary iminium ions.

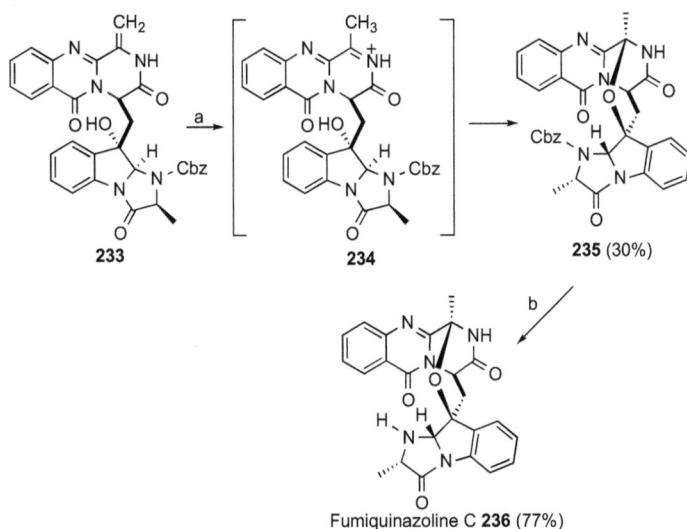

Fumiquinazoline C **236** (77%)

Reagents and Conditions: (a) CH₃CN/HOAc (100:1), reflux ; (b) H₂, Pd, 4 atm.

Scheme 46: Trapping of the tertiary *N*-acyliminium cation **234** in the synthesis of Fumiquinazoline C.

Also, an intramolecular electrophilic attack in *N*-acyliminium ion **238**, prepared by treatment of methylene compound **237** (*N*-unsubstituted) with trifluoroacetic acid, provided compound **239**, an intermediate in the synthesis of *ent*-alantrypinone **240** [131] (Scheme **47**).

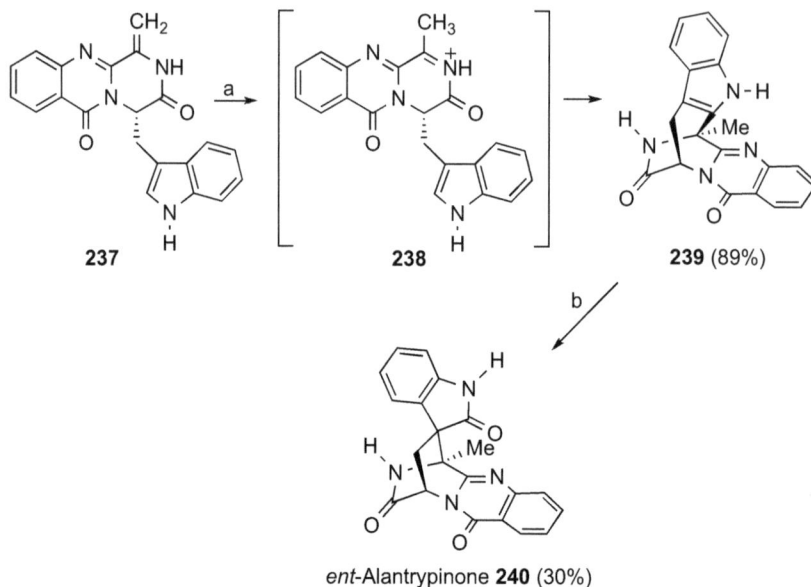

Reagents and Conditions: (a)TFA/CHCl$_3$ (1:20), reflux ; (b) NBS, THF, TFA, H$_2$O, 0 °C; (c) H$_2$, Pt/C.

Scheme 47: The *N*-acyliminium ion **238** as an intermediate in the synthesis of *ent*-alantrypinone.

These results are in contrast with the previously studied transformation of piperazinedione derivative **42** into **44** (Scheme **9**), a difference that may be rationalized by considering that the steric interactions between the two methyl groups in acyliminium species **242** (Scheme **48**) prevent its formation from **241**. Since protonation of the enamine portion in this product is unfavourable, it takes place at the N(11)-atom, permitting the coupling of the ion **243** thus formed with **241** to give **244** [129].

CONCLUSIONS

The preceding survey of the chemistry of *N*-acyliminium species derived from piperazine-2,4-diones and related compounds reflects their utility in the synthesis of a wide range of structurally diverse products, including several skeletons isolated from natural sources. We hope that this review will stimulate further applications of this fascinating chemistry.

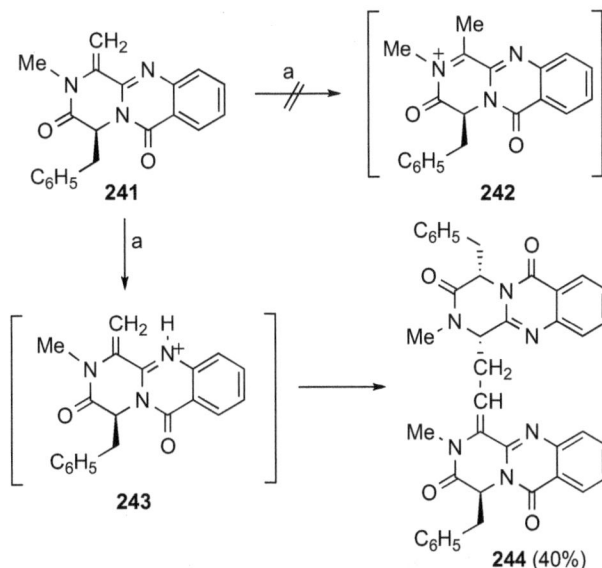

Reagents and Conditions: (a) F_3CCO_2H, reflux.

Scheme 48: Steric interactions prevent the geneation of iminium intermediate **242** in favour of imium species **243**.

ACKNOWLEDGEMENTS

We gratefully acknowledge financial support from MICINN (grant CTQ2009-12320-BQU) and UCM (Grupos de Investigación, grant GR35/10-A-920234), as well as the hard work and dedication of many coworkers, whose names can be found in the reference list.

CONFLICT OF INTEREST

The author confirms that this chapter content has no conflict of interest.

DISCLOSURE

The chapter submitted for series eBook entitled "**Advances in Organic Synthesis, Volume 5**" is an update of our article published in **CURRENT ORGANIC**

SYNTHESIS, Volume 6, Number 2, May Issue 2009, with additional text and references.

REFERENCES

[1]   Santagada, V.; Fiorino, F.; Perissutti, E.; Severino, B.; Terracciano, S.; Cirino, G.; Caliendo, G. *Tetrahedron Lett.* **2003**, *44*, 1145-1148.

[2]   Johne, S.; Groeger, D. *Pharmazie*, **1977**, *32*, 1-16.

[3]   Rajappa, S.; Natekar, M. V. *Adv. Heterocycl. Chem.* **1993**, *57*, 187-189.

[4]   Witiak, D. T.; Wei, Y. *Prog. Drug Res.* **1990**, *35*, 249-363.

[5]   Prasad, C. *Peptides* **1995**, *16*, 151-164.

[6]   See for instance: Tullberg, M.; Grøtli, M.; Luthman, K. *J. Org. Chem.* **2007**, *72*, 195-199.

[7]   Martins, M. B.; Carvalho, I. *Tetrahedron*, **2007**, *63*, 9923-9932.

[8]   Milne, P. J.; Kilian, G. The Properties, Formation, and Biological Activity of 2,5-Diketopiperazines. *Comprehensive Natural Products II: Chemistry and Biology*; Mander, L., Liu, H—W., Eds.; Elsevier: Amsterdam, 2010; Vol 5, pp 657-698.

[9]   Borthwick, A. D.; Davies, D. E.; Exall, A. M.; Hatley, R. J. D.; Hughes, J. A.; Irving, W. R.; Livermore, D. G.; Sollis, S. L.; Nerozzi, F.; Valko, K. L.; Allen, M. J.; Perren, M.; Shabbir, S. S.; Woollard, P. M; Price, M. A. *J. Med. Chem.* **2006**, *49*, 4159-4170.

[10]   Williams, R. M.; Cox, R. *Acc. Chem. Res.* **2003**, *36*, 127-139.

[11]   Scott, J. D.; Williams, R. M. *Chem. Rev.* **2002**, *102*, 1669-1730.

[12]   Borthwick, A. D. *Chem. Rev.* **2012**, *112*, 3641-3716.

[13]   González, J. F.; Ortín, I.; de la Cuesta, E.; Menéndez, J. C. *Chem. Soc. Rev.* **2012**, 41, 6902-6915.

[14]   For an example see: Bull, S. D.; Davies, S. G.; Garner, A. C.; O'Shea, M. D.; Savory, E. D.; Snow, E. J. *J. Chem. Soc. Perkin Trans. 1*, **2002**, 2442-2448.

[15]   Farloni, M.; Giacomelli, G.; Porcheddu, A.; Taddei, M. *Eur. J. Org. Chem.* **2000**, 1669-1675.

[16]   Fisher, P. M. *J. Peptide Sci.* **2003**, *9*, 9-34.

[17]   Palacin, S.; Chin, D. N.; Simanek, E. E.; MacDonald, J. C.; Whitesides, G. M.; McBride, M. T.; Palmore, G. T. R. *J. Am. Chem. Soc.* **1997**, *119*, 11807-11816.

[18]   López-Cobeñas, A.; Cledera, P.; Sánchez, J. D.; Pérez-Contreras, R.; López-Alvarado, P.; Ramos, M. T.; Avendaño, C.; Menéndez, J. C. *Synlet*t, **2005**, 1158-1160.

[19]   López-Cobeñas, A.; Cledera, P.; Sánchez, D.; López-Alvarado, P.; Ramos, M. T.; Avendaño, C.; Menéndez, J. C. *Synthesis*, **2005**, 3412-3422.

[20]   Bianco, A.; Sonksen, C. P.; Roepstorff, P.; Briand, J.-P. *J. Org. Chem.* **2000**, *65*, 2179-2187.

[21]   Dinsmore, C. J.; Beshore, D. C. *Tetrahedron* **2002**, *58*, 3297-3312.

[22]   Fukuyama, T.; Nakatsuka, S; Kishi, Y. *Tetrahedron*, **1981**, *37*, 2045-2078.

[23]   Williams, R. M.; Glinka, T.; Kwast, E.; Coffman, H.; Stille, J. K. *J. Am. Chem. Soc.* **1990**, *112*, 808-821.

[24]   Orena, M.; Porzi, G.; Sandri, S. *J. Org. Chem.* **1992**, *57*, 6532-6536.

[25]   Hayashi, Y.; Orikasa, S.; Tanaka, K.; Kanoh, K.; Kiso, Y. *J. Org. Chem.* **2000**, *65*, 8402-8405.

[26]   Pichowicz, D.; Simpkins, N. S.; Blake, A. J.; Wilson, C. *Tetrahedron* **2008**, *64*, 3713-3735.

[27] Balducci, D.; Contaldi, S.; Lazzari, I.; Porzi, G. *Tetrahedron Asymmetry* **2009**, *20*, 1398-1401.
[28] Knowles, H. S.; Hunt, K.; Parsons, A. F. *Tetrahedron Lett.* **2000**, *41*, 7121-7124.
[29] Chai, C. L. L.; Johnson, R. C.; Koh, J. *Tetrahedron* **2002**, *58*, 975-982.
[30] Sanz-Cervera, J. F.; Williams, R. M.; Marco, J. A.; López-Sánchez, J. M.; González, F.; Martínez, M. E.; Sancenón, F. *Tetrahedron* **2000**, *56*, 6345-6358.
[31] Morris, E. N.; Nenninger, E. K.; Pike, R. D.; Scheerer, J. R. *Org. Lett.* **2011**, *13*, 4430-4433.
[32] Chai, C. L. L.; Edwards, A. J.; Wilkes, B. A.; Woodgate, R. C. J. *Tetrahedron*, **2003**, *59*, 8731-8739.
[33] Huck, L.; González, J. F.; de la Cuesta, E.; Menéndez, J. C.; Avendaño, C. *Org. Biomol. Chem.* **2011**, *9*, 6271-6277.
[34] Speckamp, W. N.; Hiemstra, H. *Tetrahedron* **1985**, *41*, 4367-4416.
[35] Hiemstra, H.; Speckamp, W. N. In "The Alkaloids"; Academic Press: New York, 1988, Vol. 32, p 271.
[36] Hiemstra, H.; Speckamp, W. N. "Additions to *N*-Acyliminium Ions". In "Comprehensive Organic Synthesis", Trost, B. M.; Fleming, I. eds.; Pergamon Press: Oxford, **1991**, Vol. 2, p 1047-1082.
[37] Speckamp, W. N.; Moolenaar, M. J. *Tetrahedron*, **2000**, *56*, 3817-3856.
[38] Maryanoff, B. E.; Zhang, H-C.; Cohen, J. H.; Turchi, I. J.; Maryanoff, C. A. *Chem. Rev.* **2004**, *104*, 1431-1628.
[39] Royer, J.; Bonin, M.; Micouin, L. *Chem. Rev.* **2004**, *104*, 2311-2352.
[40] Chiou, W-H.; Mizutani, N.; Ojima, I. *J. Org. Chem.* **2007**, *72*, 1871-1882.
[41] Yazici, A.; Pyne, S. G. *Synthesis* **2009**, 339-368.
[42] Yazici, A.; Pyne, S. G. *Synthesis* **2009**, 513-541.
[43] Maryanoff, B. E.; Rebarchak, M. C. *Synthesis* **1992**, 1245-1248.
[44] Heaney, H.; Shuhaibar, K. F. *Synlett*, **1995**, 47-48.
[45] Li, W. H.; Hanau, C. E.; Davignon, A.; Moeller, K. D. *J. Org. Chem.* **1995**, *60*, 8155-8170.
[46] Lögers, M.; Overman, L. E.; Welmaker, G. S. *J. Am. Chem. Soc.* **1995**, *117*, 9139-9150.
[47] Marson, C. M.; Pink, J. H.; Smith, C. *Tetrahedron Lett.* **1995**, *36*, 8107-8110.
[48] Rigo, B.; El Ghammarti, S.; Couturier, D. *Tetrahedron Lett.* **1996**, *37*, 485-486.
[49] Yamada, H.; Aoyagi, S.; Kibayashi, C. *J. Am. Chem. Soc.* **1996**, *118*, 1054-1059.
[50] Allin, S. M.; Northfield, C. J.; Page, M. I.; Slawin, A. M. Z. *Tetrahedron Lett.* **1998**, *39*, 4905-4908.
[51] Lee, Y. S.; Lee, J. Y.; Kim, D. W.; Park, H. *Tetrahedron* **1999**, *55*, 4631-4636.
[52] Lee, Y. S.; Min, B. J.; Park, Y. K.; Lee, J. Y.; Lee, S. J.; Park, H. *Tetrahedron Lett.* **1999**, *40*, 5569-5572.
[53] Wang, H.; Ganesan, A. *Org. Lett.* **1999**, *1*, 1647-1649.
[54] Roy, B.; Pérez-Luna, A.; Ferreira, F.; Botuha, C.; Chemla, F. *Tetrahedron Lett.* **2008**, *49*, 1534-1537.
[55] Morgan, I. R.; Yazici, A.; Pyne, S. G. *Tetrahedron*, **2008**, *64*, 1409-1419.
[56] Gómez-San Juan, A.; Sotomayor, N.; Lete, E. *Tetrahedron Lett.* **2012**, *53*, 2157-2159.
[57] Liu, N.; Wang, L.; Zhang, W. *J. Org. Chem.* **2012**, *8*, 192-200.
[58] See for instance: Kubo, A.; Saito, N.; Yamato, H.; Masubuchi, K.; Nakamura, M. *J. Org. Chem.* **1988**, *53*, 4295-4310.
[59] Lee, B. H.; Clotier, M. F. *Tetrahedron Lett.* **1999**, *40*, 643-644.

[60] Ollero, L.; Mentink, G.; Rutjes, F. P. J. T.; Speckamp, N.; Hiemstra, H. *Org. Lett.* **1999**, *1*, 1331-1334.

[61] Collado, M. I.; Manteca, I.; Sotomayor, N.; Villa, M.-J.; Lete, E. *J. Org. Chem.* **1997**, *62*, 2080-2092.

[62] Kim, S.-H.; Kim, S.-I.; Cha, J. K. *J. Org. Chem.* **1999**, *64*, 6771-6775.

[63] Kim, S.-H.; Cha, J. K. *Synthesis* **2000**, 2113-2116.

[64] Moeller, K. D.; Wang, P. W.; Tarazi, Sh.; Marzabadi, M. R.; Wong, P. L. *J. Org. Chem.* **1991**, *56*, 1058-1067.

[65] Shono, T. *Tetrahedron*, **1984**, *40*, 811-850.

[66] Moeller, K. D.; Tarazi, S.; Marzabadi, M. R. *Tetrahedron Lett.* **1989**, *30*, 1213-1216.

[67] Shono, T.; Matsumura, Y.; Tsubata, K. *J. Am. Chem. Soc.* **1981**, *103*, 1172-1176.

[68] Utley, J. *Chem. Soc. Rev.* **1997**, 157-167.

[69] Magnus, P. Hulme, C. *Tetrahedron Lett.* **1994**, *35*, 8097-8100.

[70] Magnus, P.; Hulme, C.; Weber, W. *J. Am. Chem. Soc.* **1994**, *116*, 4501-4502.

[71] Boto, A.; Hernández, R.; Suárez, E. *Tetrahedron Lett.* **1999**, *40*, 5945-5948.

[72] Boto, A.; Hernández, R.; Suárez, E. *J. Org. Chem.* **2000**, *65*, 4930-4937.

[73] Boto, A.; Hernández, R.; Suárez, E. *Tetrahedron Lett.* **2000**, *41*, 2899-2902.

[74] Chai, C. L. L.; Elix, J. A.; Huleatt, P. B. *Tetrahedron Lett.* **2003**, *44*, 263-265.

[75] Chai, C. L. L.; Elix, J. A.; Huleatt, P. B. *Tetrahedron* **2005**, *61*, 8722-8739.

[76] Ottenheijm, H. C. J.; Plate, R.; Nooedik, J. H.; Herscheid, J. D. M. *J. Org. Chem.* **1982**, *47*, 2147-2154.

[77] Cannizo, L. F.; Grubbs, R. H. *J. Org. Chem.* **1985**, *50*, 2316-2323.

[78] Yoshimura, J.; Sugiyama, Y.; Nakamura, H. *Bull. Chem. Soc. Jpn.* **1973**, *46*, 2850-2853.

[79] Poisel, H.; Schmidt, U. *Angew. Chem. Int. Ed.* **1971**, *10*, 130-131.

[80] Williams, R. M.; Kwast, A. *J. Org. Chem.* **1988**, *53*, 5785-5787.

[81] Badran, T. W.; Easton C. J. *Aust. J. Chem.* **1990**, *43*, 1445-1449.

[82] Williams, R. M.; Anderson, O. P.; Armstrong, R. W.; Josey, J.; Meyers, H.; Ericksson, C. *J. Am. Chem. Soc.* **1982**, *104*, 6092-6099.

[83] Williams, R. M.; Armstrong, R. W.; Dung, J.-S. *J. Am. Chem. Soc.* **1984**, *106*, 5748-5750.

[84] Williams, R. M.; Armstrong, R. W.; Maruyama, L. K.; Dung, J-S. Anderson, O. P. *J. Am. Chem. Soc.* **1985**, *107*, 3246-3253.

[85] Whitlock, C. R.; Cava, M. P. *Tetrahedron Lett.* **1994**, *35*, 371-374.

[86] Badran, T. W.; Easton C. J.; Horn, E.; Kociuba, K.; May, B. L.; Schliebs, D. M.; Tiekink, R. T. *Tetrahedron: Asymmetry* **1993**, *4*, 197-200.

[87] For a review of the captodative effect, see: Sustmann, R.; Korth, H. G. *Adv. Phys. Org. Chem.* **1990**, *26*, 131-138.

[88] Badran, T. W.; Chai, C. L. L.; Easton C. J.; Harper, J. B.; Page, D. M. *Aust. J. Chem.* **1995**, *48*, 1379-1384.

[89] Ong, C. W.; Lee, H. C. *Aust. J. Chem.* **1990**, *43*, 773-775.

[90] Murata, S.; Suzuki, M.; Noyori, R. *Tetrahedron*, **1988**, *44*, 4259-4275.

[91] Hernández, F.; Lumetzberger, A.; Avendaño, C.; Söllhuber, M. *Synlett*, **2001**, 1387-1390.

[92] González. J. F.; de la Cuesta, E.; Avendaño, C. *Tetrahedron* **2004**, *60*, 6319-6326.

[93] González. J. F.; de la Cuesta, E.; Avendaño, C. *Tetrahedron Lett.* **2003**, *44*, 4395-4398.

[94] González. J. F.; Salazar, L.; de la Cuesta, E.; Avendaño, C. *Tetrahedron* **2005**, *61*, 7447-7455.

[95]   Ortín, I.; González, J. F.; de la Cuesta, E.; Manguán-García, C.; Perona, R.; Avendaño, C. *Bioorg. Med. Chem.* **2008**, *16*, 9065-9078.

[96]   Ortín, I.; González, J. F.; de la Cuesta, E.; Avendaño, C. *Bioorg. Med. Chem.* **2010**, *18*, 6813-6821.

[97]   Fukuyama, T.; Nunes, J. J. *J. Am. Chem Soc.* **1988**, *110*, 5196-5198.

[98]   Kubo, A.; Saito, N.; Nakamura, M.; Ogata, K.; Sakai, S. *Heterocycles*, **1987**, *26*, 1765-1770.

[99]   Saito, N.; Tachi, M.; Seki, R.; Kamayachi, H.; Kubo, A. *Chem. Pharm. Bull.* **2000**, *48*, 1549-1557.

[100]   Kumar, J. R.; Zi-rong, X. *Mar. Drugs* **2004**, *2*, 123-146.

[101]   Corey, E. J.; Gin, D. Y.; Kania, R. S. *J. Am. Chem. Soc.* **1996**, *118*, 9202-9203.

[102]   Martínez, E. J.; Corey, E. J. *Org. Lett.* **2000**, *2*, 993-996.

[103]   Fukuyama, T.; Sachleben, R. A. *J. Am. Chem. Soc.* **1982**, *104*, 4957-4958.

[104]   See the use of methanesulfonic anhydride for this purpose in: Saito, N.; Tashiro, K.; Maru, Y.; Yamaguchi, K.; Kubo, A. *Chem. Soc., Perkin Trans, 1*, **1997**, 53-69.

[105]   Kubo, A.; Saito, N.; Yamauchi, R.; Sakai, S. *Chem. Pharm. Bull.* **1987**, *35*, 2158-2161.

[106]   Kubo, A.; Saito, N.; Yamato, H.; Kawanami, Y. *Chem. Pharm. Bull.* **1987**, *35*, 2525-2532.

[107]   Kubo, A.; Saito, N.; Yamato, H.; Yamauchi, R.; Hiruma, K.; Inoue, S. *Chem. Pharm. Bull.* **1988**, *36,* 2607-2614.

[108]   Fukuyama, T.; Yang, L.; Ajeck, K. L.; Sachleben, R. *J. Am. Chem. Soc.* **1990**, *112*, 3712-3713.

[109]   Dong, W.; Liu, W.; Liao, X.; Guan, B.; Chen, Sh.; Liu, Zh. *J. Org. Chem.* **2011**, *76*, 5363-5368.

[110]   Saito, N.; Tanitsu, M.; Betsui, T.; Suzuki, R.; Kubo, A. *Chem. Pharm. Bull.* **1997**, *45*, 1120-1129.

[111]   Veerman, J. J. N.; Robin, S. B.; Hue, B. T. B.; Girones, D.; Rutges, F. P. J. T.; van Maarseveen, J. H.; Hiemstra, H. *J. Org. Chem.* **2003**, *68*, 4486-4494.

[112]   Tang, Y-F.; Liu, Z-Z.; Chen, S-Z. *Tetrahedron Lett.* **2003**, *44*, 7091-7094.

[113]   González, J. F.; de la Cuesta, E.; Avendaño, C. *Tetrahedron* Lett. **2006**, *47*, 6711-6714.

[114]   González, J. F.; de la Cuesta, E.; Avendaño, C. *Tetrahedron* **2008**, *64,* 2762-2771.

[115]   Pen, J.; Mantle, P. G.; Bilton, J. N.; Sheppard, R. N. *J. Chem. Soc., Perkin Trans 1* **1992**, 1495-1496.

[116]   Numata, A.; Takahashi, C.; Matshushita, T.; Miyamoto, T.; Kawai, K.; Usami, Y.; Matsumura, E.; Inoue, M.; Ohishi, H.; Shingu, T. *Tetrahedron Lett.* **1992**, *33*, 1621-1624.

[117]   Takahashi, C.; Matsushita, T.; Doi, M.; Minoura, K.; Shingu, T.; Kumeda, Y.; Numatta, A. *J. Chem. Soc., Perkin Trans. 1* **1995**, 2345-2353.

[118]   Wong, S.-M.; Musza, L. L.; Kydd, G. C.; Kullnig, R.; Gillum, A. M.; Cooper, R. *J. Antibiot.* **1993**, *46*, 545-553.

[119]   Fujimoto, H.; Negishi, E.; Yamaguchi, K.; Nishi, N.; Yamazaki, M. *Chem. Pharm. Bull.* **1996**, *44*, 1843-1848.

[120]   Hochlowski, J. E.; Mullally, M. M.; Spanton, S. G.; Whittern, D. N.; Hill, P.; McAlpine, J. B. *J. Antibiot.* **1993**, *46*, 380-386.

[121]   Marsden, S. P.; Depew, K. M.; Danishefsky, S. J. *J. Am. Chem. Soc.* **1994**, *116*, 11143-11144.

[122]   Depew, K. M.; Marsden, S. P.; Zatorska, D.; Zatorski, A.; Bornmann, W. G.; Danishefsky, S. J. *J. Am. Chem. Soc.* **1999**, *121*, 11953-11963.

[123]   Sinclair, P.J.; Zhai, D.; Reibenspies, J.; Williams, R.M. *J. Am. Chem. Soc.* **1986**, *108*,1103-1104.

[124]   Martín-Santamaría, S.; Espada, M.; Avendaño, C.; *Tetrahedron* **1999**, *55*, 1755-1762.

[125]   Sánchez, J. D., Ramos, M. T.; Avendaño, C. *Tetrahedron Lett.* **2000**, *41*, 2745-2748.

[126]   Sánchez, J. D., Ramos, M. T.; Avendaño, C.; *J. Org. Chem.* **2001**, *66*, 5731-5735.

[127]   Heredia, M. L.; Fernández, M.; de la Cuesta, E.; Avendaño, C. *Tetrahedron: Asymmetry* **2001**, *12*, 411-418.

[128]   Heredia, M. L.; de la Cuesta, E.; Avendaño, C. *Tetrahedron: Asymmetry* **2001**, *12*, 2883-2889.

[129]   Heredia, M. L.; de la Cuesta, E.; Avendaño, C. *Tetrahedron* **2002**, *58*, 6163-6170.

[130]   Snider, B. B.; Zeng, H. *Org. Lett.* **2002**, *4*, 1087-1090.

[131]   Hart, D. J.; Magomedov, N. A. *J. Am. Chem. Soc.* **2001**, *123*, 5892-5899.

Send Orders of Reprints at bspsaif@emirates.net.ae
Advances in Organic Synthesis, Vol. 5, 2013, 355-378 355

CHAPTER 8

Advances in the Synthesis of Organometallic Amino Acids and Analogues

Poulami Jana, Sibaprasad Maity and Debasish Haldar*

Department of Chemical Sciences, Indian Institute of Science Education and Research –Kolkata, Mohanpur, West Bengal- 741252, India

Abstract: Incorporation of metal chelating amino acids in protein or peptide sequences helps protein purification due to the metal ion affinity of the protein or peptide. Structure and function of synthetic amino acids with side chain metal complexes have particular importance for extending our understanding of organometallic compounds under physiological conditions and mimic protein folding and stabilization. Chemists are now designing biomimetic organo-metallic amino acids that can form both secondary and tertiary structures through various noncovalent interactions. An overview of the design and synthesis of non-proteinogenic amino acids containing metal ions in the side chain is discussed.

Keywords: Metal chelating amino acids, amino acid analogues, side chain metal complex, synthesis, structure, ferrocenyl bis alanine, ferrocene substituted amino acids, aminoferrocenecarboxylic acids, click chemistry, acid anhydrite, α-amino acid, β-amino acids, amino acids with iron carbonyl complexes, chromium containing amino acids, amino acids containing manganese, ruthenium containing amino acids, amino acids with rhodium, amino acids with cobalt at side chain, amino acids with platinum at side chain, amino acids, palladium at side chain.

1. INTRODUCTION

The nonproteinogenic metal chelating amino acids are versatile tools for biochemistry. The nonproteinogenic amino acids present in natural products, either in native state or as fragments of complex molecules, are highly important as broad-spectrum antibiotics activity (peptaibols) [1] or display enhanced resistance against chemical [2] and enzymatic degradation [3]. This article focuses on synthesis of organometallic nonproteinogenic amino acids from transition

*Address correspondence to Debasish Haldar: Department of Chemical Sciences, Indian Institute of Science Education and Research –Kolkata, Mohanpur, West Bengal- 741252, India; Tel: +91 33 25873020; Fax: +91 33 25873119; E-mails: deba_h76@yahoo.com; deba_h76@iiserkol.ac.in

metals with amino acids and peptides, a branch of bioorganometallic chemistry [4]. The metal containing amino acids and their analogues have wide functional diversity in recognition events, folding and stability and as delivery vehicles [5]. This functional diversity is reflected in the diversity of metal ions in the side chain of the amino acids structures. In addition, they can be used as building blocks for the synthesis of new molecules or as surrogates of native amino acids in peptide mimetic to modulate their biological activities. In this context, metal-amino acids and their derivatives have attracted a great deal of attention among synthetic organic chemists and biochemists in recent years. Wolfgang Beck and co-workers have published a comprehensive review on the amino acids-transition metal ions organometallic complexes [6].

In 1929, Cremer obtained the first α amino acids containing metal ions in side chain as a carboxylate carbonyl -[Fe(CysO)$_2$(CO)$_2$]- complex [7]. Four years later the synthesis of the compound was reported by Schubert [8]. That was the first isolated and characterized organometallic α-amino acid complex. Later, in 1967 an N, S-chelate with cis-oriented carbonyl ligands was proposed on the basis of IR spectroscopic data [9]. Recently, the active center of Ni/Fe hydrogenases has been reported to contain an iron carbonyl complex **1** (Scheme 1) [10]. This review will provide a deep and general view of the development for the synthesis of amino acids and their simple analogues containing metal ion at side chains. The review is divided into sections depending on the metal ion incorporated. For each case, methods for the introduction of the metal ions are discussed. The scope of the review is limited to synthetic amino acids that have a metal complex at side chain and have been incorporated in peptide/protein synthesis. Complexes where the metal is directly attached with amine or acid groups are out of the scope of this review.

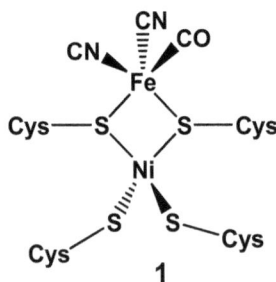

Scheme 1: Ni/Fe hydrogenases complex containing an iron carbonyl group.

2. SYNTHESIS OF IRON CONTAINING AMINO ACIDS

2.1. An Overview

The incorporation of iron in the side chain of amino acid can have a special influence on the conformational preferences of peptides with enhanced properties [11]. Ferrocene derivatives [12] have enhanced characteristics like excellent stability in water and air, favourable electrochemical properties that lead to applications in molecular receptors and sensor devices with electrochemical detection, [13] and as marker of biomolecules [14]. Moreover, about 3.3 Å inter-ring separation of ferrocene (comparable with the N···O distance in peptides β-sheets) and the rotational freedom of the cyclopentadienyl rings have allowed to use ferrocene containing amino acid as an organometallic β-turn mimetic [15] and as a molecular pair of scissors [16]. Despite the existence of many powerful methods for the synthesis of iron containing amino acids, development of truly efficient methods for their preparation in enantiomerically pure form has become of great importance.

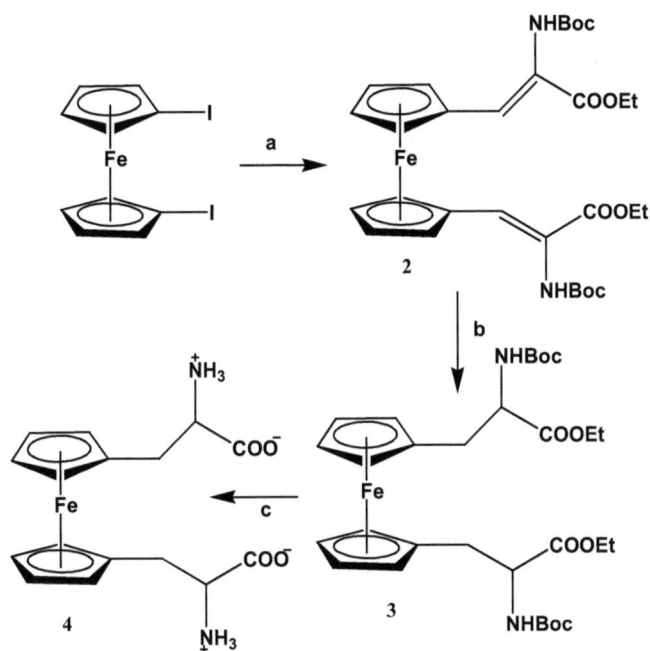

Scheme 2: 1,1′-ferrocenyl bis alanine according to Carlstrom *et al. Reagents and conditions*: (a) $CH_2=CCO_2R(NHBoc)$, $Pd(OAc)_2$, Bu_4NCl, $NaHCO_3$; (b) H_2, Rh-prophos; (c) 1. TMS-Cl/ phenol, 2. NaOH (aq).

2.2. Synthesis of 1,1′-Ferrocenyl Bis Alanine

Numerous approaches have been made in the synthesis of the amino acid 1,1′-ferrocenyl bis alanine. Carlstrom *et al.* demonstrated the synthesis of 1,1′-disubstituted ferrocenyl amino acid **4** (1,1′-ferrocenyl bis alanine) by two different routes. The optically active **4** was obtained by asymmetric homogeneous hydrogenation of the bis (didehydro amino acid) derivative **3** using [Rh((R)-prophos)(NBD)]ClO4 as catalyst followed by deprotection with TMS-Cl and phenol. The bis (didehydro amino acid) derivatives **3** were prepared in 40% yield by a palladium catalysed coupling between 1,1′-diiodoferrocene and ethyl 2-[(tert-butoxycarbonyl)amino]-acrylate **2** (Scheme 2) [17]. Alternatively racemic **3** derivatives were obtained *via* the bis (nitro ester). Bis (hydroxymethy1) ferrocene **5** was synthesized in a one-pot procedure directly from ferrocene by dilithiation followed by treatment with paraformaldehyde (Scheme 3) [17]. The key step in this synthesis was the one-step convertion (54%) of the nitro compound into the Boc-protected amino acid derivative.

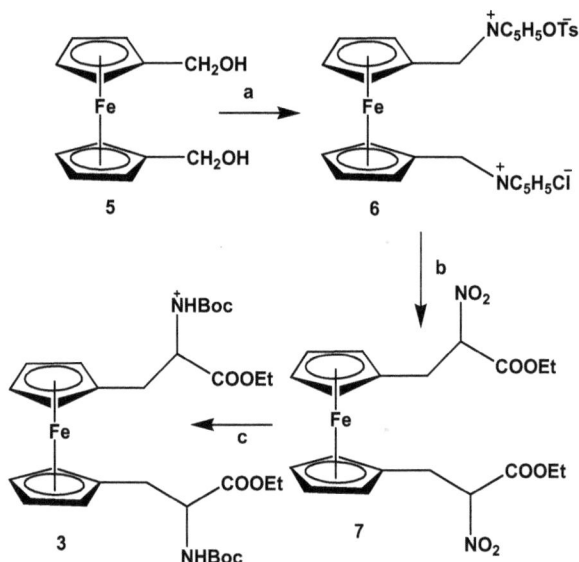

Scheme 3: Boc protected 1,1′-ferrocenyl bis alanine the one-step conversion according to Carlstrom *et al. Reagents and conditions*: (a) TsCl, Pyridine; (b) (EtO2C)O2NCHNa; (c) H2, Pd/C, Boc2O.

Basu *et al.* have reported the synthesis of optically active 1,1′-ferrocenyl bis alanine containing four different protecting groups (Scheme 4) [18]. The reaction starts with the unselective thioacetalisation of **8**. The monoprotected derivative **9** was isolated in 40% yield. The free aldehyde function was then condensed with the phosphonylglycine derivative to give the protected didehydroamino acid **10**. A second condensation reaction of **11** with phosphonylglycine derivative gave the bis armed unsaturated derivative **12**. The hydrogenation of **12** using [Rh((R,R)-DIPAMP)(COD)]BF$_4$ as a catalyst resulted in the 1,1′-ferrocenyl bis alanine containing four different protecting groups **13**.

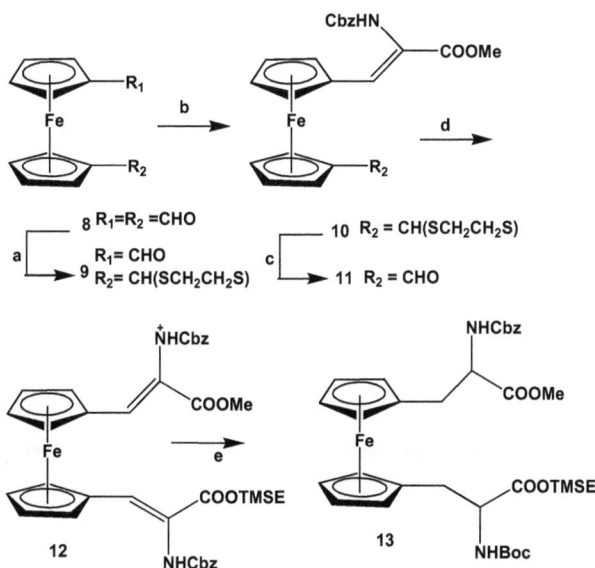

Scheme 4: 1,1′-ferrocenyl bis alanine containing four different protecting groups according to Basu *et al*. *Reagents and conditions*: (a) BF$_3$.Et$_2$O (b) TMG, (MeO)$_2$P(O)CH(NHCbz)CO$_2$Me (c) NCS/ AgNO$_3$ (d) TMG, (MeO)$_2$P(O)CH(NHBoc)CO$_2$TMSE (e) H$_2$, 4 atm, 40°C, MeOH, [Rh((R,R)-DIPAMP)(COD)]BF$_4$. TMG = 1,1,3,3-tetramethyl guanidine.

2.3. Synthesis of Ferrocene Amino Acid

Various approaches have been performed to synthesize the ferrocene substituted amino acids analogues that can be utilized as a chain reversal in peptide segment [19]. Butler and Quayle demonstrated the synthesis of ferrocene substituted amino acid by the monolithiation of 1,1'-dibromoferrocene **14** (Scheme **5**) [20]. To afford monolithiated intermediate **15** was reacted with 0.4 equivalents of *o*-benzylhydroxylamine to yield 1- bromo-1'-aminoferrocene **16** as a pale yellow crystalline solid. Lithiation of 1-bromo-1'-aminoferrocene **16** was carried out using 2.5 equivalents of *n*-butyllithium to obtained monolithiated 1-bromo-1'-

aminoferrocene **17**. The use of dry solid carbon dioxide as a quenching reagent resulted in the immediate precipitation of amino acid **18** in the form of a water-soluble salt which was removed by filtration and washed with diethyl ether.

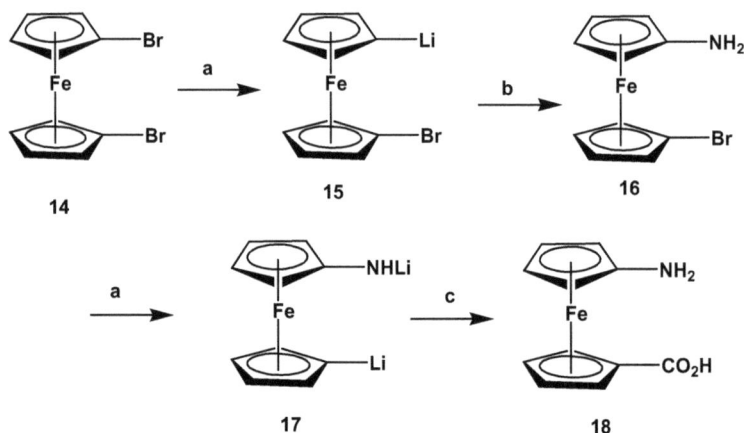

Scheme 5: Ferrocene amino acid according to Butler *et al.* [20]. *Reagents and conditions*: (a) BuLi (b) BnONH$_2$ (c) CO$_2$.

Another ferrocene amino acid (1-aminoferrocene-1′-carboxylic acid) was developed by Heinze *et al.* starting from ferrocene (Scheme 6) [21]. The iodoferrocene **21** has been synthesized from the selective mono-lithiation of ferrocene **19** with *t*-BuLi, then isolating the solid lithio-ferrocene **20** and quenching of **20** with iodine. Compound **22** was prepared in the absence of solvent directly from **21** and pre-formed Cu-phthalimide. Hydrazinolysis of **22** in ethanol gives pure **23**, which is then *N*-acetylated with acetic anhydride to form 1-(acetylamino) ferrocene **24**. The *N*-acetylamino group deactivates the substituted Cp ring, allowing the selective Friedel-Crafts acylation of **24** with 2, 6- dichlorobenzoyl chloride at the unsubstituted Cp ring giving 1-(acetylamino)-1-(2, 6-dichlorobenzoyl) ferrocene **25**. Basic hydrolysis of **25** introduces the desired COOH group seen in **26**. Removal of the acetyl protection group with hydrochloric acid furnishes the ferrocene-containing amino acid **27** as the hydrochloride.

2.4. Synthesis of *N-N*-(Ferrocene-1-Acetyl)-*L*-Lysine

Gellett *et al.* have reported a highly efficient way to synthesize *N-N*-(ferrocene-1-acetyl)-*L*-lysine **31** (Scheme 7) [22]. The synthesis begins with the preparation of

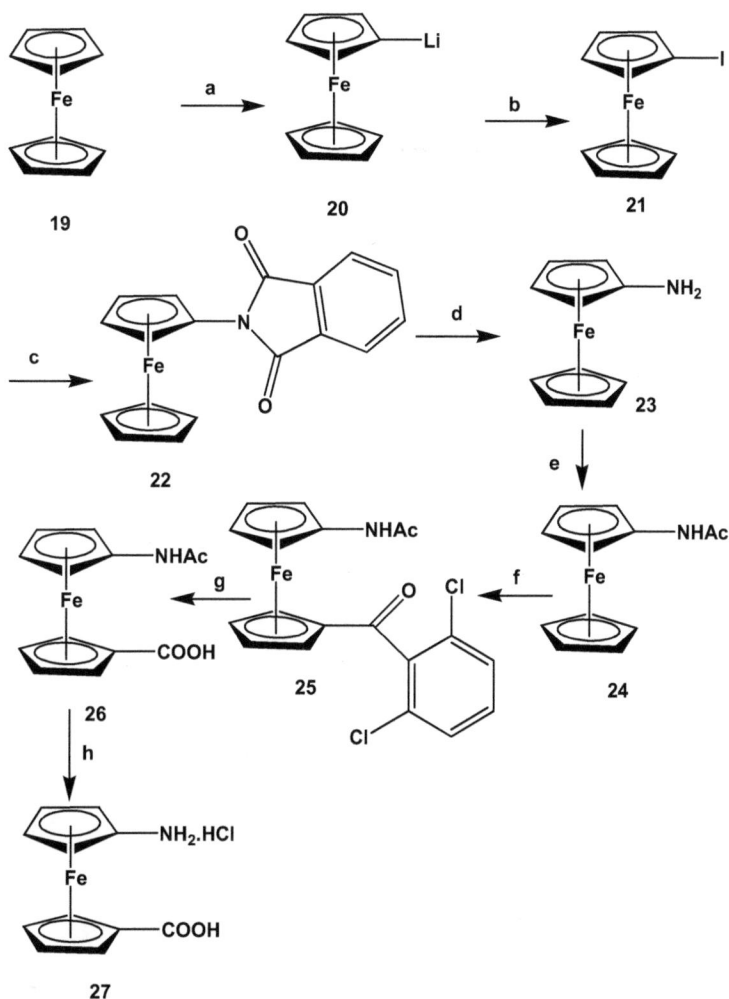

Scheme 6: Ferrocene amino acid according to Heinze *et al.* [21]. *Reagents and conditions*: (a) *t*-BuLi (b) iodine (c) Cu-phthalimide (d) NaOH, ethanol (e) acetic anhydride (f) 2, 6-dichlorobenzoyl chloride, AlCl$_3$ (g) NaOH (h) hydrochloric acid.

N^α-(1,1-dimethylethoxy-carbonyl)-*L*-lysine-tert-butyl ester **20** in two steps starting from a N^α-^{t}Boc-N^ε-(carbobenzyloxy)-*L*-lysine **28**. A coupling reaction to protect the carboxyl group followed by removal of the Cbz group from the ε-amino group of the lysine side chain. An overall yield of 70% was realized. Coupling of the free ε-amino group of **29** with the succinimido ester of ferroceneacetic acid results in a stable amide bond that attaches the ferrocene substituent onto the lysine side chain to produce compound **30** in a 73% yield. The deprotection of Boc group with TFA delivered the final product **31**.

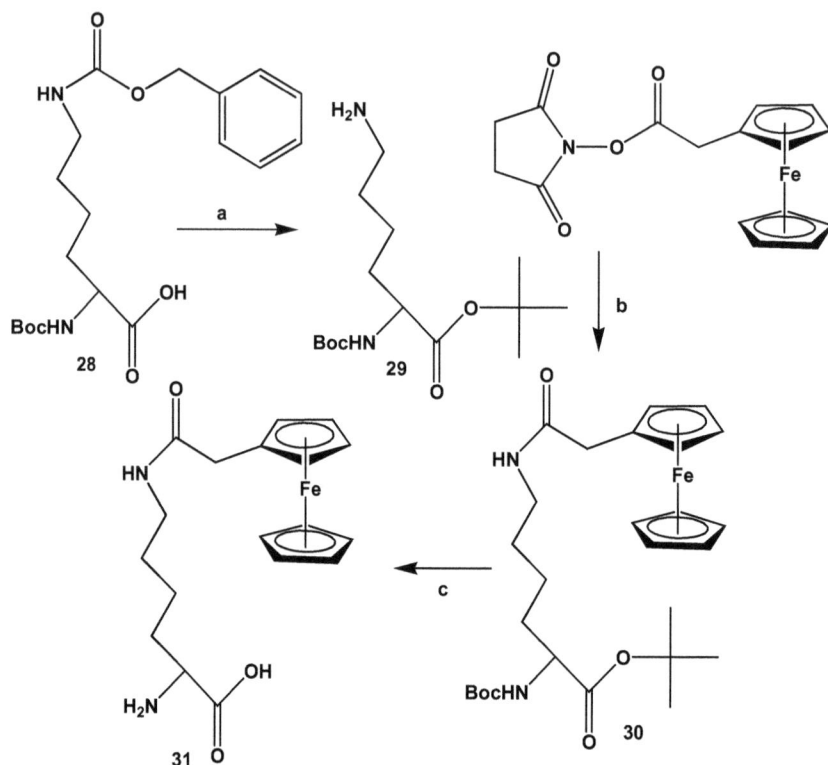

Scheme 7: *N-N-*(ferrocene-1-acetyl)-*L*-lysine according to Gellett *et al.* [22]. *Reagents and conditions*: (a) EDAC, DMAP, t-butanol, CH_2Cl_2 and H_2, Pd/C, CH_2Cl_2, MeOH (b) Na_2CO_3, CH_3CN (c) TFA, anisole, CH_2Cl_2.

2.5. Synthesis of 2-Ferrocenylvinyl-Substituted Phenylalanine

Burk and co-workers have reported the synthesis of iron containing amino acids using a versatile tandem catalysis procedure (Scheme **8**) [23]. This approach generally involves the standard Suzuki cross-coupling reactions. They have prepared enantiomerically pure (*R*)-N-acetyl methyl esters of *o*-bromo-, *m*-bromo- and *p*-bromophenylalanine **35-37,** which subsequently were employed in Pd-catalyzed cross-coupling reactions with 2-ferrocenylvinyl boronic acid derivatives to introduce 2-ferrocenylvinyl into each position of the phenylalanine ring **38-40**.

2.6. Synthesis of (*pS*)- and (*pR*)-2-Aminoferrocenecarboxylic Acids

Richards and co-workers have presented the synthesis of (*pS*)- and (*pR*)-2-aminoferrocenecarboxylic acids by lithiated ferrocenyloxazolines in conjunction with N_2O_4 as an electrophilic quench [24]. The diastereoselective lithiation of (*S*)-

2-ferrocenyl-4-(1-methylethyl) oxazoline **41**, followed by addition of N_2O_4, gave (*S*)-2-[(*pS*)-2-nitroferrocenyl]-4-(1-methylethyl) oxazoline **42** which was subsequently converted into derivatives of (*pS*)-2-aminoferrocenecarboxylic acid **44** (Scheme **9**).

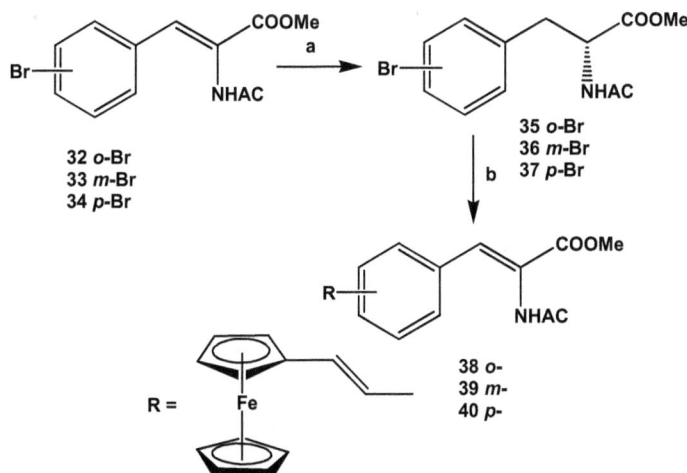

Scheme 8: Synthesis of 2-ferrocenylvinyl- substituted phenylalanine according to Bruk *et al.* [23]. *Reagents and conditions*: (a) (R,R)-Pr-DuPHOS-Rh, H_2 (b) Pd(OAc)$_2$/2 P(o-tolyl)$_3$ R-B(OH)$_2$, DME, 2 M Na$_2$CO$_3$, 80°C, 1-3 h.

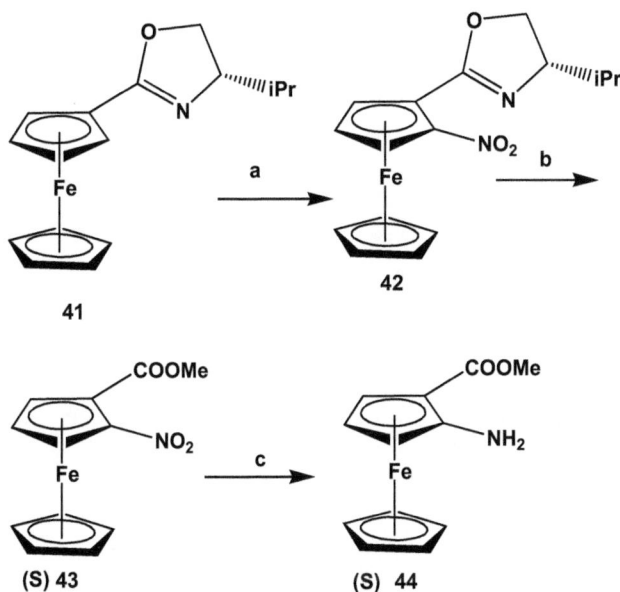

Scheme 9: Synthesis of (*pS*)-2-aminoferrocenecarboxylic acids according to Salter *et al.* [24]. *Reagents and conditions*: (a) BuLi, TMEDA, N_2O_4 (b) TFA, Ac$_2$O, NaOMe (c) PtO$_2$, H_2, EtOH.

Moreover, the enantiomers (*pR*)-2-aminoferrocenecarboxylic **48** were also obtained from **41** through use of a removable trimethylsilyl blocking group which was introduced in a one-pot procedure prior to further lithiation and addition of N_2O_4 (Scheme **10**) [24]. The resulting trisubstituted ferrocene **45**, also obtained as a single diastereoisomer, was desilylated and compound **46** was ring opened and transesterified as before to give (*R*)-**47**, and (*pR*)-2-aminoferrocenecarboxylic acid methyl ester **48** on reduction.

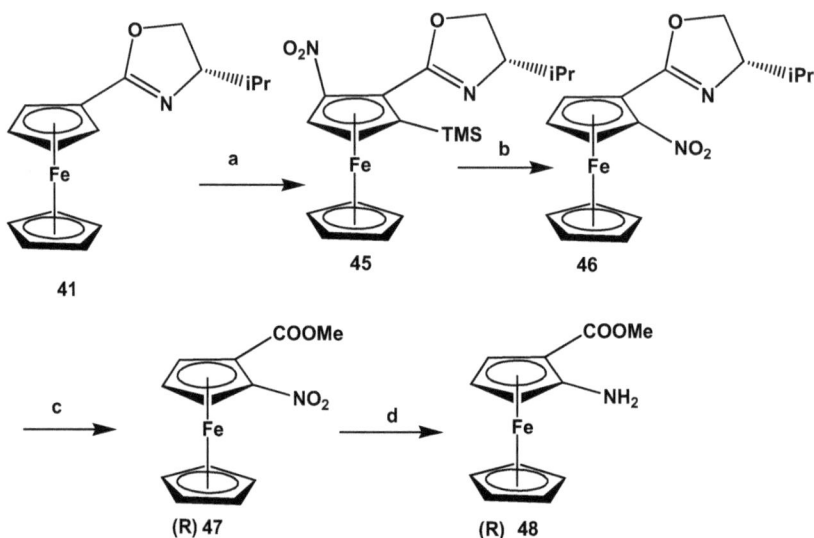

Scheme 10: Synthesis of (*pR*)-2-aminoferrocenecarboxylic acids according to Salter *et al.* [24]. *Reagents and conditions*: (a) BuLi, TMEDA; TMSCl; BuLi; N_2O_4 (b) TBAF (c) TFA, Ac_2O, NaOMe (d) PtO_2, H_2, EtOH.

2.7. Synthesis of Ferrocenyl-Amino Acids by Click Chemistry

Chandrasekaran and co-workers have been reported the synthesis of a wide range of ferrocenyl-amino acids and other derivatives in excellent yield. Diverse amino acid containing azides were synthesized and ligated to ferrocene employing click chemistry to access ferrocenyl amino acids. To synthesize ferrocene amino acids, amino acids containing azido group, which could be subjected to click reaction with ethynyl ferrocene have been developed. When azide (**50**) was treated with ethynyl ferrocene **49** in CH_3CN with CuI as catalyst and DIPEA as base (rt, 7–8 h) the corresponding protected ferrocene amino acid derivatives (**51**) was isolated in excellent yields (Scheme **11**) [25].

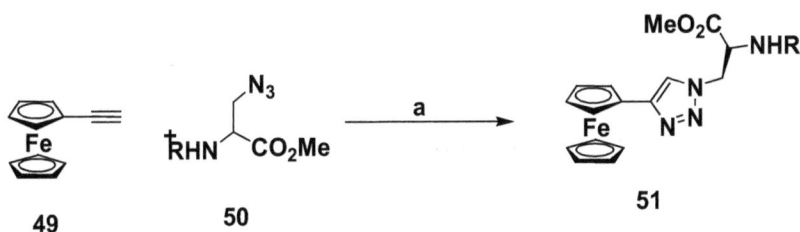

Scheme 11: Synthesis of ferrocene amino acid from amino acid derived azides according to Chandrasekaran *et al.* [25]. *Reagents and conditions*: (a) CuI, DIPEA, CH$_3$CN, rt, 7-10 h.

2.8. Synthesis of Ferrocene Amino Acid from Acid Anhydrite

Suggs and co-worker have reported ferrocene amino acid **53** from 1,10-ferrocenediacetic anhydride (**52**) with N,N-dimethylethylenediamine under an atmosphere of N$_2$ at 0 °C (Scheme **12**) [26].

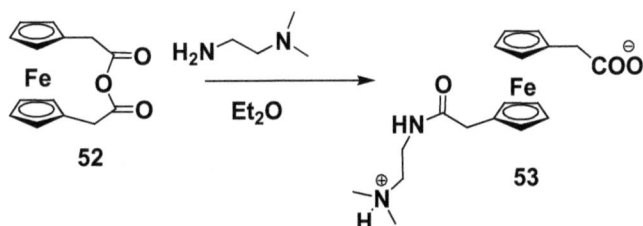

Scheme 12: Synthesis of ferrocene amino acid **53** according to Suggs *et al.* [26]. *Reagents and conditions*: N,N-dimethylethylenediamine, Et$_2$O, 0°C, N$_2$.

2.9. Synthesis of Amino Acids with Iron Carbonyl Complexes

Another type of iron containing amino acids was designed by utilizing the coordination complex of side chains π electrons. The coordination of the amino acid side chains with the organometallic complex is highly interesting due to the special properties such as relatively hydrophobic, aromatic character in some cases, unconventional shapes that offer a means of probing the stereoelectronic prerequisites for specific substrate-receptor interactions. Moreover the spectroscopic properties of the complex fragments allow them to be used as biomarkers. The carbonyl complexes of Fe and cyclobutadienylalanine **57** have been prepared by conversion of **54** into the trimethylamonium iodide derivative **55**. The **55** was then converted into **56** by treatment with diethylsodioformamido malonate. The ester **56** was hydrolysed and decarboxylated to give the desired product **57** (Scheme **13**) [27].

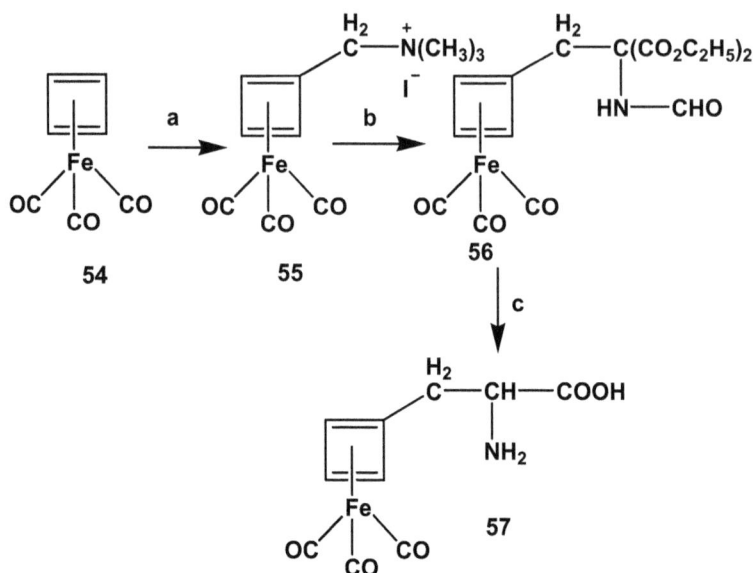

Scheme 13: Synthesis of cyclobutadienylalanine according to Brunet *et al.* [27]. *Reagents and conditions*: (a) $(CH_3)_3NCH_2I$ (b) diethylsodioformamido malonate (c) NaOH, HCl.

On the other hand the addition of π-coordinated unsaturated hydrocarbon ligands with nucleophilic glycine or alanine synthons is a versatile method for the synthesis of amino acid derivatives containing metal-carbonyl side chains. Stephenson and coworkers have reported the synthesis of **60** starting from **58** and the cationic carbonyl complexes of iron as electrophiles (Scheme **14**) [28].

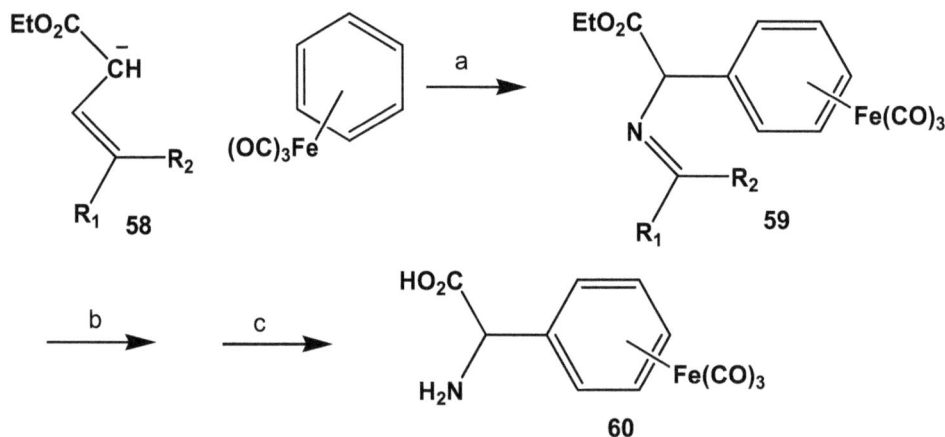

Scheme 14: Synthesis of amino acid derivatives containing metal-carbonyl side chains according to Stephenson *et al.* [28]. *Reagents and conditions*: (a) $AlCl_3$ (b) H_3O^+ (c) NaOH.

3. SYNTHESIS OF CHROMIUM CONTAINING AMINO ACIDS

3.1. An Overview

Starting in the mid-seventies, impressive achievements have been made in the synthesis of chromium carbonyl complex with amino acid side chain [29]. Actually the metal carbonyl group was introduced in tryptophan aromatic ring **61** as a protecting group to prevent oxidative destruction Scheme **15**. There are scavengers that can minimize the oxidation but electrophilic attack to the aromatic rings cannot be avoided.

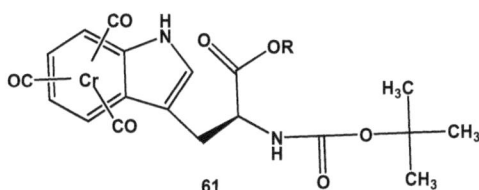

Scheme 15: Tryptophan containing $Cr(CO)_3$ in side chains according to Sergheraert *et al.* [29].

3.2. Synthesis of Amino Acids with Chromium Carbonyl Complexes

The chromium containing amino acids were mostly designed by utilizing the coordination complex of side chains π electrons with chromium carbonyl. The π-electrons of the aromatic amino acid side chains are an easy target for the introduction of transition metal complex fragments. Lavergne and co-workers have reported the synthesis of an amino acid containing a chromium carbonyl complex in its side chain **65** [30]. Methyl arylacetate–tricarbonylchromium complexes **62** formed corresponding anion **63** in presence of NaH or t-BuOK in THF at room temperature (Scheme **16**). Then the anion was readily alkylated by halogeno esters **64** in at -78 °C to give the amino acid **65** in good yield.

4. SYNTHESIS OF AMINO ACIDS CONTAINING MANGANESE

Design amino acids with manganese in side chain are highly interesting. Brunet and coworkers have reported the synthesis of amino acids containing manganese carbonyl in side chain [27]. Due to the fact that methylation of the amino group fails with cymantrene, the formamidomalonate route was not practiceable. The formyl cymantrene **67** was prepared by reaction in THF at low temperature between cymantrenyl **66** and dimethyl formamide (Scheme **17**). Compound **67** was then treated with hippuric acid to yield the azlactone **68**. Reduction of **68** using phosphorus and hydriodic acid in acetic anhydride gave **69** in 60% overall yield.

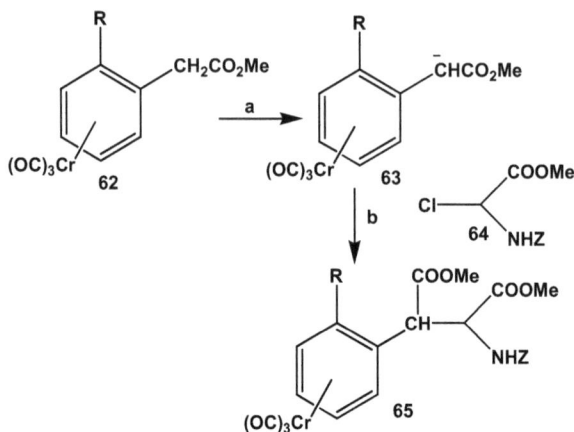

Scheme 16: Amino acids containing $Cr(CO)_3$ in side chains according to Jenhi *et al.* [30]. *Reagents and conditions*: (a) NaH (*t*-BuOK), THF, RT (b) -78 °C to RT.

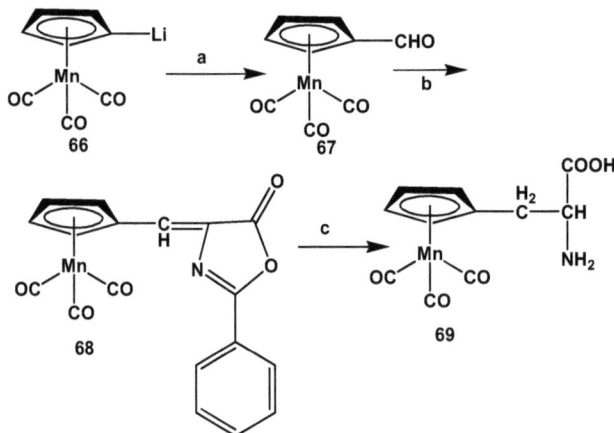

Scheme 17: Amino acids containing $Mn(CO)_3$ in side chains according to Brunet *et al.* [27]. *Reagents and conditions*: (a) DMF/THF (b) hippuric acid, acetic anhydride (c) P/HI.

5. SYNTHESIS OF RUTHENIUM CONTAINING AMINO ACIDS

5.1. An Overview

The synthesis of ruthenium containing amino acids and its derivatives have been the subject of extensive investigations because of their potential application in biological labeling. Many groups have been used to design amino acids with ruthenium in side chains. These compounds can function as "reporters" on the location of the bound molecules distributed within various media such as tissue, membranes and organs. Other potential applications include metalloimmunoassay and as radiopharmaceuticals [31].

5.2. Synthesis of Amino Acid Containing Cyclopentadienyl Ruthenium (II) Complexes

Gill and coworkers have succeeded in attaching a cyclopentadienylruthenium (II) (CpRu+) unit onto the aromatic nucleus of amino acid side chain (Scheme **18**) [32]. To a degassed solution of N-acetyl phenylalanine or tyrosine or tryptophan in 1,2-dichloroethane under a nitrogen atmosphere was added [CpRu(CH,CN),]PF, **70**. The yellow reaction mixture was stirred for 12 h at 45-50 °C. Then the solvent was removed on a rotary evaporator and the remaining dark brown residue was washed with ether to remove unreacted amino acids. The brown residue was redissolved in acetone, dried over Na_2SO_4, and decolorized with charcoal. After concentration of acetone solution, ether was added to obtain about 70% yield.

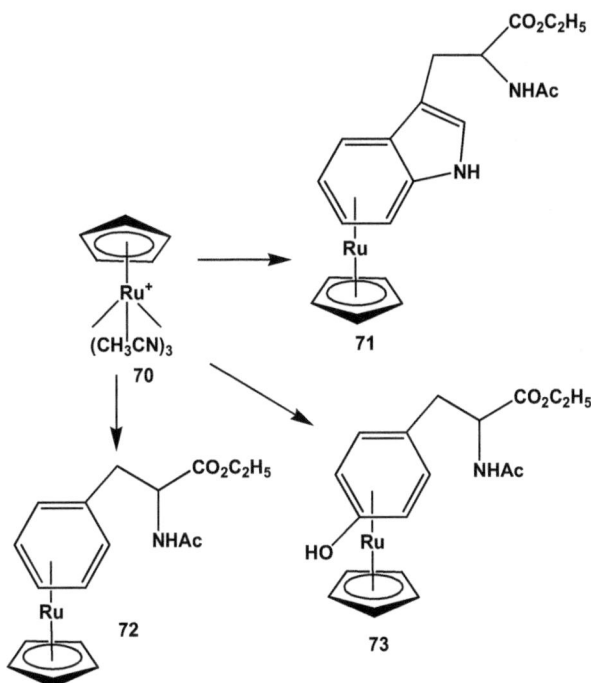

Scheme 18: Amino acids containing cyclopentadienyl ruthenium (II) complex in side chains according to Moriarty *et al.* [32].

5.3. Synthesis of Ruthenium (II) Tris (bipyridyl) Amino Acid

Kise *et. al.* have reported the synthesis of ruthenium (II) tris (bipyridyl) amino acid **79** from Boc-protected bipyridine amino acid *N-tert*-butoxycarbonyl-2-amino-3-(4,4′-methyl-2,2′-bipyridin-4-yl)propanoic acid **78**, using commercially available 4,4′-dimethyl-2,2′-bipyridine as starting material [33]. They have used the commercially available chiral phase transfer catalyst (8*S*,9*R*)-(-)-*N*-benzylcinchonidinium chloride in the reaction of 4-(bromomethyl)-4,4′-methyl-2,2′- bipyridine **74** with *N*-(diphenylmethylene)glycine *tert*-butyl ester **75**. The ee of the L isomer of 2-amino-3-(4,4′-methyl-2,2′-bipyridin-4-yl)propanoic acid **77** was 66%. The metalloamino acid **79** was produced in one step (Scheme **19**) by reaction of the Boc-protected amino acid **78** with *cis*-dichlorobis(2,2′-bipyridine)-ruthenium(II).

5.4. Synthesis of Bis(arene) Ruthenium(II) Phenylalanine

The incorporation of ruthenium in amino acids side chain has proved to be very useful for influencing the molecular recognition of enzyme active sites [34]. Sheldrick and co-workers have developed a bis(arene) ruthenium(II) phenylalanine with η^6coordination complex [35].

A solution of HpheOMe.HCl **80** in methanol was stirred with $Ag(CF_3SO_3)$ **81** for 10 minutes and the precipitated AgCl filtered off and washed with methanol (Scheme **20**). A solution of $[Ru(acetone)_3(\eta^6\text{-cymene})]^{2+}$ **83** in CH_2Cl_2 was added to the resulting solid and refluxed for 48 hours. The off-white precipitate was washed with CH_2Cl_2 and dried under vacuum to obtained 79% yield of **84**.

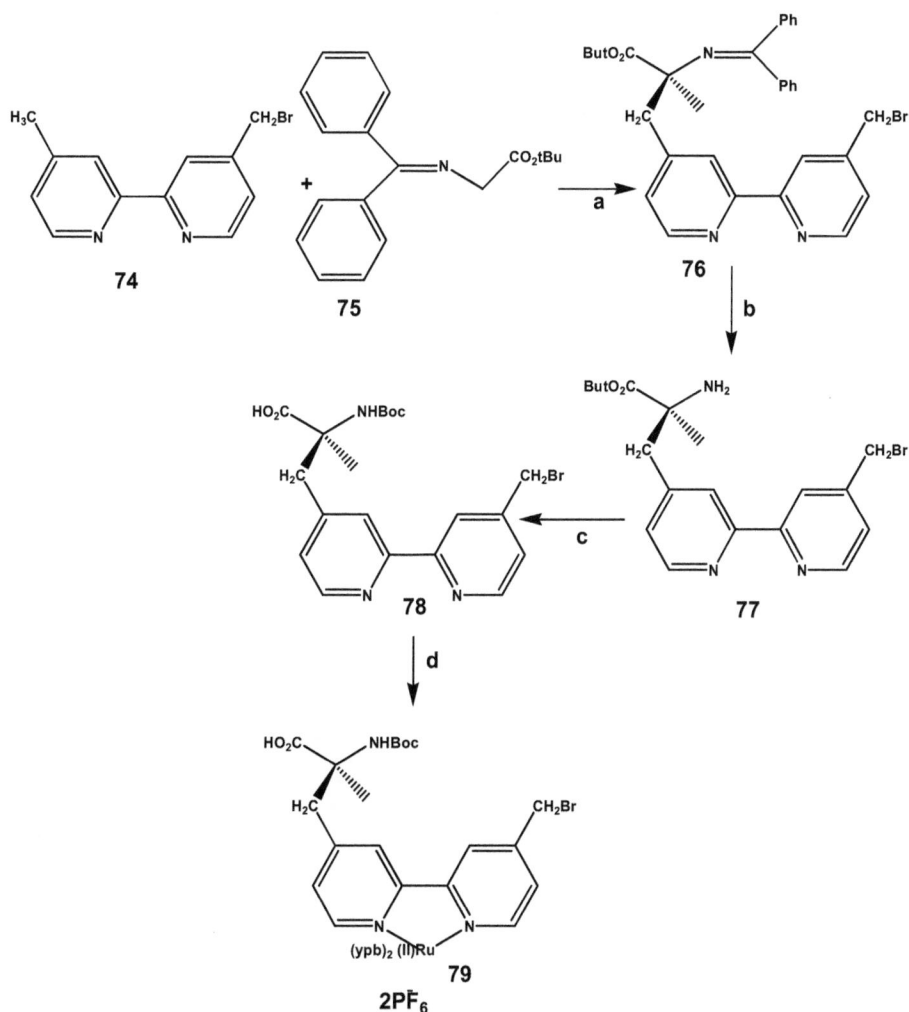

Scheme 19: Ruthenium (II) tris (bipyridyl) amino acid according to Kise *et al.* [33]. *Reagents and conditions*: (a) CsOH, -78°C, DCM, *O*(9)-allyl-N-9-anthracenyl-cinchonidinium bromide, 24 h (b) 6 N HCl, 4 h (c) di-*t*-butyl pyrocarbonate, dioxane, NaOH, 4 h (d) *cis*-Ru(bpy)$_2$Cl$_2$, 70% EtOH, CM-Sepharose, NH$_4$PF$_6$.

5.5. Synthesis of Amino Acids with Rhodium and Ruthenium

Recently, Nolte and co-worker have reported the unnatural amino acid containing thiazolylalanine (Thia) and rhodium, ruthenium at side chain [36]. These amino acids **87-90** can be obtained by addition of Ag$_2$O to the thiazolium salts in dichloromethane, containing molecular sieves to absorb the formed water. After shaking the mixture overnight, the chlorobridged metal dimers (dichloro(Cp*)rhodium(III) and dichloro(p-cymene)ruthenium(II) dimer) were added. Almost instantly, the carbene transfer reaction started, indicated by the precipitation of a white solid, likely to be silver bromide. After 24 h, the mixtures were filtered through celite and the crude products were purified by flash column chromatography. The amino acids **87-90** were obtained as orange powders in moderate yields after recrystallization from dichloromethane/pentane (Scheme **21**).

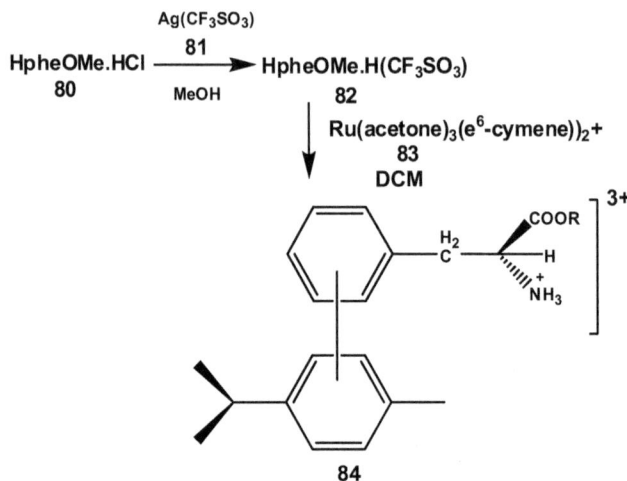

Scheme 20: Bis(arene) ruthenium(II) phenylalanine with η^6coordination complex amino acid according to Wolff *et al.* [35].

6. SYNTHESIS OF AMINO ACIDS WITH COBALT AT SIDE CHAIN

Kayser *et. al.* have reported the synthesis of amino acids with cobalt complexes at side chain, where the nucleophilic attack occured through the metal atom directly [37]. The reaction of methyl α-bromohippurate **92** with [Co (py) (dh)]⁻ **91** (bipy = 2,2′-bipyridine, py = pyridine, dh = dimethylglyoxime) affords the complexes **93** (Scheme **22**).

7. SYNTHESIS OF AMINO ACIDS WITH PLATINUM AT SIDE CHAIN

To broaden the scope of amino acids as versatile building blocks; the design, synthesis and application of novel amino acids analogues varying in the nature and stereochemistry of the side chain functional groups, is imperative. Beck and co-workers have synthesized an amino acid containing platinum at side chain

[38]. The reactions of methyl *N*-benzoyl-2-bromoglycinate **94** with $(Ph_3P)_2Pt(\eta^2\text{-}C_2H_4)$ yielded the α-metallated amino acid **95** (Scheme **23**).

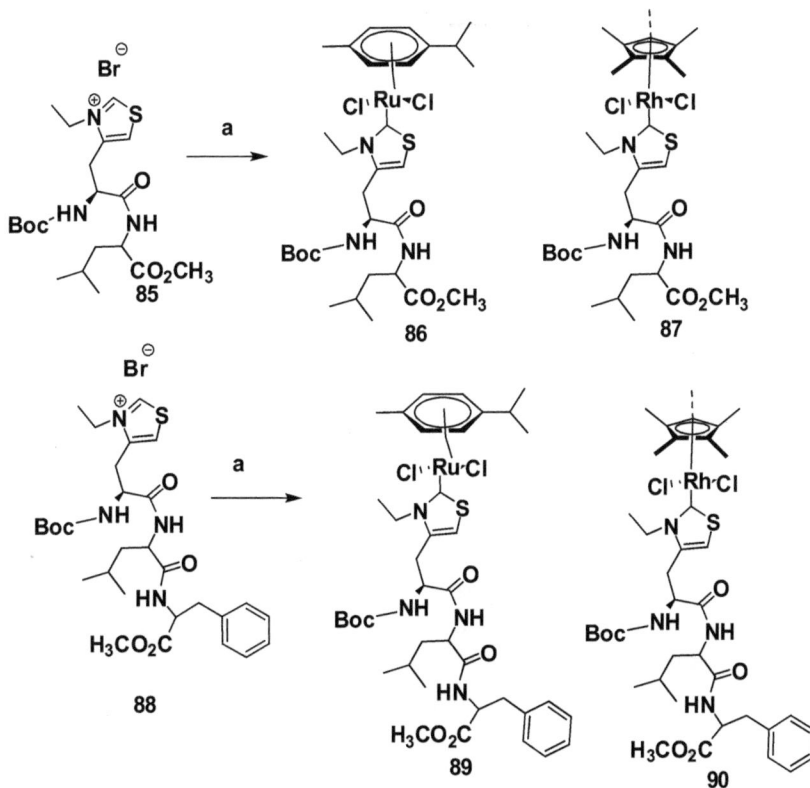

Scheme 21: Ruthenium and Rhodium containing amino acid according to Lemke *et al.* [36]. *Reagents and conditions*: (a) 2 eq. thiazolium salt, 1 eq. Ag$_2$O, 1 eq. [(p-cymene)RuCl$_2$]$_2$ or [Cp*RhCl$_2$]$_2$, CH$_2$Cl$_2$, molecular sieve, N$_2$.

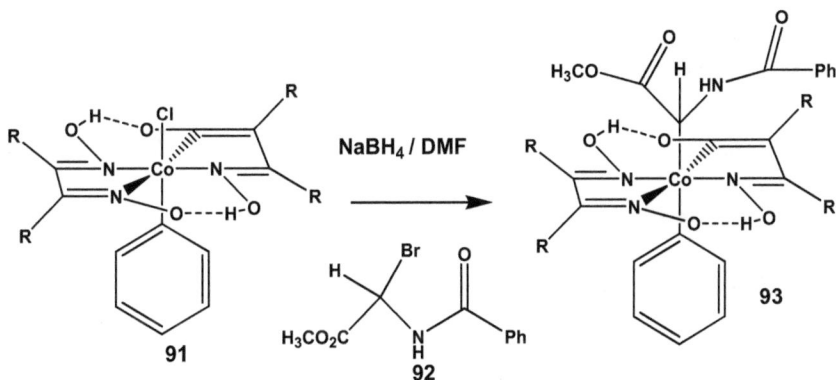

Scheme 22: Amino acids with cobalt complexes at side chain according to Kayser *et al.* [37].

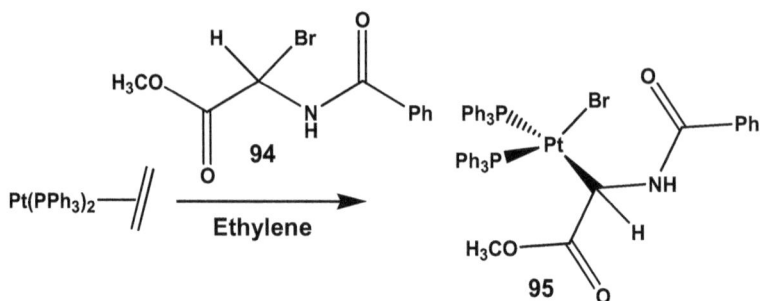

Scheme 23: Synthesis of amino acids with platinum at side chain according to Kayser *et al.* [38].

They have also reported the synthesis of amino acid containing platinum at side chain where the nucleophilic attack occured through the metal atom directly [34]. The reaction of methyl α-bromohippurate **92** with [Pt(Me)$_2$(bipy)] **96** (bipy = 2,2'-bipyridine) affords the amino acid stereo isomers **97** and **98** (Scheme **24**).

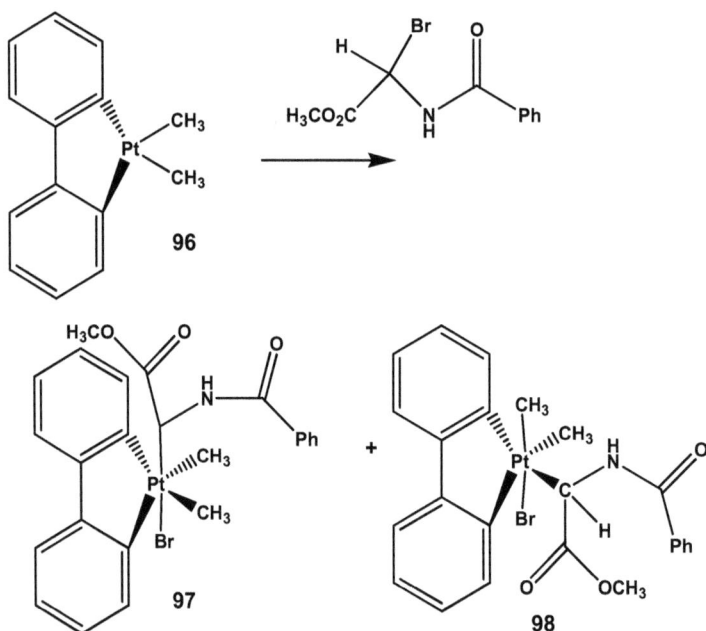

Scheme 24: Synthesis of amino acids with platinum at side chain according to Kayser *et al.* [37].

8. SYNTHESIS OF AMINO ACIDS WITH PALLADIUM AT SIDE CHAIN

The incorporation of conformationally constrained amino acids in peptides is highly efficient to provide specific conformational preferences. Beck and co-workers have used an interesting approach towards the synthesis of an amino acid containing platinum at side chain [38]. The amino acid **99** can be obtained by reaction of [Pd(dba)$_2$] (dba. dibenzylideneacetone) with methyl a-bromohippurate **92** in the presence of bipyridyl (Scheme **25**).

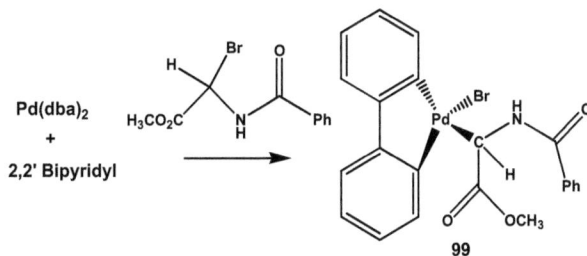

Scheme 25: Synthesis of amino acids with palladium at side chain according to Kayser *et al.* [38].

9. SYNTHESIS OF AMINO ACID SIDE CHAIN CONTAINING SODIUM

Recently, Penso and co-worker have reported the synthesis of N-tosyl derivative of tysosine (**102**) side chain containing cyclophosphazenes (CyP) crown ether Scheme **26** [39]. This compound should be exactly defined 'proton-ionizable lariat ethers' due to the presence in the molecule of a pendent amino acid group. Such a property is known to be very useful because the complexation and the release of the cation are controlled by the pH of the medium ('proton switch') and the metal ion extraction does not involve transfer of the aqueous phase anion into the organic phase. More in general, these lariat ethers could be potentially useful as 'pH-controlled active ion carriers' in liquid membranes. Scheme **26** shows the sodium ion complex with side chain ether moiety. The amino acid **103** can be obtained by reaction of L-NsTyrOMe with **101** (Scheme **26**).

CONCLUSION

The incorporation of metal ions in the side chain of conformationally constrained or flexible synthetic amino acids and analogues has led to the development of diverse therapeutic applications like an immunoassay based on carbonyl complexes as well as to a template-controlled synthesis of peptides on chiral half-sandwich complexes (synthetic ribosomes). Furthermore, the understanding of the behaviour of these organometallic amino acids under physiological conditions will help to mimic protein folding and stabilization. Therefore, the development of efficient synthetic routes to obtain large quantities of synthetic metal containing amino acids is important. This chapter shows that still there are many challenges to develop the new organometallic amino acids and analogues.

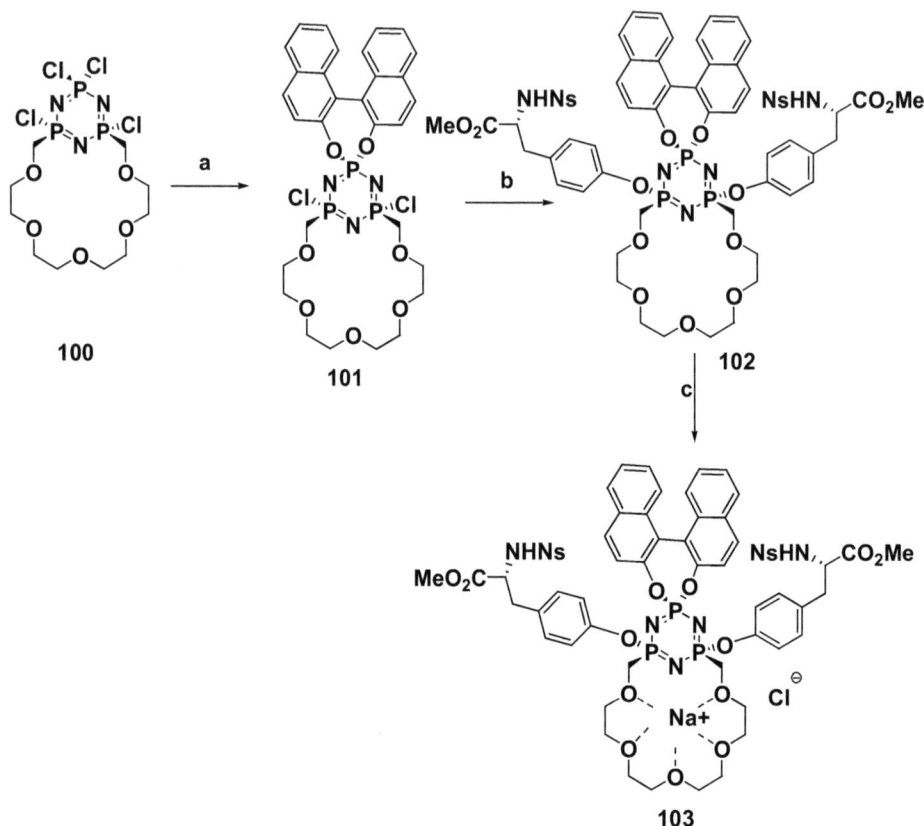

Scheme 26: Synthesis of amino acids with sodium at side chain according to Penso *et al.* [39]. *Reagents and conditions*: (a) (R)-1,10-bi-(2-naphthol), K$_2$CO$_3$, MeCN, 60°C, 21 h. (b) L-NsTyrOMe (3 equiv), NaH (5 equiv), MeCN, 40°C, 24 h, (c) NaCl.

ACKNOWLEDGEMENTS

P. Jana and S. Maity gratefully acknowledge the Council of Scientific & Industrial Research (C.S.I.R), New Delhi, India for financial assistance.

CONFLICT OF INTEREST

The author confirms that this chapter content has no conflict of interest.

DISCLOSURE

The chapter submitted for series eBook entitled "**Advances in Organic Synthesis, Volume 5**" is an update of our article published in **CURRENT ORGANIC SYNTHESIS, Volume 5, Number 1, 2008**, with additional text and references.

REFERENCES

[1]   *a*) Szekeres, A.; Leitgeb, B.; Kredics, L.; Antal, Z.; Hatvani, L.; Manczinger, L.; Vágvölgyi, C. Peptaibols and related peptaibiotics of trichoderma. A review, *Acta Microbiol. Immunol. Hung.* **2005**, *52*, 137-168; b) Degenkolb, T.; Berg, A.; Gams, W.; Schlegel, B.; Gräfe, U. The occurrence of peptaibols and structurally related peptaibiotics in fungi and their mass spectrometric identification *via* diagnostic fragment ions, *J. Pept. Sci.* **2003**, *9*, 666-678.

[2]   *a*) Polinelli, S.; Broxterman, Q. B.; Schoemaker, H. E.; Boesten, W. H. J.; Crisma, M.; Valle, G.; Toniolo, C.; Kamphuis, J. *Bioorg. Med. Chem. Lett.* **1992**, *5*, 453-458; *b*) O'Connor, S. J.; Liu, Z. *Synlett* **2003**, *14*, 2135-2138.

[3]   *a*) Almond, H. R.; Manning, D. T.; Niemann, C. *Biochemistry* **1962**, *1*, 243-247; *b*) Khosla, M. C.; Stachowiak, K.; Smeby, R. R.; Bumpus, F. M.; Piriou, F.; Lintner, K.; Fermandjian, S. *Proc. Natl. Acad. Sci. U.S.A.,* **1981**, *78*, 757-760.

[4]   *a*) Ryabov, A. D. *Angew. Chem. Int. Ed. Engl.* **1991**, *30*, 931-941; *b*) Jaouen, G.; Vessieres, A.; Butler, I. S. *Acc. Chem. Res.* **1993**, *26*, 361 - 369; *c*) Krawielitzki, S.; Beck, W. *Chem. Ber.* **1997**, *130*, 1659 - 1662; *d*) Dötz, K. H.; Ehlenz, R. *Chem. Eur. J.* 1997, *3*, 1751 - 1756,

[5]   *a*) Krämer, R. *Angew. Chem.* **1996**, *108*, 1287-1289; *b*) Wolff, J. M.; Gleichmann, A. J.; Sheldrick, W. S. *J. Inorg. Biochem.* **1995**, *59*, 219; *c*) Salmain, M.; Gunn, M.; Gorfti, A.; Top, S.; Jaouen, G. *Bioconjugate Chem.* **1993**, *4*, 425-433.

[6]   Severin, K.; Bergs, R.; Beck, W. *Angew. Chem. Int. Ed.* **1998**, *37*, 1634 – 1654.

[7]   Cremer, W. *Biochem. Z.* **1929**, *206*, 208.

[8]   Schubert, M. P. *J. Am. Chem. Soc.* **1933**, *55*, 4563 - 4570.

[9]   *a*) Tomita, A.; Hirai, H.; Makishima, S. *Inorg. Chem.* **1967**, *6*, 1746-1750; *b*) Tomita, A.; Hirai, H.; Makishima, S. *Inorg. Nucl. Chem. Lett.* **1968**, *4*, 715 - 718.

[10] *a*) de Lacey, A. L.; Hatchikian, E. C.; Volbeda, A.; Frey, M.; Fontecilla-Camps, J. C.; Fernandez, V. M. *J. Am. Chem. Soc.* **1997**, *119*, 7181-7189; *b*) Happe, R. P.; Roseboom, W.; Pierik, A. J.; Albracht, S. P. J.; K. A. Bagley. *Nature* **1997**, *385*, 126-131; *c*) Darensbourg, D. J.; Reibenspies, J. H.; Lai, C.-H.; Lee, W.-Z.; Darensbourg, M. Y. *J. Am. Chem. Soc.* **1997**, *119*, 7903 – 7904.

[11] Klok, H.-A. *Angew. Chem. Int. Ed.* **2002**, *114*, 1579-1583; *Angew. Chem. Int. Ed.* **2002**, *41*, 1509-1513.

[12] Peckham, T. J.; Elipe, P. G.; Manners, I. *Metallocenes*, Wiley-VCH, Weinheim. **1998**, pp. 723-771.

[13] *a*) Beer, P. D. *Acc. Chem. Res.* **1998**, *31*, 71-80; *b*) Beer, P. D.; Davis, J. J.; Drillsma-Milgrom. D. A.; Szemes, F. *Chem. Commun.* **2002**, 1716-1717; *c*) Ihara. T.; Nakayama,

M.; Murata, M.; Nakano, K.; Maeda, M. *Chem. Commun.* **1997**, 1609-1610; d) Kavallieratos, K.; Hwang, S.; Crabtree. R. H. *Inorg. Chem.* **1999**, *38*, 5184-5186.

[14]   a) Severin, K.; Bergs, R.; Beck, W. *Angew. Chem. Int. Ed.* **1998**, *110*, 1722-1743, *Angew. Chem. Int. Ed.* **1998**, *37*, 1634-1654; b) Brosch, O.; Weyhermüller, T.; Metzler-Nolte, N. *Inorg. Chem.* **1999**, *38*, 5308-5313.

[15]   a) Maksakov, V. A.; Ershova, V. A.; Kirin, V. P. *Russ. J. Coord. Chem.* **1996**, *22*, 399-402; b) Süss-Fink, G.; Jenke, T.; Heitz, H.; Pellinghelli, M. A.; Tiripicchio, A. *J. Organomet. Chem.* **1989**, *379*, 311 -323; c) Mani, D.; Schacht H.-T; Powell, A. K.; Vahrenkamp. H. *Chem. Ber.* **1989**, *122*, 2245 - 2251; d) Severin, K.; Sünkel. K.; Beck, W. *Chem. Ber.* **1994**, *127*, 615 - 620.

[16]   Werner, H.; Daniel, T.; Nürnberg, O.; Knaup, W.; Meyer. U. *J. Organomet. Chem.* **1993**, *445*, 229 - 235.

[17]   Carlstrom, A.-S.; Frejd, T. *J. Org. Chem.*, **1990**, *55*, 4175-4180.

[18]   Basu, B.; Chattopadhyay, S. K.; Ritzén, A.; Frejd, T. *Tetrahedron Asymm.* **1997**, *8*, 1841-1846.

[19]   Barišić, L.; Rapić, V.; Metzler-Nolte, N. *Eur. J. Inorg. Chem.* **2006**, 4019–4021.

[20]   Butler, I. R.; Quayle, S. C. *J. Organomet. Chem.* **1998**, *552*, 63-68.

[21]   Heinze, K.; Schlenker, M. *Eur. J. Inorg. Chem.* **2004**, 2974-2988.

[22]   Gellett, A. M.; Huber, P. W.; Higgins, P. J. *J. Organomet. Chem.* **2008**, *693*, 2959–2962.

[23]   Burk, M, J.; Lee, J. R.; Martinez, J. P. *J. Am. Chem. Soc.* **1994,** *116*, 10847- 10848.

[24]   Salter, R.; Pickett, T. E.; Richards, C., J. *Tetrahedron Asymm.* **1998**, *9*, 4239–4247.

[25]   Sudhir,V. Sai.; Kumar, N.Y. Phani.; Chandrasekaran, S. *Tetrahedron* **2010**, *66*, 1327–1334.

[26]   Cooper, D. C.; Yennie, C. J.; Morin, J. B.; Delaney, S.; Suggs, J. W. *J. Organomet. Chem.* **2011**, *696*, 3058-3061.

[27]   Brunet, J. C.; Cuingnet, E.; Gras, H.; Marcincal, P.; Mocz, A.; Sergheraert, C.; Tartar, A. *J. Organomet. Chem.* **1981**, *216*, 73-77.

[28]   Hudson, R. D. A.; Osborne, S. A.; Stephenson, G. R. *Synlett.*, **1996,** 845 - 846.

[29]   Sergheraert, C.; Tartar, A. *J. Organomet. Chem.*, **1982,** *240*, I63- 168.

[30]   Jenhi, A.; Lavergne, J. P.; Viallefont, P. *J. Organomet. Chem.*, **1991,** *401*, C14- C16.

[31]   Cais. M. *Meth. Enzym.*, **1983**, *92*, 445-448.

[32]   Moriarty, R. M.; Ku, Y. Y.; Gill, U. S. *J. Organomet. Chem.*, **1989,** *362*, 187-191.

[33]   Kise, K. J.; Bowler, B. E. *Inorg. Chem.* **2002**, *41*, 379-386.

[34]   Jaouen, G.; Vessieres, A.; Butler, I. S. *Acc. Chem. Res.* **1993**, *26*, 361-369.

[35]   Wolff, J. M.; Sheldrick, W. S. *J. Organomet. Chem.*, **1997,** *531*, 141-149.

[36]   Lemke, J.; Metzler-Nolte, N. *J. Organomet. Chem.*, **2011,** *696*, 1018-1022.

[37]   Kayser, B.; Nöth, H.; Schmidt, M.; Steglich, W.; Beck, W. *Chem. Ber.* **1996**, *129*, 1617-1620.

[38]   Kayser, B.; Missling, C.; Knizek, J.; Nöth, H.; Beck, W. *Eur. J. Inorg. Chem.* **1998**, 375-379.

[39]   Penso, M.; Maia, A.; Lupi, V.; Tricarico, G. *Tetrahedron* **2011**, *67*, 2096-2102.

INDEX

Monosaccharide 3, 7, 31-2, 35, 37, 130

N

N-acetylardeemin 309, 342

N-Acyliminium 310, 312, 314, 316, 318, 320, 322, 324, 326, 328, 330, 332, 334, 336-8, 340

N-acyliminium ions 309, 311, 314-15, 318, 331, 334, 340-5, 348

N-Acyliminium Ions 309, 326, 336, 339-40

N-acyliminium species 314, 321-2, 348

N-acyliminium species of type 313, 317

N-indole-2-carbonylamino esters 339-40

N-nucleophiles 138-9, 141

N-phthalimidoacetaldehyde dimethyl acetal 325

Natural products 103, 133, 149, 157, 235, 237, 253, 256-8, 266, 309-10, 319-20, 322, 326, 355

New receptors 51-2, 74

Nitrogen 34, 51, 53, 145, 314, 317-18

Nitrogen atoms 53, 56, 72, 123, 311, 329

Nitrogen nucleophiles 138-9, 144, 156

NMR 57-8, 70, 73, 75, 82, 85-6, 154, 297

Non-addition product 155

Nucleophiles 101-2, 108, 117, 124-5, 127-8, 131, 138, 147, 151, 156, 167, 169, 318, 334, 345-6

sulfur 34, 169

Nucleophilic 101-2, 112, 280, 301, 310

Nucleophilic addition 101-2, 117, 265

Nucleophilic attack 101-2, 114, 144, 146-7, 163, 323, 372, 374

Nucleophilic trifluoromethylation 281, 291, 295

O

Oligoimines 68, 70, 72, 76-7

Oligomers 61, 66, 72, 76, 83

Oligonucleotide chain 24-6, 28-30

Oligonucleotide glycoconjugates 5, 23-6, 28, 30

Q

R

S